MTP International Review of Science

Main Group Elements
Hydrogen and Groups I–IV

MTP International Review of Science

Publisher's Note

The MTP International Review of Science is an important new venture in scientific publishing, which we present in association with MTP Medical and Technical Publishing Co. Ltd. and University Park Press, Baltimore. The basic concept of the Review is to provide regular authoritative reviews of entire disciplines. We are starting with chemistry because the problems of literature survey are probably more acute in this subject than in any other. As a matter of policy, the authorship of the MTP Review of Chemistry is international and distinguished; the subject coverage is extensive, systematic and critical; and most important of all, new issues of the Review will be published every two years.

In the MTP Review of Chemistry (Series One), Inorganic, Physical and Organic Chemistry are comprehensively reviewed in 33 text volumes and 3 index volumes, details of which are shown opposite. In general, the reviews cover the period 1967 to 1971. In 1974, it is planned to issue the MTP Review of Chemistry (Series Two), consisting of a similar set of volumes covering the period 1971 to 1973. Series Three is planned for 1976, and so on.

The MTP Review of Chemistry has been conceived within a carefully organised editorial framework. The over-all plan was drawn up, and the volume editors were appointed, by three consultant editors. In turn, each volume editor planned the coverage of his field and appointed authors to write on subjects which were within the area of their own research experience. No geographical restriction was imposed. Hence, the 300 or so contributions to the MTP Review of Chemistry come from many countries of the world and provide an authoritative account of progress in chemistry.

To facilitate rapid production, individual volumes do not have an index. Instead, each chapter has been prefaced with a detailed list of contents, and an index to the 10 volumes of the MTP Review of Inorganic Chemistry (Series One) will appear, as a separate volume, after publication of the final volume. Similar arrangements will apply to the MTP Review of Physical Chemistry (Series One) and to subsequent series.

Butterworth & Co. (Publishers) Ltd.

Inorganic Chemistry Series One
Consultant Editor
H. J. Eméleus, F.R.S.
*Department of Chemistry
University of Cambridge*

Volume titles and Editors

1 **MAIN GROUP ELEMENTS—HYDROGEN AND GROUPS I–IV**
Professor M. F. Lappert, *University of Sussex*

2 **MAIN GROUP ELEMENTS—GROUPS V AND VI**
Professor C. C. Addison, F.R.S. and Dr. D. B. Sowerby, *University of Nottingham*

3 **MAIN GROUP ELEMENTS—GROUP VII AND NOBLE GASES**
Professor Viktor Gutmann, *Technical University of Vienna*

4 **ORGANOMETALLIC DERIVATIVES OF THE MAIN GROUP ELEMENTS**
Dr. B. J. Aylett, *Westfield College, University of London*

5 **TRANSITION METALS—PART 1**
Professor D. W. A. Sharp, *University of Glasgow*

6 **TRANSITION METALS—PART 2**
Dr. M. J. Mays, *University of Cambridge*

7 **LANTHANIDES AND ACTINIDES**
Professor K. W. Bagnall, *University of Manchester*

8 **RADIOCHEMISTRY**
Dr. A. G. Maddock, *University of Cambridge*

9 **REACTION MECHANISMS IN INORGANIC CHEMISTRY**
Professor M. L. Tobe, *University College, University of London*

10 **SOLID STATE CHEMISTRY**
Dr. L. E. J. Roberts, *Atomic Energy Research Establishment, Harwell*

INDEX VOLUME

**Physical Chemistry
Series One**
Consultant Editor
A. D. Buckingham
*Department of Chemistry
University of Cambridge*

Volume titles and Editors

1 THEORETICAL CHEMISTRY
   Professor W. Byers Brown, *University of Manchester*

2 MOLECULAR STRUCTURE AND PROPERTIES
   Professor G. Allen, *University of Manchester*

3 SPECTROSCOPY
   Dr. D. A. Ramsay, *National Research Council of Canada*

4 MAGNETIC RESONANCE
   Professor C. A. McDowell, *University of British Columbia*

5 MASS SPECTROMETRY
   Professor A. Maccoll, *University College, University of London*

6 ELECTROCHEMISTRY
   Professor J. O'M Bockris, *University of Pennsylvania,*

7 SURFACE CHEMISTRY AND COLLOIDS
   Professor M. Kerker, *Clarkson College of Technology, New York*

8 MACROMOLECULAR SCIENCE
   Professor C. E. H. Bawn, F.R.S., *University of Liverpool*

9 CHEMICAL KINETICS
   Professor J. C. Polanyi, F.R.S., *University of Toronto*

10 THERMOCHEMISTRY AND THERMODYNAMICS
   Dr. H. A. Skinner, *University of Manchester*

11 CHEMICAL CRYSTALLOGRAPHY
   Professor J. Monteath Robertson F.R.S., *University of Glasgow*

12 ANALYTICAL CHEMISTRY—PART 1
   Professor T. S. West, *Imperial College, University of London*

13 ANALYTICAL CHEMISTRY — PART 2
   Professor T. S. West, *Imperial College, University of London*

INDEX VOLUME

**Organic Chemistry
Series One**
Consultant Editor
D. H. Hey, F.R.S.
*Department of Chemistry
King's College, University of London*

Volume titles and Editors

1 STRUCTURE DETERMINATION IN ORGANIC CHEMISTRY
   Professor W. D. Ollis, *University of Sheffield*

2 ALIPHATIC COMPOUNDS
   Professor N. B. Chapman, *Duke University, North Carolina*

3 AROMATIC COMPOUNDS
   Professor H. Zollinger, *Swiss Federal Institute of Technology*

4 HETEROCYCLIC COMPOUNDS
   Dr. K. Schofield, *University of Exeter*

5 ALICYCLIC COMPOUNDS
   Professor W. Parker, *University of Stirling*

6 AMINO ACIDS AND PEPTIDES
   Professor D. H. Hey, F.R.S., and Dr. D. I. John, *King's College, University of London*

7 CARBOHYDRATES
   Professor G. O. Aspinall, *University of Trent, Ontario*

8 STEROIDS
   Dr. W. D. Johns, *G. D. Searle & Co., Chicago*

9 ALKALOIDS
   Professor K. Wiesner, *University of New Brunswick*

10 FREE RADICAL REACTIONS
   Professor W. A. Waters, F.R.S., *University of Oxford*

INDEX VOLUME

Inorganic Chemistry
Series One

Consultant Editor
H. J. Emeléus, F.R.S.

MTP International Review of Science

Volume 1
Main Group Elements Hydrogen and Groups I—IV

Edited by **M. F. Lappert**
University of Sussex

Butterworths · London
University Park Press · Baltimore

THE BUTTERWORTH GROUP

ENGLAND
Butterworth & Co (Publishers) Ltd
London: 88 Kingsway, WC2B 6AB

AUSTRALIA
Butterworth & Co (Australia) Ltd
Sydney: 586 Pacific Highway 2067
Melbourne: 343 Little Collins Street, 3000
Brisbane: 240 Queen Street, 4000

NEW ZEALAND
Butterworth & Co (New Zealand) Ltd
Wellington: 26–28 Waring Taylor Street, 1

SOUTH AFRICA
Butterworth & Co (South Africa) (Pty) Ltd
Durban: 152–154 Gale Street

ISBN 0 408 70256 7

UNIVERSITY PARK PRESS

U.S.A. and CANADA
University Park Press Inc
Chamber of Commerce Building
Baltimore, Maryland, 21202

> Library of Congress Cataloging in Publication Data
>
> Lappert, M   F
>   Main group elements
>
>   (Inorganic chemistry, series one, v. 1) (MTP international review of science)
>   Includes bibliographical references
>   1. Chemistry, Inorganic. 2. Chemical elements. I. Title
> QD151.2.I5 Vol. 1   546   74-160323
> ISBN 0-8391-1004-9

First Published 1972 and © 1972
MTP MEDICAL AND TECHNICAL PUBLISHING CO. LTD.
Seacourt Tower
West Way
Oxford, OX2 OJW
and
BUTTERWORTH & CO. (PUBLISHERS) LTD.

Filmset by Photoprint Plates Ltd., Rayleigh, Essex
Printed in England by Redwood Press Ltd., Trowbridge, Wilts
and bound by R. J. Acford Ltd., Chichester, Sussex

Consultant Editor's Note

The problem of keeping abreast of research literature on as broad a front as possible is one that confronts all chemists. In the past this difficulty has been met, in the main, by literature surveys and by several uncorrelated reviews of progress in certain subject areas. There are obvious inadequacies in this approach, which have become increasingly apparent in recent years. I was, therefore, grateful for the opportunity of helping to plan this new series, which has been designed to provide a comprehensive, critical survey of each of the main branches of chemistry.

This section of the MTP International Review of Science deals with progress in Inorganic Chemistry. The subject is developing at an astonishing rate and in many directions. Fortunately, however, it lends itself to a systematic treatment. Ten volumes have been prepared, three dealing with the main group elements and two with the general chemistry of the transition metals. Organometallic derivatives of the main group elements and lanthanides and actinides are covered separately, as is the subject of reaction mechanisms. The two remaining volumes on radiochemistry and solid state chemistry have been planned to avoid, as far as possible, overlap with those that have gone before.

It is a pleasure to thank the many experts who have collaborated as authors and volume editors in making this publication possible. While working to a pre-arranged over-all plan, they have been able to assess and interpret the literature in terms of their own experience in specialised fields. I believe that in this way they will not only provide a record of what has been done, but will stimulate further exploration in this fascinating branch of chemistry.

Cambridge   H. J. Emeléus

Preface

This volume provides a review of the chemistry of hydrogen and the elements of Groups I–IV, inclusive. For the present, the chemistry of Group III elements other than boron has been omitted to allow adequate coverage of the other areas. In the next issue, contributions will appear covering the period 1969–1972 inclusive on aluminium and the heavier Group III elements.

Our objective is to provide the reader with a reasonably comprehensive survey of the field for the years 1969 and 1970. In this very first issue, of what will be a biennial review, we have also drawn the reader's attention to those areas which have excited the most interest during the last five to ten years. This has clearly been a selective exercise and greatest emphasis has been laid on providing adequate secondary sources to the primary literature.

Certain aspects of the chemistry of these elements are either omitted or dealt with briefly. These, in the main, relate to organometallic (by which we mean compounds having a metal-carbon bond) and solid state chemistry, and specialised discussions on kinetics and mechanisms. Problems relating to radioactivity and nuclear chemistry are also avoided. All these areas are covered in other volumes of the review.

The various authors have, of course, been free to present the material in the manner they considered most appropriate. Some have decided to divide their chapter into a somewhat selective descriptive and critical essay, followed by a comprehensive tabular survey. This, in the editor's view, is perhaps the ideal presentation but it has not always been possible because of limitations of space.

We envisage that the book will be of greatest value to active research workers, who should find it relatively easy to obtain a comprehensive bibliography with some critical comment on their area of particular interest. By and large, the approach has concentrated on compounds rather than reaction types. Where space has been at a premium, we felt it was of greater importance to show the research worker that, for example, the structure of a particular compound had been elucidated, rather than to provide him with a detailed description of the molecular parameters.

Sussex   M. F. Lappert

Contents

Hydrogen, the alkali metals and the alkaline earth metals   1
M. H. Ford-Smith, *University of Sussex*

Beryllium, magnesium, zinc, cadmium and mercury   33
D. J. Cardin, *University of Sussex*

Boron hydrides   79
N. F. Travers, *University of Newcastle upon Tyne*

Carboranes and metallocarboranes   139
R. Snaith and K. Wade, *University of Durham*

Other aspects of boron chemistry   185
R. H. Cragg, *University of Kent*

Carbon and silicon   221
J. Simpson, *University of Otago, Dunedin*

Germanium, tin and lead   303
J. E. Drake and J. W. Anderson, *University of Windsor, Ontario*

1
Hydrogen, the Alkali Metals and the Alkaline Earth Metals

M. H. FORD-SMITH
University of Sussex

1.1	HYDROGEN	2
	1.1.1 *Introduction*	2
	1.1.2 *Interstitial hydrides of the transition metals*	2
	1.1.3 *Hydrides of the lanthanides and actinides*	3
	1.1.4 *Saline hydrides*	4
	1.1.5 *Species in solution*	5
	1.1.6 *Ions in the gas phase*	5
	1.1.7 *Miscellaneous*	6
1.2	THE ALKALI METALS	6
	1.2.1 *Introduction*	6
	1.2.2 *Solutions of the alkali metals*	8
	1.2.3 *Oxygen compounds of the alkali metals*	9
	1.2.4 *Co-ordination of the alkali metal cations by chelate ligands*	11
	1.2.5 *Solvent extraction of alkali metal ions from aqueous solution*	16
	1.2.6 *Preparation and reactions of alkali metal compounds*	17
	1.2.7 *Chemical physics*	19
	1.2.8 *Miscellaneous*	21
1.3	THE ALKALINE EARTH ELEMENTS	22
	1.3.1 *Introduction*	22
	1.3.2 *Preparations and reactions*	22
	1.3.3 *Stability constants for alkaline earth metal ions*	23
	1.3.4 *Solvent extraction*	24
	1.3.5 *Environments in the solid state*	24
	1.3.6 *Chemical physics*	25
	1.3.7 *Miscellaneous*	25

1.1 HYDROGEN

1.1.1 Introduction

Most of the chemistry of hydrogen compounds is best considered in the context of the other elements combined with hydrogen, and references to such chemistry will be found in every chapter of this and subsequent volumes. Deuterium, ^2H, and tritium, ^3H, are of chemical interest because they are used extensively in mechanistic studies of reactions in solution and, less commonly, in the gas phase. Ortho- and para-hydrogen are also sometimes used in gas kinetic studies. Aktar and Smith[1] have reviewed the work on the separation and analysis of various forms of molecular hydrogen (isotopic species and ortho- and para-hydrogen) by use of absorption and gas chromatographic techniques.

Compounds of hydrogen (hydrides) are conventionally divided into the three classes: saline hydrides, covalent hydrides and interstitial hydrides. Recent interest has been concentrated on transition-metal complexes which contain at least one hydrogen atom bonded to the metal atom (hydride complexes) and on the detailed investigation of interstitial hydrides of transition metals (binary hydrides formed by transition metals and their alloys). Hydride transition-metal complexes have been reviewed[2] as has the solubility of hydrogen in transition metals and alloys[3]. Transition-metal hydride complexes are best considered in the volume of this series dealing with transition-metal complexes because their properties are very much a function of the transition metal, its oxidation state and coordination number, and the other ligands attached to the transition metal. Hydrides of elements such as boron and silicon are discussed elsewhere in this volume.

The chemistry of metal hydrides has been the subject of two books[4].

1.1.2 Interstitial hydrides of the transition metals

A great deal of work has been done on these systems in an attempt to ascertain (a) the structure of the hydrides and, in particular, the location of hydrogen atoms in the compounds and (b) the mechanism(s) by which hydrogen diffuses through metals. Both of these topics are studied in the hope that light will be shed on the bonding in these compounds. At present it is fairly certain that hydrogen *atoms* are located in the interstices of the metal lattice (and the formation of the hydride is accompanied by some expansion of the lattice), that hydrogen atoms occupy octahedral sites in some metals and tetrahedral ones in others, and that there is considerable donation of electron density from hydrogen to the metal (i.e. hydrogen is intermediate between H^- and H^+.

An effective charge of $+0.26$ e on the diffusing species was found from studies[5] on the diffusion of H_2 generated electrolytically in pure iron in which it was found that diffusion coefficients increased linearly with over-potential over a narrow range. This value of $+0.26$ e agrees well with previous determinations.

Neutron diffraction is the only feasible method of locating directly hydrogen or deuterium in heavy-metal lattices. Russian workers[6] have confirmed and extended earlier work on the structure of Ta_2D. Fender and

Henfrey[7], by a study of the diffuse scattering of long wavelength neutrons by $NbD_{0.042}$ and $NbD_{0.079}$, have shown that at least in these compounds the deuterium atoms are randomly distributed (presumably in T_d sites) rather than present as clusters. This is an important result since it eliminates a possible explanation for observed deviations from Sievert's law, which states that the equilibrium partial pressure of H_2 should be proportional to $[H\cdot]^2$ and assumes that H atoms in the lattice do not interact (no clustering), and that the metal lattice is unperturbed. Earlier thermodynamic studies on $H_2 + V$, Nb and Ta had suggested that clustering of H atoms did occur. Recent studies have tended towards the study of interstitial hydrides which contain very small amounts of hydrogen. Ricca and Giorgi[8] have measured the pressures of H_2 and D_2 in equilibrium with their solutions in α-Hf and found that the relative partial molal entropies were essentially independent of temperature (as was found previously for Ti and Zr). This result contradicts the general assumption that dissolved H atoms occupy T_d interstices with purely vibrational motions and suggests that the H atoms are free to move around the lattice.

Studies on the rate of diffusion of H_2 through Pd and its alloys show that when palladium is alloyed with a metal such as silver[9] which forms a random substitutional alloy (Ag replacing Pd in its lattice sites) there are only small changes in diffusion coefficient and entropy of desorption, but when Pd–B alloys are formed, diffusion coefficients decrease steadily and markedly with increasing B content[10]. This is evidence that B atoms occupy interstitial sites in the alloy and that H diffuses through pure Pd via T_d holes from one O_h site to another. Diffusion coefficients are most meaningful when they are obtained for extremely low H contents of the metal. The decrease in diffusion coefficient is *not* due to an increase in activation energy. However, for Pd–Ag alloys the molar heat of absorption increases with increasing Ag concentration from 4000 cal to 8020 cal for $Pd_{0.6}Ag_{0.4}$.

After his investigations[11] into the absorption of H_2 by Pd + B and Pd + B + Ag alloys (involving measurements of electrical resistance and electrode potential as a function of solubility of H_2), Burch has written a critique[12] of the various current 'proton' models of H_2 absorption by Pd and produced his own model[13] which takes account of possible effects of lattice expansion on the solubility of H_2 in Pd and appears to give improved agreement between theory and experiment.

Although rhodium metal itself does not absorb hydrogen, Pd + Rh alloys (formed under H_2 pressures of up to 23 000 atm) absorb hydrogen[14] to give H:Pd > 1 (e.g. for 40% Rh:60% Pd, H:Pd = 1.61:1). Stoichiometric hydrides are obtained for 20% and 30% Rh alloys; this is not achieved even by pure Pd. This unique behaviour of Rh + Pd alloys is not understood.

More work has been done on the diffusion of H_2 and D_2 in tantalum[15] and changes in the structure of titanium hydride[16] at high temperatures have been investigated.

1.1.3 Hydrides of the lanthanides and actinides

These hydrides are intermediate in properties between saline and interstitial hydrides. The first reported preparation of a curium hydride[17] was achieved by treating ^{244}Cm with H_2 at 200 °C. The crystalline compound has a fcc structure and has the composition CmH_{2+x}, where $0 \leqslant x \leqslant 0.7$, although no evidence of the formation of CmH_3 was obtained.

The kinetics of formation of β-UD_3 from $U + D_2$ at various pressures and temperatures have been reported[18]. The investigators had difficulty in obtaining reproducible results because very small amounts of impurities in the D_2 had large effects on the rate of reaction. They concluded that diffusion of *molecular* D_2 across the layer of the product, UD_3, was the rate-controlling process. (This is to be contrasted with the overwhelming evidence that *atomic* diffusion is the dominant process in metals.) Ermolaev[19], from thermal decomposition and dissolution in pyrophorphoric acid experiments with uranium hydride, deduced that the hydride is properly stoichiometric UH_3 and departures from stoichiometry are due to the presence of unreacted metallic uranium in the hydride.

EuH_2, more than other rare-earth hydrides, resembles the alkaline earth saline hydrides. This is illustrated[20] by the formation of the double hydrides $LiEuH_3$ and $LiH \cdot 2EuH_2$ (established by phase-rule studies on the LiH–EuH_2 system), whereas CeH_2 forms no corresponding compounds (the m.p. of LiH is practically unchanged by the addition of CeH_2). $LiEuH_3$ was first reported in 1964 and resembles $LiBaH_3$ and $LiSrH_3$ in structure and properties. X-ray analysis is required to confirm that $LiH \cdot 2EuH_2$ is an authentic new chemical compound. Warf[21] has reviewed the chemical and physical properties of the rare-earth hydrides.

1.1.4 Saline hydrides

Messers[22] has reviewed the structures of binary and ternary hydrides and has stated that the similar values of r_{H^-} and r_{F^-} result in structural and stoichiometric similarities between hydrides and fluorides, particularly of the alkali metals, the alkaline earth elements and aluminium.

X-ray analysis of $LiH_{0.984}C_{0.016}$ (the carbon impurity occurred accidentally) showed[23] that carbon atoms occupy H^- sites, which suggests that carbon is present as the C^- ion in this compound. Heitler–London calculations[40] for solid LiH yield values for the inter-ionic distance, compressibility, etc., and the results were compared with those of earlier calculations using semi-classical and Hartree–Fock methods.

The first synthesis of $KMgH_3$ has been reported[24]. $KMg(s\text{-}C_4H_9)_2H$ dissolved in benzene was treated with H_2 at high pressure for 4 h; $KMgH_3$ was precipitated as a yellow solid. This synthesis utilised the unique solubility of $KMg(s\text{-}C_4H_9)_2H$ in benzene and avoided the use of more basic solvents such as ethers which would have competed with H^- for coordination sites. $KMgH_3$ reacts violently when exposed to the air, but is stable with respect to disproportionation although it decomposes at 250 °C. X-ray analysis proves that $KMgH_3$ is not a mixture of $LiH + MgH_2$ and indicates that it is isomorphous with $KMgF_3$, in which case Mg^{2+} will be surrounded octahedrally by six H^- ions in a Perovskite-like structure.

1.1.5 Species in solution

Detailed studies[25] involving the pulse radiolysis of oxygen-saturated aqueous solutions as a function of pH have resulted in the definitive value of $pK_a = 4.88 \pm 0.10$ for $HO_2 \rightleftharpoons H^+ + O_2^-$ and established that the perhydroxyl radical only exists in the two forms HO_2 and O_2^- in the pH range 0–13.

Deuterium sesquioxide, D_2O_3, can be generated by irradiation of acidified air-saturated water with an intense electron beam[26]. The decomposition $D_2O_3 \rightarrow D_2O + O_2$ has been studied[26] and obeys first-order kinetics. The maximum $t_{\frac{1}{2}}$ of 139 s at 0 °C is obtained in 0.027 M–$DClO_4$ and the maximum concentration of D_2O_3 obtained was 5×10^{-4} M. The decomposition exhibits a large kinetic isotope-effect ($k(H_2O_3)/k(D_2O_3) = 6$) and it is possible therefore to obtain higher concentrations of the deuterium species.

Claxton and Symons[27] have described a simple model in which H· is bonded to the oxygen atom of a water molecule with similar characteristics to that of an ordinary hydrogen bond to explain the strong absorption at 200 nm assigned[28] to H· in H_2O.

A considerable amount of work has been done on the nature of the hydrated proton in solution and in solid hydrates. It has been suggested[29] that $H_5O_2^+$ with the structure $[H_2O...H...OH_2]^+$ with $l(O—O) = 2.45$ Å may be a form of the hydrated proton in solution on the basis of calculations in which the properties of $H_5O_2^+$ are predicted by extrapolation from those of the isoelectronic HF_2^- ion and which predict that $H_5O_2^+$ is stable with respect to $H_3O^+ + H_2O$ by 10 kcal mol^{-1}.

Evidence from x-ray analysis of hydrates is usually not unequivocal about the nature of the proton environment since it is sometimes difficult to decide whether or not a water molecule is hydrogen bonded to the proton. The presence of H_3O^+ has been established in $HClO_4 \cdot H_2O$ [30, 31], $HCl \cdot H_2O$ [32] and $H_2SO_4 \cdot H_2O$ [33]. $H_5O_2^+$ has been found in $HCl \cdot 2H_2O$ [34], $HCl \cdot 3H_2O$ [35] and $HClO_4 \cdot 2H_2O$ [36]. The $H_5O_2^+$ ion is characterised by the linking of pairs of H_2O molecules by a very short hydrogen bond (O—O ≈ 2.43 Å) with a possibly symmetrical potential energy curve. In $HClO_4 \cdot 2H_2O$, but not in the case of the two HCl hydrates, the $H_5O_2^+$ has a centre of symmetry.

In $(C_{14}H_{16}N_2)^{2+} 2Br^- \cdot H_2O \cdot \frac{1}{3}HBr$ (1,1'-tetramethylene-2,2'-bipyridylium dibromide·$H_2O \cdot \frac{1}{3}HBr$) and $[H_3O][O_3S \cdot C_6H_3(OH) \cdot CO_2H] \cdot 2H_2O$ (the trihydrate of 5-sulphosalicylic acid), it is possible[37, 38] that the proton is triply hydrated as $H_7O_3^+$ although it is not certain whether it is best to describe the structures as $H_7O_3^+$ or $H_5O_2^+ + H_2O$.

The ion $(Et_2O)_nH^+$ has recently been prepared[48] in combination with MCl_4^- ions (where M = Al, Fe, Ga, or In); n varies from 1.5 to 8, depending on M and the method of purification of the compounds. Water was rigorously excluded in the preparations and the compounds are viscous oils which behave as strong 1:1 electrolytes in nitromethane solution.

1.1.6 Ions in the gas phase

Many papers are published on species such as H_2^+, LiH^+, etc., but these are likely to be of interest only to theoretical chemists. An estimate of the proton affinity (P.A.) of H_2 (4.2 eV < P.A.(H_2) < 4.7 eV) resulted from a study of the reactions between H_3^+ and N_2, CO, CO_2, N_2O, NO, CH_4, C_2H_2, H_2O and NH_3 by use of the flowing afterglow technique[39]. For each of these mixtures only protonation was observed, but in other systems more complicated reactions occurred. The majority of the information obtained was of a kinetic nature but the upper and lower limits for the P.A. of H_2 were derived from the observation that proton transfer occurs from H_3^+ to N_2 but not to O_2.

H_n^+ ions have been detected by mass spectrometry[41] and Easterfield and Linnett[42] have calculated the stabilities and shapes of some of the species by use of the FSGTO method. The H_n^+ ions are best considered as H_2 molecules clustered around a H_3^+ nucleus, i.e., $H_5^+ = [H_3^+ \cdot H_2]$ and $H_9^+ = [H_3^+ \cdot (H_2)_3]$ and the bonding between the ion and molecules arises from polarisation of the molecule by the positive charge. The H_3^+ ion is an equilateral triangle in shape with l(H—H) = 0.88 Å; and for $H_3^+ \rightarrow H_2 + H^+$, $\Delta H^0 = +110$ kcal mol^{-1}; while for $H_5^+ \rightarrow H_3^+ + H_2$, $\Delta H^0 = 9$ kcal mol^{-1}. Similar calculations[42,43] were performed on Li$^+$(H_2)$_n$ ions and BeH$_2^{2+}$ which suggested that LiH$_2^+$ should be triangular and that LiH$_2^+$, LiH$_4^+$, LiH$_6^+$, etc., should be stable with respect to dissociation to LiH$_{12}^+$; e.g. for LiH$_2^+ \rightarrow$ Li$^+ + H_2$, $\Delta H^0 = 12$ kcal mol^{-1}; and for LiH$_4^+ \rightarrow$ LiH$_2^+ + H_2$, $\Delta H^0 = 11.5$ kcal mol^{-1}. It was predicted that BeH$_2^{2+}$ should be stable with respect to dissociation into H_2 (BeH$_2^{2+} \rightarrow$ Be$^{2+} + H_2$) but unstable with respect to dissociation into H$^+$(BeH$_2^{2+} \rightarrow$ BeH$^+ + H^+$)[42,43]. Some of these predictions concerning Li(H_2)$_n^+$ species have now been observed by mass spectrometry[44]. Almost identical results for LiH$_2^+$ were calculated independently by use of similar techniques[45] but this series of calculations included Li$_2$H$^+$ (predicted to be linear, symmetrical and stable by 63 kcal mol^{-1} with respect to LiH + Li$^+$) and Li$_3^+$.

1.1.7 Miscellaneous

Pure H_2S and D_2S have been prepared by the hydrolysis of aluminium sulphide ($Al_2S_3 + 6D_2O \rightarrow 2Al(OD)_3 + 3D_2S$)[46] and very accurate thermodynamic measurements were made on these species. D_2S is less volatile than H_2S below -48 °C and more volatile above this temperature. (P_D/P_H varies from 0.9848 at -78 °C to 1.0144 at 30 °C.)

The kinetics of the reaction $WF_6 + 3H_2 \rightarrow W + 6HF$ have been studied[47] and are thought to represent a mixture of homogeneous and heterogeneous pathways.

1.2 THE ALKALI METALS   Lithium, Sodium, Potassium, Rubidium, Caesium, Francium (Group IA)

1.2.1 Introduction

The chemistry of the alkali metals is essentially the chemistry of the M$^+$ ions in the solid state, in solution and in the gas phase. Interest in them is often confined to the study of size effects since the size differences in the series Li$^+$ (0.60), Na$^+$ (0.95), K$^+$ (1.33), Rb (1.48), and Cs$^+$ (1.67) often give rise to significantly different chemistry. (The numbers in parenthesis are Pauling's Ionic radii expressed in Å.) Rubidium and caesium salts are sometimes prepared in order to obtain a solid compound of a large anion of doubtful thermal stability, although cations such as NMe$_4^+$ are now more often used for this purpose. The 'stabilisation' of large anions by large cations can be explained simply[50] by use of the relationship between lattice energy and the radii of the ions making up the lattice.

There appears to be no current interest in the chemistry of francium, although it is often identified as a product in nuclear reactions. The longest lived isotope of francium is ^{223}Fr which has a half-life of only 21 min. All the chemistry of francium indicates that its properties are those expected by extrapolation from the Li–Cs series.

The crystal structures of rubidium and caesium salts are reported very frequently, partly due to the fact that these are 'heavy metals' and their presence simplifies the problem of interpreting x-ray diffraction data.

In recent years interest in the alkali metals has centred on (a) the nature of solutions of the alkali metals (and other metals to a lesser extent) in liquid ammonia; this work has been extended to include other solvents such as amines; (b) coordination compounds in which a variety of ligands form complexes with the M$^+$ ions; (c) the role of the ions in biological systems; (d) solvent extraction of MX compounds from water into non-aqueous solvents; (e) the preparation and properties of oxygen compounds such as superoxides and ozonides; and (f) the chemical physics of a number of small molecules in the gas phase and in inert matrices. In addition, almost all physical-chemical studies of systems involving ions usually include the alkali metal cations.

Reviews have been published on the structures and properties of alkali metal–ammonia solutions[51, 52] and two collections of papers on the same subject have appeared more recently[53, 54].

The proceedings of an international symposium devoted to the alkali metals have been published[55] and these include a number of papers which deal with ammonia solutions.

The role of alkali metal ions in biological systems has been reviewed by Williams[56] and by von Hippel and Schleich[57]. The preparation, properties and structures of peroxides, superoxides and ozonides[58], and ammoniates and amides[59] of (among other elements) the alkali metals have also been reviewed.

An article by Brewer and Brackett[60] on the dissociation energies of the gaseous alkali halides contains a valuable survey of experimental data on these species in the solid state as well as in the gas phase and Fajans[61] has attempted to clarify terms such as 'degree of polarity', '% ionic character', etc., with particular reference to gaseous alkali fluorides and SrO and BaO.

As part of a general review of nuclear magnetic resonance studies of ions in solutions, Hinton and Amis[62] devote a section to the ^{1}H n.m.r. shifts of water produced by salts such as M$^+$ClO$_4^-$, M$^+$NO$_3^-$ and M$^+$X$^-$.

Topics that will be omitted from the present review include such subjects as Molecular Beam Experiments with alkali metal atoms, the study of defect and doped alkali halide crystals and the use of alkali metal salts in molten salt studies where it is felt that the interest of the inorganic chemist in the alkali metal component is likely to be very small.

1.2.2 Solutions of the alkali metals

Early work in this field was confined to the study of metal–liquid ammonia solutions and involved chiefly the measurement of magnetic, electrical and spectroscopic properties as a function of concentration of the metal and efforts to obtain thermodynamic data for such solutions.

E.S.R. spectroscopy has proved to be an exceptionally powerful tool in these investigations[63] since it enables evidence to be obtained about the location of any *unpaired* electrons in solution. One difficulty in producing a satisfactory account of metal solutions in liquid ammonia arises from the great variation in properties (electrical conductivity, visible absorption spectrum, etc.) that accompanies changes in concentration; the concentration can be varied from very dilute solutions to extremely concentrated ones and thus it is unlikely that any one model will fit the whole concentration range. In addition, significant differences are found between solutions of different metals in liquid ammonia. Golden, Guttman and Tuttle[64], in 1966, summarised the evidence relating to dilute solution of the alkali metals and produced a model comprising the species M^+, M^-, S^- and M^+M^- and M^+S^- ion pairs to explain the observations. They also obtained values for the heats of solutions of the metals in liquid ammonia[65]. Evers and Longo[66] considered in detail the mechanism of conduction by dilute lithium solutions. The study[67] of ^1H n.m.r. spectra of alkali metal *salts* in liquid ammonia has yielded information about M^+ solvation, ion pairing and hydrogen bonding and has, therefore, helped in the consideration of solutions of metals in liquid ammonia which almost certainly contain M^+ ions.

Usually all studies on metals dissolved in liquid ammonia are made on thermodynamically unstable systems (though their rates of decomposition may be very small). Kirschke and Jolly[68], however, showed that it was possible to obtain solutions *at equilibrium* when alkali metal amides were dissolved in liquid ammonia in the presence of H_2 and that at equilibrium there was a measurable concentration of solvated electrons (measurable by e.s.r. and optical spectroscopy). They were able to measure directly K_s and ΔH_s^0 for the equilibria:

$$e_{am}^- + NH_3 \rightleftharpoons NH_2^- + \tfrac{1}{2}H_2 \quad K = 5 \times 10^4$$
$$Na^+ + e_{am}^- + NH_3 \rightleftharpoons NaNH_2(s) + \tfrac{1}{2}H_2 \quad K = 3 \times 10^9$$

Studies on liquid ammonia were extended to include alkali metals dissolved in primary amines[69] and in amine–ammonia mixtures[70]. An elegant experiment used an optically active amine [(−)1,2-diaminopropane] as solvent for potassium metal[71]. The lack of observed optical rotation at absorption bands associated with the solvated electron is evidence that the solvated electron is *not* closely associated with M^+, since M^+ is presumably in an optically rotating chelate environment.

Other solvents such as hexamethylphosphoramide[72] have also received attention and x-ray diffraction showed[73] that solid $Li(NH_3)_4$ consisted of close-packed $Li(NH_3)_4$ units in which the ammonia molecules were tetrahedrally disposed round Li.

More recently, sound velocity measurements in concentrated metal–ammonia solutions as a function of metal concentration show that velocity decreases steadily with increasing concentration for Na,K and Ba but that a minimum is observed with lithium and calcium[74, 75]. This anomalous behaviour by lithium and calcium is very probably associated with compound formation by these two metals, an explanation supported by the fact that these two metals are already known from phase-rule studies to be anomalous in this respect. The solubility of lithium in liquid ammonia as a function of temperature was measured incidentally in the course of these experiments and the results differ significantly from data published previously.

The Hall coefficient and electrical conductivity of solutions of Li in liquid ammonia have been measured[76] over the concentration range 1.2–14 mol % metal. The solutions behave as liquid metals above 9 mol % and below this concentration departures from 'metallic' behaviour occurs, while at concentration near 3 mol % the influence of the 'free' electron is no longer seen.

The e.s.r. spectrum of very dilute solutions of potassium in tetrahydrofuran (THF) consists of two quartets and a singlet[77] and dilution with THF increased the relative intensity of the singlet. The two quartets are identified as arising from unpaired electrons associated with ^{39}K and ^{41}K (both with a spin of $\frac{3}{2}$), while the singlet arises from an unpaired solvated electron, S^-. The K–THF system resembles the K–ethylamine system very closely. In contrast with THF as solvent, it was found[78] that cyclohexyl-18-crown-6 [see structures (10) and (11)] + THF or diethyl ester dissolved larger amounts of potassium metal to yield deep blue solutions which were comparatively stable (no visible signs of decomposition over several days at $-78\,°C$) and that the presence of the crown ligand changed the populations of various species in solution through shift of the equilibria:

$$M \rightleftharpoons M^+ + e^-$$
$$2M(s) \rightleftharpoons M^+ + M^-$$
$$M^- \rightleftharpoons M^+ + 2e^-$$

to the right. This result is readily understandable in view of the marked ability of the crown ligand to form complexes with M^+ ions. The result of the addition of the crown ligand was to give a solution in which the solvated-electron absorption band was absent and the concentration of unpaired electron spin was very low (only a small fraction of the total dissolved metal gave an e.s.r. pattern). This work, of an essentially preliminary nature, also included the similar behaviour of caesium metal in THF + crown ligand. Nicely and Dye[79] have also published the results of their investigations into the e.s.r. line-widths and solution stability of Cs in $EtNH_2$–NH_3 mixtures.

1.2.3 Oxygen compounds of the alkali metals

Oxygen combines with the alkali metals to form four well-defined series of compounds: the oxides (containing the O^{2-} ion), the peroxides (containing

O—O^{2-}), the superoxides (containing O—O^-) and the ozonides (containing O_3^-). The relative thermodynamic stabilities of these compounds (and hence the ease and method of preparation) are classic examples of the effect of changing the size of M^+. Lithium readily forms the oxide in direct reaction with oxygen, sodium gives predominantly the peroxide + a little oxide, while rubidium and caesium yield peroxides + a little superoxide. Also, lithium superoxide and ozonide have very poor thermal stability and can only be isolated at low temperatures. Most current work on these compounds is carried out in Russia.

A suspension of Li_2O_2 in Freon-12 at $-65\,°C$ gives a 45% yield of solid LiO_2 (the superoxide) when treated with ozone[80]. E.S.R. spectroscopy indicates the presence of the O_2^- ion in LiO_2 and x-ray diffraction suggests the compound has a similar structure to that of NaO_2. LiO_2 decomposes quantitatively[81] above $-35\,°C$. An investigation[82] of the reaction between lithium atoms and O_2 molecules involved the trapping of the products at high dilution in a noble gas or oxygen matrix. I.R. spectra of the product matrices indicated that LiO_2 was predominant, but $(LiO_2)_2$, LiO_2Li, and LiLiOO were also identified. LiO_2 has the shape of an isosceles triangle; the O—O stretching force constant is that obtained elsewhere for O_2^- and it is very probable that the isolated LiO_2 molecule is highly ionic with only electrostatic forces binding Li^+ to O_2^-. Similar results were obtained[83] with the matrix isolation method for $Na+O_2$, where NaO_2 (with a symmetrical triangular structure and highly ionic bonding between Na^+ and O_2^-), NaO_2Na and $(NaO_2)_2$ were identified in the reaction products.

In contrast to the instability of LiO_2, it is reported[84] that RbO_2 melts at 813 K under oxygen at atmospheric pressure without decomposition, though decomposition does occur before the melting point if the oxygen pressure is reduced. Recent work aimed at clarifying the mechanism of the production of rubidium ozonide, RbO_3, from rubidium superoxide on treatment with ozone, has shown that RbO_2 reacts rapidly with O_3 but not with oxygen atoms[85] over the temperature range -78–$20\,°C$.

$$RbO_2 + O \nrightarrow RbO_3; \quad RbO_2 + O_3 \rightarrow RbO_3 + O_2$$

This disproves the mechanism accepted previously for the production of KO_3 from KOH and O_3, which was thought to involve oxygen atoms as a necessary reactant.

Mass spectrometric studies[86] on the vapour derived from Na_2O identified the predominant species as $Na(g)$ and $O_2(g)$, with significant amounts of $NaO(g)$ and $Na_2O(g)$. Thermodynamic information obtained from these studies included heat of formation data ($\Delta H_f^0\,[NaO_{(g)}] = 24.3$ kcal mol^{-1}; $\Delta H_f^0\,[Na_2O_{(g)}] = -9.9$ kcal mol^{-1}) and bond-energy data $[D_0^0\,(NaO) = 60.3$ kcal mol^{-1}; $D_0^0\,(Na_2O) = 119.8$ kcal $mol^{-1}]$.

Phase diagrams have been published for the K–K_2O and Rb–Rb_2O systems. With potassium, no evidence was found for a lower oxide and the disproportionation reaction $2K_2O \rightarrow K_2O_2 + 2K$ was found to occur[87] at 446 °C. However with rubidium, a copper-coloured compound with composition Rb_3O was obtained which decomposed reversibly[88] at 48 °C. The disproportionation reaction of Rb_2O occurred at 410 °C; this decrease in temperature from potassium to rubidium illustrates the effect of increase in size of M^+.

1.2.4 Coordination of the alkali metal cations by chelate ligands

Until fairly recently rather few examples were known of the M^+ ions forming reasonably 'stable' coordination compounds in other than the solid state with ligands other than H_2O. For example, LiI forms complexes with ammonia, primary, secondary and tertiary amines, alcohols and SO_2; however, they all decompose on contact with water to give the $Li(OH_2)_4^+$ ion and also have a significant vapour pressure of ligand at equilibrium. The existence of a complex ion ML_n^+ in the presence of water means that the equilibrium

$$ML_n^+ + nH_2O \rightarrow M(OH_2)_n^+ + nL$$

must lie to the left and clearly for ligands such as ammonia it lies to the right. One way of describing such an equilibrium is in terms of hard and soft acids and bases; the M^+ cations are clearly hard acids and therefore form 'strong' complexes with oxygen-containing ligands.

Recently, Hands and Mercer[89] described a large number of coordination compounds in which four oxygen atoms of four Ph_3PO molecules coordinate Li^+ or Na^+ in combination with large anions such as I^-, ClO_4^-, BPh_4^- and NO_3^-. These compounds are decomposed only very slowly by water and probably represent the most stable coordination compounds of the alkali metals yet prepared with monodentate ligands. The stoichiometry of these Ph_3PO compounds can correspond to $MX \cdot 4Ph_3PO$ or $MX \cdot 5PH_3PO$ (e.g. $LiI \cdot 5Ph_3PO$, $LiBr \cdot 4Ph_3PO$) but it is very probable that both types of compound contain the $M(OPPh_3)_4^+$ ion in which the triphenylphosphine oxide molecules are tetrahedrally disposed around M^+ and, in the case of $MX \cdot 5L$ compounds, with a fifth Ph_3PO molecule 'loose' in the lattice, as shown[90] in $LiI \cdot 5Ph_3PO$. Attempts to prepare similar complexes for K, Rb and Cs salts and with lithium and sodium salts of small anions such as F^- and Cl^- were unsuccessful[89], this demonstrates the crucial importance of the size of cation and anion in determining the stability of solid compounds.

Nyholm, Truter and co-workers[91–93] have made extensive studies of the capabilities of Li^+, Na^+, K^+, Rb^+ and Cs^+ to form isolable complexes with the bidentate ligands: 8-hydroxyquinoline (4), isonitrosoacetophenone (2), 1-nitroso-2-naphthol (3), o-nitrophenol (4), o-aminobenzoic acid (5), 2,4-dinitrophenol (6), o-nitrobenzoic acid (7), salicylic acid (8), 1,10-phenanthroline (phen) (9), and 2,9-dimethyl-1,10-phenanthroline. Compounds of the general formula $ML \cdot nHL$ were obtained when MOH in EtOH was mixed with HL in EtOH for HL = (1)—(7), with $n = 0$, 1, or 2 depending on the reaction mixture, the nature of HL and the particular alkali metal, e.g., with 8-hydroxyquinoline $n = 1$ for Li, Na, K and $n = 2$ for K, Rb and Cs. To determine whether such compounds contained complex cations, conductivity of solutions in N-methyl pyrrolidone, i.r. spectra, melting points and x-ray diffraction were used as criteria. On the basis of these criteria it was concluded that the compounds formed by (1), (2) and (3) *did* contain complexed cations and should be written as $[ML(HL)_n]^0$, while the compounds formed by o-nitrobenzoic acid are acid salts with the structure $M^+ HL_2^-$. It was concluded that isolation of adducts in the solid state is favoured by an increase in the size of M^+. Mixed complexes of composition ML·HL' could also be prepared with a variety of ligands. The structures of some of these complexes are in process of being determined[93] by x-ray diffraction but it is likely that many of them will consist of polymers in order to give a coordination number of 6 or more to M^+; e.g. for ML(phen).

The single most important development in recent years in alkali metal chemistry has undoubtedly been the synthesis and investigation of complexes of the alkali metal ions with marcocyclic polyether ligands. Pederson[94], in

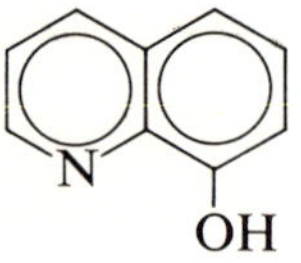

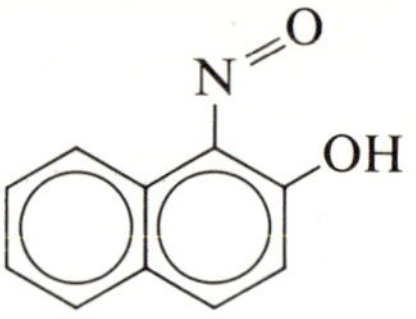

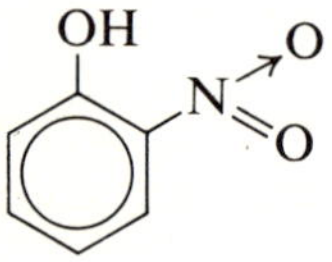

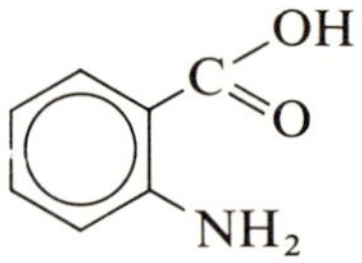

1967 published the syntheses of 33 cyclic polyethers derived from aromatic vicinal diols containing from 3 to 20 oxygen atoms. 15 of these compounds were then catalytically hydrogenated to give the corresponding saturated cyclic polyethers. He reported that many of the cyclic polyethers containing 5–10 oxygen atoms form stable complexes with a great variety of M^+, M^{2+} and M^{3+} ions. The structures of two typical crown ligands are illustrated in (10) and (11).

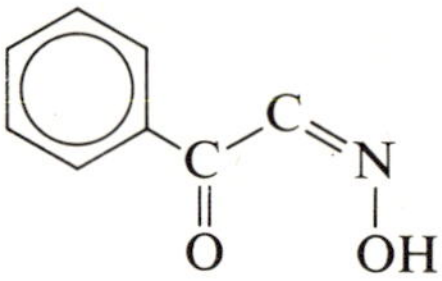

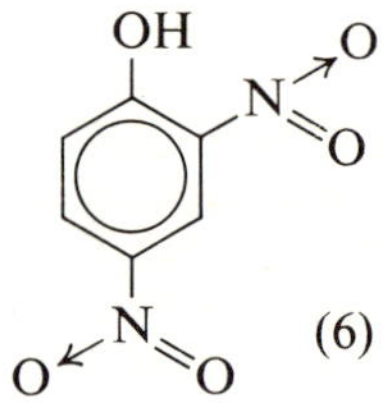

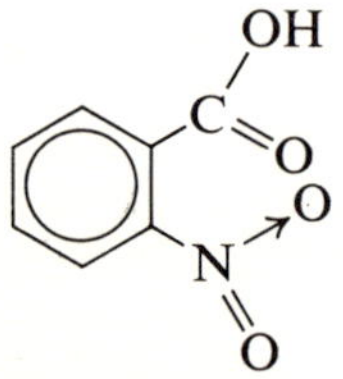

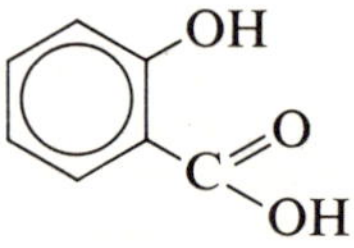

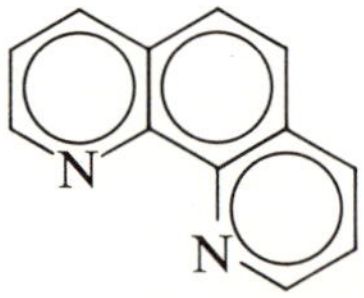

Pederson has investigated the stoichiometries of the crystalline complexes of Na^+, K^+, Rb^+, Cs^+, NH_4^+ and Ba^{2+} salts with various crown ligands in some detail[95] and found that polyether:cation ratios of 1:1, 3:2 or 2:1 could be obtained, depending on the relative sizes of the 'hole' in the poly-

Dibenzo-18-crown-6   Benzo-15-crown-5

(10)   (11)

ether and the cation and the reaction mixture used. For instance, dibenzo-18-crown-6 formed complexes of the following stoichiometry with alkali metal thiocyanates: $K^+ = 1:1$; $Rb^+ = 1:1$ and $2:1$; $Cs^+ = 2:1$ and $3:2$. By use of molecular models, he estimated the sizes of holes in the polyethers to be:

Polyether	Diameter of hole/Å	Diameter of M^+ ions/Å
All 14-crown-4	1.2–1.5	$Na^+ = 1.90$
All 15-crown-5	1.7–2.2	$K^+ = 2.66$
All 18-crown-6	2.6–3.2	$Rb^+ = 2.96$
All 21-crown-7	3.4–4.3	$Cs^+ = 3.34$

The connection between stoichiometry and size appears to be simple; if the cation is small enough to fit *inside* the ring, then 1:1 complexes result, but if the cation is too large 2:1 and 3:2 complexes are obtained and Pederson suggests that the 2:1 and 3:2 complexes may have the 'sandwich' structures:

2:1 Complex   3:2 Complex

The stabilities of complexes with crown ligands (deduced mostly from preparative work) are affected by the following factors[94]: (a) the relative sizes of the ion and the hole in the polyether ring, (b) the number of O-atoms in the polyether ring, (c) the symmetrical arrangement (or otherwise) of the O-atoms in the ring, (d) the coplanarity (or otherwise) of the O-atoms, (e) the basicity of the O-atoms, (f) steric hindrance in the rings, (g) the charge on the metal ion and (h) the tendency of M^{n+} to associate with the solvent.

The formation of these complexes can have interesting effects on solubility, e.g. dibenzo-18-crown-6 is only slightly soluble in methanol, but the solubility is greatly increased by the addition of any soluble salt of an alkali or alkaline-earth metal apart from Li^+, Mg^{2+} and Ca^{2+}. Also, almost quantitative separation of K^+ and Cs^+ was achieved by precipitating KSCN–crown solid from a mixture of KSCN and CsSCN in methanol by adding only sufficient dibenzo-18-crown-6 to complex K^+. Some of the polyethers form complexes, e.g., with KOH and $KMnO_4$, which can be extracted into aromatic hydrocarbons.

Recently, evidence has been published[96] that the order of preference of dimethyldibenzo-18-crown-6 for cations is a function of the solvent; the order $Na^+ \gg K^+ > Cs^+ > Li^+$ was observed for the fluorenyl salts in THF, but in oxetane, $K^+ > Na^+$.

Similarly, complex formation constants of sodium, potassium and caesium with the antibiotic nonactin were obtained[144] in dry acetone-d_6 and in acetone-d_6 containing up to 0.5 mole fraction of water by use of 220 MHz–H n.m.r. of the six protons in nonactin.

	Complex formation constants		
	Na^+	K^+	Cs^+
Dry acetone	7×10^4	7×10^4	1×10^4
Wet acetone	210	2×10^4	400

The difference in behaviour between dry and wet acetone can be rationalised in terms of the differences in solvation of the three ions in the solvents. The n.m.r. results also showed that the nonactin ring underwent sizeable conformation changes (though not identical ones for Na^+, K^+ and Cs^+) on formation of the complexes.

The crystal structures of four different crown complexes have been published by Truter and co-workers[97,98]: those of NaBr·(dibenzo-18-crown-6)·2H_2O, $Rb_{0.55}Na_{0.45}$NCS·(dibenzo-18-crown-6), NaI·(benzo-15-crown-5)·H_2O, and KI·(dibenzo-30-crown-10). Several interesting features are present in these compounds. The large ring systems (e.g., dibenzo-30-crown-10) can completely replace the hydration sphere of M^+ but the small rings can only do so partially and as a result in the small ring compounds the anion and/or water molecules are coordinated to M^+. The mixed $Rb^+ + Na^+$ complex contains Rb^+ and Na^+ in differing environments and the unit cell also contains a molecule of uncomplexed dibenzo-18-crown-6 which has essentially identical bond-lengths and angles but a different conformation from that of the complexed molecules[97].

The study of these crown complexes is partly motivated by a search for model compounds which can reproduce the selectivity between K^+ and Na^+ which is a characteristic of biological systems *in vivo*[56,97]. Three cyclic antibiotics have been shown to form 1:1 complexes with potassium salts, KX, and the crystal structures of the complexes have been determined: KNCS·(nonactin)[99,100], $KAuCl_4$·(valinomycin)[101], and KI·(enniatin B)[102], which all

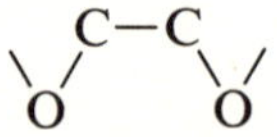

$$\diagdown_O^{C-C}\diagup_O\diagdown$$

(12)

differ in ring size and do not contain groups typical of crown ligands [structure (12)]. The coordination geometry of oxygen atoms around K^+ is very similar in KSCN·(nonactin) and KI·(dibenzo-30-crown-10). However, natural products are known which *do* contain groups as in (12), e.g. monesin[103] and the antibiotic X-537 A [104]; these ligands form complexes with alkali metal salts which are soluble in benzene, ether, etc., and insoluble in water.

Similar ligands to the crown molecules but containing some 'softer' S atoms in place of 'hard' O atoms have been synthesised and their complexes with transition-metal ions such as Co^{2+} and Ni^{2+} investigated[105, 106].

Diazopolyoxa-macrobicyclic ligands with similar properties to the crown compounds (cryptates) have been synthesised[107], their complexes with M^+

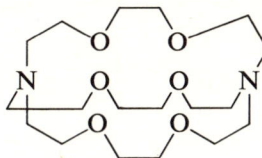

Cryptate ligand

(13)

and M^{2+} ions prepared and measurements of equilibrium constants reported[108] [see structure (13)]. The crystal structure of $RbSCN \cdot (cryptate) \cdot H_2O$ shows[109] the ligand to have the *endo-endo* conformation with Rb^+ coordinated by six oxygen and two nitrogen atoms in a polyhedron approximating to a bicapped trigonal prism. Cryptate complexes of alkali metal salts show the expected solubility in $CHCl_3$, methanol, acetone, etc. Preliminary determinations of equilibrium constants for the reaction (cryptate) $+ M^{n+}, mH_2O \rightleftharpoons (M^{n+} \cdot cryptate) + mH_2O$ yielded the following values for log K:

$$K^+ = 5.1; Rb^+ = 3.7; Na^+ = 3.6; Li^+ \approx 0.0; Cs^+ \approx -0.7$$

This demonstrates quantitatively the crucial importance of ionic size in determining the thermodynamic stabilities of these complexes.

Interesting complexes of the alkali metals with tripyrrene-α-carboxylic acid (H_3L) of stoichiometry MH_2L have been prepared which, it is thought, have a helical arrangement of several tripyrrenic acid molecules round the metal ion[110].

Russian workers have measured optical rotatory dispersion curves for chelate complexes[145] of Li^+, K^+, Cs^+ (and other ions) with optically active N-alkyl-α-phenyl-ether-dithiocarbamate ligands, and found that increasing the radius of the metal ion resulted in increased molecular rotation of the complex.

An interesting compound of aluminium, $[Na(C_4H_8O)_2]_2[Al(CH_3)_2C_{10}H_8]_2$ has been shown, from x-ray diffraction data[111], to contain $Na \cdot (THF)_2^+$ species as counter ions, with a rather complicated coordination environment around Na^+.

The variety of coordination number for M^+ ions in the solid state is well illustrated by the summary of some recent x-ray crystallographic publications. The expected pattern is seen to emerge in which the larger ions Rb^+ and Cs^+ commonly have large coordination numbers, and the smaller ions Li^+ and Na^+ are commonly 4- and 6-coordinated, respectively. Only rarely do the coordination polyhedra correspond to regular structures such as tetrahedra and octahedra. It must be borne in mind that the assignment of coordination is often a rather arbitrary process of deciding which atoms to include or exclude on the basis of observed inter-atomic distances.

Coordination numbers of M^+ *in recent publications*

Lithium:
    4: $LiI \cdot 2(NH_2)_2CO$ (Ref. 112); $Li_4H_2Si_2O_7$ (Ref. 126)
       $LiOAc \cdot 2H_2O$ (Ref. 138); NH_4LiSO_4 (Ref. 130)
 4+5: $Li_6Si_2O_7$ (Ref. 127)
    6: $Li_3Fe(MoO_4)_3$ (Ref. 117); $LiCrS_2$ (Ref. 141)

Sodium:
    6: $NaY[GeO_4]$ (Ref. 137); $NaAgS_2O_3 \cdot H_2O$ (Ref. 120)
       $Na_2CO_3 \cdot 10H_2O$ (Ref. 140); Na_2CO_3 (Ref. 132)
       $Na_2HAsO_4 \cdot 7H_2O$ (Ref. 121); $NaC_4H_3O_4$ (Ref. 123)
       $Na_6(AsO_4)_2NaF \cdot 19H_2O$ (Ref. 125); $Et \cdot COS_2Na \cdot 2H_2O$ (Ref. 135)
       $KNaThF_6$ (Ref. 118); $Rb_{0.55}Na_{0.45} \cdot NCS \cdot$(dibenzo-18-crown-6) (Ref. 97)
       $NaI \cdot$(benzo-15-crown-5)$\cdot H_2O$ (Ref. 98); $NaC_5H_3N_4O_2 \cdot 4H_2O$ (Ref. 139)
 6+7: $Mn_7Na_{12}(SO_4)_{13} \cdot 15H_2O$ (Ref. 119); Na_2ZrSiO_5 (Ref. 136)
    7: $Na_2CO_3 \cdot H_2O$ (Ref. 128)
    8: $NaIO_4$ (Ref. 122); $NaBr \cdot$(dibenzo-18-crown-6)$\cdot 2H_2O$ (Ref. 98)

Potassium:
 4+5: $K_2C_4H_2O_4 \cdot 2H_2O$ (Ref. 124)
    5: $KOCH_3$ (Ref. 143)
    6: $K_2ZrSi_2O_7$ (Ref. 115); $KZr_2(PO_4)_3$ (Ref. 134)
       $KI \cdot$(enniatin B) (Ref. 102); $KAuCl_4 \cdot$(valinomycin) (Ref. 101)
    8: $KNCS \cdot$(nonactin) (Ref. 99,100)
    9: $KNaThF_6$ (Ref. 118)
 9+10: K_2SeO_4 (Ref. 114)
  10: $KI \cdot$(dibenzo-30-crown-10) (Ref. 98)
  11: $K_2Ca_2Mg(SO_4)_4 \cdot 2H_2O$ (Ref. 116)

Rubidium:
    5: $RbOCH_3$ (Ref. 143)
    7: $Rb_{0.55}Na_{0.45}NCS \cdot$(dibenzo-18-crown-6) (Ref. 97)
    7: Rb salt of 5,5-dimethyl-1,3-cyclohexanedione-2-sulphuric acid (Ref. 133)
 7+8: $Rb_2[S_2C_2(CN)_2] \cdot H_2O$ (Ref. 129)
    8: $RbNCS \cdot$(cryptate)$\cdot H_2O$ (Ref. 109); $RbHCO_3$ (Ref. 131)
       $RbI \cdot$(thiourea)$_4$ (Ref. 113)

Caesium:
    5: $CsOCH_3$ (Ref. 143)
    8: $CsCl \cdot$(thiourea)$_4 \cdot H_2O$ (Ref. 113); $CsF \cdot$(thiourea)$_4 \cdot 2H_2O$ (Ref. 113)
10+11: $Cs_4Mg_3F_{10}$ (Ref. 142)

1.2.5 Solvent extraction of alkali metal ions from aqueous solution

A considerable amount of work relating to solvent extraction of alkali metal ions with non-aqueous solvents has been published. Much of this originates in a need for convenient practical methods of separating the alkali metals from other elements and from each other, with particular application to the separation of elements in used reactor fuels. Russian workers have also been investigating the possible use of solvent extraction for separating the lithium isotopes. The chemistry of solvent extraction is very closely related to the chemistry of complex formation, with the interest centred on the formation of complex ions in two different liquid phases.

Extraction of alkali metals into nitrobenzene in the presence of a halogen and a halide (or pseudo-halide)[146] can give quite good separations of the alkali metals, but high extraction exchange constants (of the order of 10^4 between Cs^+ and Li^+) can be obtained using dipicrylamine[147] or 2,4-dinitro-*N*-picryl-1-naphthylamine[148]. Maksimovic and Kyrs[149] have also studied in detail the extraction of Cs^+, Rb^+ and K^+ into nitrobenzene and 2-nitropropane by use of $AuCl_4^-$ and $FeCl_4^-$ as the counter ions and obtained values of the enthalpy and entropy of extraction. They found that the extraction of Cs^+ into the organic solvent is accompanied by an unfavourable entropy change and they attribute this to the fact that the lightly solvated Cs^+ ion is replaced on extraction into the organic solvent by the strongly solvated H^+ ion.

Russian workers, investigating the extraction of Rb^+ and Cs^+ into benzene, nitrobenzene, carbon tetrachloride, etc. solutions of phenol as a function of pH, found quite small single-stage separation factors

$$(\beta_{Cs}/\beta_{Rb} = 5.1)\ [150]$$

Measurements of $^6Li/^7Li$ separations in solvent extraction by use of lithium enriched with ^6Li have been reported for the extraction of lithium halides by isopentyl alcohol[151] and of lithium ions with such extractants as trioctylphosphine oxide, di-isoamyl methyl phosphonate, tributyl phosphate and acetylacetone in isopentyl alcohol[152]. Single-stage isotope effects in the range 1.005–1.010 were typical and these could well produce useful separations in counter-current extraction. The distribution of the carbonates and acetates of Na^+, K^+ and Rb^+ between water and a variety of immiscible alcohols have been measured[153].

1.2.6 Preparation and reactions of alkali metal compounds

Cs^+ was used as a large counter-ion in the first isolation of a difluorochlorate salt: $CsF + ClO_2F \rightarrow CsClO_2F_2$. The compound is reported to be thermally stable at 25 °C, but at 80 °C it decomposes to reactants. It is an extremely reactive material, fuming in moist air and reacting with all possible i.r. matrices and mulls. In this preparation, dry finely divided CsF was prepared by the thermal decomposition, *in vacuo*, of caesium heptafluoro-isopropoxide[155].

Again, Cs^+ and Rb^+ were used as large cations in the isolation of solid $M[Al(ClO_4)_4(H_2O)_2]$, in which it is suggested, from i.r. spectra, that perchlorate oxygen atoms are coordinated to Al [156]. The Rb^+ and Cs^+ salts of $[Al(ClO_4)_4(H_2O)_2]^-$ were obtained as colourless crystalline powders by addition of $AlBr_3$ to $RbClO_4$ or $CsClO_4$ in anhydrous $HClO_4$. (The coordination water molecules came from dehydration of $HClO_4$ or were produced in the main reactions.) Cs_2SnF_6 and Rb_2SnF_6 are much more soluble in liquid HF than in water. Phase-rule studies show the existence of $Rb_2SnF_6 \cdot 4HF$ and $Cs_2SnF_6 \cdot 4HF$ solvates[157]. The preparation of compounds of the type $M^IM^{II}Cl_3$ (where M^I = Rb or Cs and M^{II} = Ca, Sr, Cd, Ni) has been described, together with a review of previous work in this field[158].

Very high yields (>90%) of $Na_2S_2O_4$ were obtained when Na–Hg and SO_2 were reacted together in $HCONMe_2$, or Me_2SO; lower yields resulted when other organic solvents were used. High concentrations of radical-ion reaction intermediates were detected in some of the solvents[159]. A convenient preparative method for lithium amalgam is addition of small pieces of lithium metal to a mixture of mercury and refluxing p-cymene[160].

K_3NbO_4 can be obtained[161] from $K_2O + Nb$ at 260 °C or from liquid $K + Nb_2O_5$, NbO_2, or NbO at 600 °C. Alkali metal salts of MnO_2^- and MnO_4^{3-} can be prepared in a variety of ways, e.g., $RbOH + Mn_2O_3 \rightarrow RbMnO_2$; $MnCO_3 + Na_2CO_3 \rightarrow NaMnO_2$; $MnCO_3 + K_2CO_3 \rightarrow K_3MnO_4$; $RbO_2 + MnO_2 \rightarrow Rb_3MnO_4$. $KMnO_2$ and $RbMnO_2$ are extremely sensitive to air, the manganese is readily oxidised to higher oxidation states by oxygen[162].

The preparation (by addition of magnesium to an alkali metal dissolved in liquid ammonia), structural characteristics and properties of double amides of the formula $M_2Mg(NH_2)_4$ (where M = Na,K,Rb,Cs) have been described[163]. Thermal decomposition to the double imide, e.g. $Rb_2Mg(NH_2)_4 \rightarrow Rb_2Mg(NH)_2 + 2NH_3$ and subsequently to Mg_3N_2 (with Na and Rb) or KMgN (the double nitride), was studied and, in general, increasing size of M^+ was reflected in enhanced thermal stability of the amide.

The use of $LiAl(PH_2)_4$ as a phosphinating agent for the preparation of phosphinosilanes and germanes[164] and the preparation and reactions[165] of what may be $LiAl(AsH_2)_4$ have been described.

A new convenient preparation of alkali metal cyanates from the corresponding cyanide by oxidation with H_2O_2 or cumene hydroperoxide gives high yields of pure crystalline cyanates[166]. The kinetics and thermodynamics of the $LiBr(s) + NH_3(g) \rightleftharpoons LiBr \cdot NH_3(s)$ system over the temperature range 80–130 °C have been described[167].

The synthesis and crystal growth of $LiSrH_3$ and $LiEuH_3$ from the hydrides[168] and phase rule studies on the $LiH–EuH_2$ system[20] suggest that these are genuine compounds which are not grossly non-stoichiometric and probably have perovskite-type structures.

Dilts and Ashby[169] have examined in detail the reactions between trimethylamine, in a variety of solvents, and a number of ternary hydrides, including $LiBH_4$, $LiAlH_4$, and $NaAlH_4$. To some extent they have re-examined and clarified (and sometimes disproved) earlier work. With $LiAlH_4$ in Et_2O and Me_3N they found a reversible formation of the amine solvate precipitate, $LiAlH_4 \cdot NMe_3$. In the presence of excess NMe_3, the solvate then reacted very slowly to give the alane adduct, $AlH_3 \cdot 2NMe_3$ with an overall reaction:

$$3LiAlH_4 + 4Me_3N \rightarrow Li_3AlH_6 + 2AlH_3 \cdot 2NMe_3$$

$LiBH_4$ only yielded the solvate $LiBH_4 \cdot NMe_3$ while $NaAlH_4$ gave neither a solvate nor the alane adduct. The authors, as a result of their experiments with many other hydrides, conclude that there is correlation between ionic/covalent character in the ternary hydride and behaviour towards NMe_3. Thermal decomposition of $NaAlH_4$ and $KAlH_4$ to give Na_3AlH_6 and K_3AlH_6 and, at higher temperatures, NaH, H_2, Al and KH, H_2 and Al, respectively, has been studied by differential thermal analysis[170]. Evidence suggested that the steps: $KAlH_4 \rightarrow KH + Al + \frac{3}{2}H_2$, $\Delta H = +26$ kcal mol^{-1} and $2KH + KAlH_4 \rightarrow K_3AlH_6$, were responsible for the production of the new compound K_3AlH_6, which was also prepared by direct reaction of K, Al and H_2.

The main thermal decomposition reactions of Na_2SO_4 and Na_2CO_3, over the temperature range 500–1200 K, detected mass spectrometrically[171], are $Na_2SO_4 \rightarrow Na_2O + SO_2 + \frac{1}{2}O_2$ and $Na_2CO_3 \rightarrow Na_2O + CO_2$, although other species such as SO, C, CO and O were formed by secondary decomposition of SO_2 and CO_2.

Thermal stability varies in the following order: $MIO_4 < MBrO_3 < MClO_3 < MIO_3 < MClO_4$, where M = K, Rb or Cs. Apart from $MClO_4$ and MIO_3, the first steps in the decomposition reactions are exothermic, which explains the explosive behaviour of many of the salts[172].

Mass spectrometric study of the vapour obtained from Na_2O showed[173] that the predominant species were Na(g) and O_2(g), together with small amounts of NaO(g) and Na_2O(g). The data obtained yielded the following heat of formation and bond-energy data:

$$\Delta H_f^0[\text{NaO(g)}] = 24.3 \text{ kcal mol}^{-1}$$
$$\Delta H_f^0[\text{Na}_2\text{O(g)}] = -9.9 \text{ kcal mol}^{-1}$$
$$D_0^0(\text{NaO}) = 60.3 \text{ kcal mol}^{-1}$$
$$D_0^0(\text{Na}_2\text{O}) = 119.8 \text{ kcal mol}^{-1}$$

Addison and Davies have investigated in detail[181] the kinetics of the reaction of gaseous N_2 with molten Li, and Li–Hg or Li–Na mixtures. Unlike Na, the addition of Hg causes profound changes in the reactivity of Li. With 1:1 Li:Hg, no reaction occurs and with mixtures richer than this in Li, only sufficient reaction to yield Li_3N occurs to convert the mixture to 1:1 Li:Hg stoichiometry.

1.2.7 Chemical physics

A most important series of papers[174–176] by Kebarle and co-workers report the results of measurements and calculations of the heats and entropies of hydration of the alkali metal and halide ions in the gas phase. By use of a mass spectrometer they measured equilibrium constants for reactions of the type

$$M^+(H_2O)_{n-1} + H_2O \rightleftharpoons M^+(H_2O)_n$$

at different temperatures for all the alkali metals and for $n = 1-\sim 6$. Typical results obtained were:

$$M^+ + H_2O \rightleftharpoons M^+(H_2O)$$

	Li$^+$	Na$^+$	K$^+$	Rb$^+$	Cs$^+$
$-\Delta H^0$ (kcal mol-1)	(34.0)	24.0	17.9	15.9	13.7
$-\Delta S^0$ (cal deg^{-1} mol^{-1})	(23.0)	21.5	21.6	21.2	19.4
$-\Delta G^0$ (kcal mol-1)	(25.5)	17.6	11.4	9.6	7.9
and for Na$^+$(H$_2$O)$_{n-1}$ + H$_2$O $\rightleftharpoons$ Na$^+$(H$_2$O)$_n$					
$n-1,n$	0,1	1,2	2,3	3,4	4,5	5,6
$-\Delta H^0_{n-1,n}$	24.0	19.8	15.8	13.8	12.3	10.7
$-\Delta S^0_{n-1,n}$	21.5	22.2	21.9	25.0	28.1	26.0
$-\Delta G^0_{n-1,n}$	17.6	13.2	9.3	6.3	3.9	2.9

The authors were able to calculate[174] values for $\Delta S^0_{0,1}$ which were in excellent agreement with experiment. In general, the trend in $\Delta S°$ with increasing n results from the loss of freedom in the complexes due to crowding of ligands around the M^+ ion. The authors also attempted to calculate values of $\Delta H^0_{n-1,n}$ by calculating the potential energies of the complex ions by use of a purely electrostatic model, which gave:

P.E. of cluster = EDIP + EPOL + EDIS + RDIP + REL

where EDIP = ion−permanent dipole attraction, EPOL = ion−induced dipole attraction EDIS = attractive van der Waals forces between ion and H_2O and between H_2O and H_2O, RDIP = dipole–dipole repulsion and REL = electronic repulsion between ion and H_2O(H_2O–H_2O electronic repulsion is neglected). This provides a treatment similar to the text book treatments of lattice energy calculations in ionic solids, which had already been applied[177] to the hydration of K^+. The absolute values for $\Delta H^0_{n-1,n}$ calculated in this manner were much larger than the experimental values but the authors felt that a comparison between experiment and calculation of relative values was significant and indicated that there was significant covalent bonding in $Li^+(H_2O)$ and $Na^+(H_2O)$, i.e. dative bonding from the lone pair of the oxygen to the lowest empty orbital of M^+.

In comparing their experimental values of $\Delta H^0_{n-1,n}$ with published tabulations of ion heats of hydration[178] the authors find that their results cannot be reconciled with Latimer and Pitzer's heats of hydration (which, e.g., give larger heats of hydration to negative rather than positive isoelectronic ions, Br^- = 81.4 kcal (g-ion)$^{-1}$ Rb^+ = 69.2 kcal (g-ion)$^{-1}$ but can be reconciled quite well with Slanski and Randle's values. Extrapolation of the gas-phase data to moderately high values of n (≈ 10) gives an asymptotic approach to the Slanski and Randle data, which suggests that the heats of hydration of M^+ and X^- ions are determined essentially by the heats of attachment of the first 8–12 water molecules.

Burton and Daly[179] have published the results of MO studies of ion hydration for Li^+, Na^+, Be^{2+} and Mg^{2+} in which they calculated hydration energies and sizes and force constants for the ions treated as point charges. They obtained reasonable results for charge distribution and size and, for Be^{2+} and Na^+, heats of hydration.

Della Monica and Senatore[180] have calculated so-called 'solvated radii' for M^+ and M^{2+} ions in non-aqueous solutions from Stokes's radii for these ions (obtained from limiting equivalent conductivity and viscosity data). For all solvents but CH_3CN and sulpholane they obtain the order $Li^+ > Na^+ > K^+ > Rb^+ > Cs^+$ (as for water), while $Na^+ > Li^+$ in CH_3CN and sulpholane.

Kollman, Liebman and Allen[182] have examined the question of 'lithium bonding', by a review of the scanty experimental data available on A_2Li^- or ALi···B complexes and then a calculation on various systems and a comparison of their results with analogous H-bonded systems. For example, they calculate that a F—F distance of 3.57 Å should give the minimum energy for HF···LiF with a stabilisation energy with respect to HF and LiF of 13.5 kcal mol^{-1} (taking only s-orbitals into account) or 16.2 kcal mol^{-1} (by use of s- and p-orbitals). A similar calculation for LiF···HF yielded F—F = 2.38 Å and stabilisation energy = 24.7 kcal mol^{-1}., and the result that $Li^+HF_2^-$ is quite a good description of the bonding in this system, while $HFLi^+F^-$ is a better description of the first system. Their calculations also suggest that LiF·LiX mixed crystals might be described as $Li_2F^+ \cdot X^- \cdot$ (with Li_2F^+ linear by analogy with Li_2O), particularly if X^- is a large non-complexing anion such as SbF_6^-. Indeed, they observe that Cs_2F^+ may have already been detected in the conducting solution obtained by mixing CsF and ClF_3, where $Cs^+ClF_4^-$ can be eliminated as the cause of the conductivity[183].

Calculations have been performed on LiH_2^+, Li_2H^+ and Li_3^+ species[45] and on the cluster compounds[42, 43], $Li(H_2)_n^+$, which was followed by their mass spectrometric detection[44]. These cluster compounds have considerable stability in the gas phase, with dissociation energy for the process $Li(H_2)_n^+ \rightarrow Li(H_2)_{n-1}^+ + H_2$ in the region of 12–5 kcal mol^{-1}. Heitler–London calculations have been reported[40] for crystalline LiH, and SCF–LCAO–MO calculations for gaseous KCl, LiBr, NaBr and RbF [184] which yield reasonable values of dissociation energies, molecular force-constants and electric dipole moments.

Vapours of halides of alkali metals have received considerable attention. Mass spectrometric, torsion effusion, and molecular beam studies have yielded thermodynamic data[185] for gaseous MF, M_2F_2 and M_3F_3 species; these provide information for reactions such as $MF(s) \rightarrow MF(g)$, $MF(g) \rightarrow M^+(g) + F^-(g)$, $M_2F_2(g) \rightarrow 2MF(g)$ and $M_3F_3(g) \rightarrow 3MF(g)$. The i.r. spectra of the species in the vapour over LiF were obtained by means of the matrix isolation technique[186]. LiF, Li_2F_2 and Li_3F_3 were identified and it is probable that Li_2F_2 has a planar cyclic structure, Li_3F_3 is also probably cyclic. After their earlier work on CsOH vapour[187], Lide and co-workers have measured the microwave spectra of RbOH and RbOD, with a high-temperature spectrophotometer; they deduce that RbOH, like CsOH, is a linear molecule[188]. Cs_2O is polar (from measurement of electric deflection of molecular beams) and therefore presumably the molecule must be non-linear[189], while compressibility data for metal vapours indicate the presence of M_4 molecules in addition to the well-established M_2 species[190].

1.2.8 Miscellaneous

Considerable industrial interest has led to a continuation of work on lithium electrodes in propylene carbonate[191, 192, 201, 202] and the thermodynamics of solutions of alkali metal salts in propylene carbonate[193]. The melting point[194] of $^6LiNO_3$ is 0.03 °C higher than that of $^7LiNO_3$, while the transition to a fcc structure occurs 0.08 °C lower for $^6Li_2SO_4$ than for $^7Li_2SO_4$.

Electrical conductivity and density measurements of molten mixtures of UCl_4 and MCl are interpreted[195] in terms of M^+ and UCl_6^{2-} ions, with the stability of UCl_6^{2-} increasing with increasing size of M^+. Measurements of transport numbers in molten BeF_2 + LiF mixtures show that the mobility of Be^{II} is zero, while that of Li^+ is similar to that in $LiCl + PbCl_2$ melts[196].

Contrary to earlier work, close agreement with Debye–Hückel theory is reported for the solubilities of $KClO_4$, $RbClO_4$ and $CsClO_4$ in the presence of 1:1 electrolytes[197].

The thermodynamic complexity constant has been obtained for the ATP–K^+ complex by direct measurement of K^+ activities in solution[198].

^{23}Na n.m.r. relaxation time measurements for aminocarboxylic acid complexes of Na^+ (such as EDTA, HEEDTA) gave kinetic and thermodynamic information about the complexes in solution[199], while n.m.r. studies of solid Na_3P suggest[200] that the compound should be written as either $2Na[Na^{2-}P^{2+}]$ or $2Na^+[Na^-P^-]$.

Very careful re-examination of the $Na + K$ solid–liquid phase diagram has revealed that Na_2K is the only intermetallic compound formed by these elements[203].

1.3 THE ALKALINE EARTH ELEMENTS Calcium, Strontium, Barium and Radium (Group IIA)

1.3.1 Introduction

No recent chemistry of radium has been reported. The alkaline earth elements usually appear in research publications as 'typical' M^{2+} ions, often in contrast to 'untypical' M^{2+} transition metal ions. There has been a recent upsurge of interest in the biological roles of M^{2+} ions, with attempts to explain the thermodynamics of their complexes, reviews by Williams[56] and by Sigel and McCormick[204] on this aspect of alkaline earth chemistry have been published recently.

The gaseous molecules of Group IIA diatomic oxides[205] and triatomic dihalides[206] are favoured objects of thermochemical and chemical-physical research and results have been summarised in two review articles.

Alkaline earth phosphates have interesting optical properties and these together with preparative methods, thermodynamic information, and crystal-structure analyses, were reviewed[207] in 1961.

1.3.2 Preparations and reactions

Russian workers, making use of the unique solubility of $Me_4N^+O_2^-$ in liquid ammonia, have prepared calcium superoxide, $Ca(O_2)_2$, by mixing solutions of $Me_4N^+O_2^-$ and $Ca(NO_3)_2$ in liquid ammonia[208]. Investigations of the systems $MgO + ZnO$, $CaCO_3 + ZnO$, and $BaCO_3 + ZnO$, in CO_2 or O_2 atmospheres at elevated temperatures[209] produced evidence for the production of barium zincate only, no doubt because Ba^{2+} is larger than Ca^{2+} and Mg^{2+}. Strontium and barium thioborates, which analyse as $M_2B_2S_4$ and $M_2B_2S_5$, have been prepared and characterised[210]. Again the barium compounds are more stable than the strontium ones as regards decomposition.

Calcium, strontium and barium hydrazides, $M(NNH_2)$, have been prepared for the first time[211], by gently heating the nitrides with anhydrous hydrazine in ether. Strontium and barium hydrazides were also obtained directly from the metal. It is suggested that, being less reactive than the Group IA hydrazides, these hydrazides might become useful reagents. I.R. spectra establish that these compounds should be written as $M(N—NH_2)$ rather than $M(NH \cdot NH)$.

Maass has described and characterised a large number of strontium and barium phosphides[212]. Phase diagrams for the Sr–P and Ba–P systems were obtained and the crystal structure of Ba_3P_2 published. Compounds analysing as MP_2, M_4P_5, MP, M_4P_3, M_3P_2 and M_2P were obtained for both barium and strontium; barium in addition forms BaP_3, $BaP_{1.82}$ and $Ba_{1.1}P$. The phosphides are reactive compounds; in general they decompose in air and react vigorously with water. Many of them have highly polymerised structures and Ba in Ba_3P_2 is coordinated by six phosphorus and five barium atoms, with some negative charge located on P. The bonding appears to be intermediate between ionic and metallic in character.

The preparation of $M^+CaCl_3^-$ and $M^+SrCl_3^-$ has been described and the compounds compared to other $ABCl_3$ compounds[158].

1.3.3 Stability constants for alkaline earth metal ions

The Chemical Society's Special Publication[49] *Stability Constants* contains comprehensive data on the Group IIA M^{2+} ions. Stability constants for M^{2+} ions are frequently published, often after the synthesis of a new ligand; occasionally very accurate determinations or redeterminations are reported for well-known ligands. For example, the stability constants of MgF^+ and CaF^+ have been determined at various ionic strengths by use of a fluoride-sensitive electrode to further the understanding of the role of F^- in sea water.

No evidence for any stereospecific complex formation was obtained from accurate measurements of complexity constants for Mg^{2+}, Ca^{2+}, Sr^{2+} and Ba^{2+} with optically active and racemic ligands[214], when identical K values (within the limits of experimental error) were obtained for the optically active and racemic ligands: $(+)$pdta and $(\pm)$pdta, and $(+)$dimedta and $(\pm)$dimedta.

Earlier phase studies led Russian workers[215] to measure stability constants for Mg^{2+} and Ca^{2+} with urea. Hexamethylenediaminetetra-acetic acid (H_4Y) forms complexes of the types $[MHY]^-$, $[MY]^{2-}$ and $[M_2Y]$ with Mg^{2+}, Ca^{2+}, Sr^{2+} and Ba^{2+}, the corresponding equilibrium constants have been measured[216], and the results compared with those for similar ligands such as EDTA.

Martin and Uhlig[217] determined ΔH calorimetrically for complex formation by Mg^{2+}, Ca^{2+} and Sr^{2+} (and other M^{2+} ions) with the tetradentate ligands N-(2-carboxyphenyl)iminodiacetic acid (H_3A) and N-(2,5-dicarboxyphenyl)iminodiacetic acid (H_4B), for which stability constants had been previously published.

The obvious importance of phosphates as sequestering agents has led to a systematic investigation into trends in stability constants of Ca^{2+} complexes with phosphates. It was found that complex stability increases with increasing chain length and is also a function of the atoms which connect the phosphorus atoms[218].

The stability constants of Mg^{2+} and Ca^{2+} with six carboxylate anions have been reported[219]. The 'usual' order of stability was observed and the close relationship between pK_a of the ligands and stability constants found was evidence that the potentially multidentate ligands (acetoxyacetic acid and *N*-acetylglycine) were in fact monodentate.

Complex formation by alkaline earth cations with acidic polysaccharides is considered to be important in the cellular structures of plants. Recent work[220] suggested that there was no detectable complexing between Ca^{2+} and uronic acid monomers, but Gould and Rankin[221] have now shown that complexes *are* formed and have measured the stability constants for Mg^{2+} and Ca^{2+} with glucuronate, galacturonate, α-methyl glucuronosidate and α-methyl galacturonosidate by three different methods.

Much of the work on solvent extraction of alkaline earth metals also yields stability constant data and this is discussed in the following Section.

1.3.4 Solvent extraction

Very poor extraction of Ca^{2+}, Sr^{2+} and Ba^{2+} is achieved by dodecyliminodipropionic acid[222]. By choosing the correct pH it is possible to separate Ba^{2+}, Sr^{2+}, or Ca^{2+} from, e.g. scandium by extraction into methylisobutyl ketone with thenoyltrifluoroacetone[223]. Insoluble compounds of the type $M(H_2L)_2$ are formed between $M(OH)_2$ in H_2O and tripyrrene-α-carboxylic acid (H_3L) in $CHCl_3$. The distribution of these compounds between $CHCl_3$ and H_2O has been measured[110]. A detailed attempt has been made to investigate synergism, by use of the extraction of Sr^{2+} as the model system[224]. The extraction of strontium by a variety of di-n-alkylphosphoric acids showed that extraction increases with increasing chain length and also yielded stability constant data for the aqueous phase[225], while its extraction by dioctylphenylsulphonyl phosphoramidate has been studied by different workers[226].

Work with the cyclic polyether ligands (crown and cryptate) included the alkaline earth as well as the alkali metal ions and yielded similar results relating to sizes etc.[95, 108] (see Section 1.2.4). Preliminary determinations of stability constants with the cryptate ligand have been reported ($Ba^{2+} > Sr^{2+} > Ca^{2+}$) and, such is the stability of the barium complex, it was noted[108] that an aqueous solution of cryptate slowly dissolved $BaSO_4$. One of the natural products of which the crown ligands may be good models is the antibiotic X-537 A. The crystal structure of the barium complex of this antibiotic shows that the 9-coordinated barium atom is effectively 'sandwiched' between two antibiotic anions[227].

1.3.5 Environments in the solid state

A large number of crystal structures containing alkaline earth ions have been determined with barium featuring prominently as a 'heavy metal'. A selection of recent results which illustrates the variety of coordination environments about the M^{2+} ion follows. An interesting example of changing structure with changing atomic number is provided by $CaSi_2$, $SrSi_2$ and $BaSi_2$, all prepared directly from the elements and all having quite different structures;

Si_4 tetrahedra exist in $BaSi_2$, a three-dimensional Si_n network occurs in $SrSi_2$, while two-dimensional Si_n corrugated sheets are found[228] in $CaSi_2$.

	Coordination	Compound	Reference
Calcium			
	O_6	$CaCu_2O_3$	229
	O_6	$Ca_6(Si_2O_7)(SiO_4)(OH)_2$	230
	O_7	$CaO2Al_2O_3$	231
	O_7	$CaCl_2\cdot(glycylglycylglycine)\cdot3H_2O$	232
	O_7+O_8	$CaHSaO_4$	233
	O_8	$K_2Ca_2Mg(SO_4)_4\cdot 2H_2O$	116
	O_8	$CaBr_2\cdot 4C_4H_7NO_2$	240
Strontium			
	S_8	Sr_2GeS_4	234
Barium			
	N_9	$Ba(N_3)_2$	235
	O_9	Ba^{2+} salt of X-537A	227
	O_9	$\alpha\text{-}Ba[AlO(OH)_2]_2$	236
	O_{10}	$Ba_3Si_4Nb_6O_{26}$	237
	$O_{10}+O_6$	$Ba_3(VO_4)_2$	238
	O_{12}	$BaC_2O_4H_2C_2O_4\cdot 2H_2O$	239

1.3.6 Chemical physics

The equilibria involving the mono- and di-halides of the alkaline earth metals in the gas phase have been studied by mass spectrometry[241]. Bond-energy data for MCl and MCl_2 derived from these studies (which show some discrepancies with values obtained from electronic spectra) when compared with simple electrostatic calculations confirm that only the magnesium compounds are significantly covalent.

I.R. studies of CaF_2, SrF_2 and BaF_2 trapped in solid krypton matrices, from precise measurements of isotopic shifts of vibrational modes, suggest[242] that the apex angles decrease from 158 degrees with MgF_2 to 100 degrees with BaF_2. Rambidi has claimed that the data can equally well be rationalised on the basis of unsymmetrical linear structures[243]. The thermodynamic properties of calcium[244] and barium[245] ammoniates, $M(NH_3)_6$, have been determined and yield values for the gas-phase formation of the $M(NH_3)_6^{2+}$ species. Sound velocity measurements for liquid ammonia solutions of calcium and barium have been correlated with the fact that calcium but not barium forms a solid compound on freezing[75].

Calcium dissolves in liquid $CaCl_2$ according to the equation[246]: $Ca+Ca^{2+} \rightleftharpoons 2Ca^+$ or $Ca \rightleftharpoons Ca^{2+}+2e$, but not $Ca \rightleftharpoons Ca^0$ or $Ca+Ca^{2+} \rightleftharpoons (Ca_2)^{2+}$. (The 2e represents electrons behaving like F-centres, with thermodynamic properties similar to Cl^-.)

1.3.7 Miscellaneous

Phase rule data for methanol solvates of Mg, Ca, Sr and Ba perchlorates[247] is taken to be evidence that coordination number increases from six for

Mg^{2+} to eight for Ba^{2+}, although early work on equivalent conductivity indicates that the solvated radii of the alkali metal ions in methanol are approximately equal[180]. Free energies of transfer of $BaCl_2$ and $SrCl_2$ from H_2O to aqueous methanol have been measured by use of amalgam electrodes; calcium amalgam electrodes do not operate reversibly[248].

References

1. Aktar, S. and Smith, H. A. (1964). *Chem. Rev.*, **64**, 261
2. Green, M. L. H. and Jones, D. J. (1965). *Advan. Inorg. Chem. Radiochem.*, **7**, 115; Ginsberg, A. (1965). *Progr. Transition Metal Chem.*, **1**, 111
3. Ebisuzaki, Y. and O'Keefe, K. (1967). *Progr. Solid State Chem.*, **4**, 187
4. Metal Hydrides, Ed. by Müller, W. M. (1968). (New York: Academic Press); Mackay, K. M. (1966). *Hydrogen Compounds of the Metallic Elements* (London: Spon)
5. Wach, S. and Miodownik, A. P. (1970). *Trans. Faraday Soc.*, **66**, 2334
6. Petrunin, V. F., Somenkov, V. A., Shil'shstein, S. Sh. and Chertkov, A. A. (1970). *Kristallografiya*, **15**, 171
7. Fender, B. E. F. and Henfrey, A. W. (1970). *J. Chem. Phys.*, **52**, 3250
8. Ricca, F. and Giorgi, T. A. (1970). *J. Phys. Chem.*, **74**, 143
9. Holleck, G. L. (1970). *J. Phys. Chem.*, **74**, 503
10. Allard, K. D., Flanagan, T. B., and Wicke, E. (1970). *J. Phys. Chem.*, **74**, 298
11. Burch, R. and Lewis, F. A. (1970). *Trans. Faraday Soc.*, **66**, 727
12. Burch, R. (1970). *Trans. Faraday Soc.*, **66**, 736
13. Burch, R. (1970). *Trans. Faraday Soc.*, **66**, 749
14. Flanagan, T. B., Baranowski, B. and Majchrzak, S. (1970). *J. Phys. Chem.*, **74**, 4299
15. Züchner, H. and Wicke, E. (1969). *Z. Phys. Chem. (Frankfurt)*, **67**, 154
16. Azarkh, Z. M. and Gavrilov, P. I. (1970). *Kristallografiya*, **15**, 275
17. Bansal, B. M. and Damien, D. (1970). *Inorg. Nucl. Chem. Lett.*, **6**, 603
18. Alire, R. M., Mueller, B. A., Peterson, C. L. and Mosley, J. R. (1970). *J. Chem. Phys.*, **52**, 37
19. Ermolaev, M. I. (1970). *Russ. J. Inorg. Chem.*, **15**, 313
20. Mikheeva, V. I. and Kost, M. E. (1969). *Proc. Acad. Sci. Chem. Sect.*, **189**, 941
21. Warf, J. C. (1970). *Allg. Prakt. Chem.*, **21**, 189
22. Messer, C. E. (1970). *J. Solid State Chem.*, **2**, 144
23. Atoji, M. and Kikuchi, M. (1970). *J. Chem. Phys.*, **52**, 6434
24. Ashby, E. C., Kovar, R. and Arnott, R. (1970). *J. Amer. Chem. Soc.*, **92**, 2182
25. Behar, D., Czapski, G., Rabani, J., Dorfman, L. M. and Schwarz, H. A. (1970). *J. Phys. Chem.*, **74**, 3209
26. Bielski, B. H. J. (1970). *J. Phys. Chem.*, **74**, 3213
27. Claxton, T. A. and Symons, M. C. R. (1970). *Chem. Commun.*, 379
28. Nielson, S. O., Pagsberg, P., Rabani, J., Christensen, H. and Nilsson, G. (1969). *J. Phys. Chem.*, **73**, 1029
29. Beecham, A. F., Hurley, A. C., Mackay, M. F., Maslen, V. W. and Mathieson, A. McL. (1968). *J. Chem. Phys.*, **49**, 3312
30. Lee, F. S. and Carpenter, G. B. (1959). *J. Phys. Chem.*, **63**, 279
31. Nordman, C. E. (1962). *Acta Crystallogr.*, **15**, 18
32. Yoon, Y. K. and Carpenter, G. B. (1959). *Acta Crystallogr.*, **12**, 17
33. Taesler, I. and Olovsson, I. (1968). *Acta Crystallogr.*, **B24**, 299
34. Lundgren, J-O. and Olovsson, I. (1967). *Acta Crystallogr.*, **23**, 966
35. Lundgren, J-O. and Olovsson, I. (1967). *Acta Crystallogr.*, **23**, 971
36. Olovsson, I. (1968). *J. Chem. Phys.*, **49**, 1063
37. Derry, J. E. and Hamor, T. A. (1970). *Chem. Commun.*, 1284
38. Mootz, D., Altenburg, H., Fayos, J. and Wunderlich, H. (1969). *Acta Crystallogr.*, **A25**, S105
39. Burt, J. A., Dunn, J. L., McEwan, M. J., Sutton, M. M., Roche, A. E. and Schiff, H. I. (1970). *J. Chem. Phys.*, **52**, 6062
40. Fischer, C. R., Dellin, T. A., Harrison, S. W., Hatcher, R. D., and Wilson, W. D. (1970). *Phys. Rev.*, **B,1**, 876

41. Clampitt, R. and Gowland, L. (1969). *Nature (London)*, **223**, 815
42. Easterfield, J. and Linnett, J. W. (1970). *Chem. Commun.*, 64
43. Easterfield, J. and Linnett, J. W. (1970). *Nature (London)*, **226**, 142
44. Clampitt, R. and Jefferies, D. K. (1970). *Nature (London)*, **226**, 141
45. Ray, N. K. (1970). *J. Chem. Phys.*, **52**, 463
46. Clarke, E. C. W. and Glew, D. N. (1970). *Can. J. Chem.*, **48**, 764
47. Golovanov, Yu. N., Krasovskii, A. I., Chuzhko, R. K. and Kirillov, I. V. (1970). *Chem. Abstr.*, **72**, 71077
48. Clark, R. J. H., Crociani, B. and Wassermann, A. (1970). *J. Chem. Soc. A*, 2458
49. Martell, A. and Sillen, L. G., Eds. (1964). *Stability Constants, Special Publication No. 17*, (London: Chemical Society)
50. Johnson, D. A. (1968). *Some Thermodynamic Aspects of Inorganic Chemistry* (New York: Cambridge University Press)
51. Das, T. P. (1962). *Advan. Chem. Phys.*, **4**, 303
52. Jolly, W. L. (1959). *Progr. Inorg. Chem.*, **1**, 235
53. Le Poutre, G. and Sienko, M. J., Eds. (1964). *Metal–Ammonia Solutions* (New York: W. A. Benjamin)
54. *Metal–Ammonia Solutions 1970*, (1971). (London: Butterworth)
55. *The Alkali Metals:* An International Symposium held at Nottingham, July 1966. Special Publication No. 22 (1967). (London: The Chemical Society)
56. Williams, R. J. P. (1970). *Quart. Rev. Chem. Soc.*, **24**, 331
57. von Hippel, P. H. and Schleich, T. (1969). *Accounts Chem. Res.*, **2**, 257
58. Vannerberg, N-G. (1962). *Progr. Inorg. Chem.*, **4**, 125
59. Juza, R. (1964). *Angew. Chem. Int. Ed. Engl.*, **3**, 471
60. Brewer, L. and Brackett, E. (1961). *Chem. Rev.*, **61**, 425
61. Fajans, K. (1967). *Struct. Bonding*, **3**, 88
62. Hinton, J. F. and Amis, E. S. (1967). *Chem. Rev.*, **67**, 367
63. Catterall, R. and Symons, M. C. R. (1966). *J. Chem. Soc. A*, 13
64. Golden, S., Guttman, C. and Tuttle, T. R. (1966). *J. Chem. Phys.*, **44**, 3791
65. Tuttle, T. R., Guttman, C. and Golden, S. (1966). *J. Chem. Phys.*, **45**, 2206
66. Evers, E. C. and Longo, F. R. (1966). *J. Phys. Chem.*, **70**, 426
67. Allred, A. L. and Wendricks, R. N. (1966). *J. Chem. Soc. A*, 778
68. Kirschke, E. J. and Jolly, W. L. (1967). *Inorg. Chem.*, **6**, 855
69. Catterall, R., Symons, M. C. R. and Tipping, J. W. (1966). *J. Chem. Soc. A*, 1529
70. Catterall, R., Symons, M. C. R. and Tipping, J. W. (1967). *J. Chem. Soc. A*, 1234
71. Dowley, P. V., Gillard, R. D., Mitchell, P. R. and Price, M. G. (1968). *J. Chem. Soc. A*, 2502
72. Catterall, R., Stodulski, L. P. and Symons, M. C. R. (1968). *J. Chem. Soc. A*, 437
73. Mammano, N. and Sienko, M. J. (1968). *J. Amer. Chem. Soc.*, **90**, 6322
74. Bowen, D. E. (1969). *J. Chem. Phys.*, **51**, 1115
75. Bridges, R., Ingle, A. J. and Bowen, D. E. (1970). *J. Chem. Phys.*, **52**, 5106
76. Nasby, R. D. and Thompson, J. C. (1970). *J. Chem. Phys.*, **53**, 1970
77. Catterall, R., Slater, J. and Symons, M. C. R. (1970). *J. Chem. Phys.*, **52**, 1970
78. Dye, J. L., Debacker, M. G. and Nicely, V. A. (1970). *J. Amer. Chem. Soc.*, **92**, 5226
79. Nicely, V. A. and Dye, J. L. (1970). *J. Chem. Phys.*, **53**, 119
80. Vol'nov, I. I., Tokareva, S. A., Belevskii, V. N. and Klimanov, V. I. (1967). *Bull. Acad. Sci. USSR*, 1369
81. Bakulina, V. M., Takareva, S. A. and Volnov, I. I. (1967). *J. Struct. Chem.*, **8**, 980
82. Andrews, L. (1969). *J. Chem. Phys.*, **50**, 4288
83. Andrews, L. (1969). *J. Phys. Chem.*, **73**, 3922
84. Tsentsiper, A. B. and Dobrolyubova, M. S. (1968). *Bull. Acad. Sci. USSR*, 164
85. Tsentsiper, A. B. and Dobrolyubova, M. S. (1970). *Bull. Acad. Sci. USSR*, 644
86. Hildenbrand, D. L. and Murad, E. (1970). *J. Chem. Phys.*, **53**, 3403
87. Natola, F. and Touzain, P.(1970). *Can. J. Chem.*, **48**, 1955
88. Touzain, P. (1969). *Can. J. Chem.*, **47**, 2639
89. Hands, A. R. and Mercer, A. J. H. (1968). *J. Chem. Soc. A*, 449
90. Yasin, Y, M. G., Hodder, O. J. R. and Powell, H. M. (1966). *Chem. Commun.*, 705
91. Banerjee, A. K., Layton, A. J., Nyholm, R. S. and Truter, M. R. (1969). *J. Chem. Soc. A*, 2536
92. Banerjee, A. K., Layton, A. J., Nyholm, R. S. and Truter, M. R. (1970). *J. Chem. Soc. A*, 292

93. Layton, A. J., Nyholm, R. S., Banerjee, A. K., Fenton, D. E., Lestas, C. N. and Truter, M. R. (1970). *J. Chem. Soc., A,* 1894
94. Pederson, C. J. (1967). *J. Amer. Chem. Soc.,* **89,** 7017
95. Pederson, C. J. (1970). *J. Amer. Chem. Soc.,* **92,** 386
96. Wong, Y. H., Konizer, G. and Smid, J. (1970). *J. Amer. Chem. Soc.,* **92,** 666
97. Bright, D. and Truter, M. R. (1970). *J. Chem. Soc. B,* 1544
98. Bush, M. A. and Truter, M. R. (1970). *Chem. Commun.,* 1439
99. Kilbourn, B. T., Dunitz, J. D., Pioda, L. A. R. and Simon, W. (1967). *J. Mol. Biol.,* **30,** 559
100. Dobler, M., Dunitz, J. D. and Kilbourn, B. T. (1969). *Helv. Chim. Acta,* **52,** 2573
101. Pinkerton, M., Steinrauf, L-H. and Dawkins, P. (1969). *Biochem. Biophys. Res. Commun.,* **35,** 512
102. Dobler, M., Dunitz, J. D. and Krajewski, J. (1969). *J. Mol. Biol.,* **42,** 603
103. Agtarap, A., Chamberlain, J. W., Pinkerton, M. and Steinrauf, L.-K. (1967). *J. Amer. Chem. Soc.,* **89,** 5737
104. Westley, J. W., Evans, R. H., Williams, T. and Stempel, A. (1970). *Chem. Commun.,* 71
105. Black, D. St. C. and McLean, I. A. (1969). *Tetrahedron Lett.,* 3961
106. Rosen, W. and Busch, D. H. (1969). *Chem. Commun.,* 148
107. Dietrich, B., Lehn, J. M. and Sauvage, J. P. (1969). *Tetrahedron Lett.,* 2885
108. Dietrich, B., Lehn, J. M. and Sauvage, J. P. (1969). *Tetrahedron Lett.,* 2889
109. Metz, B., Moras, D. and Weiss, R. (1970). *Chem. Commun.,* 217
110. Plieninger, H. and Stumpf, K. (1970). *Chem. Ber.,* **103,** 2562
111. Brauer, D. J. and Stucky, G. D. (1970). *J. Amer. Chem. Soc.,* **92,** 3956
112. Verbist, J., Meulemans, R., Piret, P. and Van Meerssche, M. (1970). *Bull. Soc. Chim. Belg.,* **79,** 391
113. Boeyens, J. C. A. (1970). *Acta Crystallogr.,* **B26,** 1251
114. Kalman, A., Stephens, J. S. and Cruickshank, D. W. J. (1970). *Acta Crystallogr.,* **B26,** 1451
115. Chernov, A. N., Maksimov, B. A., Ilyukhin, V. V. and Belov, N. V. (1970). *Dokl. Akad. Nauk. SSSR,* **193,** 1293
116. Schlatti, M., Sahl, K., Zemann, A. and Zemann, J. (1970). *Chem. Abstr.,* **73,** 92455
117. Klevtsova, R. F. and Magarill, S. A. (1970). *Chem. Abstr.,* **73,** 102979
118. Brunton, G. D. (1970). *Acta Crystallogr.,* **B26,** 1185
119. Matzat, E. (1970). *Chem. Abstr.* **73,** 70757
120. Cavalca, L., Mangia, A., Palmieri, C. and Pelizzi, G. (1970). *Inorg. Chim. Acta,* **4,** 299
121. Ferraris, G. and Chiari, G. (1970). *Acta Crystallogr.,* **B26,** 1574
122. Kalman, A. and Cruickshank, D. W. J. (1970). *Acta Crystallogr.,* **B26,** 1782
123. Gupta, M. P. and Sahu, R. G. (1970). *Acta Crystallogr.,* **B26,** 1964
124. Gupta, M. P. and Sahu, B. N. (1970). *Acta Crystallogr.,* **B26,** 1969
125. Tillmanns, E. and Baur, W. H. (1970). *Naturwissenschaften,* **57,** 242
126. Völlenkle, H., Wittmann, A. and Nowotny, H. (1970). *Monatsh. Chem.,* **101,** 684
127. Völlenkle, H., Wittmann, A. and Nowotny, H. (1969). *Monatsh. Chem.,* **100,** 295
128. Dickens, B., Mauer, F. A. and Brown, W. E. (1970). *J. Res. Nat. Bur. Stand., Sect. A,* **74,** 319
129. Draeger, M. and Gattow, G. (1970). *Naturwissenschaften,* **57,** 195
130. Dollase, W. A. (1969). *Acta Crystallogr.,* **B25,** 2298
131. Kim, M. I. (1970). *Chem. Abstr.,* **72,** 304
132. Larsson, L. O. and Kierkegaard, P. (1969). *Acta Chem. Scand.,* **23,** 2253
133. Apinitis, S., Antsyshkina, A. S. and Jevins, A. (1970). *Chem. Abstr.,* **73,** 29751
134. Sljukic, M., Matkovic, B., Prodic, B. and Anderson, D. (1970). *Chem. Abstr.,* **72,** 60135
135. Mazzi, F., Tazzoli, V. and Ungaretti, L. (1970). *Chem. Abstr.,* **72,** 71763
136. Treushnikov, E. N., Ilyukhin, V. V. and Belov, N. V. (1970). *Dokl. Acad. Nauk. SSSR,* **190,** 334
137. Kuz'min, E. A., Maksimov, B. A., Ilyukhin, V. V. and Belov, N. V. (1970). *Chem. Abstr.,* **72,** 126022
138. Galigne, J. L., Mouvet, M. and Falgueirettes, J. (1970). *Acta Crystallogr.,* **B26,** 368
139. Mizuno, H., Fujiwara, T. and Tomita, K. (1969). *Bull. Chem. Soc. Jap.,* **42,** 3099
140. Taga, T. (1969). *Acta Crystallogr.,* **B25,** 2656
141. White, J. G. and Pinch, H. L. (1970). *Inorg. Chem.,* **9,** 2581
142. Steinfink, H. and Brunton, G. D. (1969). *Inorg. Chem.,* **8,** 1665

143. Weiss, E. and Alsdorf, H. (1970). *Z. Anorg. Allg. Chem.*, **372**, 206
144. Prestegard, J. H. and Chan, S. J. (1970). *J. Amer. Chem. Soc.*, **92**, 4440
145. Rukhadze, E. G., Pavlova, E. V., Dunina, V. V. and Terent'ev, A. P. (1969). *J. Gen. Chem. USSR*, **39**, 2486
146. Tribalat, S. and Grall, M. (1969). *Solv. Extr. Res., Proc. Int. Conf. Solvent Extr. Chem., 5th*, ed. Kertes, A. S. (New York: Wiley-Interscience)
147. Rais, J., Kyrs, M. and Pivonkova, M. (1968). *J. Inorg. Nucl. Chem.*, **30**, 611
148. Rais, J. and Kyrs, M. (1969). *J. Inorg. Nucl. Chem.*, **31**, 2903
149. Maksimovic, Z. and Kyrs, M. (1969). *Collect. Czech. Chem. Commun.*, **34**, 3436
150. Kuznetsova, E. M. and Medvedeva, L. V. (1969). *Russ. J. Phys. Chem.*, **43**, 1630
151. Rozen, A. M. and Mikhailichenko, A. I. (1970). *Russ. J. Phys. Chem.*, **44**, 974
152. Rozen, A. M., Mikhailichenko, A. I., Mamontova, E. P. and Khromov, Yu. F. (1970). *Russ. J. Phys. Chem.*, **44**, 977
153. Bakeev, M. I., Andamasov, R. S. and Akimova, I. P. (1970). *Chem. Abstr.*, **72**, 136974; Bakeev, M. I., Akimova, I. P. and Andamasov, R. S. (1970). *Chem. Abstr.*, **72**, 19041
154. Huggins, D. K. and Fox, W. B. (1970). *Inorg. Nucl. Chem. Lett.*, **6**, 337
155. Fraser, G. W. and Shreeve, J. M. (1967). *Inorg. Chem.*, **6**, 1711
156. Lemesheva, D. G. and Rosolovskii, V. Ya. (1969). *Bull. Acad. Sci. USSR, Chem. Sect.*, 1739
157. Tychinskaya, I. I., Yudanov, N. F. and Opalovskii, A. A. (1969). *Russ. J. Inorg. Chem.*, **14**, 1636
158. McMurdle, H. F., De Groot, J., Morris, M. and Swanson, H. E. (1969). *J. Res. Nat. Bur. Stand., Sect. A.*, **73**, 621
159. Rinker, R. G. and Lynn, S. (1969). *Ind. Eng. Chem., Prod. Res. Develop.*, **8**, 338
160. Alexander, J. and Rao, G. S. K. (1970). *J. Chem. Educ.*, **47**, 277
161. Addison, C. C., Barker, M. G. and Lintonbon, R. M. (1970). *J. Chem. Soc. A*, 1465
162. Scholder, R. and Protzer, U. (1969). *Z. Anorg. Allg. Chem.*, **369**, 313
163. Palvadeau, P. and Rouxel, J. (1970). *Bull. Soc. Chim. Fr.*, 480
164. Norman, A. D. and Wingeleth, D. C. (1970). *Inorg. Chem.*, **9**, 98
165. Anderson, J. W. and Drake, J. E. (1969). *Inorg. Nucl. Chem. Lett.*, **5**, 887
166. Nachbaur, E. (1969). *Monatsh. Chem.*, **100**, 1998
167. De Hartoulari, R. and Dufour, L. C. (1969). *Bull. Soc. Chim. Fr.*, 3017
168. Greedan, J. E. (1970). *J. Cryst. Growth*, **6**, 119
169. Dilts, D. A. and Ashby, E. C. (1970). *Inorg. Chem.*, **9**, 855
170. Dymova, T. N. and Bakum, S. I. (1969). *Russ. J. Inorg. Chem.*, **14**, 1683
171. Kosugi, T. (1970). *Chem. Abstr.*, **73**, 83428
172. Breusov, O. N., Kashina, N. I. and Revzina, T. V. (1970). *Russ. J. Inorg. Chem.*, **15**, 316
173. Hildenbrand, D. L. and Murad, E. (1970). *J. Chem. Phys.*, **53**, 3403
174. Dzidic, I. and Kebarle, P. (1970). *J. Phys. Chem.*, **74**, 1466
175. Arshadi, M., Yamdagni, R. and Kebarle, P. (1970). *J. Phys. Chem.*, **74**, 1475
176. Searles, S. K., Dzidic, I. and Kebarle, P. (1969). *J. Amer. Chem. Soc.*, **91**, 2810
177. Searles, S. K. and Kebarle, P. (1969). *Can. J. Chem.*, **47**, 2619
178. Desnoyers, J. E. *Modern Aspects of Electrochemistry*, Vol. 5, ed. by Bockris, J. O. M. (1969) (New York: Plenum Press)
179. Burton, R. E. and Daly, J. (1970). *Trans. Faraday Soc.*, **66**, 1281
180. Della Monica, M. and Senatore, L. (1970). *J. Phys. Chem.*, **74**, 205
181. Addison, C. C. and Davies, B. M. (1969). *J. Chem. Soc. A*, 1822, 1827, 1831
182. Kollman, P. A., Liebman, J. F. and Allen, L. C. (1970). *J. Amer. Chem. Soc.*, **92**, 1142
183. Lawless, E. W. and Smith, J. C. *Inorganic High Energy Oxidisers* (1968). (New York: Marcel Dekker)
184. Matcha, R. L. (1970). *J. Chem. Phys.*, **53**, 485, 4490
185. Chao, J. (1970). *Thermochim. Acta*, **1**, 71
186. Snelson, A. (1967). *J. Chem. Phys.*, **46**, 3652
187. Lide, D. R. and Kuczkowski, R. L. (1967). *J. Chem. Phys.*, **46**, 4768; Acquista, N., Abramowitz, S. and Lide, D. R. (1968). *J. Chem. Phys.*, **49**, 780
188. Matsumara, C. and Lide, D. R. (1969). *J. Chem. Phys.*, **50**, 71
189. Büchler, A., Stauffer, J. L. and Klemperer, W. (1967). *J. Chem. Phys.*, **46**, 605
190. Ewing, C. T., Stone, J. P., Spann, J. R. and Miller, R. R. (1967). *J. Phys. Chem.*, **71**, 473
191. Scarr, R. F. (1970). *J. Electrochem. Soc.*, **117**, 295
192. Meibuhr, S. G. (1970). *J. Electrochem. Soc.*, **117**, 56

193. Salomon, M. (1970). *Chem. Abstr.*, **73**, 59756
194. Jansson, B. and Lunden, A. (1970). *Z. Naturforsch.*, **25**, 697
195. Bogacz, A. and Ziolek, B. (1970). *Chem. Abstr.*, **73**, 70444
196. Romberger, K. A. and Braunstein, J. (1970). *Inorg. Chem.*, **9**, 1273
197. Guenther, W. B. (1969). *J. Amer. Chem. Soc.*, **91**, 7619
198. Rechnitz, G. A. and Mohan, M. S. (1970). *Science*, **168**, 1460
199. James, T. L. and Noggle, J. H. (1969). *J. Amer. Chem. Soc.*, **91**, 3424
200. Ossman, G. W., Silvidi, A. A. and McGrath, J. W. (1970). *J. Chem. Phys.*, **52**, 509
201. Butler, J. N., Cogley, D. R. and Synnott, J. C. (1969). *J. Phys. Chem.*, **73**, 4026
202. Jackson, G. W. and Blomgren, G. E. (1969). *J. Electrochem. Soc.*, **116**, 1483
203. Ott, J. B., Goates, J. R., Anderson, D. R. and Hall, H. T. (1969). *Trans. Faraday Soc.*, **65**, 2870
204. Sigel, H. and McCormick, D. B. (1970). *Accounts Chem. Res.*, **3**, 201
205. Schofield, K. (1967). *Chem. Rev.*, **67**, 707
206. Brewer, L., Somayajulu, G. R. and Brackett, E. (1963). *Chem. Rev.*, **63**, 111
207. Mooney, R. W. and Aia, M. A. (1961). *Chem. Rev.*, **61**, 433
208. Vol'nov, I. I., Tokareva, S. A. and Cherkasov, E. N. (1969). *Bull. Acad. Sci. USSR*, 1933, 1947
209. McAdie, H. G. (1968). *Therm. Anal., Proc. Int. Conf. 2nd*, **2**, 717
210. Chopin, F. and Capdepuy, B. (1970). *Bull. Soc. Chim. Fr.*, 505
211. Linke, K-H. and Taubert, R. (1970). *Z. Anorg. Allg. Chem.*, **376**, 289
212. Maass, K. E. (1970). *Z. Anorg. Allg. Chem.*, **374**, 1, 11, 19
213. Elgquist, B. (1970). *J. Inorg. Nucl. Chem.*, **32**, 937
214. Advan, A. T., Irving, H. M. N. H. and Pettit, L. D. (1970). *J. Chem. Soc. A*, 2649
215. Kaganskii, I. M. and Lopatina, N. P. (1970). *Russ. J. Inorg. Chem.*, **15**, 1208
216. Gorelov, I. P. and Kolosova, M. Kh. (1969). *Russ. J. Inorg. Chem.*, **14**, 1416
217. Martin, A. and Uhlig, E. (1970). *Z. Anorg. Allg. Chem.*, **376**, 282
218. Callis, C. F., Kerst, A. F. and Lyons, J. W. (1969). *Coord. Chem. Proc. John C. Barlar, Jr. Symp.*, 223
219. Bunting, J. W. and Thong, K. M. (1970). *Can. J. Chem.*, **48**, 1654
220. Kohn, R., Furda, I., Haug, A. and Smidsrod, O. A. (1968). *Acta Chem. Scand.*, **22**, 3098
221. Gould, R. O. and Rankin, A. F. (1970). *Chem. Commun.*, 489
222. Dilello, M. C., Iacobelli, C. T., Margani, A. and Palmera, M. (1970). *Chem. Abstr.*, **72**, 104523
223. Jackson, W. M., Gleason, G. I. and Hammons, P. J. (1970). *Anal. Chem.*, **42**, 1242
224. Sekine, T. and Hasegawa, Y. (1968). *Solvent Extr. Res., Proc. Int. Conf. Solvent Extr. Chem., 5th*, 289
225. Sistkova, N. V., Kolarik, S. and Chotivka, V. (1970). *J. Inorg. Nucl. Chem.*, **32**, 637
226. Shevchenko, F. D., Kuzina, L. A. and Ageev, V. A. (1969). *Russ. J. Inorg. Chem.*, **14**, 1620
227. Johnson, S. M., Herrin, J., Liu, S. J. and Paul, J. C. (1970). *Chem. Commun.*, 72; *J. Amer. Chem. Soc.*, **92**, 4428
228. Janzon, K. H., Schäfer, H. and Weiss, A. (1970). *Z. Anorg. Allg. Chem.*, **372**, 87
229. Teske, C. L. and Müller-Buschbaum, H. K. (1969). *Z. Anorg. Allg. Chem.*, **370**, 134
230. Ganiev, R. M., Ilyukhin, V. V. and Belov, N. V. (1970). *Dokl. Acad. Nauk, SSSR*, **190**, 831
231. Goodwin, D. W. and Lindop, A. J. (1970). *Acta Crystallogr.*, **B26**, 1230
232. Van der Helm, D. and Willoughby, T. V. (1969). *Acta Crystallogr.*, **B25**, 2317
233. Ferraris, G. and Chiari, G. (1970). *Acta Crystallogr.*, **B26**, 403
234. Ribes, M., Philippot, E. and Maurin, M. (1970). *C. R. Acad. Sci., Ser. C*, **270**, 1873
235. Choi, C. S. (1969). *Acta Crystallogr. Sect. B*, **25**, 2638
236. Ahmed, A. H. M. and Glasser, L. S. D. (1970). *Acta Crystallogr.*, **B26**, 1686
237. Shannon, J. and Katz, L. (1970). *Acta Crystallogr.*, **B26**, 105
238. Susse, P. and Buerger, M. J. (1970). *Chem. Abstr.*, **73**, 102981
239. Du Sansoy, Y., Protas, J., Mutin, J. C. and Watelle, G. (1970). *Acta Crystallogr.*, **B26**, 1567
240. Roux, J. P. and Boeyens, J. C. A. (1970). *Acta Crystallogr.*, **B26**, 526
241. Hildenbrand, D. L. (1970). *J. Chem. Phys.*, **52**, 5751
242. Calder, V., Mann, D. E., Seshadri, K. S., Allavena, M. and White, D. (1969). *J. Chem. Phys.*, **51**, 2093
243. Rambidi, N. G. (1969). *J. Struct. Chem.*, **10**, 118
244. Dickman, S., Senozan, N. M. and Hunt, R. L. (1970). *J. Chem. Phys.*, **52**, 2657

245. Mast, G. and Senozan, N. K. (1970). *J. Chem. Phys.*, **53,** 1296
246. Sharma, R. A. (1970). *J. Phys. Chem.*, **74,** 3896
247. Drakin, S. I., Karapet'yants, M. Kh. and Kurmalieva, R. Kh. (1969). *Russ. J. Inorg. Chem.*, **14,** 1421
248. Feakins, D. and Willmott, A. R. (1970). *J. Chem. Soc. A,* 3121

2
Beryllium, Magnesium, Zinc, Cadmium and Mercury

D. J. CARDIN
University of Sussex

2.1	INTRODUCTION	33
2.2	BERYLLIUM AND MAGNESIUM	34
	2.2.1 *Hydrides and related compounds*	34
	2.2.2 *Derivatives with Group IV ligands and associated groups*	37
	2.2.3 *Derivatives of Group V ligands*	39
	2.2.4 *Derivatives of Group VI ligands*	42
	2.2.5 *Derivatives of Group VII ligands*	45
2.3	ZINC, CADMIUM AND MERCURY	45
	2.3.1 *Hydrides and related compounds*	45
	2.3.2 *Group II ligands, Mercury(I) and related compounds*	46
	2.3.3 *Group III ligands*	46
	2.3.4 *Group IV ligands*	46
	2.3.5 *Group V ligands*	47
	2.3.6 *Compounds with Group VI ligands and associated ions*	50
	2.3.7 *Halogen derivatives*	53
	2.3.8 *Zinc, cadmium and mercury derivatives of transition metals*	53
2.4	TABULAR SURVEY	54
ACKNOWLEDGEMENTS		68

2.1 INTRODUCTION

This chapter reviews the inorganic chemistry of the title elements over the period 1969–1970, and includes reference to some important work from 1966 to 1969. While organometallic chemistry does not form a major part (it is the subject matter of Volumes 4 and 6 of this series), compounds having metal–carbon bonds are included where appropriate.

The chapter is divided into two parts: a descriptive text, which is followed by a tabular survey. The tables, which are ordered according to compounds, provide a brief description of the material of each paper cited, and are intended to be comprehensive. Limitations of space demand that the text be selective; however, text and tables follow roughly the same order of presentation to simplify the task of making cross references. In both sections compounds are described as far as possible in the order in which the coordinating atoms of ligands or associated ions occur (taken Group by Group) in the periodic table.

The greatest developments recently in the chemistry of Be, Mg, Zn, Cd and Hg have probably been made in the area of organic derivatives. This is particularly true of Be and Mg, and while the major part of the following text is concerned with inorganic chemistry, this fact is inevitably reflected to some extent. Hydrido and related compounds have also been much studied, as have the basic carboxylate salts of beryllium. A vast number of coordination compounds of the Group IIB elements, particularly zinc complexes of nitrogen donors, have been prepared. Many of these form part of a series of transition-metal complexes of a given ligand; the zinc compounds are valuable reference compounds to help distinguish effects (spectroscopic, stereochemical, etc.) associated with partially filled d-shells of other metals. The thermal behaviour of salts and coordination compounds has also been widely studied.

While the organic chemistry of the Group II elements has been extensively reviewed, the inorganic chemistry is less well documented. Books concerned with the inorganic chemistry of these elements include the titles: 'The Chemistry of Beryllium'[1], 'Beryllium'[2], 'Extraction and Metallurgy of Uranium, Thorium, and Beryllium'[3], and 'Mercury' (Part B of the Gmelin volumes appeared in 1969)[4]. Review articles have been published entitled: 'Metallurgical, Physical and Chemical Properties of Beryllium'[5], 'Some New Data on the Chemistry and Biochemistry of Beryllium'[6], 'The Chemistry of the Beryllides'[7], 'Magnesylamines'[8], 'New Theories of the Chemical Nature of Luminescence Centres of Zinc Sulphide Phosphors'[9], 'The Structural Chemistry of Mercury'[10], 'Chemistry of Halogenomercurate Complexes'[11], 'Some General Aspects of Mercury Chemistry'[12].

2.2 BERYLLIUM AND MAGNESIUM

2.2.1 Hydrides and related compounds

Several recent papers describe hydrido-beryllium compounds, particularly organo-beryllium hydrides. Tertiary amine complexes of beryllium hydride, $BeH_2 \cdot NR_3$, have been prepared from the hydride and amine heated in a sealed tube, or ball-milled together, and by reaction of, e.g. Me_3N with $Et_2Be \cdot 2Et_2AlH$ [13]. The complex $Me_3N \cdot BeH_2$ is dimeric by cryoscopy and its structure probably resembles that of MeBeH, in having a BeH_2Be bridge. For this feature there is i.r. spectral evidence (band at 1340 cm^{-1}). The tmeda ($Me_2N \cdot CH_2 \cdot CH_2 \cdot NMe_2$) complex of BeH_2, which likewise has BeH_2Be bridges (i.r. band at 1350 cm^{-1}), exists as a coordination polymer (with bridging via the ditertiary amine) which appears to be formed when solvent is removed from the dimeric material[14]. It was prepared by addition of tmeda to the pyrolysis product of $Bu_2^i Be \cdot OEt_2$ which had been taken beyond the $Bu^i BeH$ stage. Secondary amines $HNMe_2$ and $HNPr_2^i$ with BeH_2 did not afford amido-beryllium compounds which could be characterised, but $Me_2N \cdot CH_2 \cdot CH_2 \cdot NHMe$ afforded $HBeNMe \cdot CH_2 \cdot CH_2 \cdot NMe_2$ and hydrogen. The amido-beryllium hydride is trimeric in benzene with $Be \cdot H \cdot Be$ bridges and is the first well-characterised compound of this type reported[15]. ZnH_2 gives an analogue, which is, however, dimeric in benzene, but MgH_2 yielded an intractable product. BeH_2 adds to the double bonds of benzophenone and benzylidene-aniline in tetrahydrofuran (THF) affording $(HBeOCHPh_2)_n$ (n = 7–8 in benzene) and $HBeNPhCH_2Ph \cdot THF$, respectively[15]. Coordination complexes of alkyl-beryllium hydrides are also known:

$$Ph_2Be + BeCl_2 + NaBEt_3H + Me_3N \xrightarrow{(Et_2O)} [Me_3N \cdot PhBeH]_2$$
(Ref. 16)

$$Me_2Be + BeBr_2 + LiH \xrightarrow{(Et_2O)} [Et_2O \cdot MeBeH]_2 \qquad \text{(Ref. 17)}$$

$(Et_2O \cdot MeBeH)_2$ resembles BeH_2 in adding to benzylidene-aniline, benzaldehyde and benzophenone, to form $(PhCH_2PhNBeMe)_2$, $(PhCH_2OBeMe)_4$ and $Ph_2CHOBeMe \cdot OEt_2$, respectively. The ^1H n.m.r. spectra of the adducts $(Me_3N \cdot EtBeH)_2$, $(Et_2O \cdot MeBeH)_2$ and $(py \cdot MeBeH)_2$ (py = pyridine) have been recorded, and a broad resonance due to the BeH_2Be bridge observed[18]. The spectrum of $(Me_3N \cdot MeBeH)_2$ was particularly interesting. Although the bridge resonance could not be seen [nor that of the analogue having $(CD_3)_3N$ rather than Me_3N], the Me_3N resonance was partially split at 25 °C. This is attributed to *cis–trans* isomerisation about the BeH_2Be bridge:

```
   Me    H    Me              Me₃N   H    Me
    \   / \  /                    \  / \  /
     Be   Be          ⇌            Be   Be
    /   \ /  \                    /  \ /  \
 Me₃N    H    NMe₃              Me    H    NMe₃
         cis                            trans
```

A variable-temperature study showed that the entropy change associated with isomerisation is 13 ± 2 cal deg^{-1} mol^{-1} [18]. In the presence of ether, EtBeH (like $NaBeEt_2H$) reacts rapidly with 1-alkenes, only slowly with 2-pentene, and not at all with 2-methyl-2-butene[17]; this marked preference for terminal olefins recalls the behaviour of hydrido-aluminium compounds. The recent preparation of $Bu^i BeH$ free from coordinated ethereal solvent (by controlled pyrolysis of $Bu_2^i Be$) has enabled a more detailed study to be made of the effect of base on olefin additions to the Be—H bond[14]. The addition of 1-pentene to $Bu^i BeH$, $Bu^i BeH \cdot OEt_2$, and $Bu^i BeH \cdot NMe_3$ under identical conditions proceeds with relative $t_{\frac{1}{2}}$ values of 1:40:110 and is markedly retarded by bases. However, even ether-free $Bu^i BeH$ does not add 2-methyl-2-butene in 7 days at 33.5 °C [14].

Alkali metal beryllium hydrides M_2BeH_4 (M = Li, Na) have been prepared by the reaction:

$$4NaEt_2BeH + BeCl_2 \rightarrow 2NaCl + 4Et_2Be + Na_2BeH_4 \qquad \text{(Ref. 19)}$$

Both compounds, unlike BeH_2, exhibit clear x-ray diffraction patterns. These compounds which are formed as an insoluble residue in the above reaction appear to be electron-deficient polymers, and not salts of the BeH_4^{2-} ion. They are more reactive (e.g., with water) than BeH_2 but much less so than LiH or NaH. Use of 3 mol of $NaEt_2BeH$ gave products having the stoichiometry '$MBeH_3$', but these were shown to be M_2BeH_4–BeH_2 mixtures (by x-ray diffraction data)[19]. The sodium hydrides NaR_2BeH (R = Pr, Pr^i, Bu^i) and related alkoxy-salts of Na and Li have been prepared[20], but these again appear to be associated by hydrogen bridges. Pyrolysis of the Pr^i and Bu^i compounds affords olefin, some dialkylberyllium and a residue of sodium beryllium hydrides. The formation of Bu_2^iBe from the isobutyl compound is interesting; evidently it does not arise from the reaction $NaBu_2^iBeH \rightleftharpoons Bu_2^iBe + NaH$ since NaH is not present in the residue. It is probably formed as follows:

$$NaBu_2^iBeH \xrightarrow{-C_4H_8} [NaBu^iBeH_2] \rightarrow \tfrac{1}{2}Bu_2^iBe + \tfrac{1}{2}Na_2BeH_4$$

although the butene was not isolated.

The heat of formation of BeH_2 has been determined in a semi-micro static bomb by combustion in oxygen[21]; the best value is -8 kcal mol^{-1}.

Attempts to prepare authentic hydrido-magnesium halides, HMgX, (X = Cl, Br, I), (e.g., hydrogenolysis and pyrolysis of Grignard reagents) have been shown to afford MgX_2 and/or MgH_2, or their etherates[22]. Differences between the x-ray diffraction patterns of MgX_2–MgH_2 mixtures and the supposed HMgX species reported previously have been shown to be due to the presence of Et_2O, and not to new hydrido-compounds. The reaction of diborane with EtMgCl affords Et_3B and $ClMgBH_4$; and even the pyrolysis of Grignard compounds known to have the 'RMgX' structure in the solid phase (e.g., $EtMgBr \cdot 2Et_2O$) did not afford HMgBr species[22]. Stable hydride complexes of magnesium alkyls and aryls have been prepared for the first time, e.g.,

$$MgPh_2 + KH \xrightarrow[(Et_2O)]{} KH \cdot 2MgPh_2$$

although Be and Zn analogues were known previously[23]. The use of Me_2Mg (as with beryllium alkyls) resulted in ether cleavage. Other compounds prepared include $KMgBu_2^sH$, and $NaMgBu_2^sHBu_2^sMg$. The former s-butyl compound affords $KMgH_3$ either H_2 at 3000 lb/in^2 in benzene or under vacuum pyrolysis (80 °C). The trihydride is isomorphous with $KMgF_3$ (x-ray powder data); each magnesium therefore has an octahedral environment[24]. $LiAlH_4$ reduces $MgEt_2$ to MgH_2 under a variety of conditions and stoichiometries ($LiAlH_4:Et_2Mg$; 1:2, 2:3, 1:1, and 2:1), and no evidence for EtMgH (see above) or $HMgAlH_4$ was found[25]. $Mg(AlH_4)_2$ can be prepared by the reaction:

$$2NaAlH_4 + MgCl_2 \xrightarrow[(THF)]{} Mg(AlH_4)_2 + 2NaCl$$

(Ref. 26)

Earlier reports that the bromide could be used in this preparation could not be repeated; the reaction affords $BrMgAlH_4$. Compounds of the latter type can also be prepared from $NaAlH_4$ and MgX_2 (X = Cl, Br) in equimolar ratios. Mg(AlH$_4$)$_2$ is insoluble in ether and THF, although previously it was said to be soluble (see Ref. 26).

Replacement of two boron atoms of dodecaborane by the isoelectronic BeC, can lead, in principle, to a series of beryllium substituted boranes. The first such compound has been prepared from (3)-1,2-B$_9$C$_2$H$_{13}$ with Me$_2$Be in ether[27]. The product was too air and moisture sensitive for elemental analysis, but spectra suggested that the product was B$_9$BeC$_2$H$_{11}$·OEt$_2$. White crystals of the more stable Me$_3$N adduct were identified (including elemental analysis and mass spectra) as 1-trimethylamine-1-beryl-2,3-dicarba-*closo*-dodecaborane(12).

There has been much recent interest in the structure of beryllium borohydride, BeB$_2$H$_8$. Earlier views favoured a symmetric top (D_{2d}) structure (1), which received support from theoretical considerations[28, 29].

(1)   (2)   (3)

This structure was made untenable by the determination of the dipole moment of BeB$_2$H$_8$ ($\sim$2.1 D) by dielectric[30], and electric deflection[31] techniques, and by the failure to observe PQR bands in the i.r. spectrum[30]. I.R. and mass spectral studies eliminated structure (2), an important point being the high stability of the mass-spectral fragment (BeB$_2$H$_4$)$^+$ [32]. Detailed gas-phase i.r. studies now suggest the remarkable structure (3), (C_{2v}) having an almost equilateral BeB$_2$ triangle, coplanar with the four terminal (B)—H hydrogen atoms, with two BeH$_2$Be bridges lying in planes at 90 degrees to the metal atoms[33]. A similar structure apparently occurs in the molecule Me$_3$N·Be(BH$_4$)$_2$ [34]. Diethylmagnesium reacts with decaborane in ether to yield MgB$_{10}$H$_{12}$·2Et$_2$O, which decomposes rapidly in air, and has a structure very similar to the zinc analogue[35]. (See below, Section 2.3.3, and Chapter 3.)

2.2.2 Derivatives with Group IV ligands and associated groups

Compounds with metal–carbon bonds are included in other sections of the text; this section describes such compounds which are not appropriately dealt with elsewhere, and those with Si and Sn.

An important result was the establishment (based on low-temperature n.m.r. spectra, ebullioscopic molecular weights, and selective precipitation) that the exchange

$$R_2Be + BeX_2 \underset{(Ether)}{\rightleftharpoons} 2RBeX \quad (R = alkyl, X = halogen)$$

takes place, and that the equilibrium favours the mixed species[36, 37]. Lack of reaction between R$_2$Be and BeX$_2$ species reported previously is thought to be due to impure samples and halogenation of ether solvent. An equilibrium of this type is thus clearly established for Be, Mg, Zn and Hg. The first spectroscopic (n.m.r.) observation of Me$_2$Mg and MeMgBr in diethyl ether solutions is further compelling evidence for the equilibrium, which, again, appears to be greatly in favour of MeMgBr [38].

The first high resolution ^{9}Be n.m.r. studies, by Kotz, Schaeffer and Clouse[39], were carried out on predominantly ionic compounds of beryllium. They revealed little variation in chemical shifts, but showed $J(^9$Be—^{19}F) in tetrafluoroberyllates, although the magnitude of the coupling (33 Hz) was low compared with other fluorine couplings (except to carbon). More recent studies have revealed ^{9}Be chemical shifts sensitive primarily to local diamagnetic factors, in particular to changes in inductive effect of groups bonded to Be [40]. Among the compounds examined, the trimeric bis(dimethylamido)beryllium showed signals corresponding to both 3- and 4-coordination, in the expected ratio (2:1).

A recent i.r. spectral study[41] of gaseous dicyclopentadienylberyllium Cp$_2$Be reports Be—Cp stretching frequencies at 416 and 858 cm^{-1}. This is in harmony with the asymmetrical structure first shown by an electron-diffraction study[42]. An electrostatic model predicts that the energetically most favourable configuration will have the metal atom closer to one ring, and it is suggested that the best representations for solid and gaseous Cp$_2$Be are [(CpBe)$^+$Cp$^-$] and [Cp$^-$Be^{2+}Cp$^-$], respectively[41]. A Raman and i.r. spectral study of Bu$_2^t$Be is consistent with a monomeric structure, having a linear C—Be—C skeleton[43]. This behaviour, which contrasts with that of the lower alkyls, was confirmed by molecular-weight data. Electron-impact studies of the dialkyls (Me, Et, Pr, Pr^i, Bu^i, Bu^t) of beryllium show ions of associated (electron deficient) species only at low source-temperatures, as expected for weakly bridged molecules[44]. Ionisation-potential data were used to obtain a bond-dissociation energy D(RBe$^+$—R) of c. 45 kcal mol^{-1}.

Alkylmagnesium fluorides have been prepared for the first time[45,46]. Syntheses involve the reaction of alkyl fluorides with magnesium in THF and a catalyst, and the fluorination of diorganomagnesiums with Et$_2$AlF (using Et$_2$Mg) or BF$_3$·OEt$_2$ (using R$_2$Mg; R = Me, Et, Ph, hexyl). For n-hexylmagnesium fluoride, ν(Mg—C) was identified at 500 cm^{-1}, and the expected ^{1}H n.m.r. triplet was observed at τ 9.45.

Compounds having magnesium–tin bonds have been prepared:

$$\text{RMgX} + \text{Bu}_3^n\text{SnH} \rightarrow \text{Bu}_3^n\text{SnMgX} \quad \text{(Ref. 47)}$$
$$(\text{R} = \text{Pr}^i, \text{Bu}^s, \text{C}_6\text{H}_{11}, \text{Bu}^t; \text{X} = \text{Cl, Br})$$

Coupling reactions with halides (BuBr, Me$_3$SiCl, and CH$_2$=CH·CH$_2$Cl) and formaldehyde afford the expected tin compounds, and provide confirmatory evidence for the composition of the Sn—Mg species[48].

The enthalpy of formation of beryllium carbide has been determined by comparison of the heats of dissolution (in hydrochloric acid at 110 °C) of the carbide and the metal, in a closed glass vessel. The value [ΔH_f^0, $_{298}$K (Be$_2$C) = -27.96 (± 0.21) kcal mol^{-1}] is the first calorimetric value for Be$_2$C [49]. A synthesis of Mg$_2$Si from the elements is reported[50]. The product, after firing at 750–800 °C under argon is stable to hot and cold water, but dissolves with decomposition in $(NH_4)_2S_2O_8$, HOAc, oxalic, tartaric and citric acid solutions.

2.2.3 Derivatives of Group V ligands

The amides and imides of Be and Mg have been prepared and characterised. $Be(NH_2)_2$, a white crystalline material, is prepared (*inter alia*) from Be and NH_3 by a high-pressure procedure at 130–370 °C. Its properties are partially dependent on the mode of preparation[51]. Heating to ~230 °C *in vacuo* affords BeNH, a pure white solid which decomposes to Be_3N_2 at 250 °C. $Mg(NH_2)_2$ and MgNH are similarly prepared; the use of high pressures reduces the reaction time from 1.5–2 years to a few hours[52]. Double amides of magnesium with the alkali metals, $M_2Mg(NH_2)_4$ (M = Na, K, Rb, Cs), are obtained from Mg and liquid ammonia solutions of M [53]. Like the simple amide, they decompose thermally *in vacuo* to corresponding imides [$M_2Mg(NH)_2$] and NH_3, and yield Mg_3N and M on further heating. (For papers on ammine complexes see Table 2.1.)

Ether complexes of beryllium thiocyanate have been prepared by addition of excess finely powdered Be to an ether solution of $(NCS)_2$ at −78 °C. Dioxan displaces ether from $Be(NCS)_2 \cdot 2Et_2O$ affording the bis(dioxan) analogue[54]. Neither $Be(NCS)_2$ nor the double salts with Cr, Rb, K and NH_4 could be obtained from aqueous solution. The latter, $MBe(NCS)_3 \cdot n$MeCN, as acetonitrile adducts, were prepared in that solvent[55].

Complexes of several nitriles have been described in a series[56] dealing with first-row transition-elements and the Group II metals. I.R. spectral assignments [including v(M—N)] have been made for acetonitrile complexes $M(MeCN)_m(M'Br_4)_n$, [M = Mg, Cu^I, Cu^{II}, Ca, Fe^{II}, Co^{II}, Ni^{II}, Zn, Cd; M' = Al, Tl, Fe, In], which were prepared by M—M' bromide transfer reactions in acetonitrile. The complexes $ML_n(SbCl_6)_2$, [M = Mg, Zn, Co^{II}, Ni^{II}, Cu^{II}; L = PrCN, Pr^iCN, Bu^tCN, PhCN] all have $n = 6$ [57].

Magnesium complexes of several N-heterocycles are reported. The complexes [MgL_6]X_2 (L = pyrazole and imidazole; X^- = ClO_4^-, BF_4^-) were prepared from the appropriate metal salt and ligand in ethanol[58]. Like the N-methylimidazole analogues[59], they are coordinated via the 'pyridine' nitrogen atom, and show no sign of metal–anion interaction (i.r. spectra). The formation constants of the Mg, Cd and Zn complexes of sulphonic acid derivatives of azo-3- and azo-5-pyrazolones have been measured. When the sulphonyl substituent lies *ortho* to the diazo group, the ligands are tridentate, forming 1:1 metal complexes, but the *meta*- and *para*-derivatives form 2:1 adducts of the bidentate ligand[60]. The 1,8-naphthyridine (and its 2,7- dimethyl derivative) complexes of the Group IIA metal perchlorates have been characterised[61]. The methyl substituents appear to exert no stereochemical effect, except in the case of the magnesium derivative, which is also exceptional in having an undistorted (ClO_4^-) anion. Both these effects can probably be attributed to the small size of the magnesium ion. The stability constants (determined potentiometrically) of Be and Cu complexes of some naphthalene derivatives are correlated with the relative stabilities of metal chelate rings with five and six atoms[62]. The authors believe that, in addition to other factors [e.g. greater stability of (conjugated) unsaturated 6-membered chelate rings], the metal size is also important; for similar bonding atoms, smaller metals should form more stable 6-membered rings. This is borne out by the greater stability of the Be complexes of naphthoic acids, relative to the Cu analogues, both of which have 6-membered chelate rings, but apparently fails for the naphtholate derivatives.

The amine complexes of organo- and diorgano-beryllium and magnesium have been studied over the past few years; the elucidation of the factors governing molecular association has been particularly interesting. Thus $Bu^t_2Be \cdot OEt_2$ reacts with tmeda to afford a crystalline 1:1 complex in which only one of the two nitrogen atoms is coordinated to the metal; n.m.r. spectra reveal a rapid exchange of nitrogen atoms[63]. The analogous compound of the larger magnesium can accommodate four ligand atoms and does not show exchange. Similarly, the N,N,N'-trimethylethylenediamine $Bu^tBe[NMe_2(CH_2)_2NMe]$ complex is monomeric; whereas the magnesium and zinc analogues are dimeric, with the typical structure, (4).

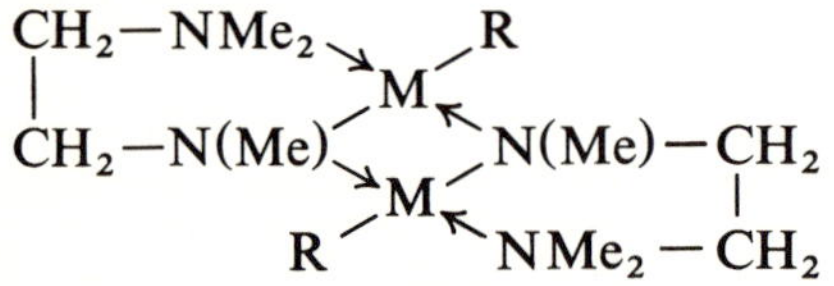

(4)

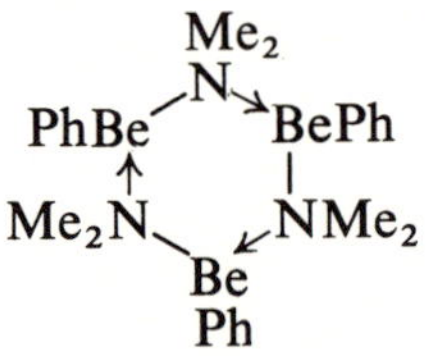

(5)

Confirmation of the steric origin of this difference lies in the fact that the replacement of Bu^t by Pr^n, Pr^i, or Ph [64] results in dimeric species. The secondary amines Me_2NH and Ph_2NH react with Ph_2Be to yield $(PhBeNMe_2)_3$ and $(PhBeNPh_2)_2$. These compounds have 3-coordinate beryllium, [see (5)] and react very rapidly with moist nitrogen or water; by contrast the phenyl derivative of (4) required dilute sulphuric acid for hydrolysis. Spectroscopic evidence for a structure analogous to (5) for $(MeBeNMe_2)_3$ (which undergoes rapid intermolecular interconversion) is reported[65]. Although this compound is trimeric in solution, $(MeBeNMe_2)_n$ is a glass at room

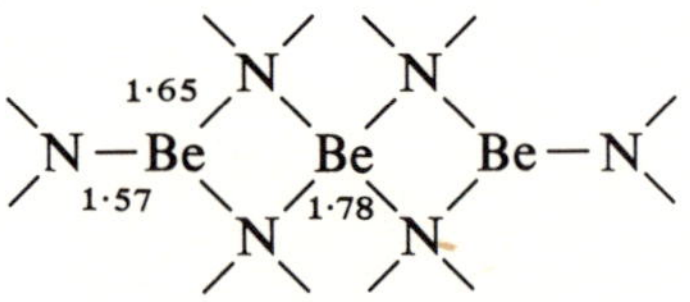

(6)

(Be—N distances in Å)

temperature. The thermodynamic control over the degree of association (entropy, angular valence strain, etc.) has been studied in the context of several compounds of the type $RBeNR'_2$ [66]. All the dimethylamido compounds were trimeric; whereas with two groups on nitrogen larger than Me, dimers resulted, with the sole exception of $(MeBeNEt_2)_3$. Some degree of association is clearly expected (presence of 2-coordinate Be and nitrogen donor-sites in monomeric $RBeNR'_2$) and entropy considerations would favour dimers. Moreover, the steric interaction of the metal and nitrogen substituents is less for dimers than trimers; accordingly dimeric species are found when large groups are present. A somewhat similar steric effect is evidently present in $[Be(NMe_2)_2]_3$. The trimers have the linear D_{2d} structure (6), in which the terminal Me_2NBe groups have coplanar C_2NBe atoms[67]. The Me—Me non-bonding distance here (3.11 Å) is very short, involving substantial distortion, and the chain is terminated (at the trimer) probably because of the steric requirements of the substituents on the nitrogen atoms of the four-membered rings. The magnesium analogue is thought to form a chain polymer[68]. The stereochemistry of the terminal $BeNMe_2$, (see also quoted bond lengths) argues strongly in favour of a Be $\leftharpoons$ N bond contribution. Other evidence for such π-bonding has been obtained spectroscopically for alkylidineamidoberyllium compounds[69]. Syntheses are shown in the following equations:

$$BeCl_2 + Ph_2C{=}NSiMe_3 \rightarrow \tfrac{1}{2}[Ph_2C{=}NBeCl]_2 + Me_3SiCl$$
$$BeCl_2 + 2Ph_2C{=}NLi \rightarrow \tfrac{1}{n}[(Ph_2C{=}N)_2Be]_n + 2LiCl$$

The dimeric species $\{[p\text{-tolyl}(Bu^t)C{=}N]_2Be\}_2$ has the structure (7), in which the i.r. absorption associated with the linear (C=N $\rightleftharpoons$ Be) group

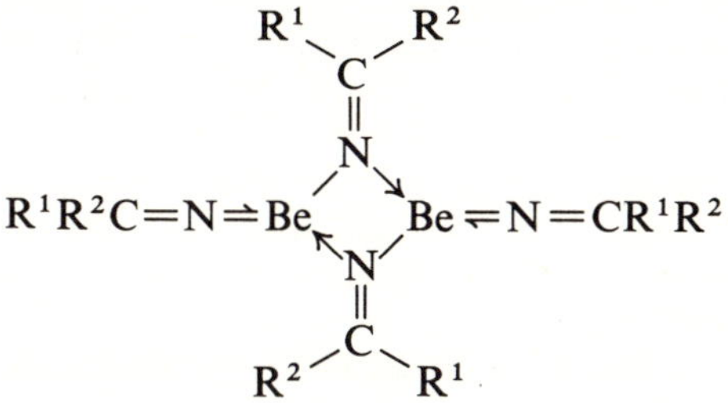

$$R^1R^2C{=}N{\rightleftharpoons}Be \underset{\substack{\| \\ C \\ R^2 \diagup \diagdown R^1}}{\overset{\substack{R^1 \diagdown \diagup R^2 \\ C \\ \|}}{\underset{N}{\overset{N}{\diagup\diagdown}}}} Be{\rightleftharpoons}N{=}CR^1R^2$$

(7)

is at 1739 cm^{-1}. The relative bond order (N $\rightleftharpoons$ M) in the series Be, B, C is nicely illustrated by comparison of this figure with ν(C=N $\rightleftharpoons$ B) $\sim$1790 cm^{-1}, and ν(C=N=C) at 1845 cm^{-1} (in $Ph_2C{=}N{=}CPh_2)^+$ [69]. The x-ray crystal study of the compound $[MeMg(NMe{\cdot}CH_2{\cdot}CH_2{\cdot}NMe_2)]_2$, which reveals a molecular geometry of the type shown in (4), is the first structural determination on a magnesium compound where a bridging electron-deficient group is replaced by an amino-group (3-electron donor)[70]. A notable feature of this structure is the steric strain evidently constraining the diamine which results in a (Me$_2$)N—Mg—N(Me) angle of 83.7 degrees (average). This is a considerable deviation from the preferred (almost regular) tetrahedral (109° 47′) angles of other chelating substituted ethylenediamine ligands.

Reduction of $TiCl_3{\cdot}3THF$ with magnesium affords a system which fixes nitrogen[71]. The stoichiometry of the initial reaction is:

$$TiCl_3{\cdot}3THF + \tfrac{5}{2}Mg + \tfrac{1}{2}N_2 \xrightarrow{\text{(THF)}} [TiNMg_2Cl_2{\cdot}THF] + \tfrac{1}{2}MgCl_2{\cdot}2THF$$

(8)

Compound (8) evolves nitrogen at 200 °C and reacts with pyridine and 2,2′-bipyridyl to form other TiMg complexes, TiNMgCl$_4$·py and TiNMg$_2$Cl$_2$·($\frac{3}{2}$bipy)THF [and also the Mg(0) complex Mg(bipy)$_3$ as violet needles]. (8) absorbs hydrogen to yield (TiNMg$_4$Cl$_3$H$_2$·2THF) which affords 1 mol of hydrogen on treatment with iodine, which may imply a structure having hydridic hydrogen. All the Ti—N—Mg species are hydrolysed with formation of NH$_3$, but none shows either $v(N_2)$, or v(metal–nitride) in the i.r. spectrum; the exact structure is uncertain[71].

The refinement of the structure determination of β-Be$_3$N$_2$ has been made, showing a sequence of hexagonal and cubic close-packed layers of nitrogen atoms. The Be atoms occupy two sites completely, and a third Be is disordered between two other (mutually exclusive) sites, all of which provide tetrahedral coordination. This is the first example of the mixed hchchchc layer structure contrasting with the fully occupied ccp (A$_2$X) and hcp (AX) types[72].

2.2.4 Derivatives of Group VI ligands

Aqueous solutions of Be^{II} contain tetrahedrally coordinated simple ions, and their polynuclear hydrolysis products [Be$_x$(OH)$_y^{(2x-y)+}$]. The combination of low coordination number and small size suggests that a relatively simple hydrolysis behaviour might be expected; no smaller ions are hydrolysed to polynuclear species. Mesmer and Baes studied the hydrolysis potentiometrically in the pH range 2–7 with Be concentrations 0.002–0.05 M[73]. At low values of the hydroxyl number the main species are Be$_2$(OH)$^{3+}$ and Be$_3$(OH)$_3^{3+}$. They found no evidence for Be(OH)$_2$, as had been suggested by Kakihana and Sillén[74], and argued that the only other species present was probably Be$_5$(OH)$_7^{3+}$. However, the exact nature of the minor products does not appear to have been agreed upon: Ohtaki[75] explained his e.m.f. data by reference to BeOH$^+$, (Be$_2$OH)$^{3+}$, Be(OH)$_2$, Be$_2$(OH)$_2^{2+}$ and Be$_3$(OH)$_3^{3+}$, and Lanza[76], his results in terms of the higher polynuclear species (see Table 2.1). The recent studies in heavy water are interpreted in terms of Be$_2$(OD)$^{3+}$, Be$_3$(OD)$_3^{3+}$, and Be(OD)$_2$ [74], but with gradual decrease in the magnitude of $\beta_{1,2}$ as the proportion of D$_2$O increased[77].

Grigor'ev and his co-workers have extensively studied the carboxylic acid salts of Be and Mg, and their ammines and amine derivatives. An i.r. spectral study of the ammine Be(OAc)$_2$·2NH$_3$ indicates the presence of the tetrahedrally coordinated metal for which the structure (9) is suggested[78]. The diammines of magnesium formate and acetate are similar; but there are non-equivalent carbonyl groups of the formate derivative of beryllium, probably arising from electronic effects rather than a difference of structure. Thermal decomposition of Be(OCOH)$_2$·2NH$_3$ affords BeO, NH$_4$(HCOO) and HCONH$_2$; the magnesium salt behaves similarly, but yields a little ammonia[79a, b]. Be(OCOH)$_2$ reacts with MeNH$_2$ (but not PrNH$_2$) affording a (polymeric) amine complex[80], which i.r. spectroscopy has shown to contain hydroxyl groups and N-methylformamide (probably decomposition products of Be(OCOH)$_2$·2NMeH$_2$)[79b]; and for which structure (10) is proposed. The amine complexes Be(OAc)$_2$·2RNH$_2$ (R = Me, Et, Pr) have similar structures, involving hydrogen-bonded six-membered rings, but unlike the formates undergo no significant ammonolysis in solution[81]. The composition of the amine complexes of basic beryllium acetate has been re-examined[82]. The reactions are best summarised by the equation:

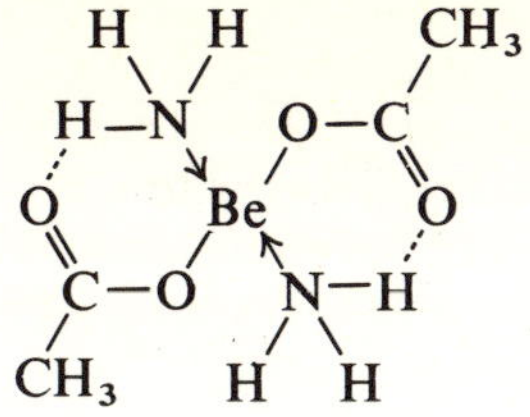

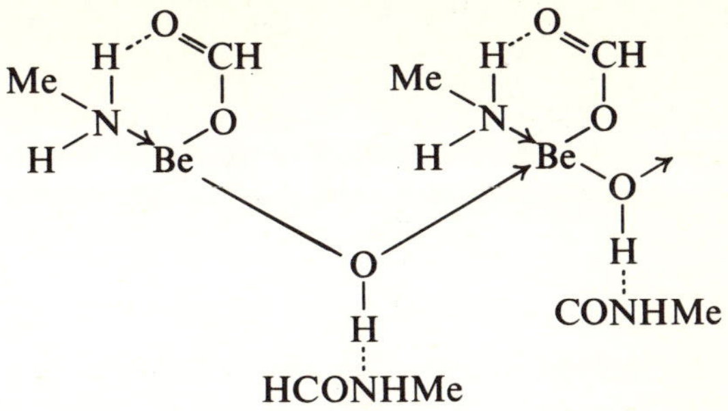

$$Be_4O(OAc)_6 + (n+2)RNH_2 \rightarrow Be_3O(OAc)_4 \cdot nRNH_2 + Be(OAc)_2 \cdot 2RNH_2$$

Products having $n = 2$ (MeNH$_2$), 3 and 6 (EtNH$_2$, PrNH$_2$, BuNH$_2$) have been characterised. Thermal decomposition of the tripropylamine complex (or the BuNH$_2$ analogue) at 100 °C affords a new beryllium oxyacetate, Be$_3$O(OAc)$_4$, which can be crystallised from pentane, affording two forms of a metastable dimer, having identical i.r. spectra and x-ray powder patterns[83]. In contrast to amines, SO$_2$ and N$_2$O$_4$ form only 'loose' addition compounds with basic beryllium acetate Be$_4$O(OAc)$_6 \cdot$2SO$_2$ and Be$_4$O(OAc)$_6 \cdot \frac{3}{2}$N$_2$O$_4$ [84a] At temperatures above 200 °C, the sublimation product of Be(OEt)(OAc) consists of the basic ethoxy-acetate Be$_4$O(OAc)$_5$OEt, which is readily soluble in organic solvents and can be recrystallised from hot octane[84b]. Pure basic mixed carboxylate salts of beryllium have been isolated for the first time with acetate and trichloroacetate ligands[85]. The compounds Be$_4$O(CH$_3$COO)$_x$(CCl$_3$COO)$_{6-x}$ (x = 5, 3, 1) were prepared by fractional crystallisation from solutions of the acetate and trichloroacetate in equilibrium, and structures have been deduced with the aid of ^{1}H n.m.r. and mass spectra. Since the anions are arranged about the Be$_4$O core in an approximately octahedral configuration, there are evident isomeric possibilities. Isomers were not isolated, however, and the n.m.r. spectra show that carboxylate scrambling occurs. The scrambling occurs faster when more trichloroacetate groups are present, and it is postulated that ionic dissociation of the less basic anion may provide the mechanism of scrambling.

Ether complexes of dialkylberyllium compounds are formed in the normal Grignard procedure for preparing the dialkyls[63]. The ether is easily removed (except from Me$_2$Be) by gentle reflux at reduced pressure, affording R$_2$Be (R = Et, Pr^n, Bu^n, Bu^i); alternative syntheses (e.g., by use of the KF salts) are not necessary. Dimethylberyllium reacts with 2-methoxyethanol affording the complex [RBeO(CH$_2$)$_2$OMe]$_4$, for which a structure with a cubic framework [like (MeZnOMe)$_4$] is suggested[86]. The same dialkyl with 8-hydroxyquinoline yields (MeBeOC$_9$H$_6$N)$_4$, probably consisting of two units similar to (11) lying parallel, and arranged such that the oxygen of one is coordinated by the Be of another. A study of the molecular aggregation of

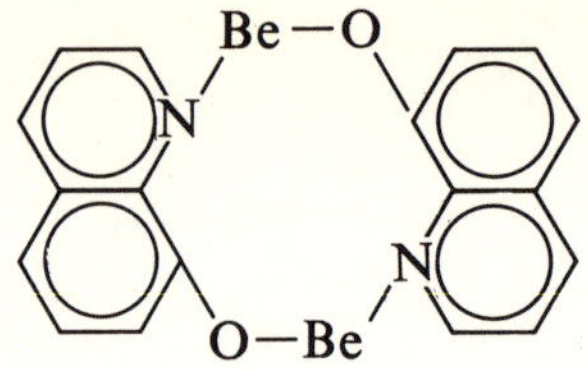

(11) 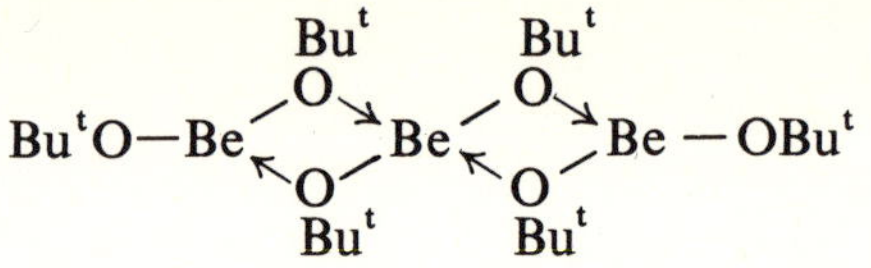   (12)

alkylberyllium alkoxides shows that they are mainly tetrameric (like many Mg, Zn and Cd analogues), unless substituents are particularly bulky[87]. Acetone reacts with (MeBeOBu^t)$_4$ affording [Be(OBu^t)$_2$]$_3$, for which structure (12) is suggested, but steric effects probably result in the analogue, [Be(OCEt$_3$)$_2$]$_2$, assuming a dimeric structure. Ether complexes of the thiomagnesium alkyls [(Bu^{t}MgSPr^i·OEt$_2$)$_2$, (MeMgSBu^t·THF)$_2$, (EtMgSBu^t·THF)$_2$] are dimeric, but the ether-free (EtMgSBu^t)$_4$ tetramer probably again has the 'cubane' based structure[88].

I.R. and Raman spectral studies on diethyl ether complexes of MgBr$_2$ and MgI$_2$ show that concentrated solutions of the iodide contain a polymer of D_{2n} symmetry. This is probably based on the MgI$_4$O$_2$ unit, with planar iodide bridges. Dilute solutions contain another species, which may contain tetrahedrally coordinated metal. The spectral evidence was not conclusive for the bromide species. However, contrary to previous reports, large changes observed in the skeletal region (for Et$_2$O) are now attributed to conformational changes of the complexed ether (greater concentration of ligands with *gauche* ethyl groups) rather than simply to coordination[89].

Several insoluble beryllium alkoxides have been prepared, either from BeCl$_2$ and the appropriate lithium alkoxide, or by alcoholysis of Be(OMe)$_2$ [90]. Reaction with acetyl halides yields the halide alkoxides Be(OR)X (X = Cl, Br; R = Me, Et, Pr, Pr^i) as ROAc adducts[91]. Various chloro-magnesium alkoxides, including several non-stoichiometric compounds, have been prepared by thermal decomposition of the chloride–alcohol adducts, MgCl$_2$·6ROH [92].

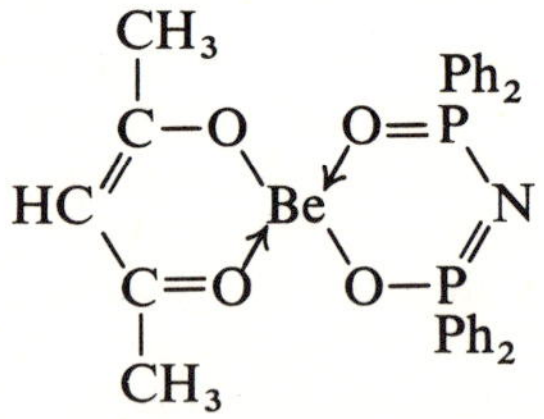

(13)

Attempts to make beryllium polymers bridged by *O*-terminal phosphonitrilic compounds resulted only in monomers[93]. Reaction of Ph$_2$P(O)NPPh$_2$OH with bis(acetylacetonato)beryllium gave compound (13) and the bisphosphonitrile-derivative. Ph$_2$P(O)(NPPh$_2$)$_3$OH likewise gave Be(C$_5$H$_7$O$_2$)[OPPh$_2$(NPPh$_2$)$_3$O] and Be[OPPh$_2$(NPPh$_2$)$_3$O]$_2$, but the structures were not established. Octamethylmethylenediphosphonic diamide forms magnesium and zinc complexes $M\{[(Me_2N)_2P(O)]_2CH_2\}_3(ClO_4)_2$, which are coordinated via both the phosphoryl oxygen atoms, to make a six-membered ring with the metal atom[94]. An x-ray crystal structure of the magnesium, cobalt(II) and copper(II) salts $M[(Me_2N)_2P(O)OP(O)(NMe_2)_2]_3$ $-(ClO_4)_2$ reveals that the three compounds have virtually identical stereochemistry: the metals are essentially octahedrally coordinated, and the chelating rings planar, suggesting that π-bonding in the ring [not including the metal (cf., the Mg compound)] could be important[95]. Hexakis(nitromethane) complexes of divalent metals, including magnesium, have been prepared, and apparently show no shift in v(N—O) in the i.r. spectrum on coordination[96].

2.2.5 Derivatives of Group VII ligands

The i.r. spectrum of vaporised MgF_2 (isolated at liquid-hydrogen temperature in Ar and Kr matrices) shows evidence of both MgF and MgF_2[97]. The spectra are interpreted in terms of a bent molecule, contrary to earlier reports. More recent work on fluorides of Mg and alkaline earth metals confirms this result and gives the following angles FM̂F: 158° (Mg); 140° (Ca); 108° (Sr); 100° (Ba). It has been suggested that the structures of these halides (and alkali-metal oxides) are best regarded as involving symmetrical but invariably linear skeletons; experimental data are lacking at present[98]. The Raman spectrum of molten $MgCl_2$ indicates octahedrally coordinated $MgCl_6^{4-}$ ions, while the liquid mixtures $MgCl_2 \cdot KCl$ and $MgCl_2 \cdot 2KCl$ have spectra consistent with the presence of $MgCl_3^-$ ions with C_{3v} symmetry[99].

The ionic-transport properties of molten BeF_2, and of mixtures with LiF, vary widely with composition, reflecting the breakdown of the BeF_2 network-type structure[100]. The mobility of beryllium ions in the LiF mixtures is ($t_{Be} = 0 \pm 0.05$) zero, reflecting the strong electrostatic forces between Be^{2+} and F^- ions. Potentiometric studies on the stability of beryllium fluoride complexes in 1 M-NaCl at 0, 25 and 60 °C have been made, with a view to identifying the expected species $B_xF_y(OH)_z^{(2x-y-z)+}$ [101]. (See also references 63–67.) No mixed species (involving either hydroxide or hydrogen ions) were detected, but the four fluoride species $y = 1$–4, are all essential for an adequate fit to the data. An x-ray examination of the new compound $Cs_4Mg_3F_{10}$, obtained from molten $CsF-MgF_2$ mixtures, has revealed a structure having corrugated sheets of octahedral MgF_6 octahedra[102]. Three octahedra, the central one of which shares two opposing faces with the other two together, form the basic unit.

2.3 ZINC, CADMIUM AND MERCURY

2.3.1 Hydrides and related compounds

Since the preparation[15] of $[ZnH(NMe \cdot CH_2 \cdot CH_2 \cdot NMe_2)]_2$ (see Section 2.2.1), a number of dialkyl- and diaryl-zinc hydride complexes have been made. The equilibrium

$$MHZnR_2 + ZnR_2 \rightleftharpoons MH(ZnR_2)_2 \ (M = Na, Li; R = alkyl)$$

has been demonstrated by n.m.r. spectroscopy. The spectrum shows a single (average) hydride resonance, occurring, surprisingly, 7 p.p.m. to low field of Me_4Si in the presence of excess ZnR_2 [103]. The 1:2 species, and the related pentafluorophenyl derivative, are believed to have Zn—H—Zn bridges[104].

2.3.2 Group II ligands, mercury (I) and related compounds

It has been shown that the rarity of Hg^I complexes is due to the tendency to disproportionate to Hg^{II}, rather than to a weak complexing ability, and a number of compounds with oxygen donor atoms have been prepared[105]. Four is a common coordination number for Hg_2^{2+}. E.S.R. spectral evidence for Cd_2^{2+} and Hg_2^{2+} has been obtained in γ-irradiated solutions containing the divalent metals[106, 107]. Monovalent cadmium ions, formed from Cd^{2+} and hydrated electrons, are powerful reducing agents. A study of reduction rates of the EDTA and nitriloacetic acid complexes suggests that both inner- and outer-sphere mechanisms occur[108] (see also Section 2.3.5).

2.3.3 Group III ligands

The interesting Zn, Cd and Hg derivatives of decaborane are discussed in Chapter 3. Zinc borohydride has been prepared, *inter alia*, from dimethoxyzinc and diborane. It forms 2:1 adducts with py, THF and trimethylamine, which contain covalently bound BH_4 groups, but ammonia and methylamine give adducts of the type $[ZnL_4][BH_4]_2$ [109].

2.3.4 Group IV ligands

Studies on the nature of the Reformatsky reagent from methyl-2-bromo-3,3-diphenylpropanoate suggest the bromozinc enolate (14) and the unsymmetrical β-lactone structure (15) for the reactive intermediates[110].

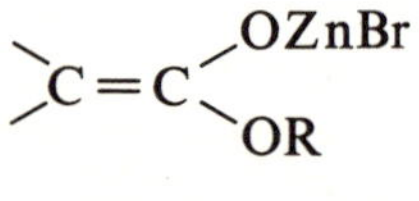

(14) 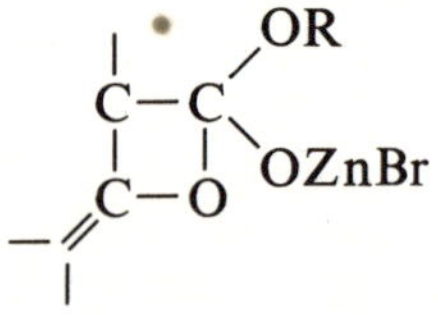   (15)

Cyclobutadienyliron tricarbonyl is reversibly mercurated by $Hg(OAc)_2$, affording a mixture containing all the possible acetoxy-mercury derivatives[111]. Germyl- and silyl-mercury compounds have been prepared from the dialkyls[112]:

$$HgR_2 + SiHCl_3 \xrightarrow{(h\nu)} (Cl_3Si)_2Hg + 2RH; \text{ (and Ge)}$$

$$HgR_2 + MeSiHCl_2 \xrightarrow{(h\nu)} (MeCl_2Si)_2Hg + 2RH$$

Reaction of $(Cl_3Si)_2Hg$ with $Mn_2(CO)_{10}$ affords $Cl_3SiMn(CO)_5$. The digermylmercury compounds react with other mercury derivatives HgX_2

and undergo exchange; when X is an electron-releasing group (e.g., Me, Pr^i), the products are stable, but are unstable when X is electronegative (e.g., C_6F_5, CH_2COOMe), and undergo immediate decomposition[113].

$$(Et_3Ge)_2Hg + X_2Hg \rightarrow [2Et_3GeHgX] \rightarrow 2Hg + 2Et_3GeX$$
$$(X = C_6F_5, CN, CH_2COOMe)$$

$(Et_3Si)_2Hg$ reacts with excess of sulphur to yield the mono-insertion product $Et_3SiSHgSiEt_3$ in good yield, but no di-sulphur compound. The germanium analogue is very unstable[114]. The new Hg—Si heterocycle (16) has been prepared by sodium-amalgam reduction of $(Me_2ClSi)_2CH_2$:

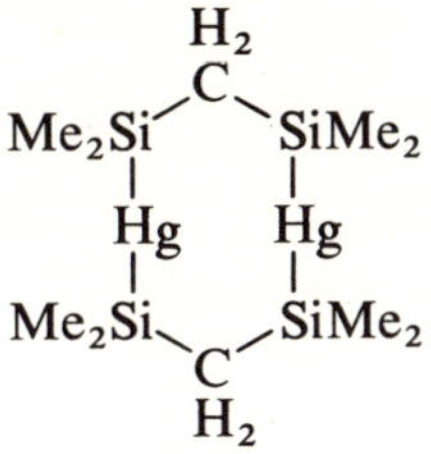

```
           H₂
            C
    Me₂Si     SiMe₂
      |         |
      Hg        Hg
      |         |
    Me₂Si     SiMe₂
            C
           H₂
```
(16)

This reaction formally resembles the synthesis of $(Me_3Si)_2Hg$ [115].

2.3.5 Group V ligands

Tricyanomethyl compounds $RM[C(CN)_3]$ (R = Ph, Et) of zinc and cadmium have been prepared from R_2M and $BrC(CN)_3$ [116]. Structures have been assigned, largely from i.r. spectral data, and involve coordination via the N of a CN group. The materials are probably polymeric, involving metal coordination by all three CN groups of a $C(CN)_3$ group.

The A_1 vibrational mode of $[Zn(NH_3)_4]I_2$ should be independent of the $^{64}Zn/^{68}Zn$ ratio since it involves no metal motion. This has been shown, and the band assignment confirmed by a Raman-polarisation study[117].

A series of nitrogen–mercury(I) compounds reveals that the system $>$N—Hg—Hg—N$<$ is more stable than previously supposed[118]. Secondary acyl acid amides react with dry, freshly prepared mercury(I) carbonate in the dark:

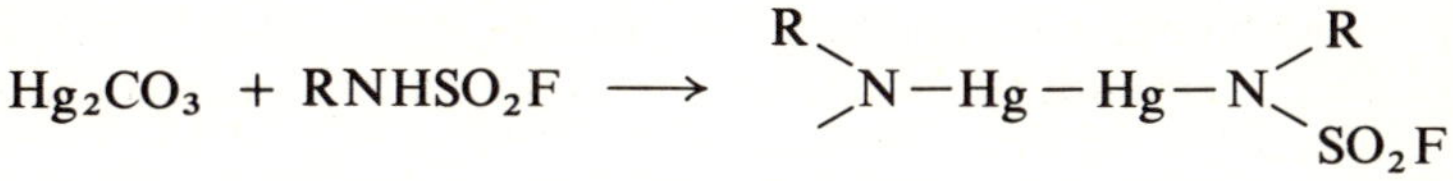

$$Hg_2CO_3 + RNHSO_2F \longrightarrow \begin{array}{c} R \\ > N-Hg-Hg-N < \\ SO_2F \end{array} \begin{array}{c} R \\ \\ \end{array}$$

R includes SO_2F, CO_2Et, CO_2Me, and $CONEt_2$. N,N-disubstituted hydrazides afford similar products with Hg_2CO_3. These compounds, $[Hg_2NAcNAc]_n$, $[Hg_2N(COCF_3)N(COCF_3)]_n$, and $[Hg_2N(COCCl_3)N(COCCl_3)]_n$ are thought, from x-ray data, to be based on chain structures, as is the ureido derivative (17), formed from N,N-bis(fluorosulphuryl)urea and Hg_2CO_3.

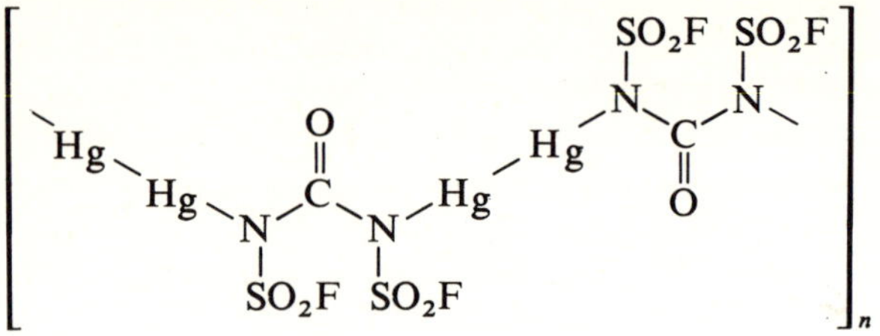

(17)

Mercury bis(sulphurdifluoridiimide) prepared from HgF_2

$$FCONSF_2 + HgF_2 \rightarrow Hg(NSF_2)_2$$

is violently hydrolysed by water to SO_2, F^- and NH_3, and decomposes thermally with formation of pure NSF[119]. The molecule, (18), has C_2 symmetry with a very nearly linear (178±1°) N—Hg—N arrangement. The short N—S distance (1.44 Å) and corresponding i.r. spectral absorption

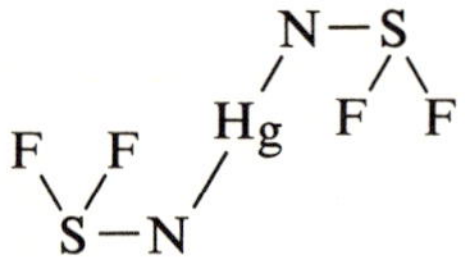

(18) 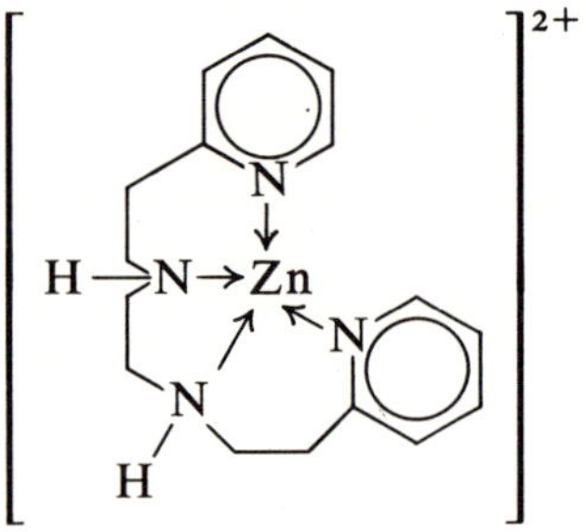   (19)

suggest a multiple N=S bond.

Steric effects of the two 'tripod' ligands $(Me_2NCH_2)_3CCH_3$ (TTN) and $(MeHNCH_2)_3CCH_3$ (TSN) have been examined by n.m.r. and i.r. spectroscopy and other physico-chemical studies of zinc complexes[120]. Both form 1:1 complexes with $ZnCl_2$ and $ZnBr_2$, but the TTN complex has 4-coordinate zinc because of the steric control exerted by the tertiary N-Me groups near the metal. The TSN complexes are relatively unhindered and the zinc atom is 5-coordinate. The zinc complex of 1,8-bis(2'-pyridyl)-3,6-diazaoctane, is of interest in having a tetrahedral N-environment provided by that ligand, (19), probably resembling the Ni analogue[121]. Trigonal prismatic coordination has been found for the first time with a Group II metal in the cis,cis-1,3,5-tris(pyridine-2-carboxaldimino)cyclohexane complex of zinc; a sketch of the ion is shown in (20). The molecular structure of the anhydrous perchlorate was determined by an x-ray diffraction study. A view along the (pseudo)-threefold axis of the trigonal prism reveals a slight (max 6.6 degrees) twist probably arising from the relief of intra- (but possibly inter-) molecular non-bonded contacts[122]. The Ni analogue shows very different x-ray powder-patterns, and it is of interest to know whether a difference of structure is due to ligand-field effects causing an octahedral arrangement in this case. The same coordination polyhedron has been found for the Fe[II], Co[II], Ni[II] and Zn complexes of the ligand shown in (21). The ^{1}H n.m.r. spectra show three pyridine-ring protons isotropically shifted in the paramagnetic Co[II] and Ni[II] complexes, and a single resonance for the HC=N proton in the diamagnetic

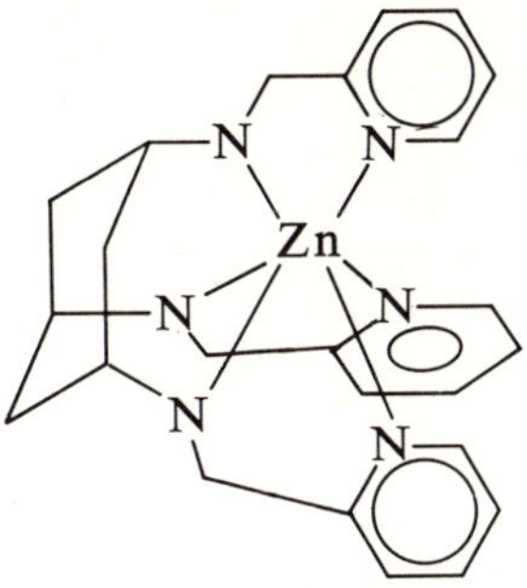

(20)

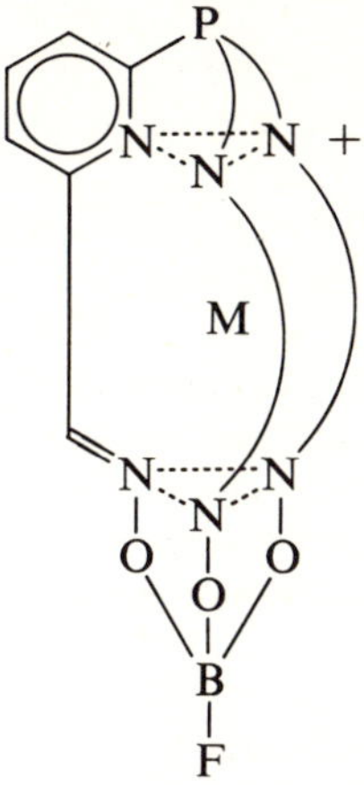

(21)

Fe[II] and Zn analogues, consistent with (at least) trigonal symmetry of the ions[123]. The electronic spectrum of the Ni complex is consistent with the presence of a C_3 axis[124].

8-Coordination has been found by x-ray examination of the zinc and cadmium complexes $ML_4(ClO_4)_2$, in which the ligand is 1,8-naphthyridine[125]. The zinc and iron compounds are the first first-row transition-metals with 8-coordinate cations.

An x-ray determination of the crystal and molecular structure of $(EtO)_2$ POHgCl reveals that all four molecules in the asymmetric unit have Cl—Hg—P angles close to 180 degrees. This has enabled a comparison of trans-effects (studied by n.m.r. spectra) to be made with analogous platinum compounds[126]. In the mercury compounds a π-bonding contribution (i.e., d_π-p_π overlap) is clearly absent. The metal–phosphorus coupling constants are 2.16–2.40 times larger in $(EtO)_2POHgCl$ than in trans-$[Pt(X)\{(PhO)_2PO\}(PBu_3)_2]$ and related compounds. A calculated ratio (from the dependence of the Fermi contact contribution upon the metal magnetogyric ratios and $|S_M(0)|^2$ is 2.1–2.6. The close agreement between the two values suggests that the bonds are similar in the two cases. Furthermore, the ratio of coupling constants in the phosphonato-complexes to the average of couplings in other complexes is 0.61 for Hg and 0.62 for Pt in harmony with the belief that the trans-influence of phosphorus is quantitatively similar in the compounds of both metals (and therefore that d_π-p_π bonding in the platinum case is too small to be of chemical significance).

Mercuration of 8-hydroxyquinoline, and its complexes with chromium(III), cobalt(III), copper(II) and lead, has been studied. The position of mercuration in the ring was determined by decomposition by DCl and n.m.r. spectral study of the deuteriated ligand. Free 8-hydroxyquinoline with a fivefold excess of mercury(II) acetate affords the mercury(II) chelate of the 5,7-dimercurated ligand. The copper complex afforded both mono-, and di-mercurated derivatives, while the cobalt and chromium derivatives yielded the 5,7-dimercurated compounds. There appeared to be no steric restriction to substitution at the 7-position, and the pyridine ring was inert throughout. The lead oxinate did not yield an (intact) mercurated derivative[127].

A number of new acetonitrile complexes $M(CH_3CN)_m X_n$, of the metals (M = Li, Mg, Ca, Mn, Fe, Co, Ni, Cu and Zn) have been prepared by chloride-ion transfer between MCl_n and $HgCl_2$ in MeCN [128]. Salts of the anions X, $(Hg_p Cl_{2p+q})^{q-}$, with $p = 1$–5, $q = 1,2$ were obtained, unsolvated, from MeCN; but in solution, $HgCl_4^{2-}$ and $HgCl_3^-$ appear to be the sole anions. From other solvents containing $HgCl_2$ and chloride donors, only $HgCl_4^{2-}$ and $HgCl_3^-$ derivatives have been obtained. The compound Et_4NHgCl_3 has tetrahedrally coordinated mercury, but the vibrational spectra of tetraethylammonium compounds with $Hg_3Cl_8^{2-}$ and $Hg_5Cl_{11}^-$ anions show both tetrahedral and linear (pseudo-octahedral) metal environments.

2.3.6 Compounds with Group VI ligands and associated ions

Hydrolysis of zinc ions in aqueous media containing NO_3^- and SO_4^{2-} has been studied[129]. Elevated temperatures were used to lessen the effect of complex formation; instability constants decrease with temperature. In addition to the species $ZnOH^+$, at higher concentrations of Zn^{2+} ($>9.75 \times 10^{-4}$ M), evidence was found for Zn_2OH^{3+}.

An x-ray study of zinc picolinate tetrahydrate reveals zinc in an octahedral environment[130]. The average zinc–ligand bond-length is 2.15 Å, compared with an average 1.97 Å in the (4-coordinate) oxyacetate of zinc; the difference is probably associated with the four antibonding electrons, $(e_g^*)^4$, of the octahedral structure.

While a number of reports of stereoselective chelation of optically active α-amino acids to metal cations have proved mistaken, chelation of the tridentate, relatively bulky histidine by Co^{II}, Ni^{II} and Zn is selective[131]. The formation of 2:1 histidine complexes favours the mixed D,L-complexes over the unmixed (e.g., L,L-species) in aqueous solution. Lack of stereoselectivity for the 2,3-diaminopropionic acid complexes suggests that selectivity is determined by the bulky imidazole group; the sense of selection implies a preference for a *trans*-arrangement of all substituents. The extent of selectivity is evidently related to the size of the metal ion, and no selectivity was found for the cadmium salts.

A ^{35}Cl and ^{37}Cl n.q.r. spectral study of complexes of $HgCl_2$ with various oxygen donors (L), suggests that 3-coordinate mercury does not occur here. The presence of two ^{35}Cl resonances implies different chlorine environments in the THF, Me_2SO (DMSO), acetophenone and benzophenone adducts. The dioxan complex, of known structure, has a similar spectrum. A chlorine-bridged dimer is in better agreement with spectral effects of changing the donor molecule than the alternative formulation $HgCl_2$ (donor)$_2 \cdot HgCl_2$ [132].

Decarboxylation of a wide range of mercury salts of aliphatic[133-135], alicyclic[136], and aromatic[137] acids, $Hg(OOCR)_2$, has been shown to afford organo-mercury salts RHgOOCR. Mercury(II) butyrate affords mainly propylmercury butyrate, ($\sim 92\%$) and CO_2, and smaller amounts of phenylmercury(II) salts (from benzene solvent), mercury(I) salts, and traces of CO, C_3H_6 and C_3H_8 [133]. The reactions are affected by solvent and initiated by u.v. light and acyl peroxides.

Bis(trifluoromethyl)nitroxide (in excess) reacts with mercury to form $Hg[(CF_3)_2NO]_2$, a white crystalline solid which decomposes into the radical and free metal at 85 °C. Metathetical reactions occur with the halides of several elements (C–halogen, Si—Cl, Ge—Cl, B—Cl, P—Cl) with transfer of the $(CF_3)_2NO$ group: e.g., BCl_3 affords $B[(CF_3)_2NO]_3$; $COCl_2$ affords $CO[(CF_3)_2NO]_2$; and PCl_5, affords $P[(CF_3)_2NO]_5$ [138]. Similar reactions of the mercury compound have been found with hydrides of silicon, germanium, nitrogen, arsenic and antimony; examples include synthesis of $Si[(CF_3)_2NO]_4$ and $As[(CF_3)_2NO]_3$ from SiH_2I_2 or SiH_3Br and AsH_3, respectively[139]. Paramagnetic nitroxide complexes have been prepared from (22) (itself

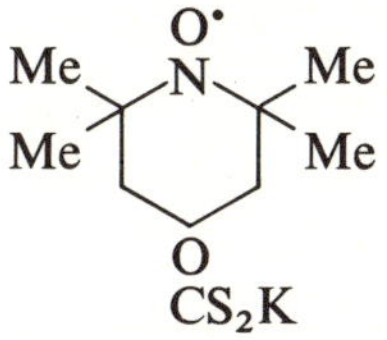

(22)

obtained from 2,2,6,6-tetramethyl-4-hydroxy-1-piperidinoxy and KOH in carbon disulphide), and their e.s.r. spectra have been recorded[140].

The reaction between $HgCl_2$ and sodium acetylacetonate [Na(acac)] affords materials of irreproducible analysis and not bisacetylacetonatomercury(II)[141]. The reaction of $HgCl_2$ and aqueous Hacac affords the carbon-bonded compound (23), whereas $Co(acac)_3$ or $Fe(acac)_3$ give (24).

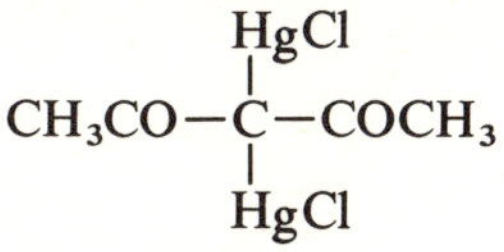

(23) 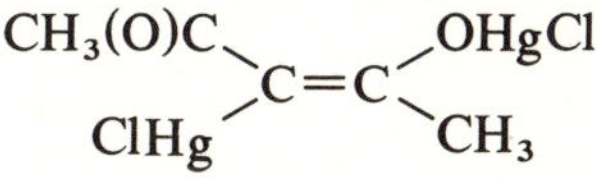   (24)

The mercury complex of 2,2,6,6-tetramethyl-3,5-heptanedione has been re-examined, and rather than the oxygen-bonded structure previously put forward, the compound, at least in solution, appears to be an equilibrium mixture of carbon- and oxygen-bonded species, with the position of equilibrium favouring the carbon-bonded species[142].

Authentic phenylmercury(II) hydroxide may be prepared by treating phenylmercury(II) acetate with sodium hydroxide solution in hot benzene; whereas the material 'methyl mercuric hydroxide' is known to be a mixture of [(MeHg)$_3$O]$^+$OH$^-$ and (MeHg)$_2$O. The phenyl analogue of the last compound (PhHg)$_2$O, may be obtained by heating the hydroxide at 100 °C under vacuum. It is stable to air and is hydrolysed by water to the hydroxide[143]. Both the hydroxide and oxide are converted by dialkylcarbonates into phenylmercury(II) alkoxides, which show a concentration-dependent equilibrium between monomers and associated species, except for PhHgOMe which was dimeric over the concentration range studied (0.0054–0.085 monomer molality)[144].

The interesting addition of sulphur to dithioarylacid chelates of zinc[145] has been further studied:

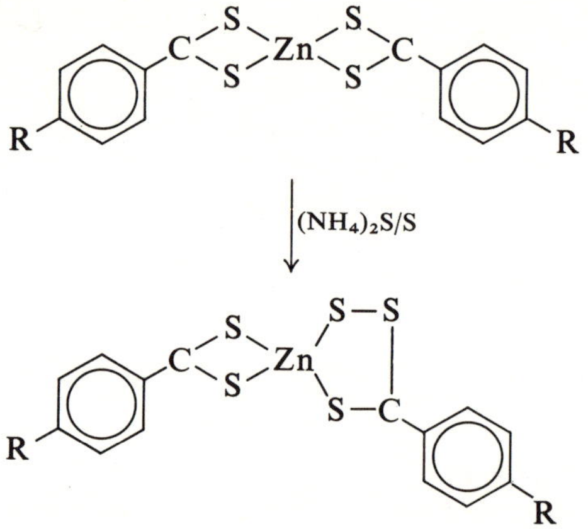

The addition can be reversed by triphenylphosphine, and the zinc perthio-complex reacts with NiCl$_2$ to yield the bis(perthio)nickel complex. A mass-spectral study[146], confirmed by n.m.r. spectral evidence[147], reveals that labelled sulphur (^{34}S) enters the nickel complex adjacent to carbon, rather than the metal, but that scrambling occurs with the zinc species, whether inter- or intra-molecularly is not known.

Zinc complexes of dialkyldithiophosphates are associated in benzene solution, but structure (25) is known for the diethyl compound, with 4-coordinate zinc. Pyridine forms 1:1 and 2:1 stable adducts; the latter

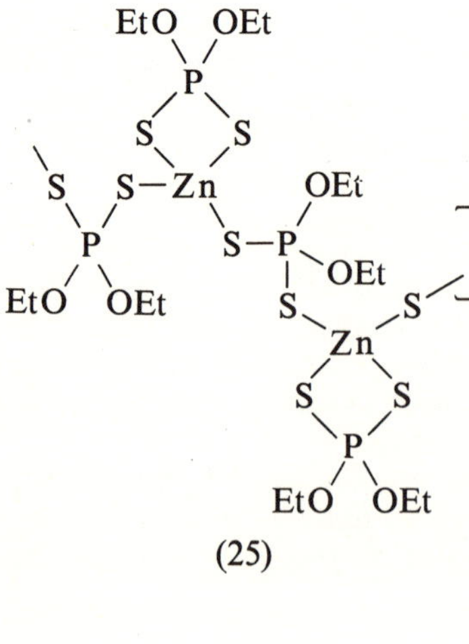

(25) 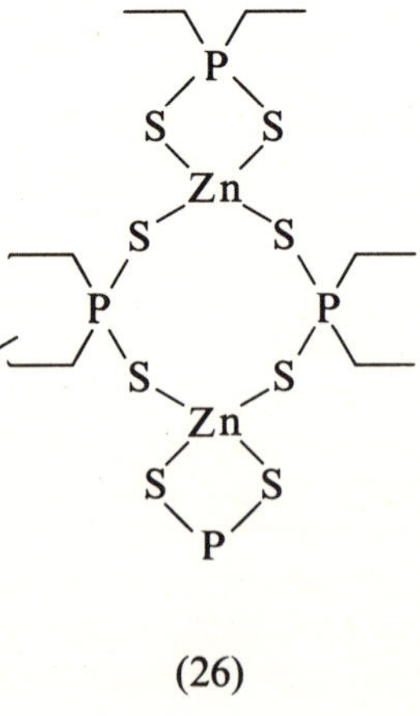   (26)

dissociates in benzene and has the *cis*-octahedral structure[148]. The crystal and molecular structure of zinc diethyldithiophosphinate also shows that the metal is 4-coordinate (26), in a distorted tetrahedral arrangement[149]. The dimeric molecules form an eight-membered ring involving two bridging dithiophosphinate groups. Replacement of sulphur by oxygen probably increases the steric requirements, and the phosphinate and monothiophosphinate complexes are linear polymers. The n-butyl phenyl phosphinate has alternate single and triple (3-atom—O—P—O—) bridging groups of the phosphinate.

The i.r. spectra of cadmium trithiocyanate[151], and hexathiocyanate[152], and of mercury thiocyanate complexes[153-155] have been studied. As expected, terminal thiocyanate groups are sulphur-bonded; but both metals form bridged species. Mercury(II) cyanide and thiocyanate form polymeric complexes with tin and titanium tetrahalides, and possible structures are discussed[153].

2.3.7 Halogen derivatives

A polarographic study of bromocadmium complexes in molten $Cd(NO_3)_2 \cdot 4H_2O$ shows best agreement with experimental data when β_2 is omitted from the scheme. The results imply that a structural change occurs at the substitution of the second bromide ion and possibly desolvation at the $CdBr_2$ stage. This suggestion is supported by evidence for the formation of a colloidal precipitate at concentrations $> 5 \times 10^{-4}$ M in bromide ions[156]. A Raman spectral study of the stepwise formation constants of the bromide complexes of zinc, cadmium and mercury also reveals anomalous behaviour for $CdBr_2$. Like some other cadmium compounds, the bromide is polymeric in aqueous solution, but the zinc and mercury analogues are monomers. There is also evidence for polymeric $CdCl_2$ in aqueous solution. $ZnBr_4^{2-}$ has tetrahedral symmetry, but $ZnBr_3^-$ is C_{3v} and $ZnBr_2$ is C_{2v}, but whether the lower symmetries of the latter species are due to hydration is not known. However, ΔH and ΔS for coordination of even the first bromide ion to zinc, and to a lesser extent cadmium, show that hydration is considerably weakened[157]. Another Raman spectral study has identified mixed bromo-chloro–mercury anions in solutions of mercury(II) and potassium chlorides to which various amounts of potassium bromide were added. Stepwise replacement of Cl by Br is thought to occur in the change $[HgCl_4]^{2-} \rightarrow [HgBr_4]^{2-}$, which was also shown to be reversible[158]. Formation constants of the (known) ions $[HgI_3]^-$ and $[HgI_4]^{2-}$ in DMSO have been measured[159].

Mass-spectroscopic examination of a mixture of zinc chloride and sodium chloride at 250–450 °C has shown the presence of $NaZnCl_3$ and $Na_2Zn_2Cl_6$, with heats of formation -53.8 and -18.3 kcal mol^{-1}, respectively[158].

2.3.8 Zinc, cadmium and mercury derivatives of transition metals

A number of references to recent work on compounds with covalent bonds between a Group IIB element and a transition metal are included in Section

2.3.8 Table 2.2. These compounds are described fully in Volume 6 of this Series.

2.4 TABULAR SURVEY

This survey is an addendum to the text and entries have been arranged under the appropriate text headings.

Table 2.1 Survey of beryllium and magnesium compounds

Compound(s)	Brief description of study	Reference
Hydrides and related compounds (see Section 2.2.1)		
BeH_2	Enthalpy of formation of BeH_2 estimated	21
NaR_2BeH, etc.	Alkali-metal dialkylkberyllium hydrides [R = Pr, Pr^i, Bu^i, and $Li(OEt_2)(Bu^i)_2BeH$, $LiBu^i_2BeH$, $Na(OEt_2)Bu^t_4Be_2H$, $Na(OMe_2)Et_2BeH$] and their thermal decomposition	20
Bu^iBeH, $Bu^t_2Be_2H_2(NMe_2C_2H_4NMe_2)$	Coordination complexes of Be hydrides and addition to olefins	14
$BeH_2 \cdot R_3N$	Tertiary amine complexes of BeH_2 prepared by direct interaction $[R_3N + (BeH_2)_n]$. (i.r.) $[R_3N = Me_3N, Et_2NMe, Et_3N, Me_2NCH_2Ph, (CH_2)_4NMe, (CH_2)_5NMe, TMEDA]$	13
MgH_2	Detailed study of the reaction between $LiAlH_4$ and Et_2Mg	25
$KMgH_3$	Preparation of $KMgH_3$ (i.r., x-ray powder).	24
$MH \cdot MgR_2$	(M = Na, K; R = Ph, Bu^s; also $NaMg_2HBu^s_4$). Preparation.	23
'HMgX'	No evidence supporting the existence of these species could be obtained	22
$Be(BH_4)_2$	I.R. and mass spectral study Structure proposed	32
$Be(BH_4)_2$	Dipole moment measured, in gas phase (dielectric cell)	30
$Be(BH_4)_2$	Dipole moment measured by electric deflection technique	31
$Be(BH_4)_2$	I.R. study. New structure proposed	33
$MeBeBH_4$	I.R., n.m.r. spectral study. Me and H are bridged; linear B—Be—Be—B structure	34
$B_9BeC_2H_{11} \cdot NMe_3$	1-Trimethylamino-1-beryl-2,3-dicarba-*closo*-dodeca-borane(12) prepared and characterised	21
$Mg(BF_4)_2 \cdot 6H_2O$	Thermal decomposition examined	161
$MgB_{10}H_{12} \cdot 2Et_2O$	Preparation and characterisation	35
$Mg(AlH_4)_2$	Detailed study of preparation (i.r., x-ray powder)	26
$MgAlB_{14}$	X-ray determination of structure	162
$MgO \cdot Al_2O_3 \cdot 7-9H_2O$	Compound prepared; shown not to be a mixture (t.g.a., x-ray, i.r.)	163
$MgO \cdot Al_2O_3 \cdot 7-9H_2O$	Thermal decomposition and products thereof examined	164

Table 2.1—continued

Compound(s)	Brief description of study	Reference

Derivatives with Group IV ligands and associated groups (see Section 2.2.2)

Compound(s)	Brief description of study	Reference
R_2Be	(R = Me, Et, Pr^n, Pr^i, Bu^i, Bu^t) Mass spectral Fragmentation behaviour, appearance and ionisation potential measurements	44
Me_2BeL, Me_2BeL_2	(L = Et_2O, Me_3N) ^{9}Be and ^{1}H n.m.r. study	40
$Me_2Be(SMe_2)_{1,2}$, $MeBeCl(SMe_2)_2$	Variable temperature n.m.r. study. (ΔH_f obtained for some species)	165
$(C_5H_5)_2Be$	I.R. study showing asymmetrical (ionic) structure	41
Be_2C	ΔH_f^0 measured	49
$Me_2Mg \cdot 2NC_7H_{13}$	Crystal and molecular structure of dimethylbis(quinuclidine)magnesium determined (x-ray)	166
$(MgC_6H_{16}N_2)_2$	Structure of bis{[2-dimethylaminoethyl(methyl)amino]methylmagnesium} determined	70
$CH_2(MgX)_2$	Geminal dimagnesium compounds prepared and some reactions investigated	167
RMgF	Preparation of organo(particularly hexyl)magnesium fluorides (R = Me, Et, Ph, C_6H_{11})	45, 46
Mg_2Si	Preparation and some chemical properties	50
$CeMg_2Si_2$	Crystal structure	168
R_3SnMgX	(R = Pr^i, Bu^s, C_6H_{11}; X = Cl, Br) Preparation[47] and properties[48], particularly reactions with halides	47, 48

Derivatives of Group V ligands (see Section 2.2.3)

Compound(s)	Brief description of study	Reference
β-Be_3N_2	Refinement of crystal structure, showing all Be atoms in tetrahedral environment, in a HCHC... layer sequence of close-packed N atoms	72
LiBeN	Preparation, crystal data and hydrolysis study	169
$Be(NH_2)_2$, BeNH	Preparation and properties, including hydrolysis	51
$BeX_2 \cdot 4NH_3$, $Mg(NO_3)_2 \cdot 6NH_3$	(X = Cl, NO_3). Preparation, thermal properties, and i.r. based structures	170
$BeX_2 \cdot 4NH_3$, $Mg(NO_3)_2 \cdot 6NH_3$	(X = Cl, ClO_4, NO_3). Compounds and some deuterated (ND_3) derivatives prepared, i.r. studied including calculation of metal–nitrogen stretching force constants (valence-field approximation)	171
$M(HCO_2)_2 \cdot 2NH_3$, $[Be(NMe_2)_2]_3$	Synthesis and thermal properties. (M = Be, Mg) x-ray structure, n.m.r. spectra and bonding	67
$[(R^1R^2C=N)_2Be]_n$	(R^1R^2 includes p-tolyl, Ph, Bu^t). Preparation, i.r. spectra and Be=N π-bonding discussed	69
$Be(NCS)_2 \cdot L_2$	(L = Et_2O, dioxan). Preparation and properties	54
$M_{1,2}Be(NCS)_{3,4} \cdot n$MeCN	Preparation, i.r., and x-ray data (powder). (M = Cs, Rb, K, NH_4)	55
$Be(OAc)_2(RNH_2)_2$	(R = Me, Et, Pr). Preparation, i.r. spectra, and structure	81
$Be(OAc)_2 \cdot 2NH_3$, $Mg(OAc)_2 \cdot 2NH_3$	Structures, from i.r. studies of acetates and analogous formates	78
$(HCOO)_6Be_4O \cdot 6PrNH_2$	Study of interaction of $MeNH_2$ and Pr^nNH_2 with Be formate and basic formate. Formate yields a polymer	80
Coordination complexes	Formation constants and thermodynamic values (ΔG, ΔH, ΔS) relating to Be complexes of adrenaline, antistine, ephedrine, benadryl and pyribenzamine	172
$Mg(NH_3)_6^{2+}$	Coordination number of magnesium in liquid ammonia shown to be 6	173

Table 2.1—continued

Compound(s)	Brief description of study	Reference
$Mg(NH_2)_2$, MgNH	Preparations and properties, including x-ray data	52
$MgM_2(NH_2)_4$	(M = Na, K, Rb, Cs). Preparation and x-ray crystal data (space group, cell dimensions), thermal decomposition	53
$Mg_3Ca_3N_4$	High temperature synthesis, properties, lattice parameters, and Debye–Scherrer diagram table	174
$MgL_6(SnCl_6)$	Magnesium (also Co, Ni, Cu, Zn) complexes (with L = PrCN, Pr^{i}CN, Bu^{t}CN, PhCN) prepared. I.R. spectral data	57
$Mg(MeCN)_m^{n+} X_n^-$	Preparation, properties, Raman spectra on compounds of magnesium and other metals. (Anions of type X = $(Hg_pCl_{2p+q})^{q-}$; $p = 1–5$, $q = 1,2$; $n = 1,2$)	175
$Mg(MeCN)_m (MBr_4)_n$	($m = 4,6$; M = Al, Tl, Fe, In; $n = 1,2$) Magnesium and other metal complexes of MeCN, prepared. Structures discussed, i.r. spectra, x-ray powder data	56
MgL_6X_2	(L = pyrazole, imidazole, N-methylimidazole; X = ClO_4, BF_4). Preparation, properties, i.r., Raman spectra, and x-ray powder data	58, 59
$MgL_{1,2}X_2$	Formation constants of complexes with sulphonic acid derivatives of azo-3- and azo-5-pyrazolinones	60
$MgL_2(ClO_4)_2(_nH_2O)$	(L = 1,8-naphthyridine, 2,7-dimethyl-1,8-naphthyridine). Preparation, i.r., ^{1}H n.m.r. spectra and structure discussed	61
$MgL_2 \cdot 4 \cdot 8py$	(L = 4,4'-bipyridyl; py = pyridine). Reaction of py with magnesium and derivatives	176
$(H_2O)Mg(TPP.)$	(TPP. = meso-tetraphenylporphin). X-ray structure	177
Mg_3P_2	Reaction with NaOH examined	178
Be^{2+}	Hydrolysis of Be^{2+} in D_2O; formation constants for principal species	77
$[Be_n(OH)_y, aq]^{(2n-y)+}$	($n = 1–3, 6$; $y = 0–4, 8, 9$). Electrometric study of aq.hydrolysis of beryllium ion.	76
$M(ClO_4)_2 \cdot 2py$, $M(ClO_4)_2 \cdot py$	(M = Be, Mg, Ca, Sr, Ba, Zn, Cd, Hg) (M = Mg, Ca, Sr). Preparation and thermal properties.	179

Derivatives of Group VI ligands (see Section 2.2.4)

Compound(s)	Brief description of study	Reference
$(BeSi_2O_x)^{n-}$	($x = 7, 8, 9$). New beryllosilicates prepared, characterised, including x-ray data	180
$Be(OR)_2$	(R = Me, Et, Pr^i, Bu^n, n-C_6H_{13}, n-C_8H_{17}). Prepared and characterised. Also double alkoxides with Li	90
$Be(OR)X \cdot nROAc$	(X = Cl, Br; $n = 0.5–1.0$). Prepared from $Be(OR)_2$ and AcX	91
$Be(HCO_2)_2$, $Mg(HCO_2)_2$	Analysis of low-frequency vibrational spectrum	181
Be oxalates	Existence of three oxalato-complexes in aq. solution demonstrated. Structure and stability	182
$Be(OH)OAc$	Preparation and properties, polymeric nature and reaction with EtOH	183
$Be(HCO_2)_2$	Decomposition reactions of weak solutions in anhydrous liquid ammonia	79b
$Be(OAc)_2$, $Be_4O(OAc)_6$	Heats of formation determined by bomb calorimetry	184

Table 2.1—continued

Compound(s)	Brief description of study	Reference
$Be_4O(OR)_6$	(R = Ac, CH_2ClCO, CH_2BrCO, C_6H_5CO). Improved synthetic procedure	185
$Be_4O(OAc)_5OEt$	Analytical data, i.r. spectra and x-ray data on compound prepared by thermal decomposition of Be(OEt)(OAc)	84b
$Be_3O(OAc)_4$	New compound prepared by thermal decomposition of $Be_3O(OAc)_4 \cdot 3PrNH_2$ or $Be_3O(OAc)_4 \cdot 3BuNH_2$. Dimer also characterised	83
$Be_4O(AcO)_x(CCl_3CO_2)_{6-x}$	Mixed basic salts examined, problem of isomerisation studied, mass and ^{1}H n.m.r. spectra.	85
$Be_3O(OAc)_4 \cdot nRNH_2$	(R = Et, Pr, Bu; n = 2, 3, 6). Stoicheiometry studied. Previous n = 8 incorrect	82
$Be(OH)(OAc) \cdot L$	(L = pyridine, $\frac{1}{2}$ dioxan). Preparation of adducts from $Be_4O(OAc)_6$. Lose L readily giving $[Be(OH)OAc]_n$	186
$Be_4O(OAc)_6 \cdot nL$	($nL = 2SO_2$, $1\frac{1}{2}N_2O_4$). Loose addition compounds reported	84a
Be β-diketone complexes	β-Diketone complexes of Be, mass spectral study	187
$(C_{14}H_{11}OS)_2Be$	Substituted β-thioxoketone Be (and other Group II metals) complexes, stability constants studied by use of ^{1}H n.m.r.	188
$(C_6H_4N_2O_2)_2Be$, $C_6H_4N_4O_4Be$	Thermal and spectral study of Cupferron and dicupferron complexes	189
$(C_7H_5O_4)_2Be$	2,4-Dihydroxybenzoic acid complex, spectrophotometric and conductometric study	190
BeL, BeL_2^{2-}	(L = 3,5-dinitrosalicylic acid). Potentiometric study of composition and formation constants	191
BeL_2	(L = 3-hydroxynaphthoic acid, 1-nitroso-2-naphthol). Potentiometric studies	62
$Be[OPPh_2NPPh_2O]_2$, $Be[OPPh_2(NPPh_2)_3O]_2$	Preparation of monomeric Be phosphonitrilates	
$S_2N_2OBeCl_2$	Structure postulated for derivatives prepared from S_4N_4 and $BeCl_2$	192
MgO	Experimental determination of ionic state	193
$Mg(O_2)_2$	X-ray powder data by Debye–Scherrer method	194
$Mg(O_2)_2$	Formation from Mg peroxide and ozone	195
$Mg(O_2)_2$	Thermal stability (100–375 °C)	196
$MgCO_3 \cdot 3H_2O$	Thermal decomposition	197
$Mg(NO_3)_2 \cdot 6H_2O$	Thermal decomposition	198
$Mg(H_2PO_4)_2 \cdot 2H_2O$	Study of dehydration products	199
$Mg(H_2PO_4)_2 \cdot 2H_2O$	Thermal dehydration products characterised using paper chromatography	200
$Mg_2As_2O_7$	Refinement of structure	201
$MgS_2O_3 \cdot 6H_2O$	Refinement of structure	202
$MgSeO_3 \cdot 6H_2O$	Thermal decomposition	203
$MgSeO_3$	Thermal stability	204
$Mg(SeCN)_2 \cdot 4L$	(L = dioxan). Preparation, i.r. study	205
$Mg(H_5TeO_6)_2$	Preparation from Mg/H_6TeO_6, and properties	206
$Mg(ClO_4)_2$	Thermal decomposition by isothermal kinetics	207
$Mg(BrO_2)_2 \cdot 6H_2O$	Preparation, x-ray and thermogravimetric study	208
$Mg(H_2O)_6H_3IO_6$	Crystal and molecular structure (x-ray)	209
$MgCrO_4$	Preparation from $MgCrO_4 \cdot 5H_2O$ under oxygen	210
$Na_2Mg_3(OH)_2(SO_4)_3 \cdot 4H_2O$	Characterised from a complex brine	211
$Mg(C_2O_4)$	Kinetics of formation of Mg oxalate by a pressure jump technique	212

Table 2.1—continued

Compound(s)	Brief description of study	Reference
$Mg(C_2O_4)\cdot 2H_2O$, $MgCl_2\cdot MgC_2O_4\cdot nH_2O$	Preparation and thermolysis of Mg oxalate and chloro-oxalates. ($n = 8, 10$)	213
$MgC_4O_6H_4\cdot 3H_2O$	Preparation of Mg and other metal tartrates	214
$MgC_4O_6H_4\cdot 3H_2O$	Thermal analyses of Mg tartrates	215
$MgCl(OR)$	(R includes Me, Et). Preparation and thermal decomposition	92
$Mg(OH)_{1.3}(OMe)_{0.7}$	Study of reaction of $Mg(OH)_2$ and MeOH. Products examined by x-ray, electron diffraction, electron microscopy, i.r. and thermal analysis	216
$Mg(MeNO_2)_6(SbCl_6)_2$	Preparation and properties. I.R.	96
$Mg(MeNO_2)_6(FeCl_4)_2$	Preparation and properties. Far i.r.	96
$Mg(ester)_6(anion)_2$	(Ester = methyl formate, ethyl acetate, diethyl malonate; anion = $FeCl_4$, $InCl_4$, $SbCl_6$). Preparation, i.r. study, [ν(metal–ligand)]	217
$Mg(C_2H_5NO)_4Cl_2\cdot 2H_2O$	N-methylformamide complexes, preparation and properties. U.V., i.r. spectra	218
$Mg(OEt_2)X_2$	(X = Br, I). I.R. and Raman spectra of complexes with almost complete vibrational assignment	89
$Mg[MePO(OPr^i)_2]_4(ClO_4)_2$	Diisopropylmethylphosphonate complexes studied. I.R. and conductance measurements	219
$Mg\{[(Me_2N)_2P(O)]_2CH_2\}_3(ClO_4)_2$	I.R., visible spectra, and conductance data, on octamethylmethylenediphosphonic diamide complexes	94
$Mg\{[Me_2N]_2P(O)OP(O)[NMe_2]_2\}_3(ClO_4)_2$	Structural study (x-ray)	95
$Mg(Bu^n_3PO)_4^{2+}$	Preparation of complexes, spectral, conductance, and x-ray powder data	220

Derivatives of Group VII ligands; miscellaneous (see Section 2.2.5)

Compound(s)	Brief description of study	Reference
BeF_2	Transference numbers and ionic mobilities on molten BeF_2 and BeF_2/LiF mixtures	100
$BeF_n^{(2-n)+}$	($n = 1$–4). Potentiometric study of aqueous fluoride solutions of Be	101
$MBeF_4\cdot 6H_2O$	M = Zn, Co^{II}, Ni^{II}. Preparation from BeF_2 and MCO_3/HF	221
$MeNH_3Cr\cdot(BeF_4)_2$, $[C(NH_2)_3]Cr(BeF_4)_2$	Properties and crystallography of the hydrated double fluoroberyllates	222
MgF_2	I.R. study, showing bent F—Mg—F skeleton	223
$MgClBr\cdot 6H_2O$	Crystal structure. (Isomorphous with $MgCl_2\cdot 6H_2O$)	224
$[TiNMg_2Cl_2\cdot THF]$	Complexes in the $TiCl_3$–Mg nitrogen fixing system	71
$BrMgMn(CO)_5$	Preparation and synthetic utility of 'transition-metal Grignard reagents'	225
Mg_3P_2	Reaction with solid NaOH	226
Mg	Reaction of vaporous Mg with O_2 and N_2	227
Be	Purification of Be by fractional sublimation of the acetylacetonate	228
$MgPbF_6$	Preparation of hexafluoroplumbates of Mg and other metals	229
$MgCl_2$	Electrolysis of the solution in liquid NH_3	230
$Mg^{2+}(Zn^{2+})$	Pressure-jump study of mechanism and kinetics of ion association	231

Table 2.2 Survey of zinc, cadmium and mercury compounds

Compound(s)	Brief description of study	Reference

Hydrides and related compounds (see Section 2.3.1)

$NaHZn(C_6F_5)_2$	Preparation and characterisation	104
$MH(ZnR_2)_n$	($M = Li, Na; n = 1,2; R = Me, Et.$)	
	Preparation, structure, n.m.r., i.r.	103
$(HZnNMeCH_2CH_2NMe_2)_2$	Preparation and characterisation	15

Group II ligands, mercury (I) and related compounds (see Section 2.3.2)

Cd_2^{2+}, etc.	E.S.R. evidence for Cd_2^{2+}, Cd^+, Hg^+ (References 106, 107); and properties of monovalent Cd^+ ions (Reference 108)	106–108

Group III ligands (see Section 2.3.3)

$(Me_4N)_2[Hg(B_{10}H_{12})_2]$	Preparation and properties of mercury derivatives of polyhedral boranes	232
$[Me_4NB_{10}H_{12}HgCl]$		
$Zn(BH_4)_2, Zn(BH_4)_2 \cdot 2L$	($L = $ THF, pyridine, NMe_3). Preparation, properties, and association of $Zn(BH_4)_2$	109

Group IV ligands (see Section 2.3.4)

$ZnNCN$	Preparation and properties, thermal decomposition	233
$C[HgOAc]_4$	Preparation and reactions	234
'Reformatsky reagent'	Structure of intermediates in Reformatsky reaction	110
$(Cl_3Si)_2Hg, (Cl_3Ge)_2Hg$	Syntheses and reactions	112
$Me_8Si_4(CH_2)_2Hg_2$	Preparation of heterocyclic Hg—Si compound	115
$Me_3SiHgGeMe_3$	Presence in equilibrium mixtures of $(Me_3Si)_2Hg$ and $(Me_3Ge)_2Hg$	235
$(Me_3Si)_2Hg$	Reactions with carbonyl compounds	236
$(Me_3Si)_2Hg$	Reactions with Sn—O and Sn—N containing compounds	237
$(Et_3Si)_2Hg, (Et_3Ge)_2Hg$	Reactions with sulphur	114
$(Me_3Si)_2Hg$	Addition to fluoro-olefins	238
$(Et_3Ge)_2Hg$	Reactions with R_2Hg ($R = CH_2CO_2Me$, CH_2COEt, Ph_F, CN)	113
$(Et_3Ge)_2Hg$	Reaction with Et_3SnX	239
Zn_3N_2	Preparation, reactions and thermal decomposition	240
Zn/Ge double nitrides	Series of Wurtzite-type oxynitrides, preparation and properties	241

Group V ligands (see Section 2.3.5)

Zn_2NX	($X = $ Cl, I). Preparation, x-ray diffraction data and thermal decomposition	242
$Zn[N(CN)_2]_2$	Reactions with methanol and H_2S	243
$RMC(CN)_3$, $PhHgC(CN)_3$	($M = $ Zn, Cd; $R = $ Ph, Et). Preparation of tricyanomethylmetal derivatives and (polymeric) structure. I.R.	116

Table 2.2—continued

Compound(s)	Brief description of study	Reference
$Zn(NH_3)_4I_2$	Metal-isotope effect on Raman spectrum, and use in assignments	117
ZnL_2X_2	(L = NH_3, ND_3, Imidazole and derivatives) Low frequency i.r. study, assignments and force constants	244
$Zn(NH_3)_2(HCO_2)_2$	Synthesis and thermal properties	79
$Cd(NH_3)_2Ni(CN)_4 \cdot 2C_6D_6$	Motion of NH_3 in molecule	245
$(Et_4N)ZnX_3$	(X = Cl, Br, I). Reactions with other N-donor molecules	246
$(Et_4N)_2ZnCl_4$	Calorimetric study of thermal behaviour, with heats and entropies of transitions	247
$M_2Zn(NH_2)_4$	(M = Na, K). Preparation, crystal data, thermal decomposition. I.R.	248
$(ZnClNH_2)$	Preparation of polymeric zinc amidochlorides	249
$Hg(NH_2)Cl$	Reaction with PCl_5	250
$Zn(NRH)_2$	(NRH = aromatic amine group). Preparation and properties	251
$Zn(N_3)_2$	Preparation, thermal decomposition of pure zinc azide	252
$K_2Zn(N_3)_4$	Preparation and x-ray powder data	253
$Zn(N_3)_2(py)_2$	X-ray crystal structure	254
$Zn(MeCN)_mX_n$	$[X = (Hg_pCl_{2p+q})^{q-}, p = 1-5, q = 1,2]$. Preparation, properties and Raman spectra	175
$M(MeCN)_m(M'Br_4)_n$	(M = Zn, Cd; M' = Al, Tl, Fe, In; n = 1,2; m = 4,6). Preparation, i.r. spectra, x-ray powder data	56
$ZnL_6(SbCl_6)_2$	(L = PrCN, Pr^{i}CN, Bu^{t}CN, PhCN). Preparation, i.r.	57
$Zn(MeCN)_4(FSO_3)_2$	Preparation, characterisation, i.r.	255
$K_2Zn(NCO)_4$, $KCd(NCO)_3$	I.R. study of cyanate complexes	256
$M(NCS)_2$	(M = Zn, Cd). Preparation and spectra (i.r., u.v.)	257
$M[Zn(NCS)_3]nH_2O$	(M = Na, K, Cs, NH_4, $C(NH_2)_3$, $\frac{1}{2}$Ba; n = 1–3). I.R. study	258
$MM'[Zn(NCS)_4]nH_2O$	(M,M' = Na, K, Cs, NH_4, $C(NH_2)_3$; n = 0–4). I.R. study	259
$M_4[Zn(NCS)_6]nH_2O$	(M = Na, K, Cs, NH_4, $C(NH_2)_3$, $\frac{1}{2}$Ba; n = 0–4). I.R. study (including compounds with mixed M)	260
$M[Zn(NCS)_4]$	I.R. study of alkaline earth metal salts, Zn and Cd analogues	261
$\mathrm{>N\!-\!Hg\!-\!Hg\!-\!N<}$	Study of stable Hg^I–nitrogen compounds, structures, preparations. I.R. and x-ray data	118
Hg_2I_2	Mössbauer spectra of intercalation compounds with BN	262
$Hg(NSF_2)_2$	Preparation, properties, x-ray structure	105
HgL_2	(L = series of amides and amidines, also compounds HgLCl and HgLOH)[263], Cd[264]	263, 264
$Zn(sal\!-\!R)_2nH_2O$	(Complexes of substituted salicylaldehydes and 2-aminopyridine derivatives). Preparation, characterisation	265
$Zn(plp,L)^{2-}$ $Zn(plp,L)(L)^{3-}$ $Zn(plp,L)_2^{6-}$	Equilibrium studies on pyridoxalphosphate (= plp) glycine and α-alanine (= L) zinc complexes	266
$ZnLX_2$	(L = terdentate Schiff base and related ligands). Stereochemistry of 5-coordinate complexes. I.R.	267
ZnL_2	(L = triazine derivatives). Structural study	268

Table 2.2—continued

Compound(s)	Brief description of study	Reference
$Zn(PhCOCHCMeNR)_2$	(R = Me, H). N.M.R. study of stereochemistry of bis(chelate) complexes	269
$Zn(C_{15}H_{13}N_2)_2$	Spectroscopic study of complexes of malonaldehydedianil	270
ML_2	Stereochemical study of Zn and Ni complexes of 1-azolyl-3-methyl-5-tolyl-formazans	271
ZnL_2	Crystal structure of zinc salicylal-o-anisidinate	272
$Hg(NSO)_2$	Preparation, characterisation, I.R., mass spectra	273
$[HgN(SO_3)_2]^-$	Thermal decomposition of salts	274a, b
$(p\text{-}NO_2C_6H_4NH_2)_2CdCl_2$	Structure determination	275
$M'L_4X_2$	(L = pyridine, γ-picoline; X = trihaloacetate). Preparation, i.r., u.v. M' = Mn, Co, Sn, Ni, Cu	276
$ZnLCl_2$	[L = 2-methylthioaniline, 3,3'-bis(methylthio) benzidine]. Preparation, i.r., n.m.r.	277
$Hg(RH)(OH)H_2O$	[RH_2 is phenylsulphonyl-α(or β)-alanine]. Preparation and properties. Also the related complexes of Zn, Cd	278
$Zn(diamine)_2(ONO)_2$	Diamine complexes (N,N'-diethylethylenediamine and others) of zinc nitrite. Preparations, i.r.	279
$(MeRNCH_2)_3CCH_3ZnX_2$	(R = H, Me; X = Cl, Br). Preparation and study of steric effects of ligands, n.m.r., i.r.	120
$(Pr^i_2NH)_2MX_2$	(M = Zn, Cd, Hg). Formation constants studied using an electrode reversible with respect to Pr^i_2NH	280
ML_1L_2	(M = Zn, Co^{II}); L_1L_2 = ethylenediamine, histamine, serine). Stability constants for mixed complexes from titration data	281
$CH_2NH_2CO_2H/ZnSO_4$	Ultrasonic study of the system in aqueous solution. Formation of 1:1 and 1:4, Zn glycine complexes	282
ML_2X_2	(M = Zn, Cd; L = AcNHNH$_2$; X = Cl, Br). Preparation, properties	283
ZnL_2X_2	(L = PhNHNH$_2$; X = Cl, Br, I). Dipole moments and structure	284
$CdCl_2L$	(L = benzoquinone guanylhydrazone thiosemicarbazone). Preparation and properties	285
Piperazine–Cd^{2+}	Formation of complexes in aqueous solution	286
$[ZnL]^{2+}$	[L = 1,8-bis(2'-pyridyl)-3,6-diazaoctane]. Preparation. I.R.	121
$[Znpy_{1-4}](ClO_4)_2$	Formation and composition. X-ray data, u.v., potentiometric study	287
$[Mpy_6](ClO_4)_2$ $[M'py_2](ClO_4)_2$	(M = Cd, Hg; M' = Zn, Cd, Hg). Thermal study	288
ML_nX_2	(M = Zn, Cd; L = pyridine, α-, β-, γ-picoline; n = 2–6; X = Cl, Br, I, OCN). Thermal decomposition	289
$Znpy_3(NO_3)_2$	Similar Cd, Hg, complexes. Thermal analysis. I.R.	290
$ZnL_{1,2}X_2$	(L = pyridine and derivatives; X = Cl, Br, I, OAc). Also similar Cd, Hg complexes. Preparation, structures, i.r.	291
$ZnL(acac)_2$	(L = substituted pyridines, quinoline). Structures, i.r.	292
$ZnL_x(NCO)_2$	Structures of cyanates and azides. (L = cyanopyridines; x = 2,4). I.R.	293

Table 2.2—continued

Compound(s)	Brief description of study	Reference
CdL_2Cl_2	(L = picolinic, nicotinic, isonicotinic acids). I.R. spectral study	294
ZnL_2Cl_2	(L = pyridine carboxylic acid, aminobenzoic acid). Study of reaction of ligand and $ZnCl_2$, $HgCl_2$	295
CdL_2X_2	(L = picolines). Kinetics of thermal decomposition	296
ZnL_2X_2	(L = lutidines). Preparation, characterisation. I.R.	297
ML_2X_2	(L = nicotinamide; X = Cl, $\frac{1}{2}SO_4$; M = Zn, Cd). I.R. study	298
ML_2Cl_2	(L = nicotinamide; M = Zn, Cd, Hg). Preparation, structure based on i.r.	299
ML_2X_2	(M = Zn, Cd; L = morpholine; X = OAc, NCS). Preparation, i.r.	300
ZnL_2SO_4	(L = morpholine). Preparation, u.v., visible spectra	301
$HgL_{1,2}X_2$	(L = morpholine; X = Cl, Br, I). Preparation, i.r.	302
$HgL_{1,2}X_2$	(L = pyridine, o-, m-, p-toluidine; X = Cl, Br). Preparation, i.r.	303
$M(dipy)_2(NCO)_2$ $M(phen)_2(NCO)_2$	Preparation and structures. I.R. (M = Zn, Cd) (M = Zn)	304
ZnL_nX_2	(L = 2,2'-bipy; n = 1,2,3; X = Cl, Br) Preparation, i.r.	305
$Zn(dipy)_2(NCSe)_2$	Preparation and spectral study of diamine zinc complexes	306
$Cd(phen)_3^{2+}$	Stability in water–propanol and water–acetone solutions	307
ZnL_3	Stability constants for 1:1, 1:2, 1:3 complexes of 8-hydroxyquinoline-5-sulphonic acid (= L)	308
$ZnLCl_2$	(L = 1,5-diazanaphthalene). Preparation and structure	309
$ML_2(ClO_4)_2 \cdot nH_2O$	(M = Zn, Cd; L = 1,8-diazanaphthalene). Preparation, and stereochemical study, i.r.	61
$ZnLSO_4 \cdot 2H_2O$	(L = 8-amino-2-methylquinoline). Preparation and structures. Visible, u.v., i.r.	310
ML_2	[M = Pb, Co^{III}, Cr^{III}; L = mercurated 8-hydroxyquinoline]. Preparation, n.m.r., and characterisation	127
$ZnLClO_4$	[L = cis,cis-1,3,5-tris(pyridine-2-carboxaldimino) cyclohexane] x-ray structure (trigonalpyramid)	122
$MLCl_3$	(M = Zn and others; L = N-methyltriethylenediaminium). Far i.r. study	311
ZnL	(L = octahydroxyanthraquinonecyanine). Synthesis, electrical properties, i.r.	304
MLX_2	(M = Zn, Cd, etc.; L = hexamethylenetetramine). Preparation, structures, i.r.[313] preliminary x-ray data[314]	313, 314
ML_2X_2	(M = Zn, X = OAc; M = Hg, X = Cl; M = Cd, X = NO_3; L = antipyrine). Conductometric study	315, 316
HgL, HgL_2	(L = substituted tetrazoline-5-thiones). Preparations, tentative structures, i.r.	317
Zn^{2+}, Cd^{2+}, Hg^{2+}/L	[L = ethylbis(4-sodio-5-tetrazolylazo)acetate]. Study of interaction of cations with L, instability constants	318
ZnL_2	(L = 2-pyrrolecarbaldehyde imine). Preparation and dipole moments	319

Table 2.2—continued

Compound(s)	Brief description of study	Reference
$ML_{1,2}X_2$	(L = sulphonic acid derivatives of azo-3- and azo-5-pyrazolinones). Formation constants, coordination numbers	60
ZnL_2X_2	(L = substituted benzothiazoles). Preparation and characterisation	320
ZnL_2X_2	(L = alkyl substituted thiazoles; X = Cl, Br, I). Preparation, coordination, i.r.	321
Zn, $CdL_{4,6}X_2$	(L = pyrazole, imidazole, N-methylimidazole; X = BF_4, ClO_4). Preparation, x-ray powder, i.r., Raman	58, 59, 322
ZnL_2	(L = 3,5-dimethyl-1-thiocarbamoyl-pyrazole). Preparation, coordination sites defined	323
ZnL_2X_2	(L = antipyrine, and 4-dimethylamino derivatives; X = Cl, NCS, NO_3). Preparation, coordination	324
ZnL_2	(L = 1,5-dibenzimidazolylformazans). Preparation, spectra, dipole moments	325
Cd, $Hg(biuret)_2Cl_2$	I.R. spectral study of stereochemistry, force constants	326
HgL_2Cl_2	(L = 5-alkyl-5-nitrotetrahydro-1,3-oxazines). Preparation	327
(Zn)	Preparation of complexes of (1-aryl-3-isoindolyl) (1-aryl-3-pseudoisoindolenylidine)arylmethane, and bromo- and iodo-derivatives	328
ZnL	(L = tricarbonylchromium complexes of *meso*-tetraphenylporphin). Preparation, properties, i.r., u.v., visible spectra	329
(Zn)	Kinetics of incorporation of Zn (and Cu) into *meso*-tetrapyridylporphin	330
(Zn, Cd, Hg)	Stability constants, and Hammett ρ-constants for porphin complexes	331, 332
(Zn)	Incorporation of Zn into water soluble porphyrin	333
(Zn)	Zn complex of a new macrocyclic hexadentate ligand	334
MLX_2	(M = Zn, Cd, Hg; L = 2,2′,6′,2″-terpyridyl; X = halogen, ClO_4, NO_3). Preparation, coordination, i.r.	335
(Zn)	Complexing behaviour of 2-aminomethylpyrrole	336
$Zn_3L_2 \cdot 2MeOH$	(L = tripyrrene-α-carboxylic acids). Preparation, characterisation, i.r., u.v.	337
(Zn, Cd, Hg)	5-coordination with complexes of a variety of potentially bi- and tri-dentate N-donors	338
Zn	Encapsulation reactions with hexadentate ligands giving *non*-octahedral symmetry	123
$(Ph_3P)_2MX_2$	(M = Zn, Cd, Hg; X = Cl, Br, I) Assignments of metal–P frequencies in i.r.	339
$(Ph_3E)_2HgX_2$, $[Ph_3EHgX_2]_2$	(E = As, P; X = SCN, NO_3, ClO_4). Preparation, structures, i.r.	340
$Cd(diphos)X_2$	[diphos = ethylenebis(diphenylphosphine), tetramethylenebis(diphenylphosphine)] Preparation, properties	341
$Hg(diphos)X_2$	[diphos = acetylenebis(diphenylphosphine); X = Cl, Br, I]. Preparation, structure, Raman, i.r.	342
$HgCl(EtO)_2PO$	X-ray crystal structure, $^{199}Hg-^{31}P$ coupling constants in related compounds and the *trans*-influence in Hg^{II}	126
$HgX_2(R_2PR')$	(R = Me, Ph, p-tolyl; R′ = o-carboxyphenyl). Preparation, coordination and structure, i.r.	343

Table 2.2—continued

Compound(s)	Brief description of study	Reference

Compounds with Group VI ligands and associated ions (see Section 2.3.6)

Compound(s)	Brief description of study	Reference
Zn^{2+}	Hydrolysis constants and enthalpies for Zn^{2+} at 40–90 °C	129
$Cd(H_3IO_6)\cdot 3H_2O$	Crystal and molecular structure	344
Zn	Structural studies on glycolato-, lactato-, and mandelato-complexes of Zn. U.V., i.r., x-ray powder data	345
$ZnC_{10}H_{14}O_4$	Physico-chemical studies on Zn camphorate	346
$Zn(C_7H_6O_2N)_2$	Preparation, structure of bis(anthranylates) of Zn, Cd; i.r.	347
$Zn(C_6H_4O_2N)_2(H_2O)_4$	Crystal structure of Zn picolinate tetrahydrate	130
(Hg)	Complexes of o-hydroxy-o-carboxyformazans with Hg	348
$Zn(C_6H_9N_3O_2)_2$	Stereoselectivity in histidine-complex formation	131
$ZnCl_2\cdot 2L$	Complexes with Zn, Cd, Hg of 2(1H)-tetrahydro-pyrimidinone. Preparation, properties, structures, i.r.	349
$(CdCl_2\cdot 2L)_3$		
$HgCl_2\cdot L$		
$M(LH)_n\cdot mH_2O$	(LH_2 = maleic acid; M = Zn, Cd, etc.). Preparation and thermal properties	350
$NH_4\cdot M^{II}L_{n+1}$	(L = 1-furoyl-3,3,3-trifluoroacetone; M = Zn, Cd, etc.). Preparation and structures	351
$ZnL\cdot 2H_2O$	(L = bromanilic acid). Thermal properties and spectra, i.r.	352
$ZnL_2\cdot 2H_2O$	(L = lawsone, juglone). Thermal properties, spectra	353
$ZnL_2(NO_3)_2$	(L = 2,6-dimethyl-4-pyrone). Preparation, coordination, i.r.	354
HgX_3^-	(X = Cl, Br, I). ^{35}Cl, ^{37}Cl n.q.r. study, structures	132
$Zn(ester)_6X_2$	(ester = methyl formate, ethyl acetate, diethyl malonate). Preparation, i.r.	217
Zn^{2+}	Study of complexation by ethoxy- and ethyl-thioacetate in 1.0 M–ClO_4^-	355
ZnL_2	(L = 2-ethoxy-4-formylphenol, 2-methoxyphenol). Preparation, i.r. spectra	356
(Zn, Cd)	Study of chelates formed by 1-nitroso-2-naphthol-7-sulphonic acid and 2-nitroso-1-naphthol-6-sulphonic acid	357
$Cd(dioxan)X_2$	(X = Cl, Br, I, NCS). Preparation, structure, i.r.	358
$CdLX_2$	(L = 1,3- or 1,4-dioxan, THF; X = Cl, Br, I). Preparation, structures, i.r.	359
ZnL_2X_2	(L = 2-methyl-benzoxazole). Preparation and structure.	360
ML_2X_2	(M = Zn, Cd; L = N-methylformamide; X = Cl, I). Preparation, properties, i.r.	218
$C_6H_4N_4O_4M$	(M = Zn, Cd, Hg). Thermal and spectral study of Cupferron and dicupferron complexes. X-ray data	189
$Hg(O_2CR)_2$	Decarboxylation studies of mercury salts.	133–136
$ZnL_2X_2, CdLX_2, HgLX_2$	Complexes of pyridine-N-oxides. Preparation and structures, i.r.	361, 362–365
$ZnL_5(ClO_4)_2$	(L = 4-ethoxypyridine-N-oxide). Preparation, spectra[366], and related nitrate complexes[367]	366, 367
$ZnL_4(ClO_4)_2\cdot nH_2O$	(L = acridine-N-oxide). Preparation, properties	368
$Hg[ON(CF_3)_2]_2$	Preparation[138], and reactions[139], especially H-abstraction	138, 139
ML_2Cl_2	(M = Zn, Cd; L = paramagnetic xanthate). Preparation, e.s.r.	140

Table 2.2—*continued*

Compound(s)	Brief description of study	Reference
$Zn(MeNO_2)_6X_2$	(X = $FeCl_4$, $SbCl_6$). Preparation, properties, i.r.	96
$Hg(acac)_2$	Preparation and structures of acetylacetone derivatives. N.M.R.[141, 142], i.r.[142], Zn, Cd complexes	141, 142
$(C_{14}H_{11}OS)_2Zn$	Substituted β-thioxoketone complex. Stability constants, n.m.r.	93
$ZnLCl_2$	(L = ethylenedi-iminobisacetylacetone). Preparation and structure, i.r., u.v., n.m.r.	369
$M(SO_2R)_2 2H_2O$	(R = Ph, p-tolyl). Coordination of sulphinate to Zn, Cd; i.r.	370
$RHgO_2SR'$	(R = Ph, p-tolyl = R'). Preparation, structures[371]; (R = Me = R'), preparation, structure, and reactions[372]	371, 372
$PhHgOH$, $(PhHg)_2O$	Structures, preparation, molecular association	143, 144
$Zn(OR)_2$	(R = Me, Et, Pr^i, Bu, Bu^t, octyl). Alkoxides and double alkoxides with Li. Preparation, reactions, i.r.	373
Zn(glycerol-alkoxides)	Glycerol alkoxides. Preparation, reactions, i.r., visible	374
$M(OH)_n(OMe)_m$	Non-stoicheiometric alkoxide-hydroxides; preparation, x-ray data, i.r., thermal properties	216
$Zn[(Me_2N)_3PO]_2Cl_2$	Preparation and properties[375], and related derivatives[376]	375, 376
$Zn\{[(Me_2N)_2PO]_2CH_2\}_3(ClO_4)_2$	Preparation, structure, i.r., visible	94
$Cd[MePO(OPr^i)_2]_4(ClO_4)_2$	Preparation, structure, i.r.	219
Na_2CdO_2	Mixed Cd oxides. Preparation, properties	377
CdO	Thermal decomposition of hydrates[378], thermal analysis of $CdCl_2$ reaction[379], thermal analysis, i.r. on hydroxy species of Cd, Zn[380]	378–380
$Cd_2(OH)_3X$	(X = Cl, Br, I). Formation of basic halides, thermal, x-ray data	381
HgO	Reaction with Na_2O_2, thermal study	382
Zn_2SiO_4, Zn_2GeO_4	Crystal structures	383
$Cd(NO_3)_2$	Basic nitrates prepared[384], thermal study of hydrate[385], i.r. data and preparation of nitrite[386]	384–386
NH_4ZnPO_4	Preparation, thermal decomposition	387
MSO_4	Thermal study of dehydration at pressure, Zn[388], thermal dissociation of basic sulphate, Zn[389], thermal analysis of basic sulphate, Cd[390]	388–390
$CdSO_3$	Thermolyses and oxidation	391
$CdS_2O_3·2H_2O$	Preparation	392
$CdCrO_4$	Thermal decomposition of Cd, Zn chromate[393], and of Cd chromate[394]	393, 394
$HgCr_2O_4$	Preparation	395
$CdMoO_4$	Conditions for preparation	396
$K_2ZnW(CN)_4(OH)_4$	Suggested by amperometric titrations	397
$ZnO·V_2O_3$	Oxidation at high temperatures	398
$Zn(ReO_4)_2·6H_2O$	Preparation and properties	399
Salts of Organic Ligands	Zn acetate pyrolysis[400], solubility of Zn oxalate[401] preparation of complex Hg oxalates[402], thermal decomposition of Zn, Cd oxalates[403], thermal behaviour of Zn, Cd tartrates[215], preparation of Cd tartrate and citrate[404], preparation of Zn, Cd, Hg tartrates[214], preparation of Cd benzoate iodide[405], boromucates of Zn, Cd[406], stability constants of Cd, Hg propylenediaminetetracetate[407]	400–407

Table 2.2—*continued*

Compound(s)	Brief description of study	Reference
$M(CS_2NH_2)_2$	Zn, Cd dithiocarbamates, preparation, ΔH_f [408] Zn, Cd, Hg oxidation[409], Zn compounds, i.r., n.m.r. of 5-coordinate complexes with N-donors[410]	408–410
$M(C_7H_5OS)_2$	Preparation and structures of thiobenzoates of Zn, Cd, Hg	411
$Zn(C_7H_5S_2)_2$	S-addition and abstraction in Zn dithiobenzoate	145–147
$Zn_2L\cdot 3H_2O$	(L = ethanediylidenetetrathiotetracetic acid). Synthesis, properties, i.r.	412
(Zn, Cd, Hg)	Thiourea(tu) derivatives. Thermochemistry of Zn, Cd, Hg compounds[413], stability constants for mixed Cd halide tu compounds[414], coordination of tu derivatives to Cd, Hg, and i.r.[415], preparation and thermal properties of cyano–Cd–tu complexes[416]	413–416
(Zn, Cd, Hg)	Thiophosphate and related derivatives. Zn, Cd derivatives $M_3(PS_4)_2$ [417], stereochemistry of bis(dialkyldithiophosphates) of Zn [418], preparation of bis(O,O-dimethyldithiophosphate) complexes of Zn, Cd, and n.m.r. [418], crystal and molecular structure of $Zn(Et_2PS_2)_2$ [149]	148–150, 417–418
$Zn(C_5H_7S_2)_2$	Preparation and properties of dithio-acetylacetonate derivative	419
$M(C_{10}H_9S_2)_2$	Preparation and properties of Zn, Cd complexes of dithiobenzoylacetone	420
$Hg(C_7H_6N_2S)\cdot Cl_2$	Preparation of 2-aminobenzothiazole Hg compounds[421], related derivatives of Zn, dipole moments[422]	421, 422
(Cd, Hg)	Thiocyanate complexes. I.R. spectra of trithiocyanate Cd compounds[423], i.r. of hexathiocyanate Cd compounds[424], $Hg(SCN)_2$ compounds with Ti, Sn tetrahalides[425], Hg pseudohalides with N-donors[426], Hg pseudohalides with N-donors and Ph_3P[427]	423–427
(Zn, Hg)	Hg, Zn complexes of lipoic acid, dihydrolipoic acid, and 1,3-dimercaptopropane[428], n.m.r. study of methionine and S-methylcysteine Hg complexes[429], cyanidexanthate mixed complexes of Hg[430]	428–430
$NSZnCl_2$, $S_2N_2ZnCl_n$	Preparation, properties, and structures postulated, i.r.	192
$Zn(C_{10}H_{20}N_2S_2)_3(ClO_4)_2$	Coordination of tetraethyldithio-oxamide to transition metals, structure	431
ML_2X_2	(L = N,N,N',N'-tetraalkylthiuram disulphide; M = Zn, Cd, Hg; X = halogen). N.M.R. study of stereochemistry, restricted rotation	432
$Zn(S_2AsMe_2)_2$	Zn dithiocacodylates. Synthesis and i.r.	433
$Hg_3S_2SiF_6$	Preparation and x-ray data	434
$(S_2Hg_3)SiF_6$	Preparation, structure and i.r.	435
$ClHg\cdot SC(=NNHPh)N=NPh$	Reaction, properties, stability constant (dithizonate)	436
$Hg_2B_2S_5$	Preparation, characterisation	437
$ML_{1,2}X_2$	(M = Zn, Cd, Hg; L = Me_3SbS; X = Cl, Br, I). Preparation, structure, i.r.	438
ZnSe	Preparation and ΔH_f	439
$ZnSeO_3$	Preparation, thermal decomposition	440
$Hg(OAc)_2\cdot 2HgS$ $Hg(OAc)_2\cdot 2HgSe$	Preparation and properties	441

Table 2.2—continued

Compound(s)	Brief description of study	Reference
$M[(PhO)_2PSSe]$	(M = Zn, Cd, Hg). Preparation, properties, i.r., u.v.	442
ZnL_nX_2	(L = selenosemicarbazone; X = Cl, n = 1; X = NO_3, n = 2). Preparation, structure	443
(Zn, Cd)	Selenocyanate complexes	444

Halogen derivatives (see Section 2.3.7)

MX_2	(M = Zn, Cd, Hg; X = halogen). Kinetics of I exchange between ZnI_2 and BuI^{445}, gas-phase Raman spectra of Zn, Hg mixed halides[446], polarography of bromo-Cd complexes in molten aqueous melts[156] preliminary x-ray study of $CdBr_2 \cdot 2OC(NH_2)_2$ [447], Raman study of formation of bromide complexes of Zn, Cd, Hg [157] n.q.r. study of HgI_2 dioxanates[448]	445–448
$MX_n^{(n-2)-}$	(M = Zn, Cd, Hg; X = halogen). $ZnCl_4^{2-}$ and $ZnCl_3^{3-}$ [449], mass spectrometric study showing $NaZnCl_3$ and Na_2ZnCl_4 [450], Raman study of Hg chloro- and bromo-anions in solution[451], formation constants of HgI_3^-, HgI_4^{2-} in DMSO [159] and kinetics of isotope exchange of Hg and I in Ag_2HgI_4 [452]	159, 449–452
(Zn, Cd, Hg)	Complex halide anions. Thermal decomposition of Zn, Cd tetrafluoroborates[161], hexafluoroplumbates of Zn, Cd, Hg, preparation and x-ray data[156], $ZnBeF_4$, preparation and thermal decomposition[140]	140, 156, 161

Zinc, cadmium and mercury derivatives of transition metals (see Section 2.3.8)

$M[Mn(CO)_5]_2$	(M = Zn, Cd). Convenient preparation and derivatives	453
$Hg[Co(CO)_4]_3^-$	Preparation and characterisation[454], of material previously reported to be $Hg[Co(CO)_4]_2$ [455]	454, 455
Carbonyl derivatives	Preparation of $M[Co(CO)_4]_2$, $M[Fe(CO)_2(C_5H_5)]_2$, $M[Cr(C_5H_5)(CO)_3]_2$, $M[Mo(CO)_3(C_5H_5)]_2$, $M[W(CO)_3(C_5H_5)]_2$, $(C_5H_5)Fe(CO)_2HgCo(CO)_4$, etc., M = Zn, Cd; i.r.	460
$(Ph_3P)_2(NO)Cl_2OsHgCl$	Preparation and x-ray structure	456
$Hg[Fe(CO)_2LNO]_2$	(L = R_3P, $(RO)_3P$, R_3As, R_3Sb). Preparation and reactions, i.r. spectra of derivatives of Fe—Hg compounds	457
$M_3[Co(CN)_6]_2$	(M = Zn, Cd). X-ray data and thermal studies	458
$Zn[Fe(CN)_5NO]5H_2O$ $Cd[Fe(CN)_5NO]$ $Hg_2[Fe(CN)_5NO]\cdot 2H_2O$	Synthesis and properties. I.R., u.v., visible spectra	459

Acknowledgements

The author wishes to thank Dr. R. Pearce for invaluable help in proof reading, Mrs. P. Keilthy for expert typing of the manuscript, and finally my wife for her patience and understanding during its preparation.

References

1. Everest, D. A. (1964). *The Chemistry of Beryllium*. (NewYork: Elsevier)
2. Darwin, F. E. and Buddery, J. H. (1968). *Beryllium*. (London: Butterworths)
3. Bellamy, R. G. and Hall, N. A. (1965). *Extraction and Metallurgy of Uranium, Thorium and Beryllium*. (New York: Pergamon Press)
4. Gmelin. (1969). *Mercury*, System Number 34, Part B, Section 4 (Conclusion of compounds). (Weinheim: Gmelin Institut, Verlag Chemie)
5. Gschneidner, K. A. (1968). *U.S. Atomic Energy Commission*, 15–1757. Available Dept. C.F.S.T.I. (1968). *Nucl.Sci.Abstr.*, **22**, 15182
6. Schubert, J. (1959). *Chimia*, **13**, 321
7. Samsonov, G. V. (1966). *Russ. Chem. Rev.*, **35**, 339
8. Petyunin, P. A. (1962). *Russ. Chem. Rev.*, **31**, 100
9. Gurvich, A. M. (1966). *Russ. Chem. Rev.*, **35**, 631
10. Grdenic, D. (1965). *Quart. Rev. Chem. Soc.*, 303
11. Deacon, G. B. (1963). *Rev. Pure. Appl. Chem.*, **13**, 189
12. Roberts, H. L. (1968). *Adv. Inorg. Chem. Radiochem.*, **11**, 309
13. Shepherd, L. H., Ter Haar, G. L. and Marlett, E. M. (1969). *Inorg. Chem.*, **8**, 976
14. Coates, G. E. and Roberts, P. D. (1969). *J. Chem. Soc. A*, 1008
15. Bell, N. A. and Coates, G. E. (1968). *J. Chem. Soc. A*, 823
16. Coates, G. E. and Tranah, M. (1967). *J. Chem. Soc. A*, 615
17. Bell, N. A. and Coates, G. E. (1966). *J. Chem. Soc. A*, 1069
18. Bell, N. A., Coates, G. E. and Emsley, J. W. (1966). *J. Chem. Soc. A*, 1361
19. Bell, N. A. and Coates, G. E. (1968). *J. Chem. Soc. A*, 628
20. Coates, G. E. and Pendlebury, R. (1970). *J. Chem. Soc. A*, 156
21. Ducros, M., Levy, R. and Mehava, G. (1970). *Bull. Soc. Chim. Fr.*, 2763
22. Ashby, E. C., Kovar, R. A. and Kawakami, K. (1970). *Inorg. Chem.*, **9**, 317
23. Ashby, E. C. and Arnott, R. C. (1970). *J. Organometal. Chem.*, **21**, P29
24. Ashby, E. C., Kovar, R. A. and Arnott, R. C. (1970). *J. Amer. Chem. Soc.*, **92**, 2182
25. Ashby, E. C. and Beach, R. G. (1970). *Inorg. Chem.*, **9**, 2300
26. Ashby, E. C., Schwartz, R. D. and James, B. D. (1970). *Inorg. Chem.*, **9**, 325
27. Popp, G. and Hawthorne, M. F. (1968). *J. Amer. Chem. Soc.*, **90**, 6553
28. Armstrong, D. R. and Perkins, P. G. (1968). *Chem. Commun.*, 352
29. Gundersen, G. and Haaland, A. (1968). *Acta Chem. Scand.*, **22**, 867
30. Nibler, J. W. and McNabb, J. (1969). *Chem. Commun.*, 134
31. Nibler, J. W. and McNabb, J. (1970). *J. Amer. Chem. Soc.*, **92**, 2920
32. Cook, T. H. and Morgan, G. L. (1969). *J. Amer. Chem. Soc.*, **91**, 774
33. Cook, T. H. and Morgan, G. L. (1970). *J. Amer. Chem. Soc.*, **92**, 6493
34. Cook, T. H. and Morgan, G. L. (1970). *J. Amer. Chem. Soc.*, **92**, 6487
35. Greenwood, N. N. and Travers, N. F. (1968). *J. Chem. Soc. A*, 15
36. Ashby, E. C., Sanders, J. R. and Carter, J. H. (1967). *Chem. Commun.*, 997
37. Sanders, J. R., Ashby, E. C. and Carter, J. H. (1968). *J. Amer. Chem. Soc.*, **90**, 6385
38. Ashby, E. C., Parris, G. and Walker, F. (1969). *Chem. Commun.*, 1464
39. Kotz, J. C., Schaeffer, R. and Clouse, A. (1967). *Inorg. Chem.*, **6**, 620
40. Kovar, R. A. and Morgan, G. L. (1970). *J. Amer. Chem. Soc.*, **92**, 5067
41. McVicker, G. B. and Morgan, G. L. (1970). *Spectrochim. Acta*, **26A**, 23
42. Almeningen, A., Bastiansen, O. and Haaland, A. (1964). *J. Chem. Phys.*, **40**, 3434
43. Coates, G. E., Roberts, P. D. and Downs, A. J. (1967). *J. Chem. Soc. A*, 1085
44. Chambers, D. B., Coates, G. E. and Glockling, F. (1970). *J. Chem. Soc. A*, 741
45. Ashby, E. C., Yu, S. H. and Beach, R. G. (1970). *J. Amer. Chem. Soc.*, **92**, 433
46. Ashby, E. C. and Nackashi, J. A. (1970). *J. Organometal. Chem.*, **24**, C17
47. Lahournere, J. C. and Valade, J. (1970). *J. Organometal. Chem.*, **22**, C3
48. Lahournere, J. C. and Valade, J. (1970). *Compt. Rend.*, **270C**, 2080

49. Blachnik, R. O. G., Gross, P. and Hayman, C. (1970). *Trans. Faraday Soc.*, **66,** 1058
50. Dvorina, L. A., Popova, O. I and Derenovskaya, N. A. (1969). *Porosh. Met.*, **9,** 29
51. Jacobs, H. and Juza, R. (1970). *Z. Anorg. Chem.*, **370,** 248
52. Jacobs, H. and Juza, R. (1970). *Z. Anorg. Chem.*, **370,** 254
53. Palvadeau, P. and Rouxel, J. (1970). *Bull. Soc. Chim. Fr.*, 480
54. Novoselova, A. V., Pochkaeva, T. I. and Tamm, N. S. and Tomashchik, A. D. (1969). *Zh. Neorg. Khim.* **14,** 605
55. Pochkarev, T. I., Mikhaeva, L. M., Grigor'ev, A. I. and Garem, A. (1970). *Zh. Neorg. Khim.*,**15,** 87
56. Reedjik, J., Vervelde, J. B. and Groeneveld, W. L. (1970). *Rec. Trav. Chim., Pays-Bas,* **89,** 42, and references therein
57. Zuur, A. P., Reintjes, A. H. L. and Groeneveld, W. L. (1970). *Rec. Trav. Chim., Pays-Bas,* **89,** 385
58. Reedjik, J. (1969). *Rec. Trav. Chim., Pays-Bas,* **88,** 1451
59. Reedjik, J. (1969). *Inorg. Chim. Acta,* **3,** 517
60. Snavely, F. A. and Sweigert, D. A. (1969). *Inorg. Chem.,* **8,** 1659
61. Bodner, R. L. and Hendricker, D. G. (1970). *Inorg. Chem.,* **9,** 1255
62. Sathe, R. M. and Shetty, S. Y. (1970). *J. Inorg. Nucl. Chem.,* **32,** 1383
63. Coates, G. E. and Roberts, P. D. (1968). *J. Chem. Soc. A,* 2651
64. Coates, G. E. and Tranah, M. (1967). *J. Chem. Soc. A,* 236
65. Bell, N. A., Coates, G. E. and Emsley, J. W. (1966). *J. Chem. Soc. A,* 49
66. Coates, G. E. and Fishwick, A. H. (1967). *J. Chem. Soc, A,* 1199
67. Atwood, J. L. and Stucky, G. D. (1969). *J. Amer. Chem. Soc.,* **91,** 4426
68. Coates, G. E. and Ridley, D. (1967). *J. Chem. Soc. A,* 56
69. Summerford, C., Wade, K. and Wyatt, B. K. (1969). *Chem. Commun.,* 61; (1970). *Idem, J. Chem. Soc. A,* 2016
70. Magnuson, V. and Stucky, G. D. (1969). *Inorg. Chem.,* **8,** 1427
71. Yamamoto, A., Ookawa, M. and Ikeda, S. (1969). *Chem. Commun.,* 841
72. Hall, D., Gurr, C. E. and Jeffrey, G. A. (1969). *Z. Anorg. Chem.,* **369,** 108
73. Mesmer, R. E. and Baes, C. F. (1967). *Inorg. Chem.,* **6,** 1951
74. Kakihana, H. and Sillén, L. G. (1956). *Acta Chem. Scand.,* **10,** 985
75. Ohtaki, H. (1967). *Inorg. Chem.,* **6,** 808
76. Lanza, E. (1969). *Rev. Chim. Minerale,* **6,** 653
77. Kakihana, H. and Maeda, M. (1969). *Bull. Chem. Soc. Japan,* **42,** 1458; (1970). **43,** 109
78. Grigor'ev, A. I. and Pogodilova, E. G. (1969). *Zh. Strukt. Khim.,* **10,** 43; Grigor'ev, A. I., Pogodilova, E. G. and Novoselova, A. V. (1965). *Zh. Neorg. Khim,* **10,** 772; (1969). **14,** 4
79. Pogodilova, E. G., Grigor'ev, A. I. and Novoselova, A. V. (1969). *Zh. Neorg. Khim,* (a) **14,** 913; (b) **14,** 2062
80. Pogodilova, E. G. Grigor'ev, A. I. and Novoselova, A. V. (1969). *Zh. Neorg. Khim.,* **14,** 2701
81. Grigor'ev, A. I., Pogodilova, E. G. and Sipachev, V. A. (1970). *Zh. Neorg. Khim.,* **15,** 1757
82. Grigor'ev, A. I. and Novoselova, A. V. (1970). *Zh. Neorg. Khim.,* **15,** 2309
83. Grigor'ev, A. I., Sipachev, V. A. and Novoselova, A. V. (1970). *Dokl. Acad. Nauk. SSSR,* **192,** 808
84. Grigor'ev, A. I., Sipachev, V. A. and Novoselova, A. V. (1969). *Dokl. Acad. Nauk. SSSR,* (a) **185,** 95; (b) **189,** 318
85. Wynne, K. J. and Bauder, W. (1969). *J. Amer. Chem. Soc.,* **91,** 5920; (1970). *idem, Inorg. Chem.,* **9,** 1985
86. Coates, G. E. and Fishwick, A. H. (1968). *J. Chem. Soc. A,* 640
87. Coates, G. E. and Fishwick, A. H. (1968). *J. Chem. Soc. A,* 477
88. Coates, G. E. and Heslop, J. A. (1968). *J. Chem. Soc. A,* 631
89. Wieser, H. and Krueger, P. J. (1970). *Spectrochim. Acta,* **26A,** 1349
90. Arora, M. and Mehrotra, R. C. (1969). *Indian J. Chem.,* **7,** 399
91. Mehrotra, R. C. (1969). *J. Less-Common Metals,* **17,** 181
92. Turova, N. Ya., Turevskaya, E. P. and Novoselova, A. V. (1969). *Dokl. Acad. Nauk. SSSR,* **186,** 358
93. Paciorek, K. L. and Kratzer, R. H. (1966). *Inorg. Chem.,* **5,** 538
94. Lannert, K. P. and Joesten, M. D. (1969). *Inorg. Chem.,* **8,** 1775
95. Joesten, M. D., Hussain, M. S. and Lenhert, P. G. (1970). *Inorg. Chem.,* **9,** 151

96. Driessen, W. L. and Groeneveld, W. L. (1969). *Rec. Trav. Chim., Pays-Bas,* **88,** 491; 620
97. Mann, D. E., Calder, G. V., Seshadri, K. S., White, D. and Linevsky, M. J. (1967). *J. Chem. Phys.,* **46,** 1138, and references therein
98. Rambidi, N. G. (1969). *J. Strukt. Khim,* **10,** 131, and references therein
99. Balasubrahmanyam, K. (1966). *J. Chem. Phys.,* **44,** 3270
100. Romberger, K. A. and Braunstein, J. (1970). *Inorg. Chem.,* **9,** 1273
101. Mesmer, T. E. and Baes, C. F. (1969). *Inorg. Chem.,* **8,** 618
102. Steinfink, H. and Brunton, G. D. (1969). *Inorg. Chem.,* **8,** 1665
103. Kubas, G. J. and Shriver, D. F. (1970). *J. Amer. Chem. Soc.,* **92,** 1949
104. Kubas, G. J. and Shriver, D. F. (1970). *Inorg. Chem.,* **9,** 1951
105. Potts, R. A. and Allred, A. L. (1966). *Inorg. Chem.,* **5,** 1066
106. Eachus, R. S., Marov, I. and Symons, M. C. R. (1970). *Chem. Commun.,* 633
107. Eachus, R. S., Symons, M. C. R. and Yandell, J. K. (1969). *Chem. Commun.,* 979
108. Meyerstein, D. and Mulac, W. A. (1970). *Inorg. Chem.,* **9,** 1762
109. Nöth, H., Wiberg, E. and Winter, L. P. (1969). *Z. Anorg. Chem.,* **370,** 209
110. Vaughan, W. R. and Knoess, H. P. (1970). *J. Org. Chem.,* **35,** 2394
111. Amiet, G., Nicholas, K. and Pettit, R. (1970). *Chem. Commun.,* 161
112. Bettler, C. R., Sendra, J. C. and Urry, G. (1970). *Inorg. Chem.,* **9,** 1060
113. Vyazankin, N. S., Kruglaya, O. A., Petrov, B. I., Egorochkin, A. N. and Khorshev, S. Y. (1970). *Zh. Obshch. Khim.,* **40,** 1279
114. Gladyshev, E. N., Andreivichev, V. S., Vyazankin, N. S. and Razuvaev, G. A. (1970). *Zh. Obshch. Khim.,* **40,** 939
115. Bettler, C. R. and Urry, G. (1970). *Inorg. Chem.,* **9,** 2372
116. Köhler, M. and Müllman, E. V. (1970). *Z. Anorg. Chem.,* **373,** 222
117. Takemoto, J. and Nakamoto, K. (1970). *Chem. Commun.,* 1017
118. Breitinger, D., Brodersen, K. and Jürgen, L. (1970). *Chem. Ber.,* **103,** 2388
119. Glemser, O., Mews, R. and Roesky, H. W. (1969). *Chem. Ber.,* **102,** 1523
120. Kasowski, W. J. and Bailar, J. C. (1969). *J. Amer. Chem. Soc.,* **91,** 3212
121. Philip, A. T., Casey, A. T. and Thompson, C. R. (1970). *Aust. J. Chem.,* **23,** 491
122. Gillum, W. O., Huffman, J. C., Streib, W. E. and Wentworth, R. A. D. (1969). *Chem. Commun.,* 843
123. Parks, J. E., Wagner, B. E. and Holm, R. H. (1970). *J. Amer. Chem. Soc.,* **92,** 3, 500
124. Sarneski, J. E. and Urbach, F. L. (1966). *Chem. Commun.,* 1025
125. Bodner, R. L. and Hendricker, D. G. (1970). *Inorg. Nucl. Chem. Lett.,* **6,** 421
126. Bennett, J., Pidcock, A., Waterhouse, C. R., Coggon, P. and McPhail, A. T. (1970). *J. Chem. Soc. A,* 2094
127. Kline, R. J. and Wardeska, J. G. (1969). *Inorg. Chem.,* **8,** 2153
128. Reedijk, J. and Groeneveld, W. L. (1969). *Rec. Trav. Chim. Pays-Bas,* **88,** 655, and references therein
129. Nikolaeva, N. M. (1969). *Zh. Neorg. Khim.,* **14,** 936
130. Lumne, P., Lundgren, G. and Mark, W. (1969). *Acta Chem. Scand.,* **23,** 3011
131. Morris, P. J. and Martin, R. B. (1970). *J. Inorg. Nucl. Chem.,* **32,** 2891
132. Brill, T. B. and Hugus, Z. Z. (1970). *Inorg. Chem.,* **9,** 984
133. Ol'dekop, Yu. A., Maier, N. A., Erdmann, A. A. and Dzhomidava, Yu.A. (1970). *Zh. Obshch. Khim.,* **40,** 300
134. Ol'dekop, Yu. A., Maier, N. A., Erdmann, A. A. and Stanovaya, S. S. (1970). *ibid.,* **40,** 305
135. Ol'dekop, Yu. A., Maier, N. A., Erdmann, A. A. and Dzhomidava, Yu. A. (1970). *ibid.,* **40,** 637
136. Ol'dekop, Yu. A., Maier, N. A. and But'ko, Yu. D. (1970). *ibid.,* **40,** 641
137. Ol'dekop, Yu.A., Maier, N. A., Erdmann, A. A. and Potopova, L. L. (1970). *Vesti Acad. Navuk Belarus, SSSR, Ser. Khim. Navuk.,* **3,** 69
138. Emeléus, H. J., Shreeve, J. M. and Spaziante, P. M. (1969). *J. Chem. Soc. A,* 431
139. *Idem.* (1969). *J. Inorg. Nucl. Chem.,* **31,** 3417
140. Kirichenko, L. N. and Medzhidov, A. A. (1969). *Isv. Acad. Nauk., SSSR, Ser, Khim.,* 2849
141. Bonati, F. and Minghetti, G. (1970). *J. Organometal. Chem.,* **22,** 5
142. Flateau, K. and Musso, H. (1970). *Angew. Chem. Int. Ed.,* **9,** 379
143. Bloodworth, A. J. (1970). *J. Organometal. Chem.,* **23,** 27
144. *Idem.* (1970). *J. Chem. Soc. C,* 2051

145. Fackler, J. P., Coucouvanis, D., Fetchin, J. A. and Seedel, W. C. (1968). *J. Amer. Chem. Soc.*, **90,** 2784
146. Fackler, J. P., Fetchin, J. A. and Smith, J. A. (1970). ibid., **92,** 2910
147. Fackler, J. P. and Fetchin, J. A. (1970). ibid., **92,** 2912
148. Dakterniks, D. R. and Graddon, D. P. (1970). *Aust. J. Chem.*, **23,** 1989
149. Calligario, M., Nardini, G. and Ripamonti, A. (1970). *J. Chem. Soc. A*, 714
150. Giordano, F., Randaccio, L. and Ripamonti, A. (1969). *Acta Crystallogr.*, **25B,** 1057
151. Kharitonov, Y. Y., Tsintsadze, G. V. and Tsivadze, A. Y. (1970). *Zh. Neorg. Khim.*, **15,** 949
152. Idem. (1970). ibid., **15,** 1811
153. Jain, S. C. and Rivest, R. (1969). *Can. J. Chem.*, **47,** 2209
154. Idem. (1969). *Inorg. Chim. Acta*, **3,** 552
155. Idem. (1970). ibid., **4,** 291
156. Lovering, D. G. and Alner, D. J. (1970). *Chem. Commun.*, 570
157. Macklin, J. W. and Plane, R. A. (1970). *Inorg. Chem.*, **9,** 821
158. Saraf, J. R., Aggarwal, R. C. and Prasad, J. (1970). *Bull. Chem. Soc. Jap.*, **43,** 264
159. Peterson, R. J., Lingane, P. J. and Reynolds, W. L. (1970). *Inorg. Chem.*, **9,** 680
160. Bloom, H., O'Grady, B. V. and Anthony, R. G. (1970). *Aust. J. Chem.*, **23,** 843
161. Ostrovskaya, T. V. and Amirova, S. A. (1970). *Zh. Neorg. Khim.*, **15,** 657
162. Matkovitch, V. I. and Economy, J. (1970). *Acta Crystallogr.*, **26B,** 616
163. Mukherjee, S. G. and Samaddar, B. N. (1969). *Indian J. Chem.*, **7,** 183
164. Mukherjee, S. G. and Samaddar, B. N. (1969). *Indian J. Chem.*, **7,** 521
165. Kovar, R. A. and Morgan, G. A. (1969). *J. Amer. Chem. Soc.*, **91,** 7269
166. Toney, J. and Stucky, G. D. (1970). *J. Organometal. Chem.*, **22,** 241
167. Bertini, F., Graselli, P., Zubiani, G. and Cainelli, G. (1970). *Tetrahedron*, **26,** 1281
168. Zmii, O. F. and Gladyshevskii, E. I. (1970). *Kristallografiya*, **15,** 939
169. Brice, J. F., Motte, J. R. and Streiff, R. (1969). *Compt. Rend.*, **269C,** 910
170. Sipachev, V. A., Grigor'ev, A. I. and Novoselova, A. V. (1969). *Zh. Strukt. Khim.*, **10,** 1031
171. Grigor'ev, A. I., Evseeva, N. K. and Sipachev, V. A. (1969). *Zh. Strukt. Khim.*, **10,** 469
172. Chawla, I. D. and Andrews, A. C. (1969). *J. Inorg. Nucl. Chem.*, **31,** 3809
173. Harrison, L. W. and Swift, T. J. (1970). *J. Amer. Chem. Soc.* **92,** 1963
174. David, J. and Lang, J. (1969). *Compt. Rend.*, **269C,** 771
175. Reedijk, J. and Groeneveld, W. L. (1969). *Rec. Trav. Chim. Pays-Bas*, **88,** 655
176. Leuhder, K. and Orfert, I. (1970). *Z. Chem.*, **10,** 32
177. Timkovitch, A. and Tulinsky, A. (1969). *J. Amer. Chem. Soc.*, **91,** 4430
178. Ugai, Ya. A. and Gukov, O. Ya. (1969). *Zh. Neorg. Khim.*, **14,** !463
179. Chudinova, L. I. (1969). *Zh. Prikl. Khim.*, **42,** 189
180. Goryachev, A. A. and Ignatev, O. S. (1970). *Zh. Neorg. Khim.*, **15,** 1614
181. Grigor'ev, A. I., Sipachev, V. A. and Pogodilova, E. G. (1970). *Zh. Strukt. Khim.*, **11,** 458
182. Couturier, Y. and Faucherre, J. (1970). *Bull. Soc. Chim. Fr.*, 1323
183. Grigor'ev, A. I., Sergeev, B. N. and Novoselova, A. V. (1969). *Zh. Neorg. Khim.*, **14,** 883
184. Kirpichev, E. P. and Rubstov, Yu. I. (1969). *Zh. Fiz. Khim.*, **43,** 2029
185. Hardt, H. D. (1969). *Z. Anorg. Chem.*, **368,** 87
186. Grigor'ev, A. I., Sergeev, B. N. and Novoselova, A. V. (1969). *Zh. Neorg. Khim.*, **14,** 112
187. Patel, K. S., Rinehardt, K. L. and Bailar, J. C. (1970). *Org. Mass. Spectrosc.*, **3,** 1239
188. Uhlemann, E., Thomas, P., Klose, G. and Arnold, K. (1969). *Z. Anorg. Chem.*, **364,** 153
189. Bottei, R. S. and Schneggenburger, R. G. (1970). *J. Inorg. Nucl. Chem.*, **32,** 1525
190. Gupta, S. L. and Soni, R. N. (1969). *J. Indian Chem. Soc.*, **46,** 561
191. Duke, S. S. and Dhindsa, S. S. (1969). *Z. Naturforsch.*, **24B,** 967
192. Bannister, A. J. and Padley, J. S. (1969). *J. Chem. Soc. A*, 658
193. Sanger, P. A. (1969). *Acta Crystallogr.*, **25A,** 694
194. Bakulina, V. M., Tokareva, S. A., Latysheva, E. I. and Vol'nov, I. I. (1970). *Zh. Strukt. Khim.*, **11,** 158
195. Vol'nov, I. I., Tokareva, S. A., Belevskii, V. N. and Latysheva, E. I. (1970). *Izv. Acad. Nauk. SSSR, Ser Khim.*, 516
196. Vol'nov, I. I. and Latysheva, E. I. (1970). *Izv. Acad. Nauk. SSSR, Ser. Khim.*, 13
197. Iwai, Shinichi; Murotani, Hiroshi; Morikawa, Hedeki (1969). *Yogyo Kyokai Shi*, **77,** 411
198. Berg, L. G., Borukhar, I. A. and Saibova, M. T. (1970). *Usb. Khim. Zh.*, **14,** 32
199. Shchegrov, L. N. and Pechkovskii, V. V. (1970). *Zh. Prikl. Khim.*, **43,** 10
200. Pechkovskii, V. V., Shchegrov, L. N. and Shul'man, A. S. (1969). *Zh. Neorg. Khim.*, **14,** 53

201. Calvo, C. and Neclakantan, K. (1970). *Can. J. Chem.*, **48,** 890
202. Baggio, S., Amzel, L. M. and Becka, L. N. (1969). *Acta Crystallogr.*, **25B,** 2650
203. Pechkovskii, V. V., Berezina, M. I. and Pinaev, G. F. (1969). *Vesti Acad. Nauk. Belaruss. SSSR., Ser. Khim. Nauk.*, 118
204. Muldagaleeva, R. A., Pashinkin, A. S. and Buketov, E. A. (1969): *Tr. Khim.—Met. Inst. Acad. Nauk. Kaz. SSSR.*, **9,** 11
205. Skopenko, V. V., Tsintsadze, G. V. and Alasaniya, R. M. (1969). *Ukr. Khim. Zh.*, **35,** 1317
206. Biryukov, V. P. and Styunkel, T. B. (1969). *Isv. Vysch. Ucheb. Zaved. Khim. Technol.*, **12,** 223
207. Acheson, R. J. and Jacobs, P. W. M. (1970). *J. Phys. Chem.*, **74,** 281
208. Diament, R. and Sediey, M. (1969). *Compt. Rend.*, **268C,** 1243
209. Bigoli, F., Lanfredi, A. M. M., Tiripicchio, A. and Camellini, M. T. (1970). *Acta Crystallogr.*, **26B,** 1075
210. Courtine, P., Charcossit, H. and Cord, P.P. (1969). *Bull. Soc. Chim. Fr.*, 57
211. Barczuk, V. J. (1970). *J. Inorg. Nucl. Chem.*, **32,** 2782
212. Lin Chin-Tung and Bear, J. L. (1969). *J. Inorg. Nucl. Chem.*, **31,** 263
213. Walter-Levy, L. and Perrotey, J. (1970). *Bull. Soc. Chim. Fr.*, 1697
214. Kubecova, K. and Frei, V. (1969). *Coll. Czech. Chem. Comm.*, **34,** 141
215. Frei, V. and Ederova, J. (1969). *Coll. Czech. Chem. Comm.*, **34,** 1304
216. Kubo, T., Uchida, K., Tsubosaki, K. and Hashimi, F. (1970). *Kogyo Kagaku Zasshi,* **73,** 75
217. Driessen, W. L., Groeneveld, W. L. and Van der Ney, F. N. (1970). *Rec. Trav. Chim. Pays-Bas,* **89,** 353
218. Mackay, R. A. and Poziomek, E. J. (1969). *J. Chem. Eng. Data,* **14,** 271
219. Karayannis, N. M., Owens, C. and Pytlewski, L. L. (1969). *J. Inorg. Nucl. Chem.*, **31,** 2059
220. Karayannis, N. M., Mikulski, C. M., Pytlewski, L. L. and Labes, M. M. (1970). *Inorg. Chem.*, **9,** 582
221. Crouzet, A. and Allonard, S. (1969). *Bull. Soc. Fr. Mineral. Cristallogr.*, **92,** 388
222. Lari-Lavassani, A., Cot, L. and Aviners, C. (1970). *Compt. Rend.*, **270C,** 1973
223. Calder, V., Mann, D. E., Seshadri, K. S., Allavena, M. and White, D. (1969). *J. Chem. Phys.*, **51,** 2093
224. Goodyear, J. and Sali. (1969). *Acta Crystallogr.*, **25B,** 2664
225. Burlitch, J. M. and Ulmer, S. W. (1969). *J. Organometal. Chem.*, **19,** P21
226. Ugai, Ya. A. and Gukov, O. Ya, (1969). *Zh. Neorg. Khim.*, **14,** 1463
227. Breakspere, R. J. and Gregg, S. J. (1969). *J. Chem. Soc. A,* 1613
228. Berg, E. W. and Shendrikar, A. D. (1969). *Anal. Chim. Acta*, **44,** 159
229. Homann, R. and Hoppe, R. (1969). *Z. Anorg. Chem.*, **368,** 271
230. Quinn, R. K. and Lagowski, J. J. (1970). *Inorg. Chem.*, **9,** 414
231. Macri, G. and Petrucci, S. (1970). *Inorg. Chem.*, **9,** 1009
232. Greenwood, N. N. and Sharrocks, D. N. (1969). *J. Chem. Soc. A,* 2334
233. Galochkina, G. M., Coryunova, N. A., Seifer, G. B., Vailpolin, A. A. and Kharatinov, Yu. Ya., (1970). *Isv. Acad. Nauk. SSSR., Neorg. Mater,* **6,** 486
234. Matteson, D. S., Castle, R. B. and Larson, G. L. (1970). *J. Amer. Chem. Soc.*, **92,** 231
235. Bennett, S. W., Clase, H. J., Eaborn, C. and Jackson, R. A. (1970). *J. Organometal. Chem.*, **23,** 403
236. Neumann, W. P. and Neumann, G. (1970). *J. Organometal. Chem.*, **25,** C59
237. Mitchell, T. N. and Neumann, W. P. (1970). *J. Organometal. Chem.*, **22,** C25
238. Fields, R., Haszeldine, R. N. and Hubbard, A. F. (Mrs.). (1970). *Chem. Commun,* 647
239. Kruglaya, O. A., Petrov, B. I. and Vyazankin, N. S. (1970). *Izv. Acad. Nauk. SSSR, Ser. Khim.*, 2413
240. Lyutaya, M. D. and Vilovenko, S. A. (1968). *Khim. Fiz. Nitridov,* 101, (Chem. Abstr., **71,** 27873k.)
241. Maunage, M., Lang, J. and Lefebvre, A. (1970). *Compt. Rend.*, **270C,** 2052
242. Marchand, R. and Lang, J. (1970). *Compt. Rend.*, **270C,** 540
243. Rembarz, G., Fischer, E., Röber, K. C., Ohff, R. and Crahmer, H. (1969). *J. für Prakt. Chem.*, **311,** 889
244. Perchard, C. and Novak, A. (1970). *Spectrochim. Acta,* **26A,** 871
245. Miyamoto, F. (1969). *Inorg. Chim. Acta,* **3,** 511
246. Guru, S., and Rao, D. V. R. (1969). *J. Indian Chem. Soc.*, **46,** 308
247. Melia, T. P. and Merrifield, R. (1970). *J. Chem. Soc. A,* 1166

248. Brisseau, L. and Rouxel, J. (1969). *Compt. Rend.*, **268C**, 2308
249. Volova, L. M., Ovechkin, E. K. and Ostroushko, V. I. (1969). *Izv. Acad. Nauk. SSSR, Neorg. Mater.*, **5**, 1298
250. Horn, H. G. and Becke-Goehring, M. (1969). *Naturwiss.*, **56**, 137
251. Ciobanu, A., Ronan, L., Bostan, M., Voiculiscu, N. and Giurgiu, D. (1970). *Rev. Roum. Chim.*, **15**, 783
252. Yoganarasimhan, S. R. and Jain, R. C. (1969). *Indian J. Chem.*, **7**, 808
253. Krischner, H. and Fritzer, H. P. (1970). *Z. Anorg. Chem.*, **376**, 162
254. Agrell, I. (1970). *Acta Chem. Scand.*, **24**, 1247
255. Milne, J. B. (1970). *Can. J. Chem.*, **48**, 75
256. Tsivadze, A. Ya, Tsintsadze, G. V., Kharitonov, Yu. Ya., Golub, A. M. and Mamulashvili, A. M. (1970). *Zh. Neorg. Khim.*, **15**, 934
257. Flint, C. D. and Goodgame, M. (1970). *J. Chem. Soc. A*, 442
258. Kharatinov, Yu. Ya., Tsintsadze, G. V. and Tsivadze, A. Yu. (1970). *Zh. Neorg. Khim.*, **15**, 390
259. *Idem.*, ibid., 1513
260. *Idem.*, ibid., 1196
261. *Idem.*, ibid., 710
262. Freeman, A. G. and Larkindale, J. P. (1969). *J. Chem. Soc. A*, 1307
263. Gould, R. O. and Sutton, H. M. (1970). *J. Chem. Soc. A*, 1184
264. *Idem.*, ibid., 1439
265. Yamada, S. and Yamanouchi, K. (1969). *Bull. Chem. Soc. Jap.*, **42**, 2562
266. Felty, W. L., Ekstrom, C. G. and Leussing, D. L. (1970). *J. Amer. Chem. Soc.*, **92**, 3006
267. Bamfield, P., Price, R. and Miller, R. G. J. (1969). *J. Chem. Soc. A*, 1447
268. Garnovskii, A. D., Tertov, B. A., Osipov, O. A., Burgkhin, V. V. and Kostromina, L. S. (1969). *Zh. Strukt. Khim.*, **10**, 339
269. Gerlach, D. H. and Holm, R. H. (1970). *Inorg. Chem.*, **9**, 588
270. Richards, C. P. and Webb, G. A. (1969). *J. Inorg. Nucl. Chem.*, **31**, 3459
271. Garnovskii, A. D., Bednyagina, N. P. and Kuznebova, L. I. (1969). *Zh. Neorg. Khim.*, **14**, 1576
272. Shkol'nikova, L. M., Obodovskaya, A. E. and Shugam, E. A. (1970). *Zh. Strukt. Khim.*, **11**, 54
273. Verbeek, W. and Sundermeyer, W. (1969). *Angew. Chem. Int. Ed.*, **8**, 376
274. Bolte, M. and Capestan, M. (1970). *Bull. Soc. Chim. Fr.*, 865 *(b)*, idem., ibid., 868
275. Ablov, A. V., Volodina, G. F. and Ivanova, Y. Ya., (1970). *Kristallografiya*, **15**, 968
276. Amasa, S., Brown, D. H. and Sharp, D. W. A. (1969). *J. Chem. Soc. A*, 2892
277. Dunski, N. and Crawford, T. H. (1969). *J. Inorg. Nucl. Chem.*, **31**, 2073
278. Ghosh, N. N. and Bhattacharya, A. (1969). *J. Indian Chem. Soc.*, **46**, 1040
279. Goodgame, D. M. L., Hitchman, M. A. and Marsham, D. F. (1970). *J. Chem. Soc. A*, 1933
280. Magearu, V. and Popa, G. (1969). *Rev. Roum. Chim.*, **14**, 1399
281. Perrin, D. D. and Sharma, V. S. (1969). *J. Chem. Soc. A*, 2060
282. Prakesh, S., Prakesh, O., Jain, S. K. and Chaturredi, C. V. (1969). *Roczniki Chem.*, **43**, 37
283. Kharatinov, Yu. Ya. and Machkhoshvili, R. I. (1969). *Zh. Neorg. Khim.*, **14**, 3181
284. Le Mau Quang, Novakovskii, M. S. and Orlov, V. D. (1969). *Zh. Strukt. Khim.*, **10**, 75
285. Green, I. and Neamtu, M. (1969). *Rev. Chim. Miner.*, **6**, 1133
286. Popa, G. and Mageaur, V. (1969). *Rev. Roum. Chim.*, **14**, 1387
287. Khairy, E. M. and Khater, M. M. (1970). *J. Electroanal. Chem. Interfac. Electrochem.*, **24**, 195
288. Chudinova, L. I. (1969). *Zh. Prikl. Khim.*, **42**, 189
289. Liptay, G., Burger, K., Moscari-Fulop, E. and Prubszky, I. (1970). *J. Therm. Anal.*, **2**, 25
290. Ouellette, T. J. and Haendler, H. M. (1969). *Inorg. Chem.*, **8**, 1777
291. Ahuja, I. S. and Rastogi, P. (1970). *J. Chem. Soc. A*, 2161
292. Mohaptra, B. K. and Rao, D. V. R. (1970). *Z. Anorg. Chem.*, **372**, 332
293. Nelson, J. and Samuel, M. (1969). *J. Chem. Soc. A*, 1597
294. Alyaviya, M. K. and Teplyakova, Z. M. (1970). *Zh. Neorg. Khim.*, **15**, 958
295. Azizov, M. A., Rashkes, Y. V., Khamikov, K. K., Ibatov, S. I. and Khamraev, A. D. (1970). *Zh. Strukt. Khim.*, **11**, 650
296. Fater, D., Fater, S., Teodorescu, A. and Segal, E. (1970). *Rev. Roum. Chim.*, **15**, 855
297. Mohaptra, B. K. and Rao, D. V. R. (1970). *Indian J. Chem.*, **8**, 564

298. Konovalov, N. V., Maslinnikova, I. S. and Shemyakin, V. N. (1970). *Zh. Neorg. Khim.*, **15,** 1993
299. Paul, R. C., Arora, H. and Chadha, S. L. (1970). *Inorg. Nucl. Chem. Lett.*, **6,** 469
300. Ahuja, I. S. and Rastogi, P. (1970). *Indian J. Chem.*, **8,** 88
301. Ahuja, I. S. (1969). *Inorg. Chim. Acta,* **3,** 110
302. Ahuja, I. S. and Rastogi, P. (1969). *Inorg. Nucl. Chem. Lett.*, **5,** 255
303. Ahuja, I. S. and Rastogi, P. (1970). *J. Inorg. Nucl. Chem.*, **32,** 2085
304. Golub, A. M., Tsintsadze, G. V. and Mamulashvili, A. M. (1969). *Kiev. Gos. Univ. Kiev.*, **14,** 3013
305. Ghose, S. N. (1969). *Inorg. Nucl. Chem. Lett.*, **5,** 841
306. Alasoniya, R. M., Skopenko, V. V. and Tsintsadze, G. V. (1969). *Ukr. Khim. Zh.*, **35,**
307. Nazarova, L. V. and Budu, G. V. (1970). *Zh. Neorg. Khim.*, **15,** 1261
308. Simpson, R. T. and Vallee, B. L. (1969). *Inorg. Chem.*, **8,** 1185
309. Stotz, R. W., Walmsley, J. A. and Walmsley, F. (1969). *Inorg. Chem.*, **8,** 807
310. Litzow, M. R., Power, L. F. and Tait, A. M. (1970). *Aust. J. Chem.*, **23,** 1375
311. Bryson, D. J. and Nutall, R. H. (1970). *J. Inorg. Nucl. Chem.*, **32,** 2569
312. Al'yanov, M. I., Bovodkin, V. F. and Benderskii, V. A. (1970). *Isv. Vyssh. Ucheb. Zaved. Khim. Tekhnol.*, **13,** 403
313. Allan, J. R., Brown, D. H. and Lappin, M. (1970). *J. Inorg. Nucl. Chem.*, **32,** 2287
314. Tsintsadze, G. V., Mamulashvili, A. M. and Demchenko, L. P. (1970). *Zh. Neorg. Khim.*, **15,** 276
315. Joshi, D. P. and Lal, K. (1969). *J. Indian Chem. Soc.*, **46,** 484
316. Idem., ibid., 477
317. Agarwala, U. and Singh, B. (1969). *Indian J. Chem.*, **7,** 726
318. Frumina, N. S., Goryunova, N. N. and Mustafin, I. S. (1969). *Zh. Neorg. Khim.*, **14,** 1510
319. Verkhovodova, D.Sh., Osipov, O. A. and Minkin, V. I. (1969). *Zh. Neorg. Khim.*, **14,** 947
320. Duff, E. J., Hughes, M. N. and Rutt, K. J. (1969). *J. Chem. Soc. A*, 2101
321. Waver, J. A., Hambright, P., Talbert, P. T., Kang, E. and Thorpe, A. N. (1970). *Inorg. Chem.*, **9,** 268
322. Reedijk, J. (1970). *Rec. Trav. Chim. Pays-Bas*, **89,** 605
323. Podder, S. N., Dey, K. and Podder, N. G. (1970). *Indian J. Chem.*, **8,** 364
324. Bailey, R. A. and Peterson, T. R. (1969). *Can. J. Chem.*, **47,** 1681
325. Oglobina, R. I., Bednyagina, N. P. and Garnovskii, A. D. (1970). *Zh. Obshch. Khim.*, **40,** 367
326. Saito, Y., Machida, K. and Uno T. (1970). *Spectrochim. Acta*, **26,** 2089
327. Schmidt-Szalowska, A. and Urbanski, T. (1970). *Bull. Acad. Pol. Sci., Ser. Sci. Khim.*, **18,** 73
328. Svirevski, I. K. and Ivanov, C. (1969). *Dokl. Bolg. Acad. Nauk.*, **22,** 683
329. Gogan, N. J. and Siddiqui, Z. U. (1970). *Chem. Commun.*, 284
330. Hambright, P. (1970). *J. Inorg. Nucl. Chem.*, **32,** 2449
331. Kirksey, C. H., Hambright, P. and Storm, C. B. (1969). *Inorg. Chem.*, **8,** 2141
332. Kirksey, C. H. and Hambright, P. (1970). ibid., **9,** 958
333. Stein, T. P. and Plane, R. A. (1969). *J. Amer. Chem. Soc.*, **81,** 607
334. Fleischer, E. B. and Tasker, P. A. (1970). *Inorg. Nucl. Chem. Lett.*, **6,** 349
335. Douglas, J. E. and Wilkins, C. J. (1969). *Inorg. Chim. Acta,* **3,** 635
336. Henning, H. and Dante, R. (1969). *Z. Chem.*, **9,** 275
337. Plienninger, H. and Stumpf, K. (1970). *Chem. Ber.*, **103,** 2562
338. Madden, D. P., DaMota, M. M. and Nelson, S. M. (1970). *J. Chem. Soc. A*, 790
339. Deacon, G. B. and Green, J. H. S. (1966). *Chem. Commun.*, 629
340. Davis, A. R., Murphy, C. J. and Plane, R. A. (1970). *Inorg. Chem.*, **9,** 423
341. Sandhu, S. S., Dass, R. and Gupta, M. P. (1970). *Indian J. Chem.*, **8,** 458
342. Anderson, W. A., Carty, A. J. and Efraty. (1969). *Can. J. Chem.*, **47,** 3361
343. Sandhu, S. S. and Parmar, S. S. (1970). *Z. Anorg. Chem.*, **373,** 64
344. Braibanti, A., Tiripichio, A., Bigoli, F. and Pellinghelli, M. A. (1970). *Acta Crystallogr.*, **26B,** 1069
345. Fischinger, A. J., Sarapu, A. and Companion, A. (1969). *Can. J. Chem.*, **47,** 2629
346. Rathi, N. G., Jain, K. C. and Banerji, S. N. (1970). *J. Indian Chem. Soc.*, **47,** 788
347. Sandhu, S. S., Manhas, B. S., Mittal, M. R. and Parmar, S. S. (1969). *Indian J. Chem.*, **7,** 286

348. Vorontsova, L. N., Ermakova, M. I. and Latosh, N. I. (1970). *J. D.I. Mendeleev. All-Union Chem. Soc.*, **15,** 472
349. Berni, R. J., Benerito, R. R. and Jonassen, H. B. (1969). *J. Inorg. Nucl. Chem.*, **31,** 1023
350. Bhat, T. R., Mathur, B. S. and Shanker, J. (1970). *Indian J. Chem.*, **8,** 275
351. McSharry, W. O. (1969). *Nucl. Sci. Abstr.*, **23,** 34960 (U.S.A.E.C., 1969, NVO-906-85).
352. Bottei, R. S. and McEachern, C. P. (1970). *J. Inorg. Nucl. Chem.*, **32,** 2645
353. *Idem., ibid.*, 2653
354. Briggs, E. and Hill, A. E. (1970). *J. Chem. Soc. A*, 2008
355. Sandell, A. (1970). *Acta Chem. Scand.*, **24,** 1718
356. Bullock, J. L. and Jones, S. L. (1970). *J. Chem. Soc. A*, 2472
357. Mäkitie, O., Saarinen, H., Mattinen, H. and Seppovara, K. (1970). *Suomen Kemistilehti*, **43,** 340
358. Ahuja, I. S. and Rastogi, P. (1969). *J. Inorg. Nucl. Chem.*, **31,** 3690
359. Barnes, J. and Conquest, D. C. S. (1969). *J. Chem. Soc. A*, 1746
360. Duff, E. J. and Hughes, M. N. (1969). *J. Chem. Soc. A*, 477
361. Ol'dekop, Yu.A., Maier, N. A., Erdmann, A. A. and Potopova, L. L. (1970). *Vesti Acad. Nauk. Belarus. SSSR. Khim. Nauk.*, 69. (*Chem. Abstr.*, **73,** 109860b)
362. Ahuja, I. S. and Rastogi, P. (1969). *J. Chem. Soc. A*, 1893
363. *Idem., ibid.* (1970). 378
364. *Idem.* (1970). *J. Inorg. Nucl. Chem.*, **32,** 1381
365. *Idem., ibid.*, 2665
366. Karayannis, N. M., Minkiewicz, V. J., Pytlewski, L. L. and Laber, M. M. (1969). *Inorg. Chim. Acta*, **3,** 129
367. Karayannis, N. M., Sonsino, S. D., Mikulski, C. M., Strocko, M. J., Pytlewski L. L. and Laber, M. M. (1970). *ibid.*, **4,** 141
368. Popp, C. J., Nathan, L. C., McKean, T. E. and Ragsdale, R. O. (1970). *J. Chem. Soc. A*, 2394
369. Bellaart, A. C., Van den Dungen, G. J., Kuijer, M. and Verbeek, J. L. (1969). *Rec. Trav. Chim. Pays-Bas*, **88,** 1089
370. Deacon, G. B. and Cookson, P. G. (1969). *Inorg. Nucl. Chem. Lett.*, **5,** 607
371. Deacon, G. B. and Felder, P. W. (1970). *Aust. J. Chem.*, **23,** 1275
372. Salib, K. A. R. and Senior, J. B. (1970). *Chem. Commun.*, 1259
373. Mehrotra, R. C. and Arora, M. (1969). *Z. Anorg. Chem.*, **370,** 300
374. Radoslovich, E. Q., Rauphach, M., Slade, P. G. and Taylor, R. M. (1970). *Aust. J. Chem.*, **23,** 1963
375. Brini-Fritz, M., Geistel, M. M. and Pousse, A. (1969). *Compt. Rend.*, **268C,** 2040
376. Le Coz, E., Guerchais, J. E. and Goodgame, D. M. L. (1969). *Bull. Soc. Chim. Fr.*, 3855
377. Vielhaber, E. and Hoppe, R. (1969). *Rev. Chim. Miner.*, **6,** 169
378. Fahim, R. B. and Kalta, G. A. (1970). *J. Phys. Chem.*, **74,** 2502
379. Ramamurthy, P. and Secco, E. A. (1969). *Can. J. Chem.*, **47,** 1045
380. *Idem., ibid.* (1970). **48,** 2656
381. Walter-Levy, L. and Groult, D. (1970). *Compt. Rend.*, **270C,** 1966
382. Viltange, M. (1969). *Chim. Anal. (Paris)*, **51,** 378
383. Chin' Khan, Simonov, M. A. and Belov, N. V. (1970). *Kristallografiya*, **15,** 457
384. Kirakosyan, A. K. (1970). *Zh. Neorg. Khim.*, **15,** 369
385. Beeker, E. (1969). *Compt. Rend.*, **268C,** 330
386. Protsenko, P. I., Ivanova, E. M. and Protsenko, G. P. (1969). *Zh. Neorg. Khim.*, **14,** 40
387. Etiene, J. J. and Boulle, A. (1969). *Bull. Soc. Chim. Fr.*, 1534
388. Churagalov, B. R. and Kalashrikov, Ya.A., (1969). *Zh. Fiz. Khim.*, **43,** 481
389. Brigeault, J. M. and Panetici, G. (1969). *Bull. Soc. Chim. Fr.*, 1061
390. Margulis, E. V., Shokarev, M. M., Beisekeeva, L. I. and Vershiwna, F. I. (1970). *Zh. Neorg. Khim.*, **15,** 374
391. Margulis, E. V. and Grishnaukina, N. S. (1968). *Sb. Nauch. Tr. Vses. Nauch. Issled. Gorno-Met. Inst. Tsvet. Metal*, (17), 62
392. Delhez, R. and Gabelica, Z. (1969). *Ann. Soc. Sci. Bruxelles, Ser. 1*, **83,** 399
393. Kohlmuller, R. and Omaly, J. (1969). *Compt. Rend.*, **268C,** 505
394. Udupa, M. R., Hariharin, P. V. and Aravamudan, G. (1970). *Curr. Sci.*, **39,** 230
395. Lamure, J. and Colas, J. L. (1969). *Compt. Rend.*, **268C,** 57
396. Zobnina, A. N. and Kislyakov, I. P. (1970). *Isv. Vyssh. Ucheb. Zaved. Khim. Technol.*, **13,** 143

397. Kabnir-ud-Din, Knan, A. A. and Beg, M. A. (1969). *J. Inorg. Nucl. Chem.*, **31**, 3657
398. Amirova, S. A., Prokhorova, V. G., Kudryashov, V. P. and Danilova, T. G. (1969). *Zh. Neorg. Khim.*, **14**, 315
399. Akimov, B. M., Gulenko, L. N., Akhmetov, S. F. and Tuerdokhletsov, A. I. (1969). *Tr. Khim-Met. Inst., Acad. Nauk. SSSR.*, **9**, 78
400. Djega-Mariadassou, G., Kerboub, E. and Pannetier, G. (1970). *Bull. Soc. Chim. Fr.*, 1353
401. Deyrieuix, R. and Peneloux, A. ibid., 2160
402. Zagdler, A. and Czakis-Sulikowska, D. (1968). *Rocz. Chem.*, **42**, 1827
403. Pribylov, K. P. and Fazlullina, D.Sh., (1969). *Zh. Neorg. Khim.*, **14**, 660
404. Zolutukhin, V. K. (1970). *Zh. Neorg. Khim.*, **15**, 1192
405. Tsuboi, H., Katabami, T. and Fujii, S. (1970). *J. Chem. Soc. Jap., (Ind. Chem.)*, **73**, 1835
406. Svares, E., Yarovitskaya, V. I. and Ieuins, A. (1969). *Zh. Neorg. Khim.*, **14**, 95
407. Garcia, S. G., Vilchez, F. G. and Cuadrado, S. (1970). *Anales de Chim.*, **66**, 471
408. Bernard, M. A. and Berel, M. M. (1969). *Bull. Soc. Chim. Fr.*, **9**, 3064
409. Brinkhoff, H. C., Cras, J. A., Steggerda, J. J. and Willemse, J. (1969). *Rec. Trav. Chim. Pays-Bas*, **88**, 633
410. Gupta, S. K. and Srivastava, T. S. (1970). *J. Inorg. Nucl. Chem.*, **32**, 1611
411. Savant, V. V., Gopalakrishnan, J. and Patel, C. C. (1970). *Inorg. Chem.*, **9**, 748
412. Ouchi, A., Ohashi, Y. and Takeuchi, T. (1970). *Bull. Chem. Soc. Jap.*, **43**, 1088
413. Ashcroft, S. J. (1970). *J. Chem. Soc. A*, 1020
414. Fridman, Ya.D. and Danilova, T. V. (1969). *Zh. Neorg. Khim.*, **14**, 1709
415. Krishnaswami, N. and Bhargava, H. D. (1969). *Indian J. Chem.*, **7**, 710
416. Sergeeva, A. N. and Kischeva, L. A. (1970). *Ukr. Khim. Zh.*, **36**, 435 *(Chem. Abstr.* **73**, 62136q)
417. Soklakov, A. I. and Nechaeva, V. V. (1970). *Izv. Acad. Nauk. SSSR. Neorg. Mater.*, **6**, 998 *(Chem. Abstr.*, **73**, 62089b)
418. Woltermann, G. M. and Wasson, J. R. (1970). *Inorg. Nucl. Chem. Lett.*, **6**, 475
419. Furuhashi, A., Kawai, S., Watanuki, K. and Ouchi, A. (1970). *Sci. Pap. Coll. Gen. Educ. Univ. Tokyo*, **20**, 47
420. Takahashi, Y., Nakatami, M. and Ouchi, A. (1970). *Nippon Kagaku Zasshi*, **91**, 636
421. Malik, W. V., Srivastava, P. and Mehra, S. C. (1969). *J. Indian Chem. Soc.*, **46**, 486
422. Kochiu, S. G., Garnovskii, A. D., Kogan, V. A. and Osipov, O. A. (1969). *Zh. Neorg. Khim.*, **14**, 1428
423. Kharatinov, Y. Y., Tsintsadze, G. V. and Tsivadze, A. Y. (1970). *Zh. Neorg. Khim.*, **15**, 949
424. *Idem.*, ibid., 1811
425. Jain, S. C. and Rivest, R. (1969). *Can. J. Chem.*, **47**, 2209
426. *Idem.*, (1969). *Inorg. Chim. Acta*, **3**, 552
427. *Idem.*, ibid. (1970). **4**, 291
428. Brown, P. R. and Edward, J. O. (1970). *J. Inorg. Nucl. Chem.*, **32**, 2671
429. Natusch, D. F. S. and Porter, L. J. (1970). *Chem. Commun.*, 596
430. Ashurst, K. G., Finkelstein, N. P. and Rice, N. M. (1970). *J. Chem. Soc. A*, 2302
431. Hart, D. M., Rolfs, P. S. and Kessinger, J. M. (1970). *J. Inorg. Nucl. Chem.*, **32**, 469
432. Brinkhoff, H. C., Grotens, A. M. and Steggerda, J. J. (1970). *Rec. Trav. Chim. Pays-Bas*, **89**, 11
433. Casey, A. T., Ham, N. S., Mackey, D. J. and Martin, R. L. (1970). *Aust. J. Chem.*, **23**, 1117
434. Puff, H., Lorbacher, G. and Heine, D. (1969). *Naturwissenschaften.*, **56**, 461
435. Breitinger, D. and Wolf, E. (1970). *Naturwissenschaften*, **57**, 89
436. Briscoe, G. B. and Cooksey, B. G. (1969). *J. Chem. Soc. A*, 205
437. Rosso, J. C. (1968). *Compt. Rend.*, **267C**, 1609
438. Saito, T., Oteva, J. and Okawara, R. (1970). *Bull. Chim. Soc. Jap.*, **43**, 1733
439. Charlot, C., Tikhomiroff, N. and Laffitte, M. (1970). *Bull. Soc. Chim. Fr.*, 459
440. Gugel, B. M. and Oranovskaya, T. V. (1969). *Zh. Neorg. Khim.*, **14**, 329
441. Leonova, M. T. and Svividov, V. V. (1970). *Zh. Neorg. Khim.*, **15**, 1767
442. Il'ina, L. A., Zemlyanskii, N. I., Larionov, S. V. and Chernaya, N. M. (1969). *Izv. Acad. Nauk. SSSR. Ser. Khim.*, 198
443. Romanov, A. M., Ablov, A. V. and Gerbeleu, N. V. (1969). *Zh. Neorg. Khim.*, **14**
444. Skopenko, V. V. (1970). *Ukrainski Khim. Zhur.*, **36**, 129
445. Howell, B. F. and Minor, J. (1969). *J. Inorg. Nucl. Chem.*, **31**, 391

446. Beattie, I. R. and Horder, J. R. (1970). *J. Chem. Soc. A*, 2433
447. Durski, Z. and Kaliszuk, I. (1970). *Rocz. Chem.*, **44,** 689
448. Brill, T. B. (1970). *J. Inorg. Nucl. Chem.*, **32,** 1869
449. Baker, D. J., Hange, S., Gerloch, M. and Lewis, J. (1969). *J. Chem. Soc. A*, 1322
450. Saraf, J. R., Aggarwal, R. C. and Prasad, J. (1970). *Bull. Chem. Soc. Jap.*, **43,** 264
451. Bloom, H., O'Grady, B. V. and Anthony, R. G. (1970). *Aust. J. Chem.*, **23,** 843
452. Tamberg, T. (1969). *J. Inorg. Nucl. Chem.*, **31,** 377
453. Carey, N. A. D. and Noltes, J. G. (1968). *Chem. Commun.*, 1471
454. Burlitch, J. M., Peterson, R. B. and Conder, H. L. (1970). *J. Amer. Chem. Soc.*, **92,** 1783
455. Vizi-Orosz, A., Papp, L. and Marko, L. (1969). *Inorg. Chim. Acta*, **3,** 103
456. Bentley, G. A., Laing, K. R., Roper, W. R. and Waters, J. M. (1970). *Chem. Commun.*, 998
457. Casey, L. and Manning, A. R. (1970). *J. Chem. Soc. A*, 2258
458. Tribonescu, P. and Cristea, M. (1969). *Rev. Roum. Chim.*, **14,** 463
459. Ayers, J. B. and Waggoner, W. H. (1969). *J. Inorg. Nucl. Chem.*, **31,** 2045

3
Boron Hydrides

N. F. TRAVERS
University of Newcastle upon Tyne

3.1	INTRODUCTION	81
	3.1.1 *Scope*	81
	3.1.2 *Presentation*	81
	3.1.3 *Nomenclature*	82
	3.1.4 *Previous reviews*	82
3.2	BONDING IN THE BORON HYDRIDES	83
3.3	NEUTRAL BORON HYDRIDES AND THEIR DERIVATIVES	84
	3.3.1 *One boron atom*	84
	3.3.1.1 *Borane*	84
	3.3.1.2 *Substituted derivatives of borane*	85
	3.3.1.3 *Adducts of borane and of substituted derivatives of borane*	86
	(a) Carbon as donor atom	86
	(b) Nitrogen as donor atom	86
	(c) Phosphorus as donor atom	88
	3.3.2 *Two boron atoms*	89
	3.3.2.1 *Diborane(4) derivatives*	89
	3.3.2.2 *Diborane(6)*	89
	3.3.2.3 *Substituted derivatives of diborane(6)*	90
	3.3.3 *Three boron atoms: triborane(7) adducts*	91
	3.3.4 *Four boron atoms*	91
	3.3.4.1 *Tetraborane(8) and its adducts*	91
	3.3.4.2 *Tetraborane(10) and its substituted derivatives*	92
	3.3.5 *Five boron atoms*	92
	3.3.5.1 *Pentaborane(9)*	92
	3.3.5.2 *Adducts and substituted derivatives of pentaborane(9)*	94
	3.3.5.3 *Pentaborane(11)*	95
	3.3.6 *Six boron atoms*	96
	3.3.6.1 *Hexaborane(10) and its adducts*	96
	3.3.6.2 *Hexaborane(12)*	97

	3.3.7	Seven boron atoms	98
	3.3.8	Eight boron atoms	99
		3.3.8.1 Octaborane(12)	99
		3.3.8.2 Octaborane(14)	99
		3.3.8.3 Octaborane(16)	99
		3.3.8.4 Octaborane(18)	99
	3.3.9	Nine boron atoms	100
		3.3.9.1 Normal nonaborane(15)	100
		3.3.9.2 Iso-nonaborane(15) and adducts of nonaborane(13)	100
	3.3.10	Ten boron atoms	101
		3.3.10.1 Decaborane(14)	101
		3.3.10.2 Decaborane(16) and iso-decaborane(16)	103
		3.3.10.3 Decaborane(18)	103
	3.3.11	Sixteen boron atoms: hexadecaborane(20)	103
	3.3.12	Eighteen boron atoms: octadecaborane(22) and iso-octadecaborane(22)	104
3.4	ANIONIC BORON HYDRIDES AND THEIR DERIVATIVES		105
	3.4.1	One boron atom: anionic adducts of borane	105
	3.4.2	Three boron atoms: the octahydrotriborate(1—) ion	106
	3.4.3	Four boron atoms	106
		3.4.3.1 The heptahydrotetraborate(1—) ion	106
		3.4.3.2 The nonohydrotetraborate(1—) ion	107
	3.4.4	Five boron atoms: the octahydropentaborate(1—) ion	107
	3.4.5	Six boron atoms: the nonahydrohexaborate(1—) ion	108
	3.4.6	Nine boron atoms: the tetradecahydrononaborate(1-) ion	108
	3.4.7	Ten boron atoms: the dodecahydrodecaborate(2—) and the tridecahydrodecaborate(1—) ions	109
	3.4.8	Twenty boron atoms	110
		3.4.8.1 The $B_{20}H_{19}^{3-}$ and $B_{20}H_{18}^{2-}$ ions	110
		3.4.8.2 The $B_{20}H_{18}NO^{3-}$ ion	110
	3.4.9	Twenty four boron atoms	111
		3.4.9.1 The $B_{24}H_{23}^{3-}$ ion and its derivatives	111
		3.4.9.2 The $B_{24}H_{20}I_2^{2-}$ ion, a derivative of $B_{24}H_{22}^{2-}$	111
3.5	METAL–BORON HYDRIDE DERIVATIVES		111
	3.5.1	Open three-centre metal–hydrogen–boron bonding (M—H—B)	112
		3.5.1.1 M—H—B: metal tetrahydroborates	112
		3.5.1.2 M—H—B: metal octahydrotriborates	114
	3.5.2	Two-centre metal–boron bonding (M—B)	116
		3.5.2.1 M—B donor–acceptor bonding: metal–borane derivatives	116
		3.5.2.2 M—B normal covalent bonding: metal substituted pentaborane(9) and decaborane(14) derivatives	116
	3.5.3	Closed three-centre boron–metal–boron bonding (B—M—B)	117

	3.5.3.1	B—M—B: *metal–monodentate boron hydride moiety*	117
	3.5.3.2	B—M—B: *metal–bidentate boron hydride moiety*	117
3.5.4		*Multi-centre or π-metal–boron hydride bonding*	122
3.6	CATIONIC BORON HYDRIDE DERIVATIVES	122
3.7	ADDENDUM	123
3.8	TABULAR SURVEY	124
3.8.1	*Presentation*	124
	Abbreviations	124

3.1 INTRODUCTION

3.1.1 Scope

The term 'boron hydride' is often used to describe a compound of boron and hydrogen in which there are insufficient valence electrons to permit allocation of two electrons to each bond between nearest neighbour atoms. This Review includes such compounds as part of a wider classification of boron hydrides which embraces all compounds containing boron–hydrogen groups. (Certain exceptions, notably systems where the BH group forms part of a carborane, or a heterocyclic ring are discussed in Chapter 4 and 5, respectively, of this Volume). Thus, such species as Lewis-base adducts of borane (BH_3), metal tetrahydroborates, and boron cations, which contain BH groups, but would hardly be considered as 'boron hydrides' by the purist, are discussed here. The use of boron–hydrogen species in organic synthesis (e.g., hydroboration) is not dealt with in this Review.

3.1.2 Presentation

A brief explanation of the divisions within the main sections of this Review is perhaps advisable here. In Sections 3.3 and 3.4, subheadings are in accordance with the number of boron atoms bound to each other, either by two-, three- or multi-centre homonuclear bonds, or by three-centre hydrogen-bridge bonds. It was found more difficult to continue this method of presentation in Section 3.5 where a better understanding of metal–boron hydride systems is achieved by a classification which stresses the type of bonding between the metal and the boron hydride moiety. Within the confines of this classification, however, the compounds are considered in order of increasing number of borons bound to each other, and of increasing number of associated hydrogen atoms. In Section 3.6 there is discussion of boron cations, in which the boron atoms are not bound to each other by either two- or three-centre homonuclear bonds, or by hydrogen bridges, and consequently the method of presentation used in Sections 3.3 and 3.4, and to a lesser extent in Section 3.5, is abandoned.

3.1.3 Nomenclature

Rules for the nomenclature of boron compounds, approved by the Council of the American Chemical Society, were published in 1968[1]. These rules are adhered to as far as is practicable in this Review.

3.1.4 Previous reviews

The earliest review of the boron hydrides is Alfred Stock's monograph, 'Hydrides of Boron and Silicon', published in 1933[2]. Here, Stock describes how, by means of specially developed high-vacuum systems, he and his students succeeded in preparing and characterising many of the presently-known boron hydrides. The subject was then largely ignored — save for work on the lower boron hydrides by Schlesinger and his colleagues[3] — until the early 'fifties, when the search for new high-energy fuels for aircraft and rockets gave enormous impetus to a full-scale investigation of the boron hydrides in the United States, which continued for over a decade. The vast amount of information which stemmed from this research programme has only recently become available in the form of a book edited by Holzmann[4]. This book is, of course, intended to be a compilation and collation of data rather than a concise review of the subject; however, a number of authors, many of whom were active in the high-energy fuel programme, have reviewed the chemistry of the boron hydrides up until 1968, whence there is a gap until the present. Only the more recent and comprehensive reviews are discussed here. Lipscomb's treatise on 'Boron Hydrides' which was published in 1963, serves as an excellent standard textbook, and contains comprehensive discussions on the bonding, structures, and reactions of these compounds[5]. Ionic boron hydrides, and neutral boron hydrides, are accorded two chapters in a book edited by Adams[6], and Hawthorne[7] has written a chapter on them in 'The Chemistry of Boron and Its Compounds'. All aspects of 'Polyhedral Boranes' are discussed in the book by Muetterties and Knoth[8]; polyhedral borane anions are also reviewed by Todd[9a] in the latest edition of the periodical 'Progress in Boron Chemistry', which also contains a discussion on the use of isotopic labels in boron hydride chemistry by Odom and Schaeffer[9b]. A recently published source book on 'N.M.R. Studies of Boron Hydrides and Related Compounds' by Eaton and Lipscomb[10] greatly expands previous reviews[11–13] of this subject, and will be invaluable to those engaged upon the elucidation of boron hydride structures by n.m.r. techniques. Synthetic procedures for the boron hydrides have been critically reviewed by Parry and Walter[14], and details for the laboratory preparation of diborane and certain polyhedral boron hydride anions have been given in *Inorganic Syntheses*[15]. James and Wallbridge[16] have recently reviewed the subject of metal tetrahydroborates, and Schmid[17] has briefly discussed some metal–boron hydride systems in a more general article on metal–boron compounds.

As mentioned above, even the most recent reviews of the boron hydrides published during 1970 do not cover the literature further than the end of

1968. For this Review, the literature for the years 1969 and 1970 inclusive has been surveyed in detail, and Table 3.1 at the end of this Review contains references, not only to all new compounds which were prepared, but also to new data for existing compounds which were obtained during the period mentioned above. Of necessity, not all of the compounds referred to in the Table are discussed in the main body of the Review (Sections 3.2–3.6).

3.2 BONDING IN THE BORON HYDRIDES

A detailed introduction to bonding theory in boron hydrides is outside the scope of this Review, and in this Section emphasis is placed on new developments rather than on fundamental principles of the subject.

The failure of the two-centre–two-electron bond treatment adequately to explain the bonding and structures of the formally electron-deficient boron hydrides led to the formulation of a semi-quantitative theory of multi-centre bonding[5]. The concept of the three-centre–two-electron bond (simply a special case of a multi-centre molecular orbital, represented as a linear combination of atomic orbitals) when applied to those boron hydrides of low symmetry possessing polyhedral fragment structures ('basket' molecules[19]) was successful in providing an explanation of their bonding, and also in correlating many of their structural and electronic properties; however, the same treatment was less adequate when applied to the more symmetrical boron hydrides, such as B_5H_9, and to the closed polyhedral species ('cage' molecules[18]) such as the $B_{10}H_{10}^{2-}$ and $B_{12}H_{12}^{2-}$ anions, and a more general multi-centre molecular orbital description was used[5]. Recently, Kettle and Tomlinson have shown that for cage[18] and basket[19] molecules the three-centre bond, or 'face' orbital approach, leads to the same pattern of molecular orbital energy levels as does an approach which considers two-centre orbitals between neighbouring boron atoms ('edge' orbitals); moreover, in the case of cage molecules, the molecular orbital energy level scheme obtained using a 'face' orbital basis was closely related to that obtained from an 'edge' orbital basis. The same authors also showed that, in general, the three-centre bond treatment could be replaced by one which employs the more classical two-centre or 'edge' bond, and gives very similar results, effectively because the method of calculation used is a topologically correct extension of the Hückel theory to three dimensions[18–20].

In an important series of recent papers, Lipscomb has reported accurate self-consistent field (S.C.F.) wave functions for B_2H_6, B_4H_{10}, B_5H_9, and B_5H_{11} [21–23]. These wave functions yield data on electron densities, net atomic charges, and overlap populations, which are discussed in terms of previously-proposed bonding schemes; in addition, magnetic properties of the boron hydrides (^{11}B and ^1H n.m.r. chemical shifts) can be related to calculated ground-state charge densities, and chemical reactivity to charge density and frontier overlap population[22]. The same S.C.F. wave functions can be transformed to localised molecular orbitals by an objective procedure in which the orbital self-energy is maximised[21, 23]. The bonding descriptions, obtained in terms of these localised orbitals, compare favourably with those predicted by the three-centre bond theory. For example, in B_2H_6 [21] there are found two localised, three-centre bridge-hydrogen bonds between two $sp^{2.5}$-hybridised borons, and, in B_4H_{10} [23], there is a reasonably well localised direct boron–boron single bond between two sp^3-hybridised boron atoms, as predicted by the three-centre bond approach[5]. An interesting feature of the description of B_5H_{11} was the unique hydrogen atom attached to the apical boron, which showed bonding properties intermediate between those of a bridge hydrogen and a terminal hydrogen[23]. In a somewhat more approximate treatment, Adamson and Linnett calculated localised molecular orbitals for B_2H_6, B_4H_{10}, B_5H_9, and $B_{10}H_{14}$; the localised orbitals were obtained from 'extended Hückel'-type delocalised molecular orbitals, by maximising the electron population of the molecular orbitals in regions chosen by symmetry and chemical criteria[24].

Pauling has proposed a simple theory to account for the number of hydrogen atoms in a particular boron hydride. It is assumed that in the boron hydride under consideration the boron atoms lie at or near the most compact set of corners of a regular icosahedron, that each boron–boron bond requires one valence electron (and is thus a half bond, analogous to a boron–boron bond in elemental boron) and that each hydrogen atom is bonded to the boron core by one or two bonds requiring a total of 1.75 electrons (the latter case corresponding to a bridge hydrogen bond). In this way, hydrogens can be correctly accounted for in B_2H_6, B_4H_{10}, B_6H_{10}, and B_8H_{12}; in B_5H_9, and $B_{10}H_{16}$, which consist of octahedral rather than icosahedral fragments, it is necessary to distort the core of five boron atoms from the icosahedral structure to form a square-pyramid[25].

3.3 NEUTRAL BORON HYDRIDES AND THEIR DERIVATIVES

3.3.1 One boron atom

3.3.1.1 Borane

The simplest, and perhaps most elusive, boron hydride, borane, BH_3, may be described as being in equilibrium with its hydrogen-bridged dimer, diborane, B_2H_6. (The equilibrium constant for $B_2H_6 \rightleftharpoons 2BH_3$ is estimated to be 1.6×10^{-5} at 155 °C and 1 atm[7].) The remarkable ease of dimerisation of borane is a direct result of the great electron deficiency at the sp^2 boron atom, where the empty 'p' orbital, unlike that of stable BF_3, cannot be involved in π-bonding because the attached hydrogen atoms are devoid of unshared electrons. Although borane was originally detected[26] in the low-pressure pyrolysis[27] of diborane, and its molecular-beam mass spectrum obtained[26], more recent studies have utilised the weakly bound complex borane–carbonyl, $BH_3 \cdot CO$, as a source of borane.

A recent electron-impact mass spectrometric study of borane, produced by pyrolysis of borane carbonyl, did not report the mass spectrum of borane[28]; however, a number of workers have investigated the thermal decomposition of $BH_3 \cdot CO$ in flow reactors using the more sophisticated techniques of molecular-beam mass spectrometry unequivocally to identify the products[29-31]. The molecular-beam mass spectrum of borane has subsequently been reported by two independent groups[30, 31], and their results, which are in excellent agreement, support the earlier data[26] obtained from a study of the pyrolysis of diborane; moreover, the recently published molecular-beam mass spectrum of borane[31] agrees with the mass spectrum used to interpret the results obtained for the high-temperature reaction of boron carbide (B_4C) with hydrogen, thus confirming the identification and quantitative estimates of borane concentration[32].

Preliminary work[29] on the low pressure decomposition of borane carbonyl had shown that the yield of borane was low due to its destruction on the surface of the reactor; however, when borane carbonyl, diluted to a high degree in an inert carrier gas, is thermally decomposed in a fast-flow reactor, borane is produced in high yield (c. 60%) and purity (up to 84%, based on total boron hydrides)[30]. The discovery of such an efficient source of borane will undoubtedly enable many of the fast reactions of this fascinating species to be studied, and, indeed, the measurement of the absolute rate of self-association of borane molecules has already been carried out[33]. The self-association reaction, which proceeds at high efficiency, and requires no activation energy, must be considered as of the same type as radical-association and ion–molecule reactions (although borane is neither a radical nor an ion). The proven existence of borane, BH_3, and its free-radical-like reactivity (which, admittedly, is at present established only for self-association) must now become important and relevant factors to be considered in any discussion of boron hydride reaction intermediates. It is indeed conceivable that this orbitally-unsaturated species has higher homologues which would also perform as highly active intermediates.

3.3.1.2 *Substituted derivatives of borane*

In substituted derivatives of borane it has often been suggested that a non-bonding electron pair on a substituent atom or group interacts with the formally vacant 2p orbital of the boron atom, and this concept has been supported by a recent S.C.F. study in which changes of certain quantities were observed throughout the series, BH_3, BH_2F, BHF_2, and BF_3; these included essentially constant atomic charges for hydrogen and fluorine atoms, linearly increasing total positive charge of the central boron atom, and an increase of the $2p_\pi$-type population on boron[34]. The electronic energies and structure of difluoroborane, HBF_2, have also been the subject of a theoretical treatment by Armstrong[35], who found excellent agreement between the calculated and experimental structural parameters. Unfortunately, a number of different attempts to isolate difluoroborane have yielded only mixtures of HBF_2 and BF_3; separation by distillation or by gas chromatography (on a column held at $-80\,°C$) proved impossible[36]. Dichloroborane, $HBCl_2$, and hydrogen chloride were the sole products detected in the reaction of boron trichloride with hydrogen at 700–850 °C, and the standard formation enthalpy and entropy of $BHCl_2$ were reported[37].

'Surface boron hydrides' may be considered as disubstituted derivatives of borane. The high-temperature pyrolysis of Aerosil which has been impregnated with boric acid, and subsequently methylated, leads to the formation of three surface =B—H species characterised by their differing infrared absorptions. Removal of these surface boranes, by pyrolytic degassing, produces an activated boria-impregnated silica ('reactive boria') which will react with hydrogen at high temperatures to re-form the same surface boron hydrides[38].

The formal substitution of an amino group for one of the hydrogens of borane produces the compound BH_2NH_2, aminoborane, which is isoelectronic with ethylene. Attempts to prepare aminoborane have usually led to its polymerisation products, but this elusive species has recently been isolated for the first time from the radiofrequency discharge of borazine vapour in a low pressure fast flow system, followed by an immediate quench to $-196\,°C$[39]. Aminoborane was characterised by its mass spectrum (obtained using a cold inlet system at $-160\,°C$), its ionisation potential (which compared well with that found for ethylene, and also with that obtained from theoretical calculations[40] on the electronic structure of aminoborane), and its infrared spectrum (observed at various temperatures down to 4 K)[39].

3.3.1.3 Adducts of borane and of substituted derivatives of borane

The filling of the empty boron 2p orbital of borane by an electron pair is achieved by the formation of Lewis base adducts of borane of the type $L·BH_3$ (where L is a Lewis base), and a wide range of these has been prepared. Substituted derivatives of borane form adducts of the same type, and these will also be considered here. It is convenient to divide the adducts into groups depending upon the nature of the donor atom.

(a) *Carbon as donor atom* — The correlation of the photoelectron spectrum of $BH_3·CO$ with that of the free donor, CO, was found to be consistent with donation from CO to BH_3 and with considerable π back-donation to CO from BH_3[41]. Agreement of experimentally-obtained ionisation potentials, for adduct and free donor, with those calculated[42] was not as favourable as might have been expected, possibly because the calculations had underestimated the extent of back-donation in the adduct[41].

(b) *Nitrogen as donor atom* — Reaction of a large excess of nitric oxide with borane carbonyl produced nitrosylborane, $BH_3·NO$, which was detected by mass spectrometry[43]. The photoelectron spectrum of $NH_3·BH_3$, when correlated with that of free ammonia, suggests simple donor–acceptor bonding in the adduct[41]; moreover, the experimentally-determined ionisation potentials for these compounds are in good agreement with the calculated values[42].

A new high-yield synthesis, applicable to adducts of borane with ammonia, primary, secondary, and tertiary amines, di- and tri-amines, and pyridines has been developed[44]. It is based on the reaction of sodium tetrahydroborate with iodine in presence of the donor (L):

$$2NaBH_4 + 2L + I_2 \longrightarrow 2NaI + 2L \cdot BH_3 + H_2$$

The optically active aminoboranes, amphetamine-borane, desoxyephedrine-borane, and N,N-dimethylamphetamine-borane have been prepared; desoxyephedrine-borane exists as two stereoisomers, as shown by ^1H n.m.r., and one of these has been isolated in the pure form[45].

In an n.m.r. study of the intermolecular exchange between borane adducts and their corresponding free Lewis bases, Cowley and Mills observed a temperature dependence of the chemical shift of $Me_3N \cdot BH_3$ when mixed with Me_3N in benzene, but *no* temperature dependence when the same components were mixed in glyme or dichloromethane[46]. It appeared then that benzene, in contrast to other solvents, must play a specific role in assisting the exchange of $Me_3N \cdot BH_3$ with Me_3N, but a later, more thorough, investigation of the n.m.r. behaviour of $Me_3N \cdot BH_3$ and Me_3N in benzene, over a wide range of concentration and temperature, proved conclusively that there was *no* exchange between these two compounds[47]. It was found that whereas the chemical shift of Me_3N was *independent* of concentration, temperature, and added $Me_3N \cdot BH_3$, the chemical shift of $Me_3N \cdot BH_3$ was *dependent* on concentration, temperature, and on added Me_3N; in fact, the chemical shift of $Me_3N \cdot BH_3$ varies with concentration and temperature in such a manner that even if no exchange were to occur, as is indeed the case, experimental observation of the chemical shift behaviour of the two components—within certain limits of concentration and temperature—would erroneously indicate that exchange between them does occur. The explanation why the chemical shift of $Me_3N \cdot BH_3$ in benzene is dependent on concentration, temperature, and addition of Me_3N, whereas that of Me_3N is independent of such factors, may be that $Me_3N \cdot BH_3$ is solvated by benzene in a conformationally more specific way than is Me_3N, so that the magnetic anisotropy is more fully felt. An increase in temperature would reduce the anisotropy associated with the solvent sphere, and an increase of concentration, by thereby increasing the self-association of $Me_3N \cdot BH_3$ in benzene, would also reduce the specific solvation; in both cases there would be resultant downfield chemical shifts of the $Me_3N \cdot BH_3$, as observed. The introduction of magnetically isotropic constituents into the solution which could associate preferentially with the amine-borane would also give downfield chemical shifts, as is observed when Me_3N is added. Work of this nature emphasises the care which must be taken in the interpretation of chemical shifts and their temperature dependence for mixtures[47].

In a study of the chlorination of $Me_3N \cdot BH_3$ by a variety of reagents[48, 49], the new compound $Me_3N \cdot BHCl_2$ was isolated from the reaction of either thionyl chloride or mercuric chloride with $Me_3N \cdot BH_3$ [49]. The new compounds $Me_3N \cdot BHCl_2$ and $Me_3N \cdot BHI_2$ were also prepared by another group, as part of an investigation of hydrogen-bonded complexes of borane adducts; it was found that adducts of the type $L \cdot BH_3$, and $L \cdot BH_2X$ (where L is a tertiary amine, phosphine, or pyridine; and X = Cl, Br, I) formed hydrogen-bonded complexes — probably involving $OH \ldots BH_3$ and $OH \ldots BH_2$ interactions — with methanol, phenol and p-fluorophenol in carbon tetrachloride solution, whereas adducts of type $L \cdot BHX_2$, and $L \cdot BX_3^1$ (where $L = Me_3N; X = Cl, Br, I; X^1 = F, Cl, Br$) did not[50]. Long-range spin–spin coupling between hydrogen and boron can be observed in adducts of the type Me_3NBX_3 (where $X = Cl, Br$) and $Me_3N \cdot BXX_2^1$ (where $X = Cl, Br$; $X^1 = Br, Cl$, respectively), but not in the adducts $Me_3N \cdot BF_3$, $Me_3N \cdot BH_2X$ ($X = Cl, Br$), $Me_3N \cdot BHX_2$ ($X = Cl, Br$), and $Me_3N \cdot BHClBr$; such lack of coupling in the latter cases was attributed to quadrupole relaxation when the field gradient was increased by hydrogen or fluorine substitution[51].

Spielvogel has reported that reaction of the cyanotrihydroborate ion (BH_3CN^-) with anhydrous acids in tetrahydrofuran produces the solvent adduct of BH_2CN[52], and it has been shown by other workers that reaction of this species with certain nitrogen bases leads to the formation of adducts of cyanoborane $L \cdot BH_2CN$ (where $L = Me_3N$, 4-picoline, and morpholine) and $L \cdot (BH_2CN)_2$ (where L = tetramethylethylenediamine), in a manner analogous to the formation of amine-boranes from the tetrahydrofuran adduct of BH_3[53].

(c) *Phosphorus as donor atom* — New syntheses of triphenylphosphine-borane, and the previously unknown compounds bis-(diphenylphosphino)methane-bis-(borane), $Ph_2P \cdot CH_2 \cdot PPh_2 \cdot 2BH_3$, and 1,2-bis-(diphenylphosphino)ethane-bis-(borane), $Ph_2P \cdot (CH_2)_2 \cdot PPh_2 \cdot 2BH_3$, have recently been reported[44].

It is often assumed that coordination in phosphine–borane complexes is enhanced by back-donation from the BH_3 group to the empty d orbitals of the phosphorus ($2p_\pi$-$3d_\pi$ bonding); in this context, recently reported SCF molecular orbital calculations — indicating that d-orbital participation, and π back-donation, are important in PH_3O but unimportant in $PH_3 \cdot BH_3$ — are of interest[54]. A related study of the molecule $PF_3 \cdot BH_3$, in which SCF–MO calculations of the bonding scheme were correlated with the photoelectron spectrum, suggested that the degree of back-donation, although small, was probably greater than that in $PH_3 \cdot BH_3$[55]. Enhanced p_π-d_π back-bonding may account for the stabilising factor which recent thermochemical studies have shown to operate in triarylphosphine-boranes, but not in either trialkylphosphine-boranes or triphenylphosphine-trifluoroborane[56].

It has been shown by ^1H n.m.r. spectroscopy that there is no proton–deuteron exchange in the partially deuterated compounds $PH_3 \cdot BD_3$ and $PD_3 \cdot BH_3$[57], and none between the carbon and boron sites in methylphosphine-borane, but there is such an exchange between the silicon and boron sites in silylphosphine-borane[58]. Although previous ^{11}B n.m.r. studies of the adduct $P_2F_4 \cdot BH_3$ had shown no P—B coupling at $-80\,°C$ (attributed to exchange of the BH_3 moiety), a recent investigation of the n.m.r. spectra, at temperatures higher than those previously used, has verified the presence of a stable P—B bond in the adduct, and has also demonstrated that addition of excess P_2F_4 to a solution of the adduct has no appreciable effect on the spectra of either species. The ^{19}F n.m.r. spectrum contains peaks attributable to both a coordinated and an uncoordinated PF_2 group, as expected for $F_2P \cdot PF_2 \cdot BH_3$ [59]. The use of n.m.r. spectroscopy has established the existence of P—B bonding in a number of adducts of borane and of substituted borane with a variety of Lewis bases, many of which have other potential donor sites, apart from phosphorus [60-66].

Phosphine-boranes have been postulated as intermediates in the formation of polyphosphinoboranes from the reaction of secondary bis-phosphines with $Et_3N \cdot BH_3$ [67, 68].

3.3.2 Two boron atoms

3.3.2.1 Diborane(4) derivatives

Diborane(4), B_2H_4, is an unknown hydride, but derivatives of the type $LBH_2 \cdot BH_2L$ (where L is a Lewis base), in which there is a direct boron–boron bond, can be prepared. The difluorophosphine derivative[69], $(PF_2H)_2 B_2H_4$, was recently obtained from the reaction of PF_2H and pentaborane(11), $(PF_3)_2B_2H_4$ [70] from PF_3 and the dimethyl ether adduct of triborane(7), and $(Me_2NPF_2)_2B_2H_4$ [69] from Me_2NPF_2 and either pentaborane(9) or pentaborane(11); the formation of $(PMe_3)_2 \cdot B_2H_4$ from $2PMe_3 \cdot B_5H_9$ has also been reported[98].

A polymeric diborane(4) species, $(B_2H_4)_n$, was obtained by reaction of tetraborane(10) with trimethyl- or triethyl-borane[71].

3.3.2.2 Diborane(6)

X-ray studies of diborane (and of other boron hydrides) yield boron–hydrogen bond lengths which are 0.1 Å shorter than the corresponding bond lengths obtained from electron-diffraction measurements; this difference in the observed bond lengths, which is five times the largest estimated standard error in either technique, has been shown — in the case of diborane — to be due to inadequacies in the spherical atom model used to interpret the x-ray data[72]. Boron–hydrogen bond lengths, calculated from a recent high-resolution infrared study of $^{10}B_2H_6$ and $^{11}B_2H_6$, were in good agreement with those obtained by electron diffraction[73].

An experimental and theoretical comparison of the isoelectronic molecules B_2H_6 and C_2H_4, by means of photoelectron spectroscopy and near Hartree–Fock calculations, has been reported by Brundle and co-workers[74]. The photoelectron spectrum of B_2H_6 obtained in this work differed, in some respects, from a previously published spectrum[75], in particular, two sharp steps present in the previous spectrum were found to be absent, and could be assigned to a hydrogen chloride impurity[74].

Reaction of diborane with water, at $-130\,°C$, yielded the previously unknown dihydrate, $B_2H_6 \cdot 2H_2O$, which may have either of the two structures $BH_2(OH_2)^+BH_4^-$ or $H_2O \cdot BH_3$. A similar reaction with alcohols produced $B_2H_6 \cdot 2ROH$ (R = Me, Et)[76].

3.3.2.3 Substituted derivatives of diborane(6)

The difficulty in obtaining Raman spectra of terminally-substituted methyldiboranes, due to their ease of disproportionation, has lately been overcome by the use of a laser source, and Raman spectra have been reported for tetramethyldiborane, and 1,1-dimethyldiborane in the liquid phase at low temperatures[77]. Reaction of a silyl-substituted sulphonium ylide, $Me_3SiCH SMe_2$, with diborane produced an unstable adduct which decomposed with formation of the new compound, 1,2-bis-(trimethylsilylmethyl)diborane(6), $[(Me_3SiCH_2)BH_2]_2$, unobtainable by conventional alkylation reactions. This novel silyl-substituted diborane reacts with trimethylamine to form the adduct $(Me_3SiCH_2)BH_2 \cdot NMe_3$ [78].

No diborane derivatives containing bridging methyl groups have so far been prepared, and recent calculations have shown that such methyl-bridged derivatives would be much less stable than the corresponding methyl-bridged aluminium compounds[79]. A new method of preparation for the dialkylamino-bridged derivative, μ-dimethylaminodiborane(6), μ-$Me_2NB_2H_5$, has recently been described in which the dimethylamido-trihydroborate(1−) anion, $Me_2NBH_3^-$, is reacted with diborane[80]. The ^{11}B n.m.r. spectrum of μ-$Me_2 NB_2H_5$ is strongly temperature dependent due to the intramolecular exchange of bridge and terminal hydrogens, which may proceed by cleavage of a bridge-hydrogen bond, followed by rotation of the BH_3 group, and re-establishment of the bridge bond; assuming the validity of such a mechanism, the acceleration of the exchange in the presence of Lewis bases could be due to reversible nucleophilic attack of the base on the boron, thus facilitating the bridge-opening step. The exchange has recently been studied by fitting the line-shape of the ^{11}B n.m.r. spectrum as a function of temperature, and supporting evidence for the proposed mechanisms was obtained. Rate constants, enthalpies and entropies of activation for the uncatalysed exchange process (in the neat liquid and in methyl-cyclohexane) and for the catalysed exchange process (in 1,2-dimethoxyethane, and tetrahydrofuran) were determined[81]. The ^{11}B n.m.r. spectrum of the new compound, μ-dimethylaminomethyldiborane, μ-$Me_2NB_2H_4Me$, also exhibits an interesting temperature dependence, which may likewise be explained by the cleavage of a bridge-hydrogen bond[82]. The n.m.r. spectrum indicates that below −30 °C, the molecule has the normal cyclic hydrogen-bridged structure, whereas, above this temperature, it has the unbridged linear structure, $H_3B \cdot Me_2N \cdot BHMe$ (bridge opening is always in the same direction — towards unsubstituted boron). This behaviour is in contrast to that of μ-$Me_2NB_2H_5$ which does not undergo bridge cleavage until relatively elevated temperatures in the presence of weak Lewis bases[81]; furthermore, the lifetime of the unbridged, methyl-substituted species is, at least, long compared to milliseconds, unlike that of the unbridged, unsubstituted species, in which complete equilibration of the boron bound protons was observed on the n.m.r. time-scale[81].

The reaction of $[Et_4N][HS(BH_3)_2]$ with anhydrous hydrogen chloride at low temperature produced the new compound μ-mercaptodiborane,

μ-SHB$_2$H$_5$, in which the presence of a bridging SH group was confirmed by ^{11}B n.m.r. spectroscopy[83].

3.3.3 Three boron atoms: triborane(7) adducts

New adducts of triborane(7), F$_2$XP·B$_3$H$_7$ (X = F, Cl and Br), have been prepared and the relative basicities of the F$_2$XP ligands towards triborane(7) have been determined by base-displacement reactions[84].

3.3.4 Four boron atoms

3.3.4.1 Tetraborane(8) and its adducts

Boron hydrides have been classified into two series called 'stable', [B$_n$H$_{n+4}$ or B$_n$H$_n$(BH$_3$)$_2$], and 'unstable', [B$_n$H$_{n+6}$ or B$_n$H$_n$(BH$_3$)$_3$] [2, 85, 86]; it thus appears that B$_4$H$_8$, which is not isolatable in the free state, is classified as a 'stable' hydride, wheareas B$_4$H$_{10}$, which has been isolated and characterised, is an 'unstable' hydride. Such an apparent anomaly may be understood on consideration of the basis for this classification[85, 86]; knowledge of the amount and type of fragmentation in the mass spectrum of a compound often helps to assess its stability. Three criteria have been applied to distinguish between 'stable' and 'unstable' boron hydrides—the relative size of the parent peak, the profile of the mono-isotopic spectrum in the parent region, and the extent

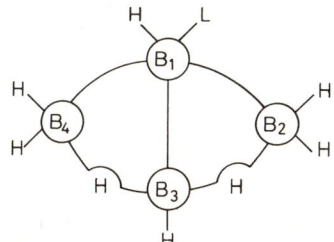

Figure 3.1 Topological representation of L·B$_4$H$_8$(L = Me$_2$N·PF$_2$, PF$_2$H)

of boron fragmentation[87]; accurate determination of this last quantity has, until recently, been somewhat difficult. Conventional instruments produce mass spectra of boron hydrides which often contain, in addition to the mass spectrum of the hydride itself, the mass spectra of pyrolysis products formed from it by contact with the hot ion source; however, the use of a molecular-beam instrument[26] eliminates such effects, and molecular-beam mass spectra can be used unequivocally to determine boron hydride fragmentation patterns, and to study their low-pressure pyrolysis behaviour. Application of the criteria discussed above to the molecular-beam mass spectrum of ^{10}B$_4$H$_8$ (obtained in a study of the pyrolysis of ^{10}B$_4$H$_8$·CO) supports the classification of B$_4$H$_8$ as a 'stable' hydride[88].

The structure of the stable adduct $(Me_2NPF_2)B_4H_8$ has been determined by x-ray crystallography[89], and, on the basis of n.m.r. spectroscopic data, the structure of $(PF_2H)B_4H_8$ is thought to be similar[90]; use of ^{19}F n.m.r. spectroscopy has disclosed the presence of two geometric isomers of $(Me_2NPF_2)B_4H_8$, but not of $(PF_2H)B_4H_8$ [90]. The adducts may be represented by a 2112 topology (Figure 3.1).

3.3.4.2 Tetraborane(10) and its substituted derivatives

The topological representation and structure[5] of tetraborane(10) are given in Figures 3.2(a) and 3.2(b), respectively. There are two types of terminal proton on the 2 and 4 boron atoms of tetraborane(10) in the crystalline state [axial(a) and equatorial(e) in Figure 3.2(b)], and the presence of these in liquid B_4H_{10} has recently been confirmed by its 220 MHz ^1H and 100 MHz ^1H n.m.r. (with multiple decoupling) spectra[91].

Reaction of (dimethyl ether)triborane(7) with 1,2-dimethyldiborane yields 2-methyltetraborane(10)[92] — also obtainable from dimethylmercury and

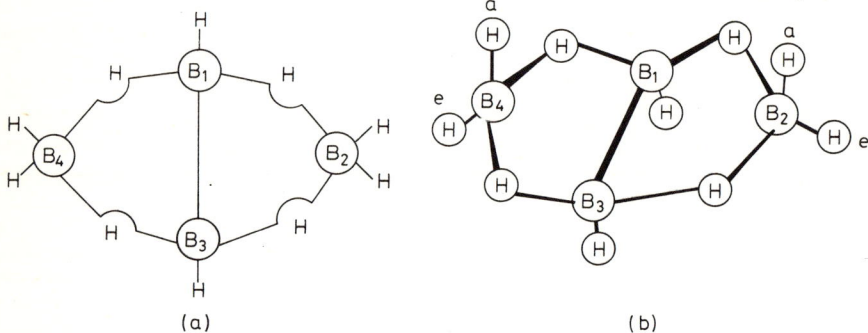

Figure 3.2 B_4H_{10} (a) Topological representation (b) Molecular structure, showing axial, 'a', and equatorial, 'e', terminal hydrogen atoms.

tetraborane[93] — and the isomeric 1,2-, 2,2-, and 2,4-dimethyltetraborane(10) derivatives[92]. Insertion of a formal Me_2B^+ moiety into a borane anion has been reported for the reaction of Me_2BCl with BH_4^-, to form 1,1-dimethyldiborane(6)[94], with $B_3H_8^-$ to form 2,2-dimethyltetraborane(10)[94], and with $B_5H_8^-$ (which is discussed in Section 3.3.5.2).

3.3.5 Five boron atoms

3.3.5.1 Pentaborane(9)

It was mentioned in Section 3.2 that pentaborane(9), which has delocalised bonding, cannot be accurately described by the localised three-centre bond approach, and, for convenience, only one possible valence-bond structure is

BORON HYDRIDES 93

shown in Figure 3.3(a); the three-dimensional structure[5] is illustrated in Figure 3.3(b).

The molecular-beam mass spectrum of B_5H_9 agreed with the spectrum obtained with a conventional instrument and verified its classification as a stable boron hydride[88]. Study of B_5H_9 pyrolysis at low pressures in a flow reactor showed that it did not decompose to give detectable amounts of volatile boron hydrides[88]; however, under more vigorous pyrolytic conditions, B_5H_9 decomposed by a first-order reaction[95].

It has been shown that the relative chemical shifts of boron and terminal hydrogen atoms in a number of pyramidal boron hydrides and carboranes can be accounted for by using a conical ring current model, in which the

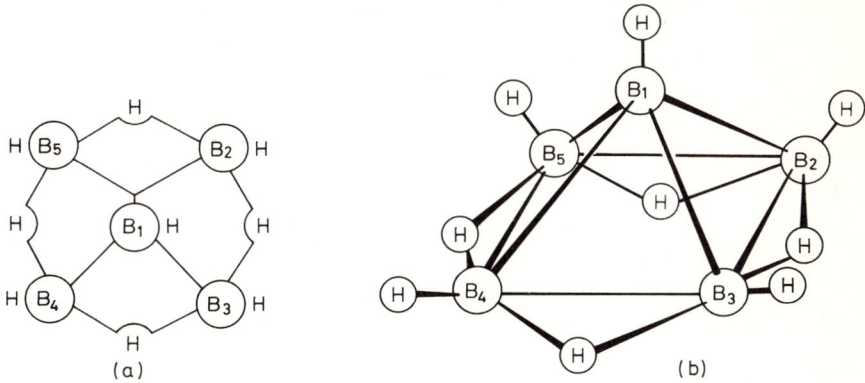

Figure 3.3 B_5H_9 (a) A valence bond structure (b) Molecular structure

pyramidal framework is treated as a cone and one or more ring current loops are placed about the conical curved surface parallel to the base[96]. In the rather exceptional case of B_5H_9, which, unlike other pyramidal boron hydrides and carboranes, appears to have a greater partial negative charge on the apex boron than on the basal borons[5, 22], placing a single ring current loop between the centroid (of the cone) and the apex leads to a result which is in good agreement with the observed difference in chemical shift between the apex and basal terminal hydrogen atoms; by placing 20 ring current loops between the apex and the base agreement is obtained between the calculated difference in boron chemical shifts (for apex and base) and that observed[96]. The difference between the ^{11}B n.m.r. spectra of B_2H_6 and B_4H_{10}, which exhibit fine structure, and the relatively featureless ^{11}B n.m.r. spectrum of B_5H_9, which displays only boron–terminal hydrogen coupling and consists of much broader lines, is not due to different relaxation rates; recent measurement of the ^{11}B and ^{10}B spin-lattice relaxation times for B_2H_6 and B_5H_9 led to the conclusion that ^{11}B line-widths should be quite narrow for both compounds, and, for B_5H_9, should be about 5 and 0.5 Hz for base and apex respectively; hence, observed line widths in B_5H_9 (>40 Hz) appear to be due to unresolved fine structure[97]. It is thus interesting that even the recently-obtained 70 MHz ^{11}B n.m.r. spectrum of B_5H_9 showed no evidence of fine structure[91], whereas the 220 MHz ^{1}H n.m.r. spectrum displayed a poorly resolved septet ($J = 35$ Hz) due to coupling of two borons to each of the equivalent basal bridge protons[101].

3.3.5.2 Adducts and substituted derivatives of pentaborane(9)

Although B_5H_9 reacts with trimethylphosphine to form the adduct $B_5H_9 \cdot 2$ PMe_3, in which one PMe_3 may be attached to an apical boron and the other to a basal boron[98], it is cleaved by Me_2NPF_2 to yield $B_2H_4(PF_2NMe_2)_2$ and a proposed adduct of triborane(5), $B_3H_5(PF_2NMe_2)_2$ [69], and also by Me_2S to yield $Me_2S \cdot BH_3$ and a resin $(Me_2S \cdot B_2H_2)_{2n}$ [99].

Recently, techniques have been described for the elucidation of the n.m.r. spectra at 70 MHz) and decoupling accessories (capable of multiple-resonance application to many of the boron hydrides, and their derivatives[91, 109, 110]; indeed, a renaissance in the use of n.m.r. spectroscopy for structural determination appears likely as access to more sophisticated spectrometers (such as 220 MHz ^{1}H n.m.r. machines, which are adaptable to obtain ^{11}B n.m.r. spectra at 70 MHz) and decoupling accessories (capable of multiple resonance decoupling) becomes more generally available. Much of the pioneering work in this area has been carried out by Onak and co-workers[100, 101], who have totally assigned the ^{1}H and ^{11}B n.m.r. spectra of a large number of new and previously known mono-, di-, and tri-substituted derivatives of pentaborane(9) by means of selective and total decoupling of both 100 MHz ^{1}H and 32 MHz ^{11}B n.m.r. spectra. High-resolution 220 MHz ^{1}H n.m.r. spectra were also obtained, which allowed geometrical conclusions to be drawn, but did not, of themselves, enable assignments to be made when several equal-area resonances were present. From the wealth of n.m.r. data available, it was found that substituent effects on the chemical shifts of contiguous, neighbouring, and *trans* borons were additive for polysubstituted pentaboranes, and that an additivity relationship also existed for the chemical shifts of bridge hydrogens in differing environments; long-range coupling ($J = 6$–7 Hz) was generally observed between the apex boron and the attached methyl hydrogens of the 1-methyl derivatives[101]. This type of coupling, which had hitherto been reported for diborane[102], and for amine adducts of boron halides[51], was not perceptible in base- methyl-substituted pentaborane derivatives, perhaps because of quadrupole broadening[101]. In the basal-substituted 2-$(BrSiH_2)B_5H_8$, the splitting of the SiH_2 resonance observed at ambient temperature was attributed to coupling with the 2-boron atom, and the temperature dependence of the resonance to quadrupolar relaxation[103].

Bridge-substituted derivatives of pentaborane(9), μ-$R_3MB_5H_8$ (R = H, Me, Et; M = Si, Ge; and R = Me; M = Sn, Pb), which isomerised in the presence of ethers to the corresponding terminal 2-substituted derivatives, were reported in 1967 by Gaines and Iorns who, on the basis of ^{1}H and ^{11}B n.m.r. spectral data, proposed a structure in which the R_3M moiety was bonded to two adjacent basal boron atoms[104]. A single crystal x-ray investigation of 1-Br-μ-$Me_3SiB_5H_7$ by Dahl[105], as yet unpublished, has verified this proposal and shown that the methyl groups of the bridging Me_3Si moiety are tetrahedrally arranged about the silicon atom, which can thus be considered to be sp^3-hybridised and bound to two basal borons by a three-centre–two-electron bond. The recently-reported[103] halosilyl derivatives, $\mu\text{-}(ClSiH_2)B_5H_8$, $2\text{-}(ClSiH_2)B_5H_8$, and $2\text{-}(BrSiH_2)B_5H_8$, probably have structures analogous to those of the corresponding silyl derivatives[104].

The bridge-substituted derivatives, $\mu\text{-}Me_2P\cdot B_5H_8$ and $\mu\text{-}MeCF_3P\cdot B_5H_8$ (of which there are two isomers), have been isolated by Burg and Heinen from the reaction of LiB_5H_8 with Me_2PCl, and $MeCF_3PCl$, respectively; confirmation that the Me_2P or $MeCF_3P$ moiety was bridging two basal borons was obtained from n.m.r. spectroscopy[106]. Recently, on the basis of ring current correlations, Onak has tentatively assigned the less-volatile isomer of $\mu\text{-}MeCF_3P\cdot B_5H_8$, A, a configuration in which the methyl group is in an axial position below the B_4 basal plane whilst the CF_3 group is equatorial and almost coplanar with the basal borons, and the more-volatile isomer, B, a configuration in which the methyl is equatorial and the CF_3 group axial[96]. In contrast to Me_2PCl and $MeCF_3PCl$, the less basic $(CF_3)_2PCl$ reacted with LiB_5H_8 to yield the *apex*-substituted $1\text{-}(CF_3)_2P\cdot B_5H_8$ [106]; interestingly, this terminally-substituted pentaborane(9) derivative reacted as a phosphine ligand with $Ni(CO)_4$, readily forming $[1\text{-}(CF_3)_2P\cdot B_5H_8]Ni(CO)_3$, whereas the bridge-substituted $\mu\text{-}MeCF_3P\cdot B_5H_8$ did not react with $Ni(CO)_4$ even under forcing conditions[107].

The compound μ-dimethylborylpentaborane(9), $\mu\text{-}Me_2B\cdot B_5H_8$, in which an Me_2B moiety appears to bridge two basal borons, has been prepared from Me_2BCl and LiB_5H_8, and has been found to isomerise in the presence of ether to 4,5-dimethylhexaborane(10); hence, this Me_2B-bridged derivative may be considered to be a stable intermediate in the formation of a B_6 pentagonal-pyramid from a B_5 tetragonal-pyramid, via a boron-insertion reaction[108].

3.3.5.3 Pentaborane(11)

The topological representation and structure[5] of pentaborane(11) are given in Figures 3.4(a) and 3.4(b) respectively. The ^1H n.m.r. spectrum of B_5H_{11}

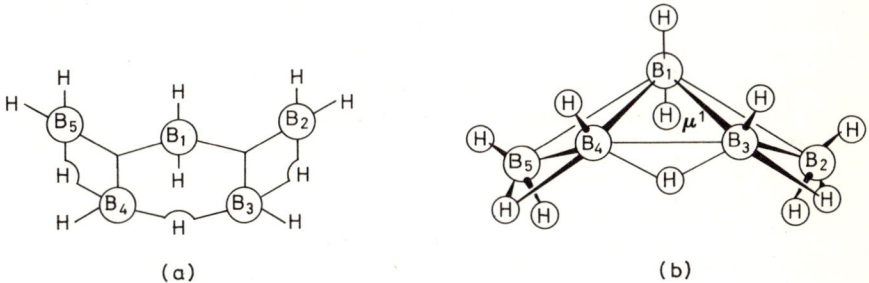

Figure 3.4 B_5H_{11} (a) Topological representation (b) Molecular structure, showing 'anomalous' hydrogen (μ^1)

has been totally assigned, and the resonance of the 'anomalous' apex hydrogen (μ^1 in Figure 3.4(b)), which lies on the symmetry plane of B_5H_{11}, has been located for the first time. The chemical shift of this hydrogen, which is unusually low not only in comparison with other bridge hydrogens in B_5H_{11} (or in other boron hydrides) but also with the apically-bonded terminal hydrogen in B_5H_{11} (or in other pyramidal boron hydrides), suggested that the anomalous hydrogen might have a greater amount of protonic character than most bridge (or terminal apical) hydrogens[91, 109, 110]. The coupling constant ($J = 55$ Hz) was somewhat higher than that found for bridge hydrogens in boron hydrides (30–40 Hz), but considerably lower than that for apical terminal hydrogens (170–180 Hz), and, in this respect, Lipscomb's proposal[23] that the anomalous hydrogen might show bonding properties intermediate between those of a bridge and a terminal hydrogen is of interest.

A molecular-beam mass spectral study of the low-pressure pyrolysis of B_5H_{11} showed that it dissociated into B_4H_8 and BH_3; although B_2H_6 and B_5H_9 were also observed, no triborane species or higher boron hydride could be detected[88].

3.3.6 Six boron atoms

3.3.6.1 Hexaborane(10) and its adducts

Two new syntheses have been described for B_6H_{10}, which had hitherto been difficult to prepare in reasonable yield[14]; Dobson and Schaeffer[111] have reported almost quantitative yields of B_6H_{10} from the limited hydrolysis of B_8H_{12}, itself obtainable via degradation of the more accessible $B_{10}H_{14}$, and Shore and co-workers[112, 113] have prepared B_6H_{10} from the reaction of diborane with LiB_5H_8.

The topology and structure[5] of hexaborane(10) are given in Figures 3.5(a) and 3.5(b) respectively. The determination of the crystal structure of B_6H_{10} (Figure 3.5(b)) has shown that there are four different types of boron atoms in the molecule—two are unique and situated on the mirror plane (1, 2), and two pairs (3 and 6; 4 and 5) are related by the mirror plane—and additionally four different types of terminal hydrogen and two types of bridge hydrogen[5]. It is hence not a little surprising that the ^{11}B n.m.r. spectrum should consist of one low-field doublet (intensity 5) and one high-field doublet (intensity 1), suggesting the presence of only two types of boron, basal and apical, respectively, and also that the 220 MHz ^1H n.m.r. spectrum and the decoupled 100 MHz ^1H n.m.r. spectrum should indicate only three types of hydrogen—terminal basal, terminal apical, and bridge basal[91]. Two possible explanations for this inconsistency are accidental degeneracy and bridge-hydrogen tautomerism; the former is discounted by the sharpness of the single resonances in the decoupled ^1H n.m.r. spectrum[91], whereas the latter is credible only if it operates by a mechanism which preserves the integrity of all five basal terminal hydrogens. A mechanism of this kind, in which *intermolecular* exchange of hydrogen occurs at the short boron–boron bond, and rapid *intra*molecular migration of the bridge hydrogens results in an equal probability of this bond being adjacent to any basal boron, has recently been proposed by Carter and Mock[114]; moreover, in a comprehensive study of hexaborane(10), which was specifically substituted with deuterium at either the bridge or the basal terminal positions, Odom and Schaeffer

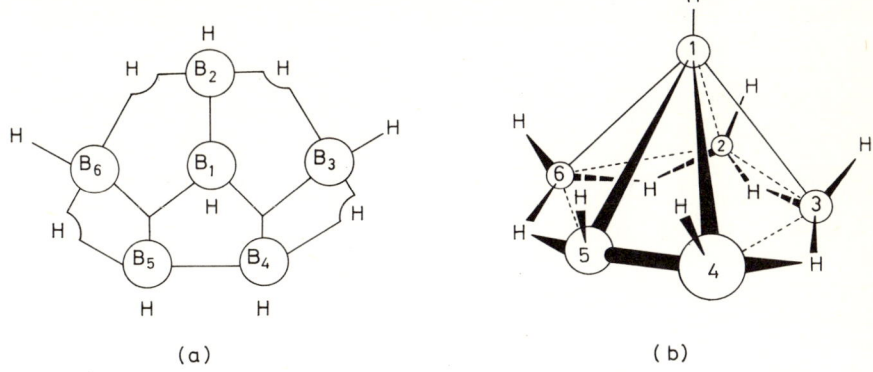

Figure 3.5 B_6H_{10} (a) Topological representation (b) Molecular structure

found that hexaborane(10) undergoes an *inter*molecular exchange at room temperature involving only the basal terminal hydrogens[115].

The broadening and collapse of the low-field doublet in the ^{11}B n.m.r. spectrum of B_6H_{10} at low temperatures without a corresponding collapse of the high-field doublet has been attributed to the shorter ^{11}B relaxation time of the basal borons[115].

Adducts of composition $B_6H_{10} \cdot L_2$ (where $L = NMe_3$, PMe_3, and PPh_3) have recently been isolated from reactions of B_6H_{10} with the appropriate Lewis base[116], and $B_6H_{10} \cdot (PF_2NMe_2)$ has been shown to be a product of the reaction of $PF_2 \cdot NMe_2$ with pentaborane(9)[69].

3.3.6.2 *Hexaborane(12)*

Determination of the structure of hexaborane(12) by x-ray crystallography has not so far proved feasible, but recent 220 MHz ^1H and 70 MHz ^{11}B n.m.r. studies[91] tend to confirm the icosahedral fragment structure Figure 3.6(b) originally proposed by Gaines and Schaeffer[117], which is represented by a 4212 topology Figure 3.6(a). It was evident from the ^1H n.m.r. spectra that there are two non-equivalent sets of hydrogens in the two BH_2 groups (1 and 4); it can be seen from Figure 3.6(b) that the 'axial' set consist of one hydrogen on B_1 and one on B_4 which protrude towards the centre of the icosahedral fragment, and the 'equatorial' set comprises of one hydrogen on B_1 and one on B_4 which are located away from the centre[91]. These BH_2- hydrogens alone are exchanged for deuteriums when hexaborane(12) reacts with deuteriodiborane at −30 °C, to form 1,1,4,4-tetradeuteriohexaborane (12), but complete deuteration of hexaborane(12) occurs on reaction at room temperature[118].

Hexaborane(12), which is moderately stable in the liquid phase at room temperature and in the gas phase at higher temperatures[118], is, however, classified as an 'unstable' boron hydride; this classification has been verified

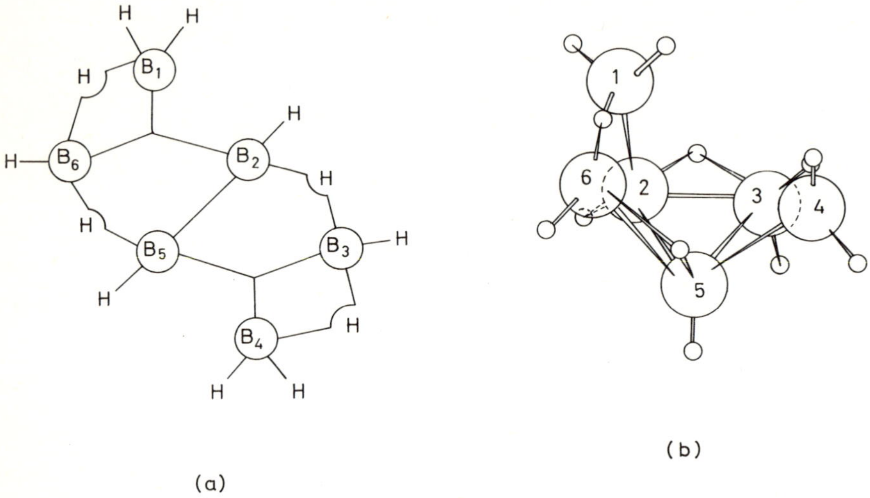

Figure 3.6 B_6H_{12} (a) Topological representation (b) Proposed structure; large circles represent borons, small circles represent hydrogens
(Reprinted from *Inorganic Chemistry* **9**, 2193 (1970) copyright by the American Chemical Society. Reprinted by permission of the copyright owner).

by the determination of the molecular-beam mass spectrum of B_6H_{12} and application of the three criteria of boron hydride stability which were discussed in Section 3.3.4.1[87].

3.3.7 Seven boron atoms

The reaction of diborane(6) and pentaborane(9) at elevated temperatures has been critically discussed by Long[119], and studied experimentally by Schaeffer[120] and co-workers; in the experimental approach, the co-pyrolysis of these two hydrides was carried out in a circulating pyrolysis system consisting of a hot zone immediately followed by a cold quench, and an impressive number of higher boron hydrides, several of which were new compounds, could thus be trapped out. One of these new compounds appeared to be a B_6 or B_7 hydride, on the basis of its volatility and mass spectrum; the instability of the compound, and its volatility and solubility which were almost identical with those of B_6H_{10}, precluded its purification and characterisation[120]. No other current reports of heptaboranes have appeared.

3.3.8 Eight boron atoms

3.3.8.1 Octaborane(12)

A new laboratory-scale preparation of octaborane(12)[5] from the decomposition of iso-B_9H_{15}, itself obtainable via degradation of the more readily accessible $B_{10}H_{14}$, has been devised by Schaeffer and co-workers[121]; it is experimentally unfortunate that B_8H_{12} should be thermally unstable above $-20\ °C$, despite its classification, on the basis of its molecular-beam mass spectrum, as a 'stable' boron hydride[87].

A study of the chemistry of B_8H_{12} has shown that it is a strong monobasic Lewis acid forming complexes of the type $B_8H_{12}\cdot L$ (where L is Et_2O, Me_3N, or MeCN, and may be bonded to the 4- boron), and that it can be converted in almost quantitative yield to hexaborane(10) by limited hydrolysis, and in good yield to n-B_9H_{15} by reaction with diborane[111].

3.3.8.2 Octaborane(14)

Small amounts of octaborane(14) have been prepared by reaction of a compound of approximate empirical composition $Me_4NB_8H_{12}$ (obtained via reaction of B_8H_{12} with either sodium hydride or sodium amalgam) with hydrogen chloride at $-78\ °C$; unfortunately, B_8H_{14} is unstable above $-30\ °C$ and its structure is unknown[111].

3.3.8.3 Octaborane(16)

The new compound, octaborane(16), although formed along with B_8H_{12} and other as yet unidentified B_8 hydrides in the co-pyrolysis of diborane and pentaborane(9), could be fairly readily purified and characterised; at present its structure is unknown[120].

3.3.8.4 Octaborane(18)

Two structures have been proposed for octaborane(18), one consists of two tetraborane fragments joined by a direct boron–boron bond at the 2,2' positions, whereas the other is a belt-line icosahedral fragment; both structures were compatible with the n.m.r. data originally available[122]. Now, however, the situation appears to have been resolved: measurement of the molecular-beam mass spectrum of B_8H_{18}, and the appearance potential of the $B_4H_9^+$ ion[123], strongly support the former structure, as does a comparison of the 220 MHz ^1H n.m.r. spectrum of B_8H_{18} with that of B_4H_{10} [110]; moreover, Lipscomb has stated that the latter structure is inconsistent with an unpublished x-ray diffraction study[10].

3.3.9 Nine boron atoms

3.3.9.1 Normal nonaborane(15)[5]

The reaction of B_8H_{12} with B_2H_6 constitutes a new method of preparation of n-B_9H_{15}[111]; an analogous reaction with $^{10}B_2H_6$ yielded n-$^{10}B^nB_8H_{15}$ and $^{10}B_2{}^nB_8H_{14}$, but B_8H_{12} and B_2D_6 gave non-specifically deuterated n-nonaborane(15) and decaborane(14)[124]. Reaction of n-B_9H_{15} with B_2D_6 produced n-nonaboranes with degrees of deuterium substitution which depended on the reaction period[124]; information on the positions of the deuteriums in these compounds may be obtainable from their ^{11}B n.m.r. spectra now that the 64 MHz ^{11}B n.m.r. spectrum of $^nB_9H_{15}$ itself has been tentatively assigned[125].

3.3.9.2 Iso-nonaborane(15) and adducts of nonaborane(13)

The reaction of hydrogen chloride with KB_9H_{14} at $-80\,^\circ C$ yields the new compound iso-nonaborane(15), an isomer of n-nonaborane(15)[121]; analogous reactions with specifically deuterated derivatives of KB_9H_{14} have led to an assignment of the ^{11}B n.m.r. spectrum of iso-B_9H_{15} in terms of C_{3v}-symmetry of the boron skeleton[126].

Above $-35\,^\circ C$, iso-B_9H_{15} decomposes with evolution of hydrogen to form B_8H_{12}, $B_{10}H_{14}$, and n-$B_{18}H_{22}$, and it seems probable that the reactive

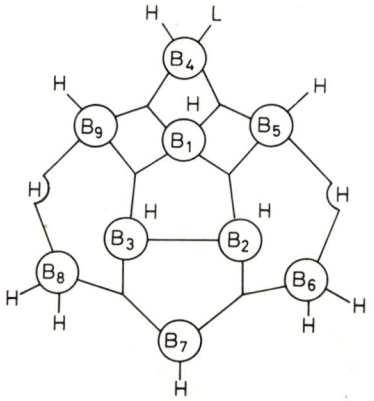

Figure 3.7 Topological representation of $B_9H_{13}L$

intermediate B_9H_{13}, formed by initial hydrogen elimination, either disproportionates or condenses to generate B_8H_{12} and $B_{10}H_{14}$, or ñ-$B_{18}H_{22}$, respectively. Such a mechanism is supported by the fact that either KB_9H_{14} or $Me_4NB_9H_{12}$ reacts with hydrogen chloride in diethyl ether at room temperature to form $B_9H_{13}\cdot OEt_2$, from which other B_9H_{13}(ligand) compounds may be obtained by ligand displacement[121]. These observations are of special interest since Lipscomb had previously considered the B_9H_{13}

(ligand) compounds as derived from a hypothetical B_9H_{15} molecule (of C_{3v} symmetry and 6330 topology), the existence of which was discounted on the basis of a rule of the topological theory relating to overcrowding of hydrogen atoms[5]. If iso-B_9H_{15} did have the 6330 topology (and at present there is no evidence either to discount or support this), the limited, but moderate, stability of the compound would serve as a measure of the importance which should be attached to those steric limitations imposed by the topological theory[121].

The alcoholysis of $B_{10}H_{12}L_2$ is an established route to $B_9H_{13}L$ compounds (where e.g., L = Me_2S); however, methanolysis of 5-Br-6,9-$(Me_2S)_2$-$B_{10}H_{11}$ did not afford the expected bromo-substituted derivative, 6-Br-4-$Me_2SB_9H_{12}$, but instead yielded the corresponding methoxy-derivative 6-MeO-4-Me_2 SB_9H_{12}, which could also be obtained by bromination of 4-$Me_2SB_9H_{13}$ and subsequent methanolysis; these reactions were reported as the first examples of direct nucleophilic replacement of a halogen by an alkoxy group in the higher boron hydrides[127]. It now appears, however, that a rearrangement *also* takes place, and the compound previously formulated as 6-MeO-4$Me_2SB_9H_{12}$ is, in fact, 7-MeO-4-$Me_2SB_9H_{12}$; as expected, this compound may also be prepared by methanolysis of 2-Br-6,9-$(Me_2S)_2B_{10}H_{11}$ [128]. (Note: the numbering system used in Figure 3.7 is in accordance with the A.C.S. recommendations[1], and not with that in the original papers [127, 128].)

3.3.10 Ten boron atoms

3.3.10.1 *Decaborane(14)*

The topological representation and structure[5] of decaborane(14) are given in Figures 3.8(a) and 3.8(b) respectively. A recent neutron diffraction study of $^{11}B_{10}D_{14}$ at $-160\,°C$ has confirmed the previously accepted structure and determined the deuterium (hydrogen) atom positions with great accuracy; the bridging deuteriums (hydrogens) were again found to be in asymmetric positions with respect to the bonding borons[129]. The distribution of bonding electrons in $^{11}B_{10}D_{14}$ has been determined from neutron and x-ray diffraction data. It was found that there were strong local accumulations of bonding electrons in normal, covalent, terminal B—D(B—H) bonds, but that there was a maximum of electron density close to the deuterium (hydrogen) atom in B—D—B (B—H—B) bridge bonds, which therefore cannot be characteristic open three-centre bonds; the electron distribution in the boron skeleton corresponded approximately to that postulated by Lipscomb[5], but only slight evidence was found for central three-centre bonds[130]. Two decaboranes, labelled with a boron isotope in specific positions, $^{10}B^nB_9H_{14}$ and $^{11}B^{10}B_9H_{14}$, have been prepared by Schaeffer and co-workers, who established, from chemical and spectral evidence, that the label was on the 6,9 and 5,7,8,10 positions[131].

Although the ^{11}B n.m.r. spectrum of $B_{10}H_{14}$ was totally assigned in 1965[132, 133], the ^1H n.m.r. spectrum has only recently been unambiguously assigned by reference to the 220 MHz ^{1}H n.m.r. spectra of 1,2,3,4-$B_{10}H_{10}D_4$, 2-Br$B_{10}H_{13}$, and $B_{10}H_{14}$ [134]. The order of the chemical shifts in the ^{1}H n.m.r. spectrum parallels that in the ^{11}B n.m.r. spectrum, with the exception of a small inversion of the 1,3 and 6,9 positions[134], and is in agreement with the order predicted by Onak from calculations which take into account both ring current contributions and positional charge differences in the molecule[135]; in this latter approach, the $B_{10}H_{14}$ icosahedral fragment is considered as two fused pentagonal-pyramids (with common atoms in the base of each pyramid) which are then treated in a manner analogous to that described for the conical ring current method as discussed in Section 3.3.5.1[96].

Czechoslovak workers have recently reported the preparation[136] and purification[137], by column chromatography, of a number of previously

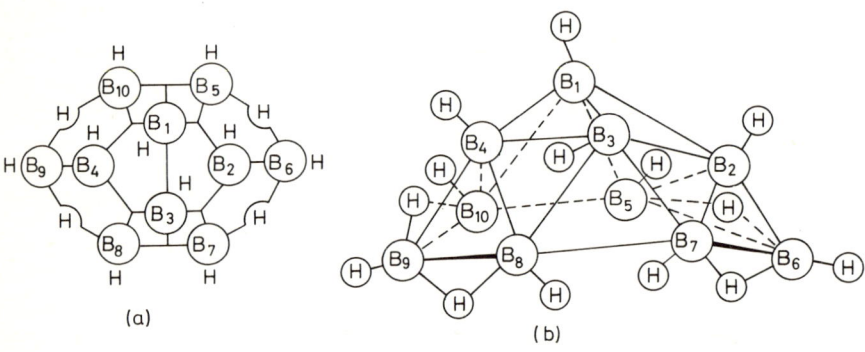

Figure 3.8 $B_{10}H_{14}$ (a) Topological representation (b) Molecular structure

unknown, or difficulty isolatable, mono-halogen substituted derivatives of decaborane; evidence for the position of substitution in the particularly interesting, but previously inaccessible, 5- and 6-halo derivatives[136] was obtained from their ^{11}B n.m.r. spectra[138] and chemical behaviour. It is apparent (from Figure 3.8(b)) that a 5-halo-decaborane, unlike an analogous 1-, 2-, or 6-substituted halo-derivative, should be resolvable into a pair of enantiomers; similarly, the corresponding 5-halo-6,9-bis-(ligand)·$B_{10}H_{11}$ derivative, which has an almost identical boron skeleton to $B_{10}H_{14}$, should also be resolvable and this has been realised in practice by the recent isolation of the two enantiomers (−) and (+)-5-Br-6,9-(Me$_2$S)$_2$·$B_{10}H_{11}$, at present the only known optical isomers in the decaborane series[139].

The use of ^{11}B n.m.r. spectroscopy has shown that 'ethoxydecaborane' (obtained by reaction of Na$B_{10}H_{13}$ with I_2 in Et_2O) is probably a 5-substituted derivative[140], and that the product from the sulphuric acid oxidation of $B_{10}H_{12}L_2$ (L = Me$_2$S, Et$_2$S) is 6,6'-($B_{10}H_{13}$)$_2$O [138], the only known oxygenated derivative of $B_{10}H_{14}$ [141]; this latter compound may be related to the recently reported ($B_{10}H_{13}$)$_3$P·Et_2O and ($B_{10}H_{13}$)POCl$_2$, in which, however, the positions of substitution are unknown[142].

3.3.10.2 Decaborane(16) and iso-decaborane(16)

Decaborane(16) has been known since 1961; its structure[5] consists of two B_5H_8 fragments joined at the 1,1' (apical) positions by a single bond. 'Iso-decaborane(16)' is, however, a new compound, isolated by Schaeffer and co-workers from the co-pyrolysis of pentaborane(9) and diborane; it appears to be quite stable thermally and does not react rapidly with air[120]. A proposed structure for iso-$B_{10}H_{16}$, in which two B_5H_8 units are joined at the 2,2' (basal) positions by a single bond, is supported by mass-spectral fragmentation data[120], the 70 MHz ^{11}B n.m.r. spectrum[110], and especially by the virtual identity of the 220 MHz ^1H n.m.r. spectrum with that of B_5H_9[110]; the results of a single crystal x-ray study, now in progress[120], are eagerly awaited.

3.3.10.3 Decaborane(18)

This new compound, which, unlike iso-$B_{10}H_{16}$, is unstable at room temperature, was isolated from the co-pyrolysis of pentaborane(9) and diborane; its structure is unknown[120].

3.3.11 Sixteen boron atoms: hexadecaborane(20)

The pyrolysis of $B_9H_{13}\cdot SMe_2$, *in vacuo*, produces the known hydrides decaborane(14), $B_{10}H_{14}$, and n-octadecaborane(22), n-$B_{18}H_{22}$, in moderate yield and also a small amount of the new boron hydride, hexadecaborane(20), $B_{16}H_{20}$; unfortunately, this low-yield (*c.* 7%) process constitutes the only known synthesis of this novel compound[143]. A mechanism for the pyrolysis has been proposed in which the initially-generated intermediate B_9H_{13} may dimerise to $B_{18}H_{26}$, which cleaves to form $B_{10}H_{14}$ and B_8H_{12} (isolation of the latter being precluded by its thermal instability[111]), may lose hydrogen to form B_9H_{11}, which then condenses to $B_{18}H_{22}$, or may lose a BH_3 group to form B_8H_{10}, which perhaps dimerises to $B_{16}H_{20}$ [143, 144]; similar mechanisms were put forward by Schaeffer and co-workers[121] to rationalise the production of $B_{10}H_{14}$, B_8H_{12}, and n-$B_{18}H_{22}$ from the decomposition of iso-B_9H_{15} (see Section 3.3.9.2).

The determination of the structure of $B_{16}H_{20}$ (Figure 3.9) by x-ray diffraction techniques has demonstrated that $B_{16}H_{20}$ is a member, along with n- and iso-$B_{18}H_{22}$[5], of that presently-select class of boron hydride molecules which can be regarded as fusions of simpler icosahedral fragment boranes; however, unlike n- and iso- $B_{18}H_{22}$, which are fusions of two identical B_{10} fragments, $B_{16}H_{20}$ represents the unsymmetrical fusion of a B_{10} icosahedral fragment with a B_8 icosahedral fragment (in which the component fragments open in opposite directions) and has *no* molecular symmetry, thus differing from all other neutral boron hydrides of known structure[144]. It is apparent that numerous such fused icosahedral fragment boron hydrides may be capable of existence; formal generation of such species, from two simple known icosahedral fragment boranes, by removal of one bridge and two terminal hydrogens from each component fragment, and reduction of the sum of boron atoms from the components by 2, leads to the conclusion that 123 fused-fragment boron hydrides (235 including enantiomers) may exist[144]. The attempted resolution of chiral $B_{16}H_{20}$ into its enantiomers unfortunately

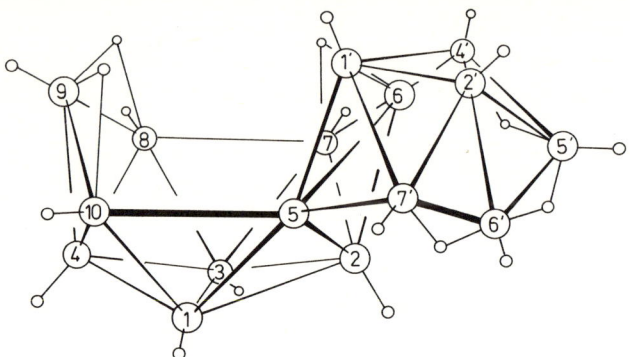

Figure 3.9 Molecular structure of $B_{16}H_{20}$; large circles represent borons, small circles represent hydrogens

failed, probably because of its sensitivity towards bases (see below); however, chiral iso-$B_{18}H_{22}$ *has* been separated into enantiomers (see Section 3.3.12)[145].

The cleavage of $B_{16}H_{20}$ by Lewis bases L (L = Me_2S, Ph_3P), to yield firstly $B_{10}H_{14}$ and then $B_{10}H_{12}L_2$, as well as other unidentified products (B_6 adducts?), is consistent with its established structure; also, like the structurally-related compounds $B_{10}H_{14}$, and n- and iso-$B_{18}H_{22}$, $B_{16}H_{20}$ is a strong acid, although it differs from $B_{18}H_{22}$ by being unstable in aqueous media[143].

3.3.12 Eighteen boron atoms: octadecaborane(22) and iso-octadecaborane(22)

The isomeric boron hydrides n- and iso-octadecaborane(22) were prepared in 1962 by hydrolysis of the hydronium ion salt of $B_{20}H_{18}^{2-}$, and separated by crystallisation, or by crystallisation of their diquaternary salts and subsequent regeneration[146, 147]; recently, a more efficient and higher yield separation of n- and iso-$B_{18}H_{22}$, using column chromatography on silica gel, has been described[145]. Three other laboratory-scale procedures, which yield moderate amounts of n-$B_{18}H_{22}$, have been reported — the thermal decomposition of $B_9H_{13}·L$ *in vacuo* (L = Me_2S)[143], or in refluxing di-n-butyl ether (L = n-Bu_2O)[121], and the protonolysis of the $B_{11}H_{14}^-$ ion, which is readily obtainable from $B_{10}H_{14}$[148].

Iso-$B_{18}H_{22}$, which possesses a chiral structure[5], has recently been separated into its two enantiomers, and thus provides the first example of optical activity in an unsubstituted boron hydride[145].

The chemistry of n- and iso-$B_{18}H_{22}$ has been quite extensively investigated by Hawthorne and co-workers, who found that both hydrides were strong diprotic acids, forming the n- and iso-$B_{18}H_{20}^{2-}$ ions with aqueous alkali, and also with sodium hydride or butyl lithium, in anhydrous solvents; moreover, unlike $B_{10}H_{14}$ and $B_{16}H_{20}$, neither n- nor iso-$B_{18}H_{22}$ reacted with Lewis bases[147].

3.4 ANIONIC BORON HYDRIDES AND THEIR DERIVATIVES

3.4.1 One boron atom: anionic adducts of borane

Lithium cyanotrihydroborate, $LiBH_3CN$, although isolated by Wittig[149] in 1951, was comparatively ignored until a recent report of its utility[150] — enhanced by its resistance to acid hydrolysis[151] — as a selective reducing agent in organic synthesis; now, somewhat more convenient preparations of $LiBH_3CN \cdot dioxan$, $NaBH_3CN$, and KBH_3CN have been described[152, 153]. Sodium isocyanotrihydroborate, $NaBH_3NC$, is produced, along with $NaBH_3CN$, in the reaction of $NaBH_4$ with HCN in T.H.F. *below* room temperature, but has not been isolated in pure form (under reflux conditions, $NaBH_3CN$ is the only product); its isomerisation to $NaBH_3CN$ is acid-catalysed and further accelerated by the presence of cyanide ions[153]. Unlike $NaBH_3CN$, $NaBH_3NC$ is unstable to aqueous acid. Full infrared, Raman, and n.m.r. spectroscopic data have been obtained for these cyano- and isocyanotrihydroborates and their corresponding deuterated derivatives, and have been used to make qualitative comparisons of their bonding with that of other borane adducts[152, 153]. The recently-reported transition metal cyanotrihydroborate complexes $(Ph_3P)_3M(NCBH_3)$ (M = Cu, Ag) and $(en)_2Ni(NCBH_3)_2$ (en = ethylenediamine) were assigned structures in which the cyanotrihydroborate ion is bonded to the metal through the nitrogen atom, but the possibility of M—$CNBH_3$ linkages, the formation of which would require a structural rearrangement[153], could not be excluded[154]; such M—$CNBH_3$ linkages are known to be present in the related complex (o-phen)$_2$Fe(CNBH$_3$)$_2$ [155].

The new anions $Me_2N(BH_3)_2^-$, $H_2P(BH_3)_2^-$, $PhHP(BH_3)_2^-$, and $HS(BH_3)_2^-$ are structurally related. The anion $Me_2N(BH_3)_2^-$, the first derivative of the unknown $H_2N(BH_3)_2^-$, is prepared by reaction of sodium hydride with μ-$Me_2N \cdot B_2H_5$, rather than from sodium tetrahydroborate and μ-$Me_2N \cdot B_2H_5$, which exist in equilibrium with $B_2H_7^-$ and $Me_2N(BH_3)_2^-$; it reacts with diborane to yield μ-$Me_2NB_2H_5$ and $B_2H_7^-$ [156]. Reaction of either phosphine-borane or phosphonium iodide with sodium tetrahydroborate yields $H_2P(BH_3)_2^-$, and phenylphosphine-borane with sodium tetrahydroborate yields $PhHP(BH_3)_2^-$ [157]; detailed studies of $H_2P(BH_3)_2^-$, and its deuterated analogues, by infrared, Raman, and n.m.r. spectroscopy have recently been carried out[158, 159]. Reaction of tetraethylammonium tetrahydroborate with liquid hydrogen sulphide produces the new compound [Et_4N][$HS(BH_3)$], a solution of which in liquid hydrogen sulphide can absorb diborane to yield the thermally unstable $[Et_4N][HS(BH_3)_2]$ [160], also obtainable from $[Et_4N][SH]$ and diborane in the same solvent[83]. The $HS(BH_3)_2^-$ anion reacts with anhydrous hydrogen chloride to form μ-mercaptodiborane, $\mu\text{-}SHB_2H_5$ (see Section 3.3.2.3)[83].

3.4.2 Three boron atoms: the octahydrotriborate(1−) ion

A new convenient high-yield synthesis of the octahydrotriborate(1−) ion, from sodium tetrahydroborate and iodine, has been described[161].

It is well known that the ^{11}B n.m.r. spectrum of $B_3H_8^-$ consists of a symmetrical nine-line multiplet, and that the ^1H n.m.r. spectrum consists of a symmetrical ten-line multiplet; these observations, which suggest that all three borons couple equivalently to all eight hydrogens, and *vice versa*, are at variance with the known structure[5] of the $B_3H_8^-$ ion, and are attributed to a rapid intramolecular hydrogen tautomerism[10]. Attempts to slow, or stop, this tautomerism have led to studies of the temperature dependence of the n.m.r. spectra of $B_3H_8^-$; it has been found that both spectra are temperature dependent, and in both cases ^{11}B—H coupling disappears at low temperatures. The ^{11}B resonance *broadens* (with consequent loss of fine structure) as the temperature is lowered, as might be expected on the basis of the decreased relaxation time of the boron nuclei[162] (cf., B_9H_{10}, Section 3.3.6.1; $B_5H_8^-$, Section 3.4.4; $B_6H_9^-$, Section 3.4.5). The ^1H multiplet collapses to a singlet at low temperatures, which continues to *sharpen* until about −90 °C (and then very slowly broadens); such behaviour is attributed to more effective boron quadrupole-induced relaxation at low temperatures, effectively decoupling boron from hydrogen[162, 163]. Lipscomb, in anticipation of such experimental observations, had previously predicted that when hydrogen is coupled to boron the ^1H resonance will become a single sharp line in the extreme of very fast relaxation of the boron quadrupolar nucleus[10]. The observation of only one singlet in the ^1H n.m.r. spectrum of $B_3H_8^-$, even at very low temperatures (−135 °C), implies that the protons are still equivalent and undergoing rapid exchange, and leads to an upper estimate (c. 8 kcal mol^{-1}) for the free energy of activation for the exchange process[162]. Partial ^{11}B—H decoupling of the n.m.r. spectra of $B_3H_8^-$ can be effected by addition of the paramagnetic manganese(II) ion at room temperature[162].

3.4.3 Four boron atoms

3.4.3.1 The heptahydrotetraborate (1−) ion

The reaction of ammonia gas with pentaborane(9) in diethyl ether at low temperatures produces the thermally-unstable $[(NH_3)_2 BH_2][B_4H_7]$, a salt of the previously unknown anion $B_4H_7^-$; the new compound was characterised by its reaction with excess hydrogen chloride to give $(NH_3)_2BH_2^+ Cl^-$, and with an equimolar amount of the same reagent to yield $(NH_3)_2BH_2^+ Cl^-$, and the presumed ether adduct of tetraborane(8), which could be converted to the known[90] $(Me_2NPF_2)B_4H_8$ by reaction with Me_2NPF_2 [164].

3.4.3.2 The nonahydrotetraborate (1–) ion

Two competing reactions occur between tetraborane(10) and ammonia in diethyl ether and, depending on the conditions, either $NH_4B_4H_9$, a salt of the new anion $B_4H_9^-$, or $[(NH_3)_2BH_2][B_3H_8]$ is the main product[116]; another route to this new anion is the reaction of pentaborane(11) with ammonia which yields $[(NH_3)_2BH_2][B_4H_9]$ [164]. (See Section 3.7.)

3.4.4 Five boron atoms: the octahydropentaborate(1–) ion

Although Onak and co-workers established the Brønsted acidity of pentaborane(9) by showing that it could be deprotonated by lithium or sodium hydride in ether solvents at room temperature, no evidence for the presence of an octahydropentaborate(1–) ion, $B_5H_8^-$, was obtained[165]; however, Gaines and Iorns[166] found that reaction of lithium alkyls with pentaborane(9) in ether at low temperatures yielded LiB_5H_8, and Geanangel and Shore[112] reported the formation of alkali metal salts of $B_5H_8^-$ by reaction of pentaborane(9) with lithium alkyls and sodium and potassium hydrides in ether solvents at low temperatures. A tetramethylammonium salt, $Me_4NB_5H_8$, obtained by reaction of Me_4NCl with LiB_5H_8 in ether at ambient temperatures appears to contain a $B_5H_8^-$ ion, which is a structural isomer of the $B_5H_8^-$ ion present in the corresponding alkali metal salts[112,113]. Recently, full details of the preparation and characterisation of the alkali metal salts of $B_5H_8^-$ have been published; unlike LiB_5H_8, the sodium and potassium salts could be isolated as microcrystalline solids of limited thermal stability[113].

It appears that the $B_5H_8^-$ ion is generated by removal of a basal bridge proton from pentaborane(9) (Figure 3.3(b)), since reaction of LiB_5H_8 with HCl or DCl forms B_5H_9 or μ-DB_5H_8 in high yield, and hence must have a very similar boron framework to that of pentaborane(9)[166]; such a conclusion is substantiated by the ^{11}B n.m.r. spectrum of $B_5H_8^-$ which is qualitatively similar to that of B_5H_9, and consists of a high-field doublet, assigned to the apical boron, and a low-field doublet, assigned to the basal borons, rendered magnetically equivalent by a rapid intramolecular tautomerism[113]. Such a tautomerism must occur by a mechanism which involves migration of *only* bridge hydrogens around the base, and does not allow rapid bridge–terminal hydrogen exchange[113]; an analogous mechanism was proposed to rationalise the magnetic equivalence of the basal borons in the ^{11}B n.m.r. spectrum of B_6H_{10} [114] (see Section 3.3.6.1.). The ^{11}B n.m.r. spectra of the alkali metal salts of $B_5H_8^-$ are temperature dependent; the low-field doublet (basal borons) collapses to a broad singlet at low temperatures whereas the high-field doublet (apical boron) is comparatively unaffected. Such behaviour may perhaps be due to the shorter ^{11}B relaxation time of the basal borons[113] (cf., B_6H_{10}, Section 3.3.6.1).

3.4.5 Six boron atoms: the nonahydrohexaborate(1−) ion

Alkali metal salts of the new nonahydrohexaborate(1−) ion, $B_6H_9^-$, can be prepared by the deprotonation of hexaborane(10) with methyllithium and sodium and potassium hydrides in ether solvents at low temperatures; the sodium and potassium salts may be isolated as microcrystalline solids[113, 167]. In all cases, hexaborane(10) was more rapidly deprotonated than pentaborane(9), as anticipated from their relative Brønsted acidities[113] (n.m.r. studies[167] have shown that in the series B_5H_9, B_6H_{10}, $B_{10}H_{14}$ Brønsted acidity increases with increasing size of the polyhedral boron framework). Hexaborane(10) is also deprotonated by liquid ammonia, or by equimolar proportions of ammonia in ether solvents at low temperatures, with presumed formation of $NH_4B_6H_9$, which may be converted to the known solid salt $Bu_4NB_6H_9$ by metathesis[116]. The $B_6H_9^-$ ion is more stable thermally, both in solution and in the solid state, than is its analogue $B_5H_8^-$ [113, 167].

Reaction of methyl lithium with basal terminal-deuterated hexaborane(10) yields only CH_4, showing that deprotonation takes place with loss of a basal bridge hydrogen; moreover, reaction of $B_6H_9^-$ with HCl and DCl generates B_6H_{10} and $\mu\text{-}DB_6H_9$ [167]. Both these observations strongly suggest that the boron framework in the anion is the same as that in hexaborane(10) (Figure 3.5(b)), and this is further supported by the close resemblance of the ^{11}B n.m.r. spectra of $B_6H_9^-$ and B_6H_{10}. The ^{11}B n.m.r. spectrum of $B_6H_9^-$, like that of $B_5H_8^-$, consists of a high-field doublet (apical boron) and a low-field doublet (magnetically equivalent basal borons), and the mechanism invoked to explain the magnetic equivalence of the basal borons is analogous to that used for $B_5H_8^-$ (see Section 3.4.4)[113, 167]. The temperature dependence of the ^{11}B n.m.r. spectrum of $B_6H_9^-$ is analogous to that of $B_5H_8^-$ and may be similarly explained[113].

3.4.6 Nine boron atoms: the tetradecahydrononaborate(1−) ion

The tetradecahydrononaborate(1−) ion, $B_9H_{14}^-$, originally obtained in 1963 by the degradation of decaborane(14)[168], has recently been prepared by the reaction of an alkali metal tetrahydroborate with pentaborane(9), which may be regarded as a new type of specific insertion into a smaller boron cage compound[169].

The solid-state structure of $B_9H_{14}^-$, recently determined by Greenwood and co-workers from a single crystal x-ray diffraction study of CsB_9H_{14}, is related to decaborane(14) by removal of the B6(9) boron atom and the formation of BH_2 groups on the three adjacent borons ($B_9H_{14}^-$ is represented topologically by Figure 3.10)[170]; thus, the boron atom arrangement, but *not* the hydrogen atom arrangement, is in agreement with Lipscomb's prediction[5] that $B_9H_{14}^-$ would have an analogous structure to that of the isoelectronic adduct $B_9H_{13}L$ (Figure 3.7, $L = H^-$). The $B_9H_{14}^-$ ion provides yet another example of a boron hydride species whose structure in the solid state differs from that in solution because of intramolecular hydrogen exchange; in contrast to the x-ray data, the n.m.r. spectra indicate that in solution the $B_9H_{14}^-$ ion has effective C_{3v} symmetry (on the n.m.r. time-scale), with nine terminal hydrogens (each of which is coupled to one boron) and five labile

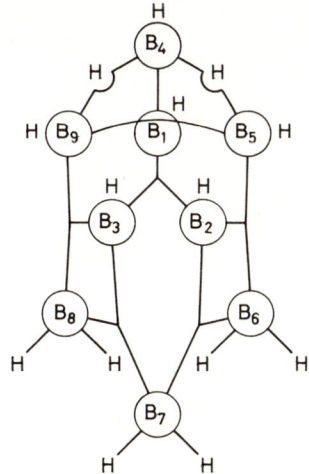

Figure 3.10 Topological representation of $B_9H_{14}^-$

hydrogens (which do *not* couple to boron)[170]. These labile hydrogens may be exchanged for deuteriums in weakly basic deuterium oxide, while in more strongly basic deuterium oxide both they and the 4, 6, and 8 terminal hydrogens are exchanged; in weakly acidic deuterium oxide only the 4, 6, and 8 terminal hydrogens are exchanged. By means of such selective deuterium labelling procedures the ^{11}B n.m.r. spectrum of $B_9H_{14}^-$ may be assigned[126].

3.4.7 Ten boron atoms: the dodecahydrodecaborate(2−) and the tridecahydrodecaborate(1−) ions

In 1958, Siegel and co-workers reported that decaborane(14) could behave as a diprotic acid towards Grignard reagents in diethyl ether with formation of the ether-solvated oil $B_{10}H_{12}(MgI)_2·2Et_2O$, which may contain the dodecahydrodecaborate(2−) ion[171]. Later work, by Greenwood and Travers, showed that decaborane behaved similarly in its reactions with magnesium, zinc, and cadmium alkyls in ether solvents with formation of the crystalline, ether-solvated compounds $MB_{10}H_{12}·2Et_2O$ (where M = Mg, Zn, Cd)[172]; almost simultaneously, Wilks and Carter reported that decaborane reacted with two moles of sodium hydride in diethyl ether, forming the unsolvated disodium salt of the dodecahydrodecaborate(2−) ion $Na_2B_{10}H_{12}$ [173].

Recently, the crystalline solvated salt $Na_2B_{10}H_{12} \cdot 2T.H.F.$ has been prepared by a similar route, and the unsolvated salt $(Me_4N)_2B_{10}H_{12}$ obtained from it by metathesis[174]. Pyrolysis of unsolvated $Na_2B_{10}H_{12}$ serves as a new method of synthesis for the $B_9H_9^{2-}$ ion, as does the pyrolysis of $Rb_2B_{10}H_{14}$ in which $Rb_2B_{10}H_{12}$ may be an intermediate[175].

Amberger and Leidl reacted decaborane(14) with lithium and sodium hydride, potassium hydroxide, and silylpotassium (SiH_3K) in ether solvents at various temperatures, and isolated a range of ether solvated and unsolvated alkali metal salts of the known tridecahydrodecaborate(1−) ion, $B_{10}H_{13}^-$; the reactivity of such salts towards Me_3SiCl, in the formation of the new compound $Me_3SiB_{10}H_{13} \cdot T.H.F.$ depended greatly on the nature of the alkali metal and of the coordinated ether, as well as on the degree of solvation[176]. The same authors also reported that 2-iododecaborane was more readily deprotonated than decaborane by alkali metal hydrides and potassium metal, with formation of alkali metal salts of the new anions $2\text{-}IB_{10}H_{12}^-$ and $2\text{-}IB_{10}H_{11}^{2-}$; although $2\text{-}IB_{10}H_{13}$ reacted with silyl-potassium to yield the $2\text{-}IB_{10}H_{12}^-$ ion, rather than $2\text{-}H_3SiB_{10}H_{13}$, the $2\text{-}IB_{10}H_{12}^-$ ion reacted with silylpotassium to form $H_3SiB_{10}H_{12}^-$ and with triphenylsilylpotassium to give $Ph_3SiB_{10}H_{12}^-$ [177].

3.4.8 Twenty boron atoms

3.4.8.1 The $B_{20}H_{19}^{3-}$ and $B_{20}H_{18}^{2-}$ ions

Electrochemical oxidation of the polyhedral ion $B_{10}H_{10}^{2-}$ in acetonitrile yields the ion $B_{20}H_{19}^{3-}$, the product of a one-electron oxidation, which can undergo a further two-electron oxidation to give $B_{20}H_{18}^{2-}$ [178]. No x-ray studies of $B_{20}H_{19}^{3-}$ have been reported, but infrared and ^{11}B n.m.r. spectral data support a structure in which an apical boron of one B_{10} polyhedron is bound, via a hydrogen bridge, to an equatorial boron of another B_{10} polyhedron[179]. An x-ray diffraction study of $(Et_3NH)_2B_{20}H_{18}$ has led to the determination of the structure of the $B_{20}H_{18}^{2-}$ ion, which consists of two B_{10} polyhedra linked by two three-centre homonuclear boron bonds[180]. U.v. irradiation of the $B_{20}H_{18}^{2-}$ ion gives a photoisomer[181], 'photo-$B_{20}H_{18}^{2-}$' in which x-ray work has shown that two B_{10} polyhedra are linked at the equatorial positions by a pair of hydrogen bridges[182].

3.4.8.2 The $B_{20}H_{18}NO^{3-}$ ion

The $B_{10}H_{10}^{2-}$ ion is oxidised by NO_2 to yield $B_{20}H_{18}NO^{3-}$ (not $B_{14}H_{12}NO^{2-}$, as previously suggested)[5]; a single crystal x-ray study[180] of the purple salt $(Et_3NH)_3B_{20}N_{18}NO$, has shown that the anion consists of two B_{10}-cages linked at the apical positions through the nitrogen atom of the NO group [although $B_{20}H_{18}NO^{3-}$ should be included in Section 3.4.7 (see Section 3.1.2), it is more appropriate to discuss it in this Section].

3.4.9 Twenty four boron atoms

3.4.9.1 The $B_{24}H_{23}^{3-}$ ion and its derivatives

Electrochemical oxidation of the polyhedral ion $B_{12}H_{12}^{2-}$ in acetonitrile yields the new anion $B_{24}H_{23}^{3-}$, the product of a one-electron oxidation of $B_{12}H_{12}^{2-}$. At present, the structure of this new anion is unknown, but it has been suggested, by analogy with the $B_{20}H_{19}^{3-}$ ion (see Section 3.4.8.1), that the two B_{12}-polyhedra are linked by a single hydrogen bridge; it appears, from its infrared spectrum, and the lack of exchange with deuterium oxide in basic conditions, that the bridging hydrogen in $B_{24}H_{23}^{3-}$ is more hydridic than that in $B_{20}H_{19}^{3-}$ [183]. There are interesting differences in the chemical behaviours of $B_{24}H_{23}^{3-}$ and $B_{20}H_{19}^{3-}$. Although $B_{24}H_{23}^{3-}$ is cleaved to $B_{12}H_{12}^{2-}$ under basic reducing conditions (sodium in liquid ammonia)[183], $B_{20}H_{19}^{3-}$ is not cleaved but instead reduced to $B_{20}H_{18}^{4-}$ [179], and such behaviour may be related to the strength of the cage–cage linkage in the B_{24} and B_{20} anions. The fact that $B_{24}H_{23}^{3-}$ is polyhalogenated without cleavage or decomposition to yield $B_{24}X_nH_{23-n}^{3-}$ (where $X = I$, $n = 2$; $X = Br$, $n = 7$, 10, 11) and $B_{24}X_nH_{22-n}^{4-}$ (where $X = Br$, $n = 11, 14, 18$; $X = Cl$, $n = 18$), whereas $B_{20}H_{19}^{3-}$ is cleaved under such conditions, may be explained by the stability of the $B_{12}H_{12}^{2-}$ ion being greater than that of the $B_{10}H_{10}^{2-}$ ion towards polyhalogenation conditions[183].

The formation of such anions as $B_{24}H_{23}^{3-}$ from $B_{12}H_{12}^{2-}$, and $B_{20}H_{19}^{3-}$ from $B_{10}H_{10}^{2-}$, may be regarded as a one-electron oxidation, followed by dimerisation, and elimination of H^+ [178, 183], and thus prompts the suggestion that such oxidative coupling reactions, at present confined to two identical polyhedral ions of the type $B_nH_n^{2-}$, might also occur between two different polyhedral ions, $B_nH_n^{2-}$ and $B_mH_m^{2-}$; moreover, perhaps coupling might be induced between polyhedral and non-polyhedral ions, as well as between different non-polyhedral ions.

3.4.9.2 The $B_{24}H_{20}I_2^{2-}$ ion, a derivative of $B_{24}H_{22}^{2-}$

Unlike its analogue, $B_{20}H_{19}^{3-}$ (see Section 3.4.8.1), the $B_{24}H_{23}^{3-}$ ion does not undergo further electrochemical oxidation, and the $B_{24}H_{22}^{2-}$ ion cannot be so obtained; however, partially-halogenated derivatives of $B_{24}H_{23}^{3-}$ are more susceptible to oxidation[183] and, in particular, $B_{24}H_{21}I_2^{3-}$ undergoes a two-electron oxidation to yield the new anion $B_{24}H_{20}I_2^{2-}$, a derivative of the unknown anion $B_{24}H_{22}^{2-}$, which may itself be regarded as an analogue of $B_{20}H_{18}^{2-}$. It has been suggested that $B_{24}H_{20}I_2^{2-}$ has a structure analogous to that of 'photo-$B_{20}H_{18}^{2-}$'[182], in which the two B_{12}-cages are linked at the equatorial positions by two hydrogen bridges[184].

3.5 METAL–BORON HYDRIDE DERIVATIVES

This section is sub-divided according to the type of bonding which is either known (from x-ray or electron-diffraction studies) or proposed (from spectroscopic and other physico-chemical studies) to exist between the metal (M) and the boron hydride moiety.

3.5.1 Open three-centre metal–hydrogen–boron bonding (M—H—B)

3.5.1.1 M—H—B: *metal tetrahydroborates*

The vapour of methylberyllium tetrahydroborate, MeBeBH$_4$, consists of a mixture of monomer and dimer, and by means of infrared investigations of normal and isotopically substituted molecules it can be shown that the monomer has a linear C..Be..B skeleton, with the carbon bound to the beryllium by a normal σ-bond, and the boron bound to the beryllium by a double hydrogen bridge, whereas the dimer has a linear B..Be..Be..B skeleton, with the two beryllium atoms bound to one another by a double methyl bridge, and to the two boron atoms by double hydrogen bridges. In solution, molecular weight and n.m.r. studies indicate that methylberyllium tetrahydroborate exists as a dimer, which has the same structure as that present in the vapour phase[185].

The structure of beryllium tetrahydroborate, 'BeB$_2$H$_8$', has been the subject of much controversy in recent years. In 1968, Haaland and co-workers reported that a new electron-diffraction study of beryllium tetrahydroborate supported a C_{2v} structure, in which the beryllium and boron atoms were located at the corners of a roughly equilateral triangle (Figure 3.11(b)), rather than one in which the heavy atoms were linear, as in the classical D_{2d} structure (Figure 3.11(a))[186]; however, bond-overlap populations, obtained from M.O. calculations, were not compatible with certain structural parameters[187], and moreover, from a consideration of total energies, it was shown that this proposed structure (Figure 3.11(b)) was less stable than the

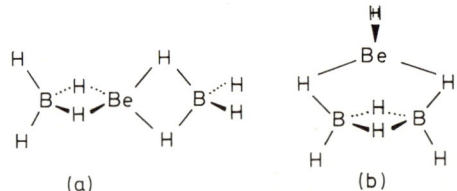

Figure 3.11 (a)–(d). Proposed structures for BeB$_2$H$_8$
((c)–(d) see opposing page)

classical structure[188]. Later, on the basis of infrared spectral evidence (indicating terminal BH$_2$-groups) and mass spectral data, Cook and Morgan proposed another structure (Figure 3.11(c)) in preference to Haaland's, which still retained the triangular arrangement of heavy atoms and was found to fit Haaland's original electron-diffraction data[189]; about the same time, Nibler and co-workers eliminated the classical linear D_{2d} structure on the basis of electron-deflection experiments (which showed the molecule had a dipole moment)[190] and gas-phase and matrix infrared studies (which indicated C_{2v} symmetry)[191], and suggested that either of the structures 3.11(b) or 3.11(c) was thus acceptable[190]. The situation was further complicated when recent electron-diffraction studies failed to reproduce the results reported in 1968 and seemed to imply that a linear structure might after all

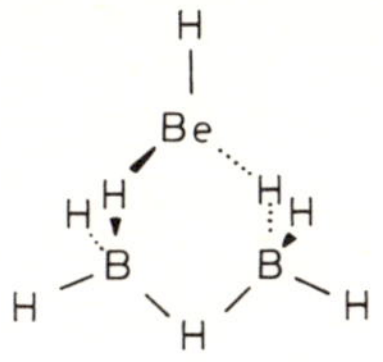

```
            H
            |
            Be
          /    ⋮
       H H    H H
       ˙B      B˙
      /  \   /  \
     H    \ /    H
           H
          (c)
```

be correct[192]. The latest report from Cook and Morgan, however, describes gas-phase infrared spectral studies of the normal and isotopically-substituted molecule, which strongly support a structure with an equilateral triangular arrangement of heavy atoms and C_{2v} symmetry (Figure 3.11(d)). This new structure differs from those previously proposed by having each boron bound to beryllium by a double hydrogen bridge, and all four bridge hydrogens in

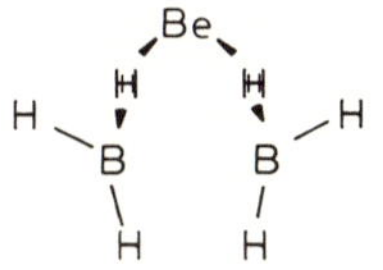

```
            Be
          /    \
        H H    H H
       ˙B      B˙
      /         \
     H           H
      \         /
       H       H
          (d)
```

a plane perpendicular to the plane of the three heavy atoms; moreover, each boron has two terminal hydrogens and all four terminal hydrogens lie in the plane of the heavy atoms. The gas phase infrared spectrum of the trimethylamine adduct of beryllium tetrahydroborate supported the proposed structure (Figure 3.11(d)), and, when compared with the infrared spectrum of beryllium tetrahydroborate, verified the absence of beryllium–hydrogen terminal bonds in that molecule[193]. Although the infrared spectrum of *unsaturated* vapours of isotopically-substituted beryllium tetrahydroborate agreed with the proposed structure, the infrared spectra of the *saturated* vapours were consistent with dimeric and trimeric structures, and the presence of such dimers and trimers may account for the recently-obtained anomalous electron-diffraction data[192, 193].

Structural investigations on other metal tetrahydroborates, apart from beryllium tetrahydroborate, have proved more conclusive; electron-diffraction studies on gaseous $Al(BH_4)_3$ [194], and x-ray diffraction studies on $Al(BH_4)_3 \cdot NMe_3$ at $-160\,°C$ [195] and $(Ph_3P)_2CuBH_4$ [196], have shown that in each case the metal atom is bonded to the boron by two hydrogen bridges. X-ray work[197] on $Zr(BH_4)_4$ at $-160\,°C$, however, indicated that the metal atom is bound to the boron by *three* hydrogen bridges and, at present, this compound is the only known example having this unique bonding arrangement. Although, as shown by x-ray and infrared spectral studies, metal tetrahydroborates exist as hydrogen-bridged species in the solid state, they undergo a rapid intramolecular exchange in solution effectively rendering the bridge and terminal hydrogens equivalent on the n.m.r. time-scale. Wallbridge and co-workers have proposed a mechanism involving a triply hydrogen-bridged intermediate [cf., $Zr(BH_4)_4$], which may be applicable both to the intramolecular exchange which occurs in crystalline $Al(BH_4)_3$·NMe_3 at room temperature (gross delocalisation of the electron density of hydrogen atoms of the BH_4 groups and large thermal motions of the boron atoms were observed in the x-ray diffraction study[195]), and to the exchange which takes place in solutions of metal tetrahydroborates at room temperature[198]. Although the x-ray study[195] indicates that the hydrogen atoms in $Al(BH_4)_3$·NMe_3 become localised in the crystal at −160 °C, n.m.r. spectra show that both this aluminium compound and $Zr(BH_4)_4$ still undergo intramolecular hydrogen-exchange at low temperatures[198]. Efforts to slow such intramolecular exchange processes in other metal tetrahydroborates have led to similar studies on the temperature dependence of their n.m.r. spectra. In the case of $(Ph_3P)_2CuBH_4$, the initially broad ^1H resonance, observed at room temperature, was found to sharpen to a singlet at −110 °C; such sharpening (loss of coupling) was attributed to efficient quadrupole-induced ^{10}B and ^{11}B spin relaxation, while the presence of a singlet indicated that rapid intramolecular hydrogen-exchange was still occurring, even at −110 °C [199]. Similar ^1H–^{11}B nuclear spin decoupling, explicable in terms of boron quadrupole relaxation, has been observed in the ^1H and ^{11}B n.m.r. spectra of the recently-reported compounds $(\pi\text{-}C_5H_5)_3ThBH_4$ [200] and $(\pi\text{-}C_5H_5)_3UBH_4$ [200–202], and of the known compound $(\pi\text{-}C_5H_5)_2Zr(BH_4)_2$; in each case, intramolecular hydrogen-exchange was still rapid at −70 °C [203] (see also $B_3H_8^-$, Section 3.4.2; $(Ph_3P)_2CuB_3H_8$, Section 3.5.1.2).

Zinc tetrahydroborate forms complexes of the type $Zn(BH_4)_2$·2L (where L = T.H.F., C_5H_5N, Me_3N), in which there are covalently-bonded BH_4 groups; however, the complexes $Zn(BH_4)_2$·4L (where L = NH_3, $MeNH_2$) are ionic and are best represented as $(ZnL_4)(BH_4)_2$ [204]. This transition from covalent to ionic bonding is also noticeable in the analogous complexes of $Al(BH_4)_3$; although $Al(BH_4)_3$·NMe_3 contains covalently-bound BH_4 groups[195], $Al(BH_4)_3$·$6NH_3$ is best formulated as $[AlH_2(NH_3)_4][BH_2(NH_3)_2][BH_4]_2$ in which there are discrete BH_4^- ions[205].

3.5.1.2 M—H—B: metal octahydrotriborates

Reaction of the hexacarbonyls of chromium, molybdenum, and tungsten, with the octahydrotriborate(1−) ion, $B_3H_8^-$, results in the formation of the anionic species $(CO)_4M·B_3H_8^-$ (where M = Cr, Mo, W)[206]. An x-ray diffraction study of the air-stable salt $[Me_4N][(CO)_4CrB_3H_8]$ has shown that, in the anion, the octahedrally-hybridised chromium atom is coordinated to two axial carbonyls, two equatorial carbonyls, and, via two hydrogen bridges, to two different boron atoms of the B_3H_8 moiety; the corresponding molybdenum and tungsten salts are isostructural[207]. In solution, the ^{11}B n.m.r.

BORON HYDRIDES

spectrum of $(CO)_4Cr \cdot B_3H_8^-$ indicates the presence of two different boron environments, and thus supports the reported solid-state structure; the ^{11}B n.m.r. spectra of the molybdenum and tungsten analogues are similar[206].

A bis-cyclopentadienyltitanium octahydrotriborate, $(\pi\text{-}C_5H_5)_2TiB_3H_8$, obtained by reaction of $(\pi\text{-}C_5H_5)_2TiCl_2$ with CsB_3H_8, was found to be volatile and readily oxidised in air; at present, the structure of this paramagnetic titanium derivative is unknown[206].

Bis-phosphinocopper(I) and bis-phosphinosilver(I) octahydrotriborates, $L_2CuB_3H_8$ and $L_2AgB_3H_8$ (where L = triphenylphosphine, or tri-p-tolylphosphine), have been prepared by reaction of a copper or a silver salt with CsB_3H_8 in the presence of the appropriate phosphine[206, 208]; bis-triphenylphosphinecopper(I) octahydrotriborate, $(Ph_3P)_2CuB_3H_8$, has also been obtained from the reaction of $(Ph_3P)_3CuCl$ with CsB_3H_8 [209]. No phosphinogold octahydrotriborates have so far been isolated, perhaps because of their inherent instability[206]. An x-ray diffraction study of $(Ph_3P)_2CuB_3H_8$ showed that the quasi-tetrahedrally hybridised copper atom was coordinated to two Ph_3P ligands, and bonded via two hydrogen bridges to two different boron atoms of the B_3H_8 group (Figure 3.12); the structural features of the complex correlated well with the results found for other bis-triphenylphosphinecopper(I) complexes and for the B_3H_8 group in $(CO)_4CrB_3H_8^-$ [210]. The structure of the corresponding silver derivative, $(Ph_3P)_2AgB_3H_8$, has not so far been determined by x-ray methods, but in the solid state, at any rate, the mode of bonding between the metal and the B_3H_8 group is probably similar to that in $(Ph_3P)_2CuB_3H_8$ [206].

Muetterties and co-workers have studied the solution-state structures of $L_2CuB_3H_8$ and $L_2AgB_3H_8$ (where L = tri-p-tolylphosphine), by means of

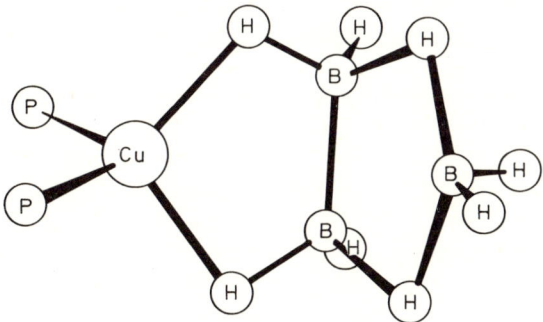

Figure 3.12 Molecular structure of $(Ph_3P)_2CuB_3H_8$; phenyl groups are not shown

^{31}P and ^{11}B n.m.r. spectroscopy and conductivity measurements. Between $-120\,°C$ and $-85\,°C$ the ^{31}P n.m.r. spectrum of $L_2CuB_3H_8$ is an AB pattern fully consistent with the known[210] solid-state structure of its triphenylphosphine analogue (Figure 3.12), but at $-60\,°C$ the AB pattern merges to a broad singlet which sharpens with further temperature increase; such behaviour suggests that an exchange process, perhaps one involving bidentate ⇌ unidentate B_3H_8 ion coordination, is equilibrating phosphine environments. Although the nature of the bonding between the copper atom and the B_3H_8 moiety in $L_2CuB_3H_8$ at room temperature is unknown, its strength is still substantial, since the broad resonance in the ^{11}B n.m.r. spectrum began to sharpen to the nonet of free $B_3H_8^-$ only after addition of 20 equivalents of free phosphine; furthermore solutions of $L_2CuB_3H_8$ in dichloromethane are non-conducting at room temperature[208, 211]. Muetterties' results, and suggested exchange mechanism, are particularly relevant to a recently-reported study of the temperature dependence of the ^1H n.m.r. spectrum of $(Ph_3P)_2CuB_3H_8$, in which the observation of five separate ^1H resonances at $-90\,°C$ was consistent with the solid-state structure of $(Ph_3P)_2CuB_3H_8$ and also with quadrupole-induced ^{11}B and ^{10}B spin relaxation[163]. In the case of $L_2AgB_3H_8$ (L = tri-p-tolylphosphine), the ^{31}P n.m.r. spectrum showed only a single ^{31}P environment at $-80\,°C$, and thus precluded any proposal regarding the precise structure of the compound in solution; although there is bonding between the silver and the B_3H_8 group, its strength appears to be less than that between copper and the B_3H_8 group, since the ^{11}B resonance is transformed to the characteristic nonet of $B_3H_8^-$ after addition of only six equivalents of free phosphine (cf., $L_2CuB_3H_8$)[208].

3.5.2 Two-centre metal–boron bonding (M—B)

3.5.2.1 M—B donor-acceptor bonding: metal–borane derivatives

These metal–boron hydride derivatives may be regarded as metal complex–borane (BH_3) adducts (cf., Section 3.3.1.3 and 3.4.1); comparatively few of them have been characterised. Adducts of transition metal complexes with borane were first prepared by Parshall, who reacted manganese and rhenium carbonyl anions with diborane and isolated salts of the anions $L(CO)_4\cdot M\cdot BH_3^-$ (where L = PPh_3, CO, and M = Mn; L = CO, and M = Re), and $(CO)_5Re(BH_3)_2^-$ [212]. Although $(\pi-C_5H_5)_2MoH_2$, $(\pi-C_5H_5)_2WH_2$, and $(\pi-C_5H_5)_2ReH$ react with BF_3 and BCl_3 to form 1:1 adducts containing metal–boron coordinate bonds, they do not form borane adducts with diborane[213]. This behaviour is surprising since the acidity of BH_3 towards amines is greater than that of BF_3, but may perhaps be explained by the transition metal hydrides being 'hard' bases which thus react with the 'hard' acids BF_3 and BCl_3 in preference to the 'soft' acid BH_3 [214]. Reaction of potassium germyl, $KGeH_3$, with diborane yields the potassium salt of the anionic stable adduct $H_3GeBH_3^-$ [215].

3.5.2.2 M—B normal covalent bonding: metal-substituted pentaborane(9) and decaborane(14) derivatives

Gaines and Iorns have reported that $NaMn(CO)_5$ and $NaRe(CO)_5$ react with 2-halo derivatives of pentaborane(9) to yield the compounds 2-[(CO)_5

Mn]B_5H_8 and 2-[(CO)$_5$Re]B_5H_8, respectively, in which n.m.r. spectral data suggests that there is a direct σ-bond between the transition metal and the boron hydride moiety[216]. The same authors obtained trialkylmetal-substituted derivatives of pentaborane(9), 2-$R_3MB_5H_8$ (where R = H, Me,Et; M = Si,Ge), by isomerisation of the corresponding basal-bridged derivatives μ-$R_3MB_5H_8$ in the presence of ether solvents (see Section 3.3.5.2); evidence for the position of terminal substitution in the compounds was given by their n.m.r. spectra[104] (see also (Section 3.3.5.2) the analogous halosilyl derivatives 2-(XH$_2$Si)B_5H_8 (X = Cl,Br))[103]. No n.m.r. data have been presented for the related trimethylsilyl-substituted derivative of decaborane(14), $Me_3SiB_{10}H_{13}$·T.H.F.[176], or for the silyl-substituted derivatives of the $B_{10}H_{13}^-$ ion, $B_{10}H_{12}SiR_3^-$ (R = H,Ph)[177] (see Section 3.4.7), and the nature of the bonding between the R_3Si group and the boron hydride moiety in these compounds is open to speculation.

3.5.3 Closed three-centre boron–metal–boron bonding (B—M—B)

3.5.3.1 B—M—B: metal–monodentate boron hydride moiety

A single crystal x-ray diffraction study of 1-Br-μ-$Me_3SiB_5H_7$ has shown that the silicon atom may be regarded as sp^3-hybridised and bound to two adjacent basal boron atoms of the pentaborane(9) framework by a three-centre B—Si—B two-electron bond[105]; other μ-$R_3MB_5H_8$ derivatives (where R = H, Me, Et; M = Si, Ge: and R = Me; M = Sn, Pb) may have analogous structures[104]. The $B_5H_8^-$ ion thus appears to be acting as a monodentate ligand at one of its basal edges. Such a concept is not new. Dobrott and Lipscomb had similarly rationalised the findings of an x-ray diffraction study of $Cu_2B_{10}H_{10}$, in which each copper atom appeared bonded to two B_{10}-polyhedra, and each polyhedron to four coppers, by considering that each Cu(I) was sp-hybridised, and that each hybrid formed a three-centre bond with an edge B—B bond of $B_{10}H_{10}^{2-}$ [217]. The structures of the air-stable compounds (Ph$_3$P)$_2$CuB$_5$H$_8$ and (Ph$_3$P)$_2$CuB$_6$H$_9$, recently reported by Brice and Shore, are unknown, and, although the (Ph$_3$P)$_2$Cu unit may bridge two adjacent basal borons (as does the Me$_3$Si unit, see above), its connection to the base by two hydrogen bridges cannot be discounted, especially since detection of B—H—B infrared absorptions was unfortunately prevented by the presence of phenylphosphine absorptions[218]; such hydrogen bridging – at present known with certainty only for metal derivatives of the lower boron hydrides (Sections 3.5.1.1 and 3.5.1.2) – has been proposed to exist in $L_2CuB_{10}H_{13}$ (L = tri-p-tolylphosphine) and in $Cl_2CuB_{10}H_{13}$ [208].

3.5.3.2 B—M—B: metal– bidentate boron hydride moiety

In this structurally well-characterised class of metal–boron hydride derivatives, the boron hydride moiety may be regarded as a bidentate ligand which coordinates to two hybrid orbitals of the metal species. It should, however, be noted that in all compounds studied so far the incorporation of the metal into the open face of the bonding boron hydride moiety gives a metal–boron icosahedral fragment in which the metal would be expected to participate more fully in delocalised bonding than is evident from the simple bidentate bonding description.

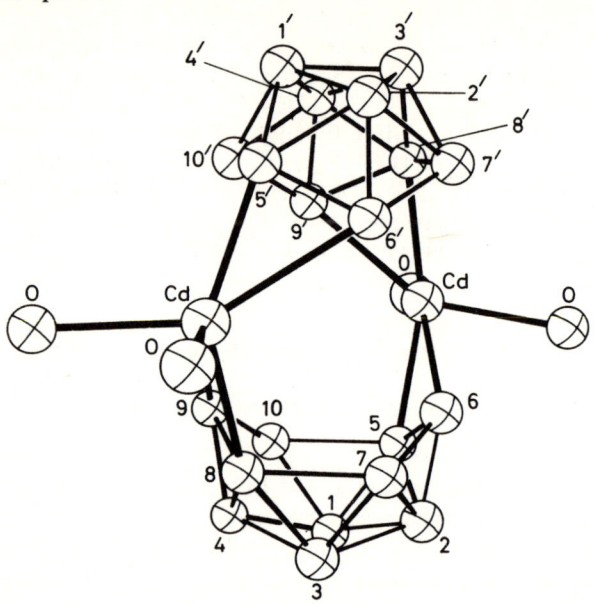

Figure 3.13 Molecular structure of $(CdB_{10}H_{12} \cdot 2Et_2O)_2$; hydrogens attached to the boron cages and ethyl groups attached to the oxygen atoms are omitted. Numbering used is the same as that for $B_{10}H_{14}$

Although Greenwood and Travers reported the metal–boron hydride derivatives $MB_{10}H_{12} \cdot 2Et_2O$ (where M = Mg, Zn or Cd) [172, 219] and $M(B_{10}H_{12})_2^{2-}$ (M = Zn, Cd, Hg) [219, 220] in 1966 and 1967, respectively, it was not then realised that these were in fact metal derivatives of the formally bidentate ligand $B_{10}H_{12}^{2-}$. A later report by Muetterties and co-workers gave details for the preparation of three classes of metal–$B_{10}H_{12}^{2-}$ derivatives, $M(B_{10}H_{12})_2^{2-}$ (where M = Co, Ni, Pd, Pt), $M(B_{10}H_{12})L_2$ (where M = Pd or Pt and L is a phosphine ligand), and $M(B_{10}H_{12})L_3^-$ (where M = Co, Rh, Ir; L = carbonyl and phosphine ligands); in addition, preliminary x-ray diffraction data were presented for $[Me_4N]_2[Ni(B_{10}H_{12})_2]$ (the cobalt analogue is isomorphous) which indicated that the $B_{10}H_{12}^{2-}$ ion interacted with the nickel as a bidentate ligand to form an 11-atom icosahedal fragment, and that fusion of two such fragments through the central metal atom generated the $Ni(B_{10}H_{12})_2^{2-}$ ion, which had C_{2h} symmetry and a square-planar arrangement about the metal atom. The bonding arrangement in the analogous palladium and platinum derivatives, and in the other two classes of transition metal–boron hydride derivatives, was proposed to be similar on the basis of n.m.r. spectra[221]. The compounds $M(B_{10}H_{12})_2^{2-}$ (M = Co, Ni,

Pd, Pt), have also been obtained by Siedle and Hill from the reaction of $Cs_2B_{10}H_{14}$ with the appropriate transition metal halide derivative[222]. Muetterties and co-workers have recently described the formation of another class of transition metal–$B_{10}H_{12}^{2-}$ derivative, $(CO)_4M(B_{10}H_{12})^{2-}$ (M = Cr, Mo, W), by reaction of aqueous base with either of the novel quasi-icosahedral derivatives $(CO)_4M(B_{10}H_{10}COH)^-$ or $(CO)_3O\underline{CMO}CB_{10}H_{10}^{2-}$ (note: although these two latter anions are produced via reaction of $NaB_{10}H_{13}$ with $M(CO)_6$ they are regarded as metal derivatives of the *carborane* anion, $B_{10}H_{10}COH^{3-}$, and are hence not discussed in detail here); the new $(CO)_4M(B_{10}H_{12})^{2-}$ ions are believed to be structurally-analogous to the 11-atom icosahedral fragment transition metal–boron hydride derivatives discussed above[223].

Recent single crystal x-ray diffraction studies in this laboratory have revealed that the previously-formulated '$CdB_{10}H_{12} \cdot 2Et_2O$' is in fact dimeric in the solid state, and has a unique structure (Figure 3.13) in which *two* cadmiums are bonded to *two* $B_{10}H_{12}$ units; each cadmium is also coordinated to two molecules of diethyl ether and the distance between the cadmiums precludes metal–metal bonding[224]. The structure of $Zn(B_{10}H_{12})_2^{2-}$ has also been elucidated (the cadmium and mercury analogues are isostructural), and has been shown to consist of two 11-atom icosahedral fragments fused through a central quasi-tetrahedrally coordinated zinc atom (Figure 3.14); formally, coordination occurs from the two-centre–two-electron B_5—B_6, and B_9—B_{10} bonds, to two of the tetrahedrally directed sp^3 hybrid orbitals of Zn^{2+}, but comparison of certain boron–boron distances in the anion with the corresponding distances in $B_{10}H_{14}$ supports the concept that the zinc atom is truly incorporated into the icosahedral boron hydride fragment, with consequent participation in the delocalised bonding[224].

The results of such x-ray diffraction investigations lead to structural proposals for previously prepared, but structurally uncharacterised, metal–boron hydride derivatives. The anion $B_{10}H_{12}AlH_2^-$, formed by reaction of trimethylamine–alane, $Me_3N \cdot AlH_3$, with decaborane(14), may be a derivative of the bidentate ligand $B_{10}H_{12}^{2-}$, although the structure of $B_{10}H_{14}GaH_2^-$, formed by an analogous reaction with trimethylamine–gallane, $Me_3N \cdot GaH_3$, remains obscure (this anion was originally proposed to be a 6,9-bridged derivative of the unknown ion $B_{10}H_{16}^{2-}$)[225]; the anion, $pyCoX_2B_{10}H_{11}py^-$ (where py = pyridine, and X = Cl, Br) may have a structure in which there is a square-pyramidal arrangement about the 5-coordinate Co^{II}, with the $B_{10}H_{11}py^-$ ion acting as a bidentate ligand in the same manner as its analogue $B_{10}H_{12}^{2-}$ [174].

Three new classes of icosahedral fragment metal–boron hydride derivatives, in which the $(Et_3P)_2Pt$ moiety is bonded to the formally bidentate anions $B_9H_{10}S^-$, $B_9H_{11}L^{2-}$ (where L is an organic base), and $B_8H_{12}^{2-}$, have been reported by Muetterties and co-workers. X-ray diffraction data for $(Et_3P)_2Pt(H)B_9H_{10}S$ indicate that the $B_9H_{10}S^-$ ion, isoelectronic with the bidentate ion $B_{10}H_{12}^{2-}$, behaves in an analogous manner. The coordination about the platinum may be regarded as quasi-5-coordinate, with distorted square-pyramidal form; one of the hybrid orbitals of the platinum may form a three-centre bond with a B—B edge bond, and another may interact similarly with a B—S edge bond (Figure 3.15). The compound $(Et_3P)_2PtB_9H_{11}L$ is proposed to be isoelectronic and isostructural with the compounds[5] $B_{10}H_{12}L_2$ (where L is an organic base), with $(Et_3P)_2Pt^{2+}$ replacing a 6(9)-BHL^{2+} unit (Figure 3.16); the $B_9H_{11}L^{2-}$ ion appears to be

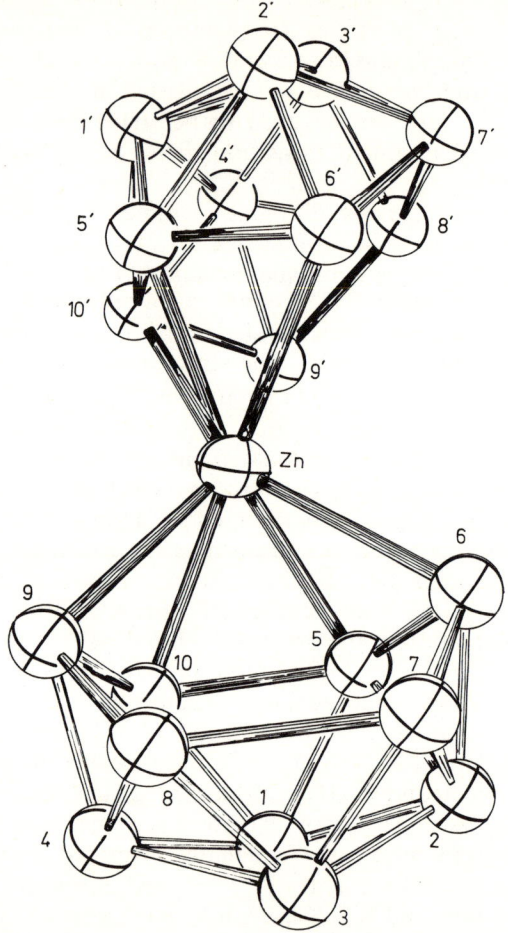

Figure 3.14 Molecular structure of $Zn(B_{10}H_{12})_2^{2-}$; only heavy atoms are shown. Numbering used is the same as that for $B_{10}H_{14}$

bidentate with two square-planar platinum orbitals forming three-centre bonds with the B_2—B_5 and B_2—B_7 edges. Alcohol degrades $(Et_3P)_2PtB_9H_{11}L$ to $(Et_3P)_2PtB_8H_{12}$, and this novel metal–$B_8H_{12}^{2-}$ compound may have a structure derived from that of its B_9 precursor (Figure 3.16) by elimination of BHL^{2+} and addition of $2H^+$ at edge-bridging positions, which would be analogous to that of $B_9H_{13}L$ [226].

The boron hydride n-$B_{18}H_{22}$ consists of two decaborane(14) fragments

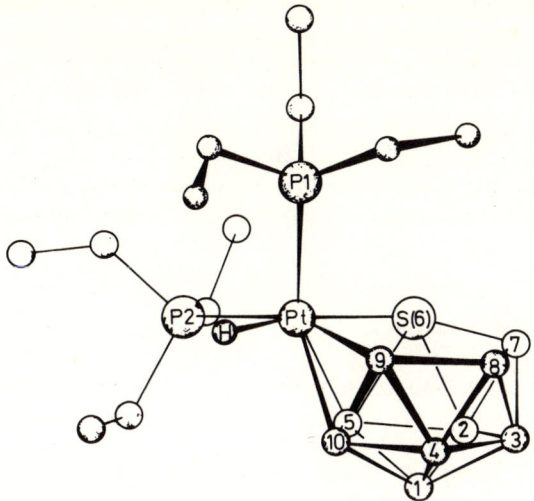

Figure 3.15 Molecular structure of $(Et_3P)_2\ Pt(H)B_9H_{10}S$; ethyl hydrogens and hydrogens attached to boron cage are not shown

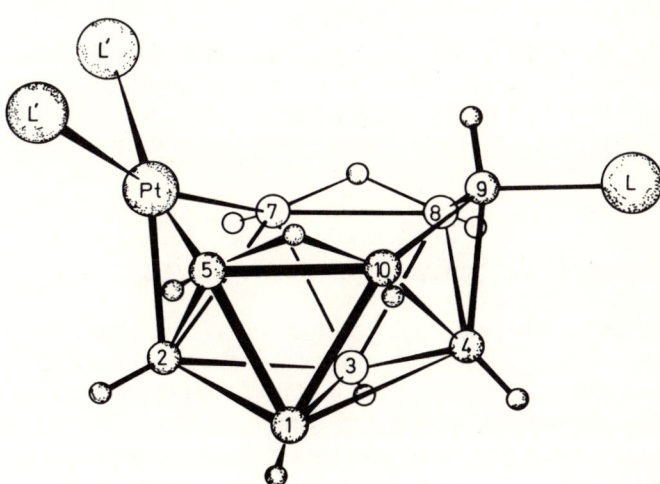

Figure 3.16 Proposed molecular structure of $(Et_3P)_2PtB_9H_{11}L$; L' represents Et_3P

which open in opposite directions and share a common edge, and its bridge-hydrogen arrangement is similar to that in decaborane(14)[5]. The anion n-$B_{18}H_{20}^{2-}$, which may be generated from n-$B_{18}H_{22}$, in the same manner as $B_{10}H_{12}^{2-}$ is from $B_{10}H_{14}$ (see Sections 3.3.12 and 3.4.7 respectively), is probably structurally similar (in the B_{10} fragment) to $B_{10}H_{12}^{2-}$, and might likewise be expected to form metal derivatives in which it acts as a bidentate ligand. Todd and co-workers have recently shown that this may indeed be the case by preparing the new complexes $(CO)_3Co$-$(n$-$B_{18}H_{20})^-$, $(Ph_3P)_2Ni(n$-$B_{18}H_{20})$, and $(Ph_2PCH_2CH_2PPh_2)Ni(n$-$B_{18}H_{20})$, by reaction of n-$B_{18}H_{20}^{2-}$ with the appropriate transition metal derivative, and by proposing that the metal atom is incorporated into the open face of the B_{10} unit of the B_{18} skeleton[227].

3.5.4 Multi-centre or π-metal–boron hydride bonding

Compounds in which the boron hydride moiety may formally donate six electrons to the metal to form a *closed icosahedral* metal–boron structure, with the metal atom in a close nearly-equivalent approach to *five* atoms of the ligand skeleton (one of which may be sulphur), were reported by Muetterties and co-workers in 1967; such metal–boron hydride derivatives $(CO)_3Re(B_{10}H_{10}S)^-$, $Fe(B_{10}H_{10}S)_2^{2-}$, $Co(B_{10}H_{10}S)_2^-$, π-$C_5H_5Co(B_{10}H_{10}S)$, $(Et_3P)_2Co(B_{10}H_{10}S)$, and $(Et_3P)_2Pt(B_{10}H_{10}S)$ were prepared by reaction of the $B_{10}H_{10}S^{2-}$ ion with the appropriate transition metal derivative. These compounds, unlike their probable structural analogues – the corresponding metal derivatives of the carborane anions $B_9C_2H_{11}^{2-}$ and $B_{10}CH_{11}^{3-}$ (see Chapter 4) – have not been studied by x-ray diffraction techniques; however, available n.m.r. data suggests that they have the structure discussed above[228].

More recently, Little has reported the preparation of the new anion $(CO)_3Mn(B_{10}H_{10}PPh)^-$, from the deprotonation of the new boron hydride $B_{10}H_{12}PPh$ (itself obtainable from $NaB_{10}H_{13}$ and $PhPCl_2$), and subsequent reaction with $BrMn(CO)_5$ [229]; the structure of this new metal derivative of the $B_{10}H_{10}PPh^{2-}$ ligand may be analogous to that of the metal derivatives of the isoelectronic ligand $B_{10}H_{10}S^{2-}$ [228].

3.6 CATIONIC BORON HYDRIDE DERIVATIVES

Large numbers of cationic boron hydride derivatives have recently been prepared (see Table 3.1), but, unfortunately, space is not available to discuss all of them here, and only new methods of preparation and some of the more novel cations will be dealt with in this section.

Perhaps the most useful method for preparing boron hydride cations is the reaction of an adduct of a monohalo- or dihalo-borane (see Section 3.3.1.3) with an amine; for example, in the simple case below displacement of iodide results in the formation of an 'unsymmetrical' boron cation[230]: $R_3N \cdot BH_2I +$ $R'_3N \rightarrow [R_3N \cdot BH_2 \cdot NR'_3]^+ I^-$ (where R = Me, R' = Et). By variation of the type of substituted borane adduct and the amine with which it reacts, a wide variety of cations may be obtained (as may be seen from Table 3.1). The interesting species $[Me_3N \cdot BH_2 \cdot NCCH_3]^+$ was prepared by reaction of $Me_3N \cdot BH_2I$ with acetonitrile, and it seems likely that acetonitrile is close to the lower limit in base strength required for such a displacement reaction (benzonitrile, a somewhat weaker base, did *not* displace iodide from $Me_3N \cdot BH_2I$); the extremely fast displacement of the CH_3CN group by pyridine, and the marginal hydrolytic stability of this novel cation, even in acidic solution, suggest that it may represent the practical limit of kinetic stability for such species[231].

Recently, Benjamin and co-workers described a new method of preparation of boron hydride cations which involves the abstraction of hydride ions from B—H-containing compounds by the reagent triphenylmethyl fluoroborate, $Ph_3C^+BF_4^-$ (which had already been extensively used, in both organic and organometallic chemistry, to prepare stable cations); by treating amine-boranes, or the tetrahydroborate ion, with this reagent in the presence of bases, several boron cations were obtained. For example, reaction of either $py \cdot BH_3$ or BH_4^- with $Ph_3C^+BF_4^-$ in the presence of pyridine (py) yields the 'symmetrical' cation $py_2BH_2^+BF_4^-$, while reaction of $Me_3N \cdot BH_3$ with the same reagent in the presence of acetonitrile generates the novel 'unsymmetrical' cation $[Me_3N \cdot BH_2 \cdot NCCH_3]^+$, which was discussed above[232].

Boron cations, in which pyridine and substituted pyridines are coordinated to a central BH^{2+} moiety, were recently reported by Ryschkewitsch and co-workers as the first examples of mononuclear 2+ species with B—H bonds; these new cations were found to be more resistant to electrophilic attack than the corresponding singly-charged boron hydride cations[233, 234].

A polymeric boron $(n+)$ ion, representing the first example of a polyboronium ion containing more than three boron atoms, has been obtained by vinyl polymerisation of the trimethylamine-(4-vinylpyridine)-di-hydroboron(1+) ion, $[Me_3N \cdot BH_2 \cdot N\langle\rangle CH=CH_2]^+$ [235].

Reaction of $[Me_3PCH_2NMe_2]^+Cl^-$ with diborane yields salts of the cationic adduct $[Me_3P \cdot CH_2 \cdot NMe_2 \cdot BH_3]^+$, the first example of a boron hydride cation with a borane (BH_3) group[236].

3.7 ADDENDUM

While this review was being typed two important papers appeared. Bond and Pinsky have reported that tetraborane(10) may be deprotonated in diethyl ether at low temperatures by either methyl lithium or lithium octahydropentaborate, LiB_5H_8, with formation of the nonahydrotetraborate(1−) ion $B_4H_9^-$ (see Section 3.4.3.2), and also that the proton is lost from a bridge position of tetraborane(10); interestingly, tetraborane(10) appears to be a weaker Brønsted acid than decaborane(14), but stronger than diborane(6) or pentaborane(9)[237]. In a related consecutive report, Johnson and Shore have shown that ammonia deprotonates tetraborane(10) at $-65\,°C$ in diethyl ether to yield the $B_4H_9^-$ ion, but at higher temperatures bridge cleavage occurs with formation of $[BH_2(NH_3)_2]B_3H_8$ (see Section 3.4.3.2). Tetraborane(10) is also deprotonated by potassium hydride in diethyl ether, and by the $B_5H_8^-$ or $B_6H_9^-$ ion in dichloromethane at low temperatures. The ^{11}B n.m.r. spectrum of $B_4H_9^-$ at low temperatures has been assigned on the basis of Lipscomb's proposed[5] 2113 structure (Figure 3.1; H^- replaces L); analogous assignments have been made for the ^{11}B n.m.r. spectra of the isoelectronic species B_4H_8L (where L = PF_2NMe_2, PF_2H)[90] (see Section 3.3.4.1). The temperature dependence of the ^{11}B n.m.r. spectrum of $B_4H_9^-$ may be explained in terms of a rapid intramolecular tautomerism which occurs above $-45\,°C$. Johnson and Shore have also reported that pentaborane(11) may be deprotonated by either ammonia or potassium hydride in ether solvents at low temperatures, with formation of the new decahydropentaborate(1 −) ion, $B_5H_{10}^-$ [238].

3.8 TABULAR SURVEY

3.8.1 Presentation

The tabular survey closely follows the order and content of the main sections of the Review. Where a reference is given for a compound *without* being followed by a comment in parentheses, the preparation and standard infrared, n.m.r., and mass spectral data are — if appropriate and available — provided in that reference. Where a reference *is* followed by a comment in parentheses the preceding remarks may still apply, but, in addition, the comment draws attention to particularly important investigations on the compound. Single crystal x-ray diffraction studies are indicated by 'x-ray' in parentheses, and detailed n.m.r., and infrared spectral studies by 'n.m.r.', and 'i.r.', respectively, in parentheses. If a boron hydride has been the subject of theoretical calculations, 'theor.' appears in parentheses behind the relevant reference; other comments should be self-explanatory.

Abbreviations

The following abbreviations have been used throughout Table 3.1.
bipy	2,2′-bipyridyl
cy	cyclohexyl
diox.	dioxan
diphos	1,2-bis-diphenylphosphinoethane
DMA	N,N,-dimethylacetamide
DMF	N,N-dimethylformamide
DMpip	N,N′-dimethylpiperazine
dppa	1,2-bis-diphenylphosphinoacetylene

en	ethylenediamine
PMDT	pentamethyldiethylenetriamine
py	pyridine
TEDA	triethylenediamine
TEED	$N,N,N,'N'$-tetraethyl-1,2-ethanediamine
THF	tetrahydrofuran
TMBD	$N,N,N,'N'$-tetramethyl-1,4-butanediamine
TMED	$N,N,N,'N'$-tetramethyl-1,2-ethanediamine
TMPD	$N,N,N,'N'$-tetramethyl-1,3-propanediamine

Table 3.1 Survey of neutral boron hydrides; anionic boron hydrides, metal-boron hydrides and cationic boron hydrides

Compound	References and Comments
One boron atom (see Section 3.3.1)	
BH_3	27,28,29,32,33; 34,251,301 (theor.); 30, 31 (molec. beam m.s.)
HBF_2	34,35 (theor.); 36; 300 (theor.: coupling constants)
$HBCl_2$	37
$HB(OMe)_2$	239 (n.m.r.: relaxation studies)
$HB(OR)_2 (R = Pr^n, Bu^n, Bu^i, Bu^s, Bu^t)$	240
Surface = BH	38
NH_2BH_2	39 (cryochem. prep.); 40 (theor.)
$OC \cdot BH_3$	31 (molec. beam m.s.); 41 (photoelectron spec.); 42 (theor.)
$ON \cdot BH_3$	43
$H_3N \cdot BH_3$	41 (photoelectron spec.); 42,250 (theor.)
$Me_2NH \cdot BH_3$	261 (H_2 elimination: kinetics)
$Me_3N \cdot BH_3$	46,47 (n.m.r.: exchange react'ns); 264 (n.m.r.; wide-line)
$Me_3N \cdot BD_3$	241 (D n.m.r.)
$PhCH_2CH(Me)NRR' \cdot BH_3$ (R = R' = H,Me; R = H,R' = Me)	45 (optical activity)
$Me_3N \cdot BH_2CH_2NMe_2 \cdot BH_3$	242
$MeSCH_2NMe_2 \cdot BH_3$	236
2-MeOpy$\cdot BH_3$	44
3,4-diMepy$\cdot BH_3$	234
$RN{=}NR \cdot BH_3$ (R = Me,Et)	243
$(CH_2)_2NH \cdot BH_3$	244; 246 (i.r.); 245 (x-ray)
$(CH_2)_2NR \cdot BH_3$ (R = C_2H_4OH, C_2H_4CN)	244
$(CH_2)_6N_4 \cdot BH_3$	247 (x-ray)
$TMED \cdot 2BH_3$; $TMPD \cdot 2BH_3$; $TMBD \cdot 2BH_3$	234
$(CH_2)_2NR \cdot 2BH_3$ (R = $C_2H_4NH_2$)	244, 246
bipy$\cdot 2BH_3$	44
$Me_3N \cdot BH_2X$ (X = Cl,Br,I)	49, 51; 262 (solvolysis: kinetics)
$Me_3N \cdot BH_2CN$; $O(CH_2)_4NH \cdot BH_2CN$; 4-Mepy$\cdot BH_2CN$; $TMED \cdot (BH_2CN)_2$	53
$Me_3N \cdot BH_2CH_2SiMe_3$	78
$Me_3N \cdot BH_2CH_2NMe_2$	242
$NH_2NH_2 \cdot BH_2$t-Bu	248
$Me_3N \cdot BHCl_2$	49, 50, 51, 262

Table 3.1—*continued*

Compound	References and comments
Me$_3$N·BHBr$_2$	50, 51, 262
Me$_3$N·BHI$_2$	50, 262
Me$_3$N·BHClBr	51
H$_3$P·BH$_3$	54 (theor.); 57 (isotopic subst.: i.r., n.m.r.)
F$_3$P·BH$_3$	55 (photoelectron spec.: theor.)
P$_2$F$_4$·BH$_3$	59 (n.m.r.)
MePH$_2$·BH$_3$; H$_3$SiPH$_2$·BH$_3$	58 (isotopic subst.: n.m.r.)
(cyclopropyl)$_3$P·BH$_3$; i-Pr$_3$P·BH$_3$	249
Ph$_3$P·BH$_3$	56 (thermochem.)
R$_2$PH·BH$_3$ (R = Me,Et,Pr)	60 (n.m.r.); 253 (magneto-optical)
R$_3$P·BH$_3$ (R = Me,Et,n-Bu)	61,64 (n.m.r.); 252, 253 (magneto-optical)
(RO)$_3$P·BH$_3$ (R = Me,Et,Pr,Bu)	61 (n.m.r.); 252 (magneto-optical)
(R$_2$N)$_3$P·BH$_3$ (R = Me,Et,Pr)	61, 62, 66 (n.m.r.); 252 (magneto-optical)
(Me$_2$N)Me$_2$P·BH$_3$	62 (n.m.r.)
Me(Et$_2$N)(i-PrO)P·BH$_3$	63 (n.m.r.); 252 (magneto-optical)
R$_2$PH·BH$_2$Et (R = Et, Pr)	60 (n.m.r.); 253 (magneto-optical)
R$_3$P·BH$_2$R' (R = Et, Bu; R' = Et, Pr)	253 (magneto-optical)
Et$_3$P·BH$_2$X; Et$_3$P·BHX$_2$ (X = Cl)	64 (n.m.r.)
Et$_3$P·BH$_2$X; Et$_3$P·BHX$_2$ (X = Br)	65 (n.m.r.)
Et$_3$P·BHClBr	65 (n.m.r.)
(Me$_2$N)$_3$P·BH$_{3-n}$Cl$_n$ (0 ≤ n ≤ 3)	66 (n.m.r.)
(Me$_2$N)Me$_2$P·BEt$_2$H;(Me$_2$N)$_3$P·BEt$_2$H	62 (n.m.r.)
Me$_2$NCH$_2$PMe$_2$·BH$_3$	236
MeSCH$_2$PMe$_2$·BH$_3$	236
Me$_2$P·CH$_2$NMe$_2$·2BH$_3$	236
Ph$_2$P·CH$_2$·PPh$_2$·2BH$_3$	44
diphos·2BH$_3$	44
C(CH$_2$PPh$_2$)$_4$·2BH$_3$	254

Two boron atoms (see Section 3.3.2)

(Me$_2$NPF$_2$)$_2$·B$_2$H$_4$	69
(PF$_3$)$_2$·B$_2$H$_4$	70
(B$_2$H$_4$)$_n$	71
B$_2$H$_6$	21, 74, 255–257 (theor.); 74, 75 (photoelectron spec.); 72 (x-ray: theor.); 73, 258 (i.r.); 97 (n.m.r.: relaxation time); 263 (acetone: kinetics); 300 (theor.: coupling constants)
B$_2$H$_6^-$	259 (e.s.r.)
B$_2$H$_6$·2ROH (R = H, Me,Et)	76
B$_2$H$_5$Br	260 (isotopic subst.: microwave spec.)
1,1-Me$_2$B$_2$H$_4$; Me$_4$B$_2$H$_2$	77 (Raman); 79 (theor.)
1,2-(Me$_3$SiCH$_2$)$_2$B$_2$H$_4$	78
μ-Me$_2$NB$_2$H$_5$	80; 81 (n.m.r.: intramolecular exchange)
μ-Me$_2$NB$_2$H$_4$Me	82
μ-SHB$_2$H$_5$	83

Three boron atoms (see Section 3.3.3)

(Me$_2$NPF$_2$)$_2$·B$_3$H$_5$	69
F$_2$XP·B$_3$H$_7$ (X = F, Cl,Br)	84

Four boron atoms (see Section 3.3.4)

B$_4$H$_8$	88 (molec. beam m.s.)
B$_4$H$_8$·CO	88 (molec. beam m.s.)

Table 3.1—*continued*

Compound	References and comments
$(Me_2NPF_2) \cdot B_4H_8$	89 (x-ray); 90 (n.m.r.)
$PF_2H \cdot B_4H_8$	90 (n.m.r.)
B_4H_{10}	265; 266 (decomposition: kinetics); 22, 23 (theor.); 91 (n.m.r.)
2-MeB_4H_9	70, 92, 93
1,2-diMe-, 2,2-diMe-, and 2,4-diMeB_4H_8	70, 92, 94

Five boron atoms (see Section 3.3.5)

B_5H_9	22, 23 (theor.); 88 (molec. beam m.s.); 91, 101 (n.m.r.); 96 (ring current); 95 (pyrolysis: kinetics); 97 (n.m.r.: relaxation time); 267 (u.v.)
$B_5H_9 \cdot 2PMe_3$	98
1-Cl; 2-Cl; 1-Br; 2-Br; 1-Me; 2-MeB_5H_8	101 (n.m.r.); 267 (u.v.)
1-I; 2-IB_5H_8	267 (u.v.)
1-$[(CF_3)_2P \cdot Ni(CO)_3] B_5H_8$	107
2MeO-B_5H_8	268
2-$XSiH_2B_5H_8$ (X = Cl, Br)	103
1,2-diMeB_5H_7	100, 101 (n.m.r.)
2,3-diMe; 1-Br, 2-Me; 2-Br, 3-Me; 1-Cl, 2-Me; 2-Cl, 1-Me; 2-Cl, 3-Me; 2-Cl, 4-MeB_5H_7; 1,2,3-triMe-; 2,3,4-triMe-; 1-Cl, 2,3-diMe-; 2-Cl, 1,3-diMe-; 2-Cl, 1,4-diMe-; 2-Cl, 3,4-diMe-; 3-Cl, 2,4-diMe B_5H_6	101 (n.m.r.)
μ-$ClSiH_2B_5H_8$	103
μ-$Me_2B \cdot B_5H_8$	108
B_5H_{11}	22, 23 (theor.); 88 (molec. beam m.s.); 91, 109, 110 (n.m.r.)

Six boron atoms (see Section 3.3.6)

B_6H_{10}	91 (n.m.r.); 96 (ring current); 114, 115 (isotopic subst.; n.m.r.)
$B_6H_{10} \cdot PF_2NMe_2$	69
$B_6H_{10} \cdot 2L$ (L = NMe_3, PMe_3, PPh_3)	116
4,5-diMeB_6H_8	108
B_6H_{12}	87 (molec. beam m.s.); 91 (n.m.r.); 118 (isotopic subst.: n.m.r.)

Eight boron atoms (see Section 3.3.7)

B_8H_{12}	87 (molec. beam m.s.)
B_8H_{16}	120
B_8H_{18}	110 (n.m.r.); 123 (molec. beam m.s.)

Nine boron atoms (see Section 3.3.39

n-B_9H_{15}	125 (n.m.r.)
iso-B_9H_{15}	126 (isotopic subst.: n.m.r.)
7-MeO, 4-$Me_2SB_9H_{12}$	127, 128

Ten boron atoms (see Section 3.3.10)

$B_{10}H_{14}$	129 (neutron diffraction); 130 (electron distrib.); 134 (n.m.r.); 135 (ring current)
2-$BrB_{10}H_{13}$	137; 134 (n.m.r.)
5-F; 5-Br; 5-I; 6-$ClB_{10}H_{13}$	136

Table 3.1—*continued*

Compound	References and comments
1-Br; 1-Cl; 2-Cl; 5-Br; 6-BrB$_{10}$H$_{13}$	137
5-OEtB$_{10}$H$_{13}$	140 (isotopic subst.: n.m.r.)
Me$_3$SiB$_{10}$H$_{13}$·THF	176
B$_{10}$H$_{13}$POCl$_2$; (B$_{10}$H$_{13}$)$_3$P·Et$_2$O	142
B$_{10}$H$_{12}$[P(CN)$_3$]$_2$	269
L$_2$B$_{10}$H$_{12}$ (L = Ph$_3$P, Ph$_2$PH, PhPH$_2$, (Me$_2$N)$_3$P,(Me$_2$N)$_2$PCl, (MeO)$_3$P, (PhO)$_3$P, Ph(Me$_2$N)$_2$P, Me$_2$NPhPCl, Me$_2$S, Et$_2$S,C$_4$H$_8$S, C$_4$H$_8$SO, (PhNH)$_2$CS, MeCN, PhCN)	270 (u.v.)
LB$_{10}$H$_{12}$ (L = diphos, TMED, TEED)	270 (u.v.)
5-Br-6,9(Me$_2$S)$_2$B$_{10}$H$_{11}$	127, 128; 139 (optical isomers)
iso-B$_{10}$H$_{16}$	120; 110 (n.m.r.)
B$_{10}$H$_{18}$	120

Sixteen boron atoms (see Section 3.3.11)

B$_{16}$H$_{20}$	144 (x-ray)

Eighteen boron atoms (see Section 3.3.12)

n-B$_{18}$H$_{22}$	148; 145 (chromat. sep.)
iso-B$_{18}$H$_{22}$	145 (optical isomers)

Compound	References and comments
BH$_4^-$	271 (theor.)
LiBH$_4$; NaBH$_4$; KBH$_4$	272 (n.m.r.: relaxation studies)
KBH$_4$	273 (isotopic subst.: i.r.)
LiBH$_4$·NMe$_3$; LiBH$_4$·TEDA	274
[C(NH$_2$)$_3$]BH$_4$	275
LiBH$_3$CN	150, 151, 152
NaBH$_3$CN; KBH$_3$CN	152, 153 (isotopic subst.: i.r., Raman)
NaBH$_3$NC	153 (isotopic subst.: i.r.)
NaBH$_3$CNBH$_3$	153
(Ph$_3$P)$_3$M(NCBH$_3$)(M = Cu, Ag)	154
(en)$_2$Ni(NCBH$_3$)$_2$	154
Me$_2$N(BH$_3$)$_2^-$	156
H$_2$P(BH$_3$)$_2^-$	157; 159 (isotopic subst.: i.r., Raman)
PhHP(BH$_3$)$_2^-$	157
HSBH$_3^-$; HS(BH$_3$)$_2^-$	160
NaBH$_2$Te$_3$	276

Three boron atoms (see Section 3.4.2)

B$_3$H$_8^-$	161; 162, 163 (n.m.r.)

Four boron atoms (see Section 3.4.3)

B$_4$H$^-$	164
B$_4$H$_9^-$	116, 164, 237, 238 (n.m.r.)

Five boron atoms (see Section 3.4.4)

B$_5$H$_8^-$	113 (n.m.r.)
B$_5$H$_{10}^-$	238

Six boron atoms (see Section 3.4.5)

B$_6$H$_9^-$	116, 167; 113 (n.m.r.)

Table 3.1—*continued*

Compound	References and comments
Eight boron atoms	
$[(NH_3)_4Zn]B_8H_8$	277 (x-ray)
Nine boron atoms (see Section 3.4.6)	
$B_9H_9^{2-}$	175
$L_4CuB_9H_{12}S$; $L_3AgB_9H_{12}S$;	
$\quad L_4AgB_9H_{12}S$; $L_2AuB_9H_{12}S$;	
$\quad L_3AuB_9H_{12}S$; $L_4AuB_9H_{12}S$	
$\quad$ (L = tri-*p*-tolylphosphine)	208 (n.m.r.)
$B_9H_{14}^-$	169; 126 (isotopic subst.: n.m.r.); 170 (x-ray; n.m.r.)
$L_4CuB_9H_{14}$; $L_4AgB_9H_{14}$; $L_4AuB_9H_{14}$	
$\quad$ (L = tri-*p*-tolylphosphine)	208 (n.m.r.)
Ten boron atoms (see Section 3.4.7)	
$B_{10}H_9NH_3^-$	278
$C_7H_6 - B_{10}H_9^-$	279
$NaB_{10}H_{13}\cdot 2.5$ diox; $Na_2B_{10}H_{12}\cdot 2THF$;	
$\quad (Me_4N)_2B_{10}H_{12}$	174
$L_4AgB_{10}H_{13}$ (L = tri-*p*-tolylphosphine)	208 (n.m.r.)
2-I-$B_{10}H_{12}^-$; 2-I-$B_{10}H_{11}^{2-}$	177
$R_3SiB_{10}H_{12}^-$ (R = H, Ph)	177
$L_4AuB_{10}H_{15}$ (L = tri-*p*-tolylphosphine)	208
Eleven boron atoms	
$L_4CuB_{11}H_{14}$; $L_4AgB_{11}H_{14}$; $L_4AuB_{11}H_{14}$	
$\quad$ (L = tri-*p*-tolylphosphine)	208 (n.m.r.)
Twelve boron atoms	
$B_{12}H_{12}^{2-}$	280 (new synthesis)
$C_7H_6 - B_{12}H_{11}^-$	279
$Cs_2B_{12}H_{11}SH\cdot H_2O$	281 (x-ray)
Twenty boron atoms (see Section 3.4.8)	
$(Et_3NH)_2B_{20}H_{18}$; $(Et_3NH)_3B_{20}H_{18}NO$	180 (x-ray)
Twenty four boron atoms (see Section 3.4.9)	
$B_{24}H_{23}^{3-}$	183 (electrochem. prep.)
$B_{24}H_{23-n}X_n^{3-}$ (X = I, n = 2; X = Br,	
$\quad n = 7, 10, 11$)	
$B_{24}H_{22-n}X_n^{4-}$ (X = Br, n = 11, 14, 18;	
$\quad$ X = Cl, n = 18)	183
$B_{24}H_{20}I_2^{2-}$	184 (electrochem. prep.)
Metal tetrahydroborates (see Section 3.5.1)	
$MeBeBH_4$	185 (isotopic subst.: i.r.)
$Be(BH_4)_2$	189, 193 (isotopic subst.: i.r.);
	190 (electron deflection); 191 (i.r.)
$Me_3N\cdot Be(BH_4)_2$	193
$MgClBH_4\cdot 2THF$ (X = Cl, Br, I)	282
$Al(BH_4)_3$	283; 284 (theor.)

Table 3.1—*continued*

Compound	References and comments
$Al(BH_4)_3 \cdot 6NH_3$	205
$Al(BH_4)_3 \cdot NMe_3$	198 (n.m.r.)
Et_2AlBH_4; $Et_2AlBH_4 \cdot L$	
($L = Me_3N$, Et_2O)	285
$(Me_2Al)_2B_2H_6$	71
$Sc(BH_4)_3 \cdot THF$	286
$(\pi\text{-}C_5H_5)_2TiBH_4$	287, 288 (isotopic subst.; i.r.)
$Zr(BH_4)_4$	198 (n.m.r.); 289 (isotopic subst.: i.r.)
$(\pi\text{-}C_5H_5)_2Zr(BH_4)_2$	200, 203 (n.m.r.); 288 (isotopic subst.: i.r.)
$(\pi\text{-}C_5H_5)_2Hf(BH_4)_2$	288 (isotopic subst.: i.r.)
L_3MHBH_4 (M = Ni, L = cy_3P)	290
(M = Ni, I = i-Pr_3P; M = Pd, L = cy_3P, i-Pr_3P)	291
$(Ph_3P)_2CuBH_4$	199 (n.m.r.)
$(dppa)_3(CuBH_4)_2$	292
$Zn(BH_4)_2 \cdot 2L$ (L = THF, py, Me_3N)	204
$Zn(BH_4)_2 \cdot 4L$ (L = NH_3, $MeNH_2$)	204
$(\pi\text{-}C_5H_5)_3ThBH_4$	200, 203 (n.m.r.)
$(\pi\text{-}C_5H_5)_3UBH_4$	200, 201, 203 (n.m.r.); 202

Metal octahydrotriborates (see Section 3.5.1)

$Me_4N[(CO)_4CrB_3H_8]$	207 (x-ray)
$(Ph_3P)_2CuB_3H_8$	163 (n.m.r.); 210 (x-ray)
$L_2CuB_3H_8$; $L_2AgB_3H_8$ (L = tri-p-tolylphosphine)	208, 211 (n.m.r.)

Other metal–boron hydride derivatives (see Sections 3.5.2, 3.5.3 and 3.5.4)

$Al_3B_3(NMe_2)_7H_5$	293
$(Ph_3P)_2CuB_5H_8$; $(Ph_3P)_2CuB_6H_9$	218
$(Et_3P)_2PtB_8H_{12}$	226
$(Et_3P)_2PtH(B_9H_{10}S)$	226 (x-ray)
$(Et_3P)_2Pt\,B_9H_{11}L$ (L = org. base)	226
$L_2AgB_9H_{12}S$ (L = tri-p-tolylphosphine)	208 (n.m.r.)
$pyCoX_2B_{10}H_{11}p\bar{y}$ (X = Cl, Br)	174
$pyBr_2FeB_{10}H_{11}py_2$	294
$Fe(B_{10}H_{11}py_2)_2(B_{10}H_{13}py)$	294
$(CO)_4MB_{10}H_{12}^-$ (M = Cr, Mo, W)	223
$M(B_{10}H_{12})_2^{2-}$ (M = Ni, Co, Pd, Pt)	222
$(Ph_3PMe)_2Zn(B_{10}H_{12})_2$	224 (x-ray)
$(CdB_{10}H_{12} \cdot 2Et_2O)_2$	224 (x-ray)
$Hg(B_{10}H_{12})_2^{2-}$	174
$Cl_2CuB_{10}H_{13}$; $L_2CuB_{10}H_{13}$ (L = tri-p-tolylphosphine)	208 (n.m.r.)
$pyBr_2FeB_{10}H_{13}py$	294
$(py_2B_{10}H_{11})FeBr\,\mu\text{-}Br_2FeBrpy_2$ $(B_{10}H_{13}py)$	294
$(CO)_3Co(n\text{-}B_{18}H_{20})^-$; $(Ph_3P)_2Ni(n\text{-}B_{18}H_{20})$; diphos·Ni($n\text{-}B_{18}H_{20}$)	227
$(CO)_3Mn(B_{10}H_{10}PPh)^-$	229

Table 3.1—*continued*

Compound	References and comments
Type $L_1L_2L_3BH^{2+}$	
py$_3$BH	233
(4-Mepy)$_3$BH	233
py·bipy·BH	234
3,4-diMepy·bipy·BH	234
3,4-diMepy·TMED·BH	233, 234
Type $L_1L_2 BHX^+$	
Me$_3$N·py·BH(CH$_2$SiMe$_3$)	78
Me$_3$N·Me$_3$P·BH(CH$_2$SiMe$_3$)	78
Type $LBHX^+$	
TMED·BHBr	295
DMpip·BHX (X = Cl, Br)	234
Type LBH_2^+	
TMED·BH$_2$	234, 295
TMPD·BH$_2$; TMBD·BH$_2$	234
bipy·BH$_2$	232
2-Me$_2$NC$_2$H$_4$py·BH$_2$	234
DMpip·BH$_2$	234
Type $L_1BH_2L_2^+$	
(i) carbon base	
Me$_3$N·BH$_2$·CH$_2$NMe$_2$H	242
Me$_3$N·BH$_2$·CH$_2$NMe$_2$CH$_2$CO$_2$Et	242
Me$_3$N·BH$_2$·CH$_2$PPh$_3$	296
Me$_3$N·BH$_2$·CH(SiMe$_3$)PMe$_3$	296
Me$_3$N·BH$_2$·CH(SiMe$_3$)AsMe$_3$	296
(ii) nitrogen base	
(Me$_2$NH)$_2$BH$_2$	48, 297
(Me$_2$NH)·BH$_2$·py	297
(Me$_2$NH)·BH$_2$·4-Mepy	297
Me$_3$N·BH$_2$·NCMe	231, 232
Me$_3$N·BH$_2$·NEt$_3$	230
Me$_3$N·BH$_2$·NMe$_2$CH$_2$CO$_2$Et	296; 299 (hydrol.)
Me$_3$N·BH$_2$·NMe$_2$CH$_2$SiMe$_3$	296
Me$_3$N·BH$_2$·NMe$_2$·CH$_2$SMe	296
Me$_3$N·BH$_2$·NH = C(OMe)Me	232
Me$_3$N·BH$_2$NMe$_2$CH$_2$BH$_2$NMe$_3$	242
Me$_3$N·BH$_2$·py	230
Me$_3$N·BH$_2$·2-Xpy (X = F, Cl, OMe)	230
Me$_3$N·BH$_2$·3-Xpy (X = Cl, Br, I)	230
Me$_3$N·BH$_2$·4-Xpy (X = CN, Me, Ph PhCH$_2$, PhCO, MeCO, MeO)	230
Me$_3$N·BH$_2$·4-vinylpy	235
Me$_3$N·BH$_2$·3,4-diMepy	230
Me$_3$N·BH$_2$·3-Me, 5-CNpy	230
Me$_3$N·BH$_2$2,4,6-triMepy	230
Me$_3$N·BH$_2$NMe$_2$C$_6$H$_5$	230
Me$_3$N·BH$_2$·NC$_9$H$_7$	230
py·BH$_2$·NEt$_3$	230

Table 3.1—*continued*

Compound	References and comments
$(py)_2BH_2$	232
$py·BH_2·2\text{-MeOpy}$	230
$py·BH_2·4\text{-Mepy}$	230
$py·BH_2·2,4,6\text{-triMepy}$	230
(iii) phosphorus/arsenic base	
$Me_3N·BH_2·PMe_3$	295
$Me_3N·BH_2·PPh_3$	295
$Me_3N·BH_2·CH_2NMe_2·BH_2PMe_3$	242
$Me_3N·BH_2·PMe_2·CH_2SiMe_3$	296
$Me_3N·BH_2·PMe_2CH_2SMe$	296
$Me_3SiCH_2Me_2N·BH_2·PMe_3$	296
$Me_3N·BH_2·AsMe_3$	295
$Et_3P·BH_2·AsMe_3$	295
(iv) oxygen base	
$Me_3N·BH_2·O$ base (base = Me_3N, py, Me_3P, Me_2S)	298
$Me_3N·BH_2·DMF$	296
$Me_3N·BH_2·DMA$	296
$Me_3SiCH_2NMe_2·BH_2·DMF$	296
$Me_3SiCH_2NMe_2·BH_2·DMA$	296
(v) sulphur base	
$Me_3N·BH_2·SMe_2$	295
Type $L_1BH_2L_2BH_2L_3^{2+}$	
$Me_3N·BH_2·TMED·BH_2NMe_3$	234, 295, 296
$Me_3N·BH_2·TMPD·BH_2NMe_3$	234
$Me_3N·BH_2·TMBD·BH_2·NMe_3$	234
$Me_3SiCH_2NMe_2·BH_2TMED·BH_2·NMe_2CH_2SiMe_3$	296
$py·BH_2·TMED·BH_2·py$	234
$py·BH_2·TMPD·BH_2·py$	234
$py·BH_2·TMBD·BH_2·py$	234
$py·BH_2·DMpip·BH_2·py$	234
$DMF·BH_2·TMED·BH_2·DMF$	296
$DMA·BH_2·TMED·BH_2·DMA$	296
Other boron hydride cations	
$Me_3P·CH_2NMe_2·BH_3^+$	236
$PMDT·BH_2N·Me_3^{3+}$	235
$(TMBD·BH_2)_n^{n+}$	234
$(Me_3N·BH_2·py\text{-}4\text{-}CHCH_2)_n^{n+}$	235

References

1. Council of the American Chemical Society (1968). *Inorg. Chem.*, **7**, 1945
2. Stock, A. E. (1933). *Hydrides of Boron and Silicon*, (Ithaca, New York: Cornell University Press)
3. Schlesinger, H. I. and Burg, A. B. (1942). *Chem. Rev.*, **31**, 1

4. Holzmann, R. T. (1967). *Production of Boranes and Related Research* (New York: Academic Press)
5. Lipscomb, W. N. (1963). *Boron Hydrides,* (New York: W. A. Benjamin)
6. Adams, R. M. (1964). *Boron, Metallo-Boron Compounds and Boranes* (New York: Interscience)
7. Muetterties, E. L. (1967). *The Chemistry of Boron and Its Compounds* (New York: John Wiley)
8. Muetterties, E. L. and Knoth, W. H. (1968). *Polyhedral Boranes* (New York: Marcel Dekker)
9a. Todd, L. J. (1970). *Progr. Boron Chem.,* **2,** 1
9b. Odom, J. D. and Schaeffer, R. (1970). *ibid.,* 141
10. Eaton, G. R. and Lipscomb, W. N. (1969). *N.M.R. Studies of Boron Hydrides and Related Compounds,* (New York: W. A. Benjamin)
11. Schaeffer, R. (1964). *Progr. Boron Chem.,* **1,** 417
12. Eaton, G. R. (1969). *J. Chem. Educ.,* **46,** 547
13. Mooney, E. F. (1969). *Ann. Rev. N.M.R. Spectroscopy,* **2**
14. Parry, R. W. and Walter, M. K. (1968). *Preparative Inorganic Reactions,* **5,** 45
15. Jolly, W. L. (1968). *Inorg. Syn.,* **11**
16. James, B. D. and Wallbridge, M. G. H. (1970). *Progr. Inorg. Chem.,* **11,** 99
17. Schmid, G. (1970). *Angew. Chem. Int. Ed. Engl.,* **9,** 819
18. Kettle, S. F. A. and Tomlinson, V. (1969). *J. Chem. Soc. A,* 2002
19. Kettle, S. F. A. and Tomlinson, V. (1969). *J. Chem. Soc. A,* 2007
20. Kettle, S. F. A. and Tomlinson, V. (1969). *Theor. Chim. Acta,* **14,** 175
21. Switkes, E., Stevens, R. M., Lipscomb, W. N. and Newton, M. D. (1969). *J. Chem. Phys.,* **51,** 2085
22. Switkes, E., Epstein, I. R., Tossell, J. A., Stevens, R. M., Lipscomb, W. N. (1970). *J. Amer. Chem. Soc.,* **92,** 3837
23. Switkes, E., Lipscomb, W. N. and Newton, M. D. (1970). *J. Amer. Chem. Soc.,* **92,** 3847
24. Adamson, G. W. and Linnett, J. W. (1969). *J. Chem. Soc. A,* 1697
25. Pauling, L. (1970). *J. Inorg. Nucl. Chem.,* **32,** 3745
26. Baylis, A. B., Pressley, G. A. and Stafford, F. E. (1966). *J. Amer. Chem. Soc.,* **88,** 2428
27. Fehlner, T. P. and Fridmann, S. A. (1970). *Inorg. Chem.,* **9,** 2288
28. Ganguli, P. S. and McGee, H. A. (1969). *J. Chem. Phys.,* **50,** 4658
29. Fehlner, T. P. and Mappes, G. W. (1969). *J. Phys. Chem.,* **73,** 873
30. Mappes, G. W. and Fehlner, T. P. (1970). *J. Amer. Chem. Soc.,* **92,** 1562
31. Herstad, O., Pressley, G. A. and Stafford, F. E. (1970). *J. Phys. Chem.,* **74,** 874
32. Steck, S. J., Pressley, G. A. and Stafford, F. E. (1969). *J. Phys. Chem.,* **73,** 1000
33. Mappes, G. W., Fridmann, S. A. and Fehlner, T. P. (1970). *J. Phys. Chem.,* **74,** 3307
34. Scwartz, M. E. and Allen, L. C. (1970). *J. Amer. Chem. Soc.,* **92,** 1466
35. Armstrong, D. R. (1970). *Inorg. Chem.,* **9,** 874
36. Cueilleron, J. and Dazard, J. (1970). *Bull. Soc. Chim. Fr.,* 1741
37. Attwood, B. and Shelton, R. A. J. (1970). *J. Less-Common Metals,* **20,** 131
38. Morterra, C. and Low, M. J. D. (1969). *Chem. Commun.,* 862; (1970). *J. Phys. Chem.,* **74,** 1297
39. Kwon, C. T. and McGee, H. A. (1970). *Inorg. Chem.,* **9,** 2458
40. Armstrong, D. R., Duke, B. J. and Perkins, P. G. (1969). *J. Chem. Soc. A,* 2566
41. Lloyd, D. R. and Lynaugh, N. (1970). *Chem. Commun.,* 1545
42. Armstrong, D. R. and Perkins, P. G. (1969). *J. Chem. Soc. A,* 1044
43. Hoffman, K. F. and Engelhardt, U. (1970). *Z. Naturforsch B,* **25,** 317
44. Nainan, K. C. and Ryschkewitsch, G. E. (1969). *Inorg. Chem.,* **8,** 2671
45. Fiaud, J. C. and Kagan, H. B. (1969). *Bull. Soc. Chim. Fr.,* 2742
46. Cowley, A. H. and Mills, J. L. (1969). *J. Amer. Chem. Soc.,* **91,** 2911
47. Ryschkewitsch, G. E. and Cowley, A. H. (1970). *J. Amer. Chem. Soc.,* **92,** 745
48. Miller, V. R., Ryschkewitsch, G. E. and Chandra, S. (1970). *Inorg. Chem.,* **9,** 1427
49. Wiggins, J. W. and Ryschkewitsch, G. E. (1970). *Inorg. Chim. Acta,* **4,** 33
50. Brown, M. P., Heseltine, R. W., Smith, P. A. and Walker, P. J. (1970). *J. Chem. Soc. A,* 410
51. Ryschkewitsch, G. E. and Rademaker, W. J. (1969). *J. Magn. Res.,* **1,** 584
52. Spielvogel, B. F., Purser, J. M. and Moreland, C. G. (1969). *Chem. Eng. News,* Nov. 10, p. 38

53. Uppal, S. S. and Kelly, H. C. (1970). *Chem. Commun.*, 1619
54. Demuynck, J. and Veillard, A. (1970). *Chem. Commun.*, 873
55. Hillier, I. H., Marriott, J. C., Saunders, V. R., Ware, M. J., Lloyd, D. R. and Lynaugh, N. (1970). *Chem. Commun.*, 1586
56. McAllister, T. and Mackle, H. (1969). *Trans. Faraday Soc.*, **65,** 1734
57. Davis, J. and Drake, J. E. (1970). *J. Chem. Soc. A*, 2959
58. Davis, J., Drake, J. E. and Goddard, N. (1970). *J. Chem. Soc. A*, 2962
59. Hodges, H. L. and Rudolph, R. W. (1970). *Abstracts 160th A.C.S. National Meeting, Chicago, Illinois*, INOR 38
60. Jugie, G., Pouyanne, J. P. and Laurent, J. P. (1969). *C. R. Acad. Sci. Ser. C.*, **268,** 1377
61. Jugie, G., Laurent, J. P. and Commenges, G. (1970). *Bull. Soc. Chim. Fr.*, 838
62. Laurent, J. P., Jugie, G. and Commenges, G. (1969). *J. Inorg. Nucl. Chem.*, **31,** 1353
63. Laurent, J. P., Jugie, G., Wolf, R. and Commenges, G. (1969). *J. Chim. Phys. Physicochim. Biol.*, **66,** 409
64. Laussac, J. P., Jugie, G. and Laurent, J. P. (1969). *C. R. Acad. Sci. Ser. C.*, **269,** 698
65. Jugie, G., Laussac, J. P. and Laurent, J. P. (1970). *Bull. Soc. Chim. Fr.*, 2542
66. Jugie, G., Laussac, J. P. and Laurent, J. P. (1970). *J. Inorg. Nucl. Chem.*, **32,** 3455
67. Korshak, V. V., Zamyatina, V. A., Solomatina, A. I., Fedin, E. I. and Petrovskii, P. V. (1969). *J. Organometal. Chem.*, **17,** 201
68. Zamyatina, V. A., Korshak, V. V., Solomatina, A. I. and Dubova, T. A. (1970). *Russ. J. Inorg. Chem.*, **14,** 831
69. Lory, E. R. and Ritter, D. M. (1970). *Inorg. Chem.*, **9,** 1847
70. Deever, W. R., Lory, E. R. and Ritter, D. M. (1969). *Inorg. Chem.*, **8,** 1263
71. Williams, R. E. and Gerhart, F. J. (1970). *Inorg. Nucl. Chem. Lett.*, **6,** 221
72. Jones, D. S. and Lipscomb, W. N. (1969). *J. Chem. Phys.*, **51,** 3133; (1970). *Acta. Crystallogr. Sect. A*, **26,** 196
73. Lafferty, W. J., Maki, A. G. and Coyle, T. D. (1970). *J. Mol. Spectrosc.*, **33,** 345
74. Brundle, C. R., Robin, M. B., Basch, H., Pinsky, M. and Bond, A. (1970). *J. Amer. Chem. Soc.*, **92,** 3863
75. Rose, T., Frey, R. and Brehm, B. (1969). *Chem. Commun.*, 1518
76. Finn, P. A. and Jolly, W. L. (1970). *Chem. Commun.*, 1090
77. Carpenter, J. H., Jones, W. J., Jotham, R. W. and Long, L. H. (1970). *Spectrochim. Acta., Part A*, **26,** 1199
78. McMullen, J. C. and Miller, N. E. (1970). *Inorg. Chem.*, **9,** 2291
79. Levison, K. A. and Perkins, P. G. (1970). *Theor. Chim. Acta.*, **17,** 1
80. Keller, P. C. (1969). *J. Amer. Chem. Soc.*, **91,** 1231
81. Schirmer, R. E., Noggle, J. H. and Gaines, D. F. (1969). *J. Amer. Chem. Soc.*, **91,** 6240
82. Dobson, J. and Schaeffer, R. (1970). *Inorg. Chem.*, **9,** 2183
83. Keller, P. C. (1969). *Chem. Commun.*, 209; *Inorg. Chem.*, **8,** 2457
84. Paine, R. T. and Parry, R. W. (1970). *Abstracts 160th A.C.S. National Meeting, Chicago, Illinois*, INOR 16
85. Ditter, J. F., Spielman, J. R. and Williams, R. E. (1966). *Inorg. Chem.*, **5,** 118
86. Ditter, J. F., Gerhart, F. J. and Williams, R. E. (1968). *Advan. Chem. Ser.*, **72,** 191
87. Steck, S. J., Pressley, G. A., Stafford, F. E., Dobson, J. and Schaeffer, R. (1970). *Inorg. Chem.*, **9,** 2452
88. Hollins, R. E. and Stafford, F. E. (1970). *Inorg. Chem.*, **9,** 877
89. La Prade, M. D. and Nordman, C. E. (1969). *Inorg. Chem.*, **8,** 1669
90. Centofani, L. F., Kodama, G. and Parry, R. W. (1969). *Inorg. Chem.* **8,** 2072
91. Leach, J. B., Onak, T., Spielman, J., Rietz, R. R., Schaeffer, R. and Sneddon, L. G. (1970). *Inorg. Chem.*, **9,** 2170
92. Deever, W. R. and Ritter, D. M. (1969). *Inorg. Chem.*, **8,** 2461
93. Miller, F. M. and Ritter, D. M. (1970). *Inorg. Chem.*, **9,** 1284
94. Gaines, D. F. (1969). *J. Amer. Chem. Soc.*, **91,** 6503
95. Bond, A. C. and Hairston, G. (1970). *Inorg. Chem.*, **9,** 2610
96. Marynick, D. and Onak, T. (1969). *J. Chem. Soc. A*, 1797
97. Allerhand, A., Odom, J. D. and Moll, R. E. (1969). *J. Chem. Phys.*, **50,** 5037
98. Denniston, M. L. and Shore, S. G. (1969). *Abstracts 158th A.C.S. National Meeting, New York, N.Y.*, INOR 104
99. Mishra, I. B. and Burg, A. B. (1970). *Inorg. Chem.*, **9,** 2188
100. Tucker, P. M. and Onak, T. (1969). *J. Amer. Chem. Soc.*, **91,** 6869
101. Tucker, P. M., Onak, T. and Leach, J. B. (1970). *Inorg. Chem.*, **9,** 1430

102. Farrar, T. C., Johannesen, R. B. and Coyle, T. D. (1968). *J. Chem. Phys.*, **49**, 281
103. Geisler, T. C. and Norman, A. D. (1970). *Inorg. Chem.*, **9**, 2167
104. Gaines, D. F. and Iorns, T. V. (1967). *J. Amer. Chem. Soc.*, **89**, 4249; (1968). *J. Amer. Chem. Soc.*, **90**, 6617
105. Dahl, L. F. and Calabrese, J. Unpublished results quoted in ref. 104
106. Burg, A. B. and Heinen, H. (1968). *Inorg. Chem.*, **7**, 1021
107. Burg, A. B. and Mishra, I. B. (1970). *J. Organometal. Chem.*, **24**, C33
108. Gaines, D. F. and Iorns, T. V. (1970). *J. Amer. Chem. Soc.*, **92**, 4571
109. Onak, T. and Leach, J. B. (1970). *J. Amer. Chem. Soc.*, **92**, 3513
110. Rietz, R. R., Schaeffer, R. and Sneddon, L. G. (1970). *J. Amer. Chem. Soc.*, **92**, 3514
111. Dobson, J. and Schaeffer, R. (1968). *Inorg. Chem.*, **7**, 402
112. Geanangel, R. A. and Shore, S. G. (1967). *J. Amer. Chem. Soc.*, **89**, 6771
113. Johnson, H. D., Geanangel, R. A. and Shore, S. G. (1970). *Inorg. Chem.*, **9**, 908
114. Carter, J. C. and Mock, N. L. H. (1969). *J. Amer. Chem. Soc.*, **91**, 5891
115. Odom, J. D. and Schaeffer, R. (1970). *Inorg. Chem.*, **9**, 2157
116. Brubaker, G. L., Denniston, M. L., Shore, S. G., Carter, J. C. and Swicker, F. (1970). *J. Amer. Chem. Soc.*, **92**, 7216
117. Gaines, D. F. and Schaeffer, R. (1964). *Inorg. Chem.*, **3**, 438
118. Collins, A. L. and Schaeffer, R. (1970). *Inorg. Chem.*, **9**, 2153
119. Long, L. H. (1970). *J. Inorg. Nucl. Chem.*, **32**, 1097
120. Dobson, J., Maruca, R. and Schaeffer, R. (1970). *Inorg. Chem.*, **9**, 2161
121. Dobson, J., Keller, P. C. and Schaeffer, R. (1968). *Inorg. Chem.*, **7**, 399
122. Dobson, J., Gaines, D. and Schaeffer, R. (1965). *J. Amer. Chem. Soc.*, **87**, 4072
123. Steck, S. J., Pressley, G. A., Stafford, F. E., Dobson, J. and Schaeffer, R. (1969). *Inorg. Chem.*, **8**, 830
124. Maruca, R., Odom, J. D. and Schaeffer, R. (1968). *Inorg. Chem.*, **7**, 412
125. Keller, P. C. and Schaeffer, R. (1970). *Inorg. Chem.*, **9**, 390
126. Keller, P. C. (1970). *Inorg. Chem.*, **9**, 75
127. Stibr, B., Plesek, J. and Hermanek, S. (1969). *Collect. Czech. Chem. Commun.*, **34**, 3241
128. Plesek, J., Hermanek, S. and Stibr, B. (1970). *Collect. Czech. Chem. Commun.*, **35**, 344
129. Tippe, A. and Hamilton, W. C. (1969). *Inorg. Chem.*, **8**, 464
130. Brill, R., Dietrich, H. and Dierks, H. (1970). *Angew. Chem. Int. Ed. Engl.*, **9**, 524
131. MacLean, D. B., Odom, J. D. and Schaeffer, R. (1968). *Inorg. Chem.*, **7**, 408
132. Pilling, R. L., Tebbe, F. N., Hawthorne, M. F. and Pier, E. A. (1964). *Proc. Chem. Soc.*, 402
133. Keller, P. C., MacLean, D. and Schaeffer, R. O. (1965). *Chem. Commun.*, 204
134. Bodner, G. M. and Sneddon, L. G. (1970). *Inorg. Chem.*, **9**, 1421
135. Onak, T. and Marynick, D. (1970). *Trans. Faraday Soc.*, **66**, 1843
136. Stibr, B., Plesek, J. and Hermanek, S. (1969). *Collect. Czech. Chem. Commun.*, **34**, 194
137. Stuchlik, J., Hermanek, S., Plesek, J. and Stibr, B. (1970). *Collect. Czech. Chem. Commun.*, **35**, 339
138. Sedmera, P., Hanousek, F. and Samek, Z. (1968). *Collect. Czech. Chem. Commun.*, **33**, 2169
139. Plesek, J., Hermanek, S. and Stibr, B. (1969). *Collect. Czech. Chem. Commun.*, **34**, 3233
140. Norman, A. D. and Rosell, S. L. (1969). *Inorg. Chem.*, **8**, 2818
141. Plesek, J., Hermanek, S. and Stibr, B. (1968). *Collect. Czech. Chem. Commun.*, **33**, 691
142. Kuznetsov, N. T. and Klimchuk, G. S. (1969). *Russ. J. Inorg. Chem.*, **14**, 1424
143. Plesek, J., Hermanek, S. and Hanousek, F. (1967). *Collect. Czech. Chem. Commun.*, **32**, 1095; (1968). *Collect Czech. Chem. Commun.*, **33**, 699
144. Friedman, L. B., Cook, R. E. and Glick, M. D. (1970). *Inorg. Chem.*, **9**, 1452
145. Hermanek, S. and Plesek, J. (1970). *Collect. Czech. Chem. Commun.*, **35**, 2488
146. Pitochelli, A. R. and Hawthorne, M. F. (1962). *J. Amer. Chem. Soc.*, **84**, 2318
147. Olsen, F. P., Vasavada, R. C. and Hawthorne, M. F. (1968). *J. Amer. Chem. Soc.*, **90**, 3946
148. McAvoy, J. and Wallbridge, M. G. H. (1969). *Chem. Commun.*, 1378
149. Wittig, G. and Raff, P. (1951). *Z. Naturforsch, B*, **6**, 225
150. Borch, R. F. and Durst, H. D. (1969). *J. Amer. Chem. Soc.*, **91**, 3996
151. Kreevoy, M. M. and Hutchins, J. E. C. (1969). *J. Amer. Chem. Soc.*, **91**, 4329
152. Berschied, J. R. and Purcell, K. F. (1970). *Inorg. Chem.*, **9**, 624
153. Wade, R. C., Sullivan, E. A., Berschied, J. R. and Purcell, K. F. (1970). *Inorg. Chem.*, **9**, 2146

154. Lippard, S. J. and Welcker, P. S. (1970). *Chem. Commun.*, 515
155. Shriver, D. F. (1963). *J. Amer. Chem. Soc.*, **85**, 1405
156. Keller, P. C. (1969). *Chem. Commun.*, 1465
157. Mayer, E. and Laubengayer, A. W. (1970). *Monatsh. Chem.*, **101**, 1138
158. Gilje, J. W., Morse, K. W. and Parry, R. W. (1967). *Inorg. Chem.*, **6**, 1761
159. Mayer, E. and Hester, R. E. (1969). *Spectrochim. Acta. A.*, **25**, 237
160. Keller, P. C. (1969). *Inorg. Chem.*, **8**, 1695
161. Nainan, K. C. and Ryschkewitsch, G. E. (1970). *Inorg. Nucl. Chem. Lett.*, **6**, 765
162. Marynick, D. and Onak, T. (1970). *J. Chem. Soc., A*, 1160
163. Beall, H., Bushweller, C. H., Dewkett, W. J. and Grace, M. (1970). *J. Amer. Chem. Soc.*, **92**, 3484
164. Kodama, G. (1970). *J. Amer. Chem. Soc.*, **92**, 3482
165. Onak, T., Dunks, G. B., Searcy, I. W. and Spielman, J. (1967). *Inorg. Chem.*, **6**, 1465
166. Gaines, D. F. and Iorns, T. V. (1967). *J. Amer. Chem. Soc.*, **89**, 3375
167. Johnson, H. D., Shore, S. G., Mock, N. L. and Carter, J. C. (1969). *J. Amer. Chem. Soc.*, **91**, 2131
168. Benjamin, L. E., Stafiej, S. F. and Takacs, E. A. (1963). *J. Amer. Chem. Soc.*, **85**, 2674
169. Savory, C. G. and Wallbridge, M. G. H. (1970). *Chem. Commun.*, 1526
170. Greenwood, N. N., Gysling, H. J., McGinnety, J. A. and Owen, J. D. (1970). *Chem. Commun.*, 505
171. Siegel, B., Mack, J. L., Lowe, J. V. and Gallaghan, J. (1958). *J. Amer. Chem. Soc.*, **80**, 4523
172. Greenwood, N. N. and Travers, N. F. (1966). *Inorg. Nucl. Chem. Lett.*, **2**, 169
173. Wilks, P. H. and Carter, J. C. (1966). *J. Amer. Chem. Soc.*, **88**, 3441
174. Greenwood, N. N. and Sharrocks, D. N. (1969). *J. Chem. Soc., A*, 2334
175. Carter, J. C. and Wilks, P. H. (1970). *Inorg. Chem.*, **9**, 1777
176. Amberger, E. and Leidl, P. (1969). *J. Organometal Chem.*, **18**, 345
177. Amberger, E. and Leidl, P. (1969). *Chem. Ber.*, **102**, 2764
178. Middaugh, R. L. and Farha, F. (1966). *J. Amer. Chem. Soc.*, **88**, 4147
179. Hawthorne, M. F., Pilling, R. L. and Stokely, P. F. (1965). *J. Amer. Chem. Soc.*, **87**, 1893
180. Schwalbe, C. H. and Lipscomb, W. N. (1969). *J. Amer. Chem. Soc.*, **91**, 194
181. Hawthorne, M. F. and Pilling, R. L. (1966). *J. Amer. Chem. Soc.*, **88**, 3873
182. DeBoer, B. G., Zalkin, A. and Templeton, D. H. (1968). *Inorg. Chem.*, **7**, 1085
183. Wiersema, R. J. and Middaugh, R. L. (1969). *Inorg. Chem.*, **8**, 2074
184. Wiersema, R. J. and Middaugh, R. L. (1970). *J. Amer. Chem. Soc.*, **92**, 223
185. Cook, T. H. and Morgan, G. L. (1970). *J. Amer. Chem. Soc.*, **92**, 6487
186. Almenningen, A., Gundersen, G. and Haaland, A. (1968). *Acta. Chem. Scand.*, **22**, 859
187. Gundersen, G. and Haaland, A. (1968). *Acta. Chem. Scand.*, **22**, 867
188. Armstrong, D. R. and Perkins, P. G. (1968). *Chem. Commun.*, 353
189. Cook, T. H. and Morgan, G. L. (1969). *J. Amer. Chem. Soc.*, **91**, 774
190. Nibler, J. W. and Dyke, T. (1970). *J. Amer. Chem. Soc.*, **92**, 2920
191. Nibler, J. W. and McNab, J. (1969). *Chem. Commun.*, 134
192. Almenningen, A., Haaland, A. and Morgan, G. L. (1969). Unpublished results; Gundersen, G. and Hedberg, K. W. (1969). Unpublished results: quoted in ref. 193
193. Cook, T. H. and Morgan, G. L. (1970). *J. Amer. Chem. Soc.*, **92**, 6493
194. Almenningen, A., Gundersen, G. and Haaland, A. (1968). *Acta. Chem. Scand.*, **22**, 328
195. Bailey, N. A., Bird, P. H. and Wallbridge, M. G. H. (1968). *Inorg. Chem.*, **7**, 1575
196. Lippard, S. J. and Melmed, K. M. (1967). *Inorg. Chem.*, **6**, 2223
197. Bird, P. H. and Churchill, M. R. (1967). *Chem. Commun.*, 403
198. Bailey, N. A., Bird, P. H., Davies, N. and Wallbridge, M. G. H. (1970). *J. Inorg. Nucl. Chem.*, **32**, 3116
199. Grace, M., Beall, H. and Bushweller, C. H. (1970). *Chem. Commun.*, 701
200. von Ammon, R., Kanellakopulos, B., Fischer, R. D. and Laubereau, P. (1969). *Inorg. Nucl. Chem. Lett.*, **5**, 219
201. von Ammon, R., Kanellakopulos, B. and Fischer, R. D. (1970). *Chem. Phys. Lett.*, **4**, 553
202. Anderson, M. L. and Crisler, L. R. (1969). *J. Organometal. Chem.*, **17**, 345
203. von Ammon, R., Kanellakopulos, B., Schmid, G. and Fischer, R. D. (1970). *J. Organometal Chem.*, **25**, Cl
204. Nöth, H., Wiberg, E. and Winter, L. P. (1969). *Z. Anorg. Allg. Chem.*, **370**, 209
205. Maybury, P. C., Davis, J. C. and Patz, R. A. (1969). *Inorg. Chem.*, **8**, 160

206. Klanberg, F., Muetterties, E. L. and Guggenberger, L. J. (1968). *Inorg. Chem.*, **7,** 2272
207. Guggenberger, L. J. (1970). *Inorg. Chem.*, **9,** 367
208. Muetterties, E. L., Peet, W. G., Wegner, P. A. and Alegranti, C. W. (1970). *Inorg. Chem.*, **9,** 2447
209. Lippard, S. J. and Ucko, D. A. (1968). *Inorg. Chem.*, **7,** 1051
210. Lippard, S. J. and Melmed, K. M. (1969). *Inorg. Chem.*, **8,** 2755
211. Muetterties, E. L. and Alegranti, C. W. (1970). *J. Amer. Chem. Soc.*, **92,** 4114
212. Parshall, G. W. (1964). *J. Amer. Chem. Soc.*, **86,** 361
213. Johnson, M. P. and Shriver, D. F. (1966). *J. Amer. Chem. Soc.*, **88,** 301
214. Pearson, R. G. (1963). *J. Amer. Chem. Soc.*, **85,** 3533
215. Rustad, D. S. and Jolly, W. L. (1968). *Inorg. Chem.*, **7,** 213
216. Gaines, D. F. and Iorns, T. V. (1968). *Inorg. Chem.*, **7,** 1041
217. Dobrott, R. D. and Lipscomb, W. N. (1962). *J. Chem. Phys.*, **37,** 1779
218. Brice, V. T. and Shore, S. G. (1970). *Chem. Commun.*, 1312
219. Greenwood, N. N. and Travers, N. F. (1967). *J. Chem. Soc., A,* 880; (1968). *J. Chem. Soc., A,* 15
220. Greenwood, N. N. and Travers, N. F. (1967). *Chem. Commun.*, 216
221. Klanberg, F., Wegner, P. A., Parshall, G. W. and Muetterties, E. L. (1968). *Inorg. Chem.*, **7,** 2072
222. Siedle, A. R. and Hill, T. A. (1969). *J. Inorg. Nucl. Chem.*, **31,** 3874
223. Wegner, P. A., Guggenberger, L. J. and Muetterties, E. L. (1970). *J. Amer. Chem. Soc.*, **92,** 3473
224. Greenwood, N. N., McGinnety, J. A. and Owen, J. D. (1970). Unpublished results; (1971). *J. Chem. Soc., A,* in press.
225. Greenwood, N. N. and McGinnety, J. A. (1965). *Chem. Commun.*, 331; (1966). *J. Chem. Soc., A,* 1090
226. Kane, A. R., Guggenberger, L. J. and Muetterties, E. L. (1970). *J. Amer. Chem. Soc.*, **92,** 2571
227. Sneath, R. L., Little, J. L., Burke, A. R. and Todd, L. J. (1970). *Chem. Commun.*, 693
228. Hertler, W. R., Klanberg, F. and Muetterties, E. L. (1967). *Inorg. Chem.*, **6,** 1696
229. Little, J. L. (1970). *Abstracts 160th A.C.S. National Meeting, Chicago, Illinois.* INOR. 11
230. Nainan, K. C. and Ryschkewitsch, G. E. (1969). *J. Amer. Chem. Soc.*, **91,** 330
231. Ryschkewitsch, G. E. and Zutshi, K. (1970). *Inorg. Chem.*, **9,** 411
232. Benjamin, L. E., Carvalho, D. A., Stafiej, S. F. and Takacs, E. A. (1970). *Inorg. Chem.*, **9,** 1844
233. Ryschkewitsch, G. E., Mathur, M. A. and Sullivan, T. E. (1970). *Chem. Commun.*, 117
234. Ryschkewitsch, G. E. and Sullivan, T. E. (1970). *Inorg. Chem.*, **9,** 899
235. Smith, G. L., Johnson, L. R. and Kelly, H. C. (1970). *Chem. Commun.*, 922
236. Lundberg, K. L., Rowatt, R. J. and Miller, N. E. (1969). *Inorg. Chem.*, **8,** 1336
237. Bond, A. C. and Pinsky, M. L. (1970). *J. Amer. Chem. Soc.*, **92,** 7585
238. Johnson, H. D. and Shore, S. G. (1970). *J. Amer. Chem. Soc.*, **92,** 7586
239. Farrar, T. C. and Tsang, T. (1969). *J. Res. Nat. Bur. Stand., Sect. A.*, **73,** 195
240. Pasto, D. J., Balasubramaniyan, V. and Wojtkowski, P. W. (1969). *Inorg. Chem.*, **8,** 594
241. Merchant, S. Z. and Fung, B. M. (1969). *J. Chem. Phys.*, **50,** 2265
242. Miller, N. E. and Reznicek, D. L. (1969). *Inorg. Chem.*, **8,** 275
243. Kaldor, A., Pines, I. and Porter, R. F. (1969). *Inorg. Chem.*, **8,** 1418
244. Akerfeldt, S., Wahlberg, K. and Hellstrom, M. (1969). *Acta. Chem. Scand.*, **23,** 115
245. Ringertz, H. (1969). *Acta. Chem. Scand.*, **23,** 137
246. Williams, R. L. (1969). *Acta. Chem. Scand.*, **23,** 149
247. Hanic, F. and Subrtova, V. (1969). *Acta. Crystallogr., B.*, **25,** 405
248. Miller, J. J. and Johnson, F. A. (1970). *Inorg. Chem.*, **9,** 69
249. Cowley, A. H. and Mills, J. L. (1969). *J. Amer. Chem. Soc.*, **91,** 2915
250. Perkins, P. G. and Stewart, J. J. (1970). *Inorg. Chim. Acta.*, **4,** 40
251. Switkes, E., Stevens, R. M. and Lipscomb, W. N. (1969). *J. Chem. Phys.*, **51,** 5229
252. Laurent, J. P. and Jugie, G. (1969). *Bull. Soc. Chim. Fr.*, 26
253. Gallais, F., Laurent, J. P. and Jugie, G. (1970). *J. Chim. Phys. Physicochim. Biol.*, **67,** 934
254. Ellerman, J. and Gruber, W. H. (1969). *Chem. Ber.*, **102,** 1
255. Frost, A. A. (1970). *Theor. Chim. Acta.*, **18,** 156
256. Maksic, Z. B. and Randic, M. (1970). *J. Mol. Struct.*, **6,** 215
257. Hensen, K. (1969). *Theor. Chim. Acta.*, **14,** 273

258. Adams, D. M. and Churchill, R. G. (1970). *J. Chem. Soc., A,* 697
259. Kasai, P. H. and McLeod, D. (1969). *J. Chem. Phys.,* **51,** 1250
260. Ferguson, A. C. and Cornwell, C. D. (1970). *J. Chem. Phys.,* **53,** 1851
261. Ryschkewitsch, G. E. and Wiggins, J. W. (1970). *Inorg. Chem.,* **9,** 314
262. Lowe, J. R., Uppal, S. S., Weidig, C. and Kelly, H. C. (1970. *Inorg. Chem.,* **9,** 1423
263. Kuhn, L. P. and Doali, J. O. (1970). *J. Amer. Chem. Soc.,* **92,** 5475
264. Yim, C. T. and Gilson, D. F. R. (1970). *Can. J. Chem.,* **48,** 515
265. Mongeot, H. and Dazord, J. (1970). *Bull. Soc. Chim. Fr.,* 2157
266. Bond, A. C. and Pinsky, M. L. (1970). *J. Amer. Chem. Soc.,* **92,** 32
267. Murphy, C. B. and Enrione, R. E. (1970). *Inorg. Chem.,* **9,** 1924
268. Gaines, D. F. (1969). *J. Amer. Chem. Soc.,* **91,** 1230
269. Kuznetsov, N. T. and Klimchuk, G. S. (1970). *Russ. J. Inorg. Chem.,* **15,** 136
270. Cragg, R. H., Fortuin, M. S. and Greenwood, N. N. (1970). *J. Chem. Soc., A,* 1817
271. Turner, A. G. (1969). *Theor. Chim. Acta.,* **14,** 350
272. Tsang, T. and Farrar, T. C. (1969). *J. Chem. Phys.,* **50,** 3498
273. Coker, E. H. and Hofer, D. E. (1970). *J. Chem. Phys.,* **53,** 1652
274. Dilts, J. A. and Ashby, E. C. (1970). *Inorg. Chem.,* **9,** 855
275. Titov, L. V. and Levicheva, M. D. (1969). *Russ. J. Inorg. Chem.,* **14,** 1522
276. Lalancette, J. M. and Arnac, M. (1969). *Can. J. Chem.,* **47,** 3695
277. Guggenberger, L. J. (1969). *Inorg. Chem.,* **8,** 2771
278. John, K. C., Kaczmarczyk, A. and Soloway, A. H. (1969). *J. Med. Chem.,* **12,** 54
279. Harmon, K. M., Harmon, A. B. and MacDonald, A. A. (1969). *J. Amer. Chem. Soc.,* **91,** 323
280. Harzdorf, C., Niederprum, H. and Odenbach, H. (1970). *Z. Naturforsch. B.,* **25,** 6
281. Shiro, M., Aono, K. and Watanabe, H. (1970). *Chem. Ind. (London),* 564
282. Ewerling, J. and Noth, H. (1970). *Z. Naturforsch. B.,* **25,** 780
283. Mikheeva, V. I. and Zapolskii, S. V. (1970). *Russ. J. Inorg. Chem.,* **15,** 326
284. Levison, K. A. and Perkins, P. G. (1970). *Rev. Roum. Chim.,* **15,** 153
285. Davies, N., Smith, C. A. and Wallbridge, M. G. H. (1970). *J. Chem. Soc., A,* 342
286. Morris, J. H. and Smith, W. E. (1970). *Chem. Commun.,* 245
287. Bercaw, J. E. and Brintzinger, H. H. (1969). *J. Amer. Chem. Soc.,* **91,** 7301
288. Davies, N., James, B. D. and Wallbridge, M. G. H. (1969). *J. Chem. Soc., A,* 2601
289. Davies, N., Saunders, D. and Wallbridge, M. G. H. (1970). *J. Chem. Soc., A,* 2915
290. Green, M. L. H., Munakata, H. and Saito, T. (1969). *Chem. Commun.,* 1287
291. Munakata, H. and Green, M. L. H. (1970). *Chem. Commun.,* 881
292. Carty, A. J. and Efraty, A. (1969). *Inorg. Chem.,* **8,** 543
293. Hall, R. E. and Schram, E. P. (1969). *Inorg. Chem.,* **8,** 270
294. Greenwood, N. N. and Schick, H. (1969). *Chem. Commun.,* 935
295. Smith, G. L. and Kelly, H. C. (1969). *Inorg. Chem.,* **8,** 2000
296. Miller, N. E., Reznicek, D. L., Rowatt, R. J. and Lundberg, K. R. (1969). *Inorg. Chem.,* **8,** 862
297. Miller, V. R. and Ryschkewitsch, G. E. (1970). *J. Amer. Chem. Soc.,* **92,** 1558
298. Miller, N. E. (1969). *Inorg. Chem.,* **8,** 1693
299. Miller, N. E. (1970). *J. Amer. Chem. Soc.,* **92,** 4564
300. Cowley, A. H. and White, W. D. (1969). *J. Amer. Chem. Soc.,* **91,** 1917
301. Fruhbeis, H. and Seelig, F. F. (1970). *Z. Naturforsch. A.,* **25,** 816

4
Carboranes and Metallocarboranes

R. SNAITH and K. WADE
University of Durham

4.1	INTRODUCTION	140
4.2	STRUCTURAL TYPES AND NOMENCLATURE	141
4.3	THE DICARBA-*closo*-DODECABORANES, $C_2B_{10}H_{12}$	145
	4.3.1 *Preparation and isomerisation of the dicarba-closo-dodecaboranes*	145
	4.3.2 *Some properties of the dicarba-closo-dodecaboranes*	147
	4.3.3 *Halogenation of the dicarba-closo-dodecaboranes*	147
	4.3.4 *C-Organo-substituted dicarba-closo-dodecaboranes*	149
	4.3.5 *C-Carboranyl derivatives of some metals and metalloids*	151
	4.3.5.1 *General features*	151
	4.3.5.2 *Some silicon compounds*	151
	4.3.5.3 *Carboranyl-sulphur compounds*	152
	4.3.5.4 *Carboranylmercurials*	154
	4.3.5.5 *Transition metal derivatives*	154
4.4	OTHER DICARBA-BORANES, $C_2B_mH_{m+2,4,\text{or }6}$	155
	4.4.1 *The dicarba-undecaboranes, $C_2B_9H_{11}$ and $C_2B_9H_{13}$*	155
	4.4.2 *Dicarba-boranes containing six, seven, or eight boron atoms*	157
	4.4.2.1 *The arachno-carborane $C_2B_7H_{13}$*	157
	4.4.2.2 *The preparation of $C_2B_6H_8$, $C_2B_7H_9$, and $C_2B_8H_{10}$*	158
	4.4.2.3 *The structures of $C_2B_6H_8$, $C_2B_7H_9$, and $C_2B_8H_{10}$*	159
	4.4.2.4 *Some reactions of $C_2B_7H_9$ and $C_2B_8H_{10}$*	159
	4.4.3 *Dicarba-boranes containing three, four, or five boron atoms*	160
	4.4.3.1 *Preparation of the lower dicarba-boranes*	160
	4.4.3.2 *1,2-Dicarba-nido-pentaborane(7), $C_2B_3H_7$*	162
	4.4.3.3 *2,3-Dicarba-nido-hexaborane(8), $C_2B_4H_8$*	162
	4.4.3.4 *The dicarba-closo-pentaboranes, $C_2B_3H_5$*	162
	4.4.3.5 *The dicarba-closo-hexaboranes, 1,2- and 1,6- $C_2B_4H_6$*	163
	4.4.3.6 *2,4-Dicarba-closo-heptaborane, $C_2B_5H_7$*	163

4.5	CARBORANES WITH OTHER THAN TWO SKELETAL CARBON ATOMS		163
	4.5.1	*Carboranes containing one skeletal carbon atom*	164
		4.5.1.1 *Monocarba-closo-hexaborane(7)*, CB_5H_7	164
		4.5.1.2 *Monocarba-nido-hexaborane(9)*, CB_5H_9	164
		4.5.1.3 *The CB_9, CB_{10} and CB_{11} carboranes*	165
	4.5.2	*Carboranes containing three skeletal carbon atoms; 2,3,4-$C_3B_3H_7$*	166
	4.5.3	*Carboranes containing four skeletal carbon atoms; $C_4B_2H_6$, $C_4B_6H_{10}$ and $C_4B_7H_{11}$*	166
4.6	METALLOCARBORANES AND RELATED SPECIES		167
	4.6.1	*Introduction*	167
	4.6.2	*Main group metallocarboranes*	168
		4.6.2.1 *Species formally derived from $C_2B_9H_{11}^{2-}$*	168
		4.6.2.2 *Species formally derived from $C_2B_4H_6^{2-}$*	169
		4.6.2.3 *Species formally derived from $CB_{10}H_{11}^{3-}$*	169
	4.6.3	*Transition metal complexes of the dicarbollide ion, $C_2B_9H_{11}^{2-}$*	170
		4.6.3.1 *Bis(dicarbollyl) complexes $[M(C_2B_9H_{11})_2]^{n-}$*	170
		4.6.3.2 *Mixed ligand dicarbollyl complexes $(C_2B_9H_{11})ML_x$*	172
	4.6.4	*Other transition metal-carborane complexes*	174
		Complexes of the dicarbacanastide ion (3,6)-1,2-$C_2B_8H_{10}^{4-}$	174
		4.6.4.2 *Complexes of the dicarbazapide ion $C_2B_7H_9^{2-}$*	174
		4.6.4.3 *Complexes of the ions $C_2B_6H_8^{2-}$ and $C_2B_6H_8^{4-}$*	176
		4.6.4.4 *Complexes of the ion $C_2B_4H_6^{2-}$*	176
		4.6.4.5 *Complexes of the ion $C_2B_3H_7^{2-}$*	176
		4.6.4.6 *Complexes of the ion $CB_{10}H_{11}^{3-}$*	177
		4.6.4.7 *Complexes of the ions $CB_9H_{10}P^{2-}$ and $CB_9H_{10}As^{2-}$*	178
		4.6.4.8 *Complexes of the ion $C_3B_3H_6^-$*	179

4.1 INTRODUCTION

Carboranes are those mixed hydrides of carbon and boron in which both carbon and boron atoms feature in the (electron-deficient) molecular skeleton. The first examples to be prepared and characterised were made during the big expansion in boron hydride research in the 1950s, when derivatives of the icosahedral carborane, $C_2B_{10}H_{12}$, were obtained from experiments aimed at the synthesis of organo derivatives of decaborane(14) for possible use as high-energy fuels[1]. The early findings of American and Russian carborane researchers were released for publication in 1963, since when there has been considerable activity in the area, almost exclusively on the part of groups in the U.S.A. and Russia.

Reviews charting progress abound[1-15, 170, 171]. Whereas Bobinski[1] briefly outlined the accidental discovery of icosahedral carboranes, Onak[2], Stanko et al.[3], Isslieb et al.[4], Hawthorne[5], Köster and Grassberger[6] and Williams[7, 8] have given more detailed accounts of varying emphasis. Annual developments have received prominent treatment in 'Annual Reports of the London Chemical Society' and in 'Organometallic Chemistry Reviews, Section B' and a comprehensive survey is provided by Grimes' book[9]. Other reviews have dealt with specific aspects, e.g., rearrangement reactions[10], polymers incorporating carboranes[11], labelled isotope[170] and n.m.r.[12] studies on boron hydrides including carboranes, carboranyl derivatives of various elements[13] and metallocarboranes[14, 15, 171].

The recent origin and extremely rapid growth of carborane chemistry have made it unrealistic to use a specific earlier review as a basis for the present survey. Instead, the major features that have emerged will be outlined, in sufficient detail to illustrate particularly those respects in which carboranes differ from other substances. The relatively straightforward derivative chemistry of icosahedral carboranes is accordingly covered in less detail than some other aspects, although leads into the copious literature of the subject are provided.

4.2 STRUCTURAL TYPES AND NOMENCLATURE

Known carboranes, and carboranes known in the form of derivatives, are listed in Table 4.1. Like all known higher boranes and cage borane anions, all known carboranes adopt structures which fall into one or other of the following three categories: (a) *closo*-, (b) *nido*-, and (c) *arachno*- structures[16].

(a) *Closo*- structures are adopted by species of formula $B_nH_n^{2-}$, $CB_{n-1}H_n^-$ and $C_2B_{n-2}H_n$ (where n is an integer in the range 5–12), i.e. by neutral compounds $C_aB_mH_{m+2}$ in general ($a = 0, 1$ or 2). In these, the $n(= a+m)$ skeletal carbon and boron atoms occupy the corners of the triangular-faced polyhedra shown in Figure 4.1, in which the symmetries of the polyhedra and skeletal atom numbering conventions are also indicated. Each skeletal atom has a single terminal hydrogen attached. The systematic names for these compounds recommended by the American Chemical Society Nomenclature Committee[17] indicate the positions and numbers of the carbon atoms, the closed cage structure, and the overall number of cage atoms in that sequence. For example, the three possible isomers of icosahedral carboranes of formula $C_2B_{10}H_{12}$ are 1,2-, 1,7- and 1,12-dicarba-*closo*-dodecaborane, respectively (see Figure 4.3). The prefix '*closo*-' replaces the earlier '*clovo*-' which was dropped because of confusion with '*chloro*-'.

(b) *Nido*- (Greek for 'nest') structures are adopted by substances of formulae $C_aB_mH_{m+4}$ ($a = 0 \rightarrow 4$). The polyhedra shown in Figure 4.1 again form the basis for the structures of these compounds, but with one important difference from *closo*- species: invariably, one cage corner is unoccupied by a skeletal atom. The n skeletal atoms occupy all but one of the corners of the $(n+1)$-cornered polyhedron. Examples include the complete series, from $a = 0$ to $a = 4$, of compounds $C_aB_mH_{m+4}$ containing six skeletal atoms $(a+m = 6)$. Their pentagonal pyramidal structures are shown in Figure 4.2.

(c) *Arachno*- (Greek for 'cobweb') structures are adopted by substances of formulae $C_aB_mH_{m+6}$ (a lies in the range $0 \rightarrow 6$). In these, the n skeletal atoms define all but two of the corners of the $(n+2)$-cornered polyhedron. Though several boranes (B_4H_{10}, B_5H_{11}, etc.) fall into this category, neutral carborane examples are at present few (e.g., $C_2B_7H_{13}$), though *arachno*- anions

Table 4.1 Known carboranes and carborane anions

Number of skeletal bonding electron-pairs	Polyhedron on which skeleton is based (Figure 4.1)	Closo- species	Nido- species (see also Table 4.2)	Arachno- species (see also Table 4.2)
13	Icosahedron	$CB_{11}H_{12}^-$ $C_2B_{10}H_{12}$ (Figure 4.3) $(B_{12}H_{12}^{2-})$	$CB_{10}H_{14}, CB_{10}H_{13}^-$ $CB_{10}H_{11}^{3-}$ (Figure 4.8) $C_2B_9H_{13}, C_2B_9H_{12}^-,$ $C_2B_9H_{11}^{2-}$ (Figure 4.5) $C_4B_7H_{11}$	$CB_9H_{11}^{5-}$ $C_2B_8H_{10}^{4-}$ (Figure 4.11)
12	Octadecahedron	$CB_{10}H_{11}^-$ $C_2B_9H_{11}$ (Figure 4.5) $(B_{11}H_{11}^{2-})$	$C_4B_6H_{10}$ $(B_{10}H_{14})$	$C_2B_7H_{13}$ (Figure 4.6) $C_2B_7H_{12}^-$ $C_2B_7H_{11}^{2-}$
11	Bicapped Archimedean antiprism	$CB_9H_{10}^-$ $C_2B_8H_{10}$ (Figure 4.6) $(B_{10}H_{10}^{2-})$	$C_2B_7H_9^{2-}$ (Figure 4.12)	$C_2B_6H_8^{4-}$ (Figure 4.12)
10	Tricapped trigonal prism	$C_2B_7H_9$ (Figure 4.6) $(B_9H_9^{2-})$	$C_2B_6H_8^{2-}$ (Figure 4.12)	
9	Dodecahedron	$C_2B_6H_8$ (Figure 4.6) $(B_8H_8^{2-})$		
8	Pentagonal bipyramid	$C_2B_5H_7$ (Figure 4.7) $(B_7H_7^{2-})$	$CB_5H_9, C_2B_4H_8$ $C_2B_4H_7^-, C_2B_4H_6^{2-},$ $C_3B_3H_7, C_3B_3H_6^-$ $C_4B_2H_6; (B_6H_{10})$ (Figure 4.2)	$C_2B_3H_7^-$ (B_5H_{11})
7	Octahedron	$CB_5H_7, CB_5H_6^-$ $C_2B_4H_6$ (Figure 4.7) $(B_6H_6^{2-})$	$C_2B_3H_7$ (Figure 4.7) (B_5H_9)	(B_4H_{10})
6	Trigonal bipyramid	$C_2B_3H_5$ (Figure 4.7)		$(B_3H_8^-)$

Some familiar boranes and borane anions are listed in parentheses

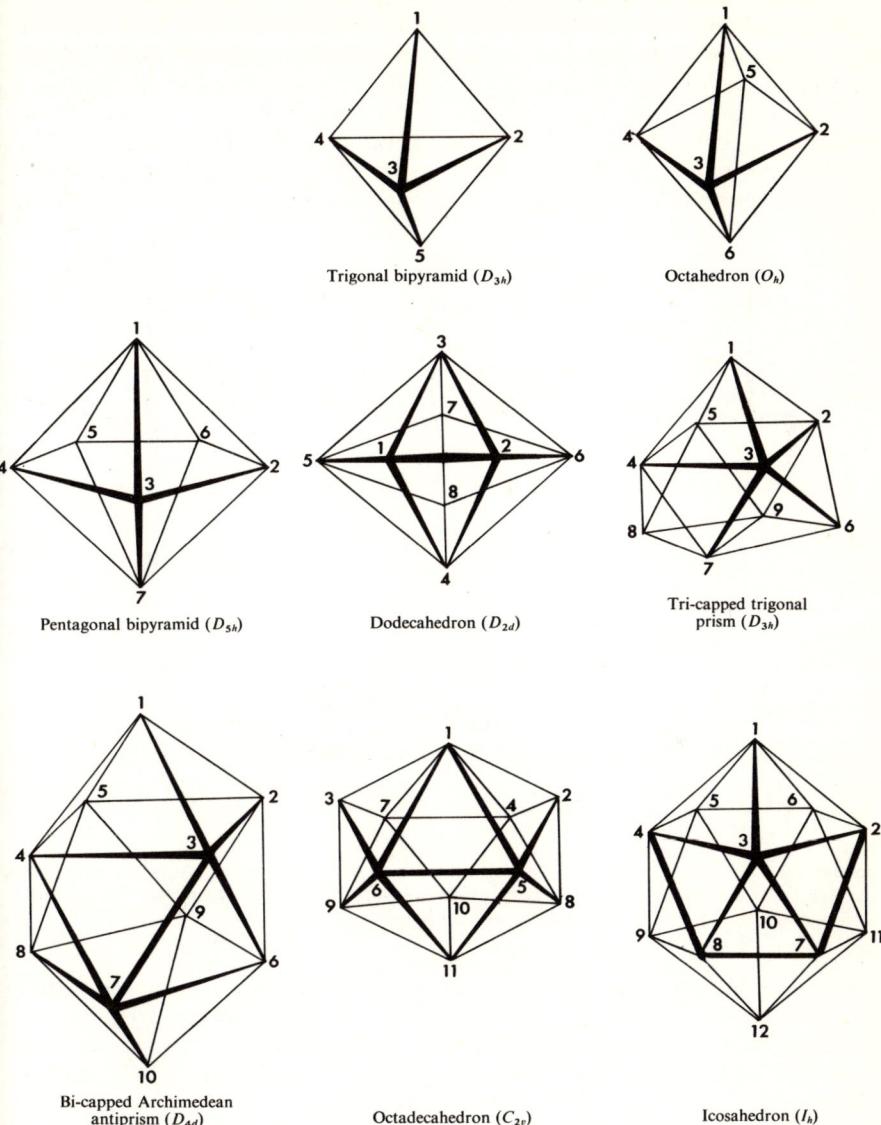

Figure 4.1 The polyhedra upon which the structures of all known carboranes may be regarded as based

$C_2B_6H_8^{4-}$ and $C_2B_8H_{10}^{4-}$ are known in the form of transition metal complexes, the structures of which are illustrated below.

A rationalisation of these three types of carborane structure follows from a consideration of the number of skeletal bonding electrons they contain. Molecular orbital treatments of the *closo-* borane species $B_nH_n^{c-}$ have shown that, for all the polyhedra shown in Figure 4.1, a closed shell electronic configuration is reached in the case of the anion $B_nH_n^{2-}$ [18, 19], i.e. *closo-* species with n skeletal atoms have $(n+1)$ bonding molecular orbitals. The isoelectronic *closo-* carboranes with n skeletal carbon and boron atoms likewise require $(n+1)$ pairs of skeletal bonding electrons for a closed shell

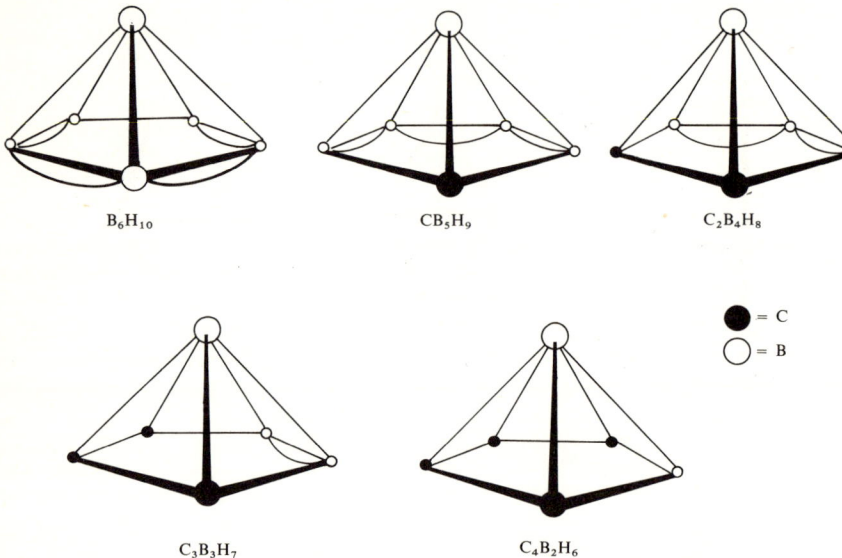

Figure 4.2 The *nido-*carboranes formally related to B_6H_{10}. (All have eight pairs of electrons bonding the six cage atoms together. Hydrogen bridges are represented by curved lines; terminal hydrogens are not shown in this and subsequent figures)

electronic configuration. *Nido-* species, on the other hand, all of which may be regarded as formally derived from hypothetical species $B_nH_n^{4-}$, have $(n+2)$ pairs of skeletal bonding electrons with which to hold together their n skeletal atoms, i.e., the appropriate number for the $(n+1)$- cornered polyhedron. *Arachno-* species, formally derived from hypothetical anions $B_nH_n^{6-}$, have $(n+3)$ pairs of skeletal bonding electrons to bind their n skeletal atoms together, i.e., the appropriate number for the $(n+2)$-cornered polyhedron.

Of these different types of carborane, the dicarba-*closo*-boranes $C_2B_mH_{m+2}$ are those compounds which have been by far the most thoroughly studied, and are given greatest prominence in this review. The icosahedral compounds having $m = 10$ are dealt with first, as their chemistry is by now well documented: sections dealing with the smaller *closo-* compounds and some

CARBORANES AND METALLOCARBORANES

nido- dicarba species follow. Species containing other than two carbon atoms are dealt with next, and the final section is devoted to compounds containing another element in the cage, the metallocarboranes.

4.3 THE DICARBA-*CLOSO*-DODECABORANES $C_2B_{10}H_{12}$

These compounds, the highest known members of the series $C_2B_mH_{m+2}$, contain the icosahedral C_2B_{10} skeleton, which is so thermally stable and survives attack by such a wide range of reagents as to allow these substances to have an extensive derivative chemistry, and indeed to have prompted their incorporation in thermally stable polymers[11]. They are dealt with before smaller dicarba-*closo*-boranes because of their importance and because they serve as key intermediates through which a number of other carboranes can be prepared.

4.3.1 Preparation and isomerisation of the dicarba-*closo*-dodecaboranes

The crucial cage-forming reaction through which the dicarba-*closo*-dodecaboranes are prepared is that between decaborane and an acetylene in the presence of a Lewis base, which is usually of the π-acid type (e.g., RCN, R_2S, R_3P) [2–9, 20]:

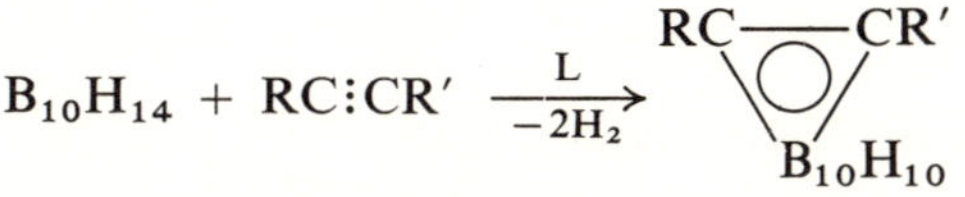

$$B_{10}H_{14} + RC\vcentcolon CR' \xrightarrow[-2H_2]{L} \underset{B_{10}H_{10}}{RC\text{---}CR'} \quad c.\ 70\% \text{ yield}$$

The substituents R and R' can be hydrogen, alkyl, aryl or organofunctional groups not containing active hydrogen, which would cause degradation of the decaborane. Use of substituted decaboranes leads to B-substituted carboranes, which are also accessible through another cage-completing reaction, that between the *nido*- anion $C_2B_9H_{11}^{2-}$ and boron halides RBX_2 (R = F, Br, NR_2, organo) [21, 22]:

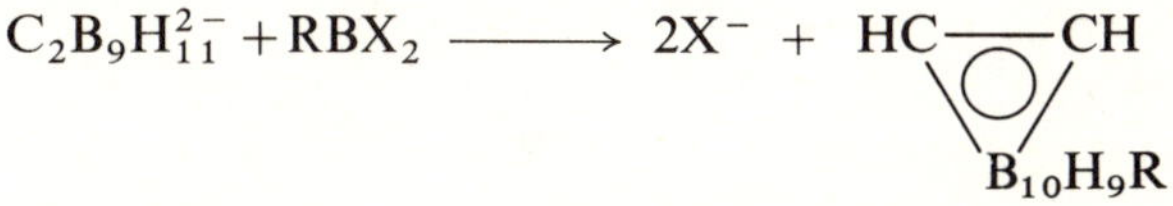

$$C_2B_9H_{11}^{2-} + RBX_2 \longrightarrow 2X^- + \underset{B_{10}H_9R}{HC\text{---}CH}$$

The *nido*- anion $C_2B_9H_{11}^{2-}$ is itself prepared by degradation of $C_2B_{10}H_{12}$, however, so this does not provide an alternative route to the parent species. The introduction of other substituents is described below.

1,2-Dicarba-*closo*-dodecaborane, commonly referred to as 'ortho-carborane' ('barene' in Russian papers) is unaffected by heat up to about 470 °C when it isomerises smoothly and in good yield (c. 90%) to give the *meta*-isomer, 1,7-$C_2B_{10}H_{12}$ (Figure 4.3): the yield is almost quantitative if the *ortho*-isomer is treated at 600 °C for about 30 s. If held longer at this higher temperature, or at 700 °C for a few seconds, it gives the *para*-isomer,

1,12-$C_2B_{10}H_{12}$, though in only 20% yield as general cage degradation sets in [23].

$$1,2\text{-}C_2B_{10}H_{12} \xrightarrow{480-500\,°C} 1,7\text{-}C_2B_{10}H_{12} \xrightarrow{600\,°C} 1,12\text{-}C_2B_{10}H_{12}$$

Though the isomer with the carbon atoms furthest apart, the *para*-isomer, is that which is thermodynamically most stable, it is apparently significant that it is not obtained in high yield by such isomerisations, one likely mechanism for which involves a cubo-octahedral intermediate of the type shown in Figure 4.4 [10]. Such an intermediate would allow originally adjacent carbon

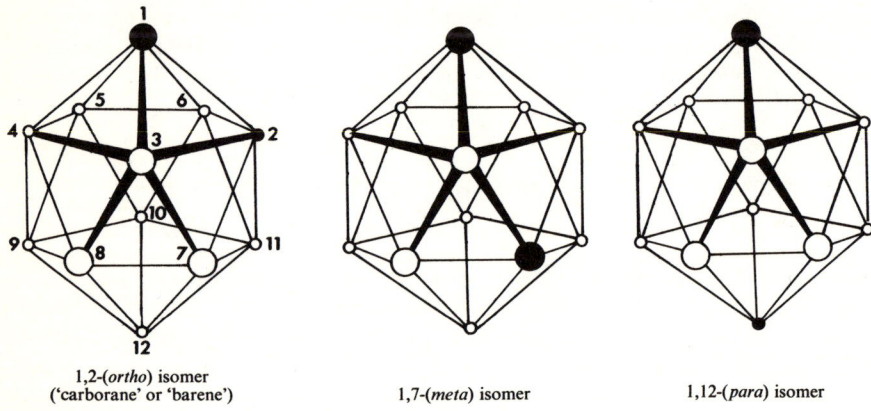

Figure 4.3 The three dicarba-*closo*-dodecaboranes, $C_2B_{10}H_{12}$

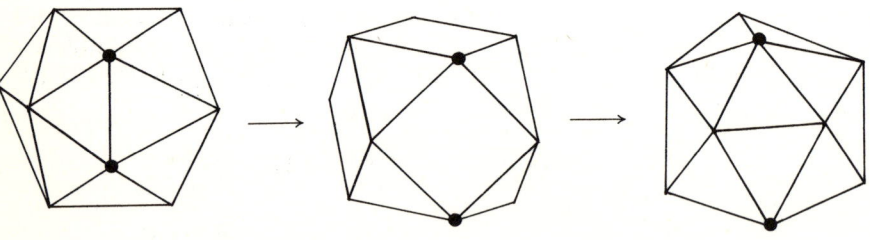

Figure 4.4 A possible route from 1,2-$C_2B_{10}H_{12}$ to 1,7-$C_2B_{10}H_{12}$ via an intermediate with cubo-octahedral geometry (an example of a diamond square-diamond' mechanism)

atoms to move to *meta*-positions by a 'diamond-square-diamond' mechanism but would not allow further isomerisation to the *para*-isomer. For this latter isomerisation, it is possible that one section of the cage rotates with respect to the remainder. Detailed studies[24-27] on the rearrangement of substituted carboranes consistent with such mechanisms have been surveyed by Lipscomb[10, 27]. These thermal isomerisations can be reversed chemically, e.g.[28],

$$1,7\text{-}C_2B_{10}H_{12} \xrightarrow[\text{(ii) } H_2O]{\text{(i) 2Na}} 16\% \; 1,2\text{-}C_2B_{10}H_{12}$$

$$1,12\text{-}C_2B_{10}H_{12} \xrightarrow{\text{Na/liqd.NH}_3} Na_2\,[C_2B_{10}H_{12}] \xrightarrow{\text{KMnO}_4} 1,7\text{-}C_2B_{10}H_{12}$$

The extra pair of electrons in the intermediate anion $C_2B_{10}H_{12}^{2-}$ may well cause distortion to a *nido*-structure based on a 13-cornered polyhedron, facilitating this isomerisation.

4.3.2 Some properties of the dicarba-*closo*-dodecaboranes

All three icosahedral carboranes are chemically very stable, remarkably so in comparison with the decaborane, $B_{10}H_{14}$, used in their synthesis. They are stable to hydrolysis, to strong oxidising agents such as potassium permanganate and hydrogen peroxide, and to reducing agents. However, the *ortho*- and *meta*-isomers, though not the *para*, are susceptible to nucleophilic attack by, for example, methoxide ion[29] or hydrazine[30], which attack one of the cage boron atoms adjacent to both carbons and remove it from the cage, generating the dicarba-*nido*-dodecahydroundecaborate(-1) ion $C_2B_9H_{12}^-$. This cage degradation reaction provides the route into metallocarboranes containing the *nido*-ligand $C_2B_9H_{11}^{2-}$, and also to lower *closo*-dicarba species.

X-ray crystallographic studies on three brominated *ortho*-carboranes[31], the 9,12-dibromo-, 8(or 10),9,12-tribromo-, and 8,9,10,12-tetrabromo-compounds have established the effectively icosahedral geometry of their skeletons, comparable to that of the *closo*-dodecahydrododecaborate(2-) anion $B_{12}H_{12}^{2-}$ [32]. Average cage edge distances are as follows: C—C, 1.64 Å; C—B3(6), 1.71 Å; C—B4(5), 1.73 Å; B—B, 1.77 Å; cf. B—B for $B_{12}H_{12}^{2-}$, 1.77 Å.

With thirteen pairs of skeletal bonding electrons shared between the thirty edges of the icosahedron, these distances correspond roughly to two-centre bonds of order one-half. Comparable distances in 9,10-dibromo-*meta*-carborane $C_2B_{10}H_{10}Br_2$ are C—B, 1.69 Å; B—B, 1.78 Å [33]. The cage diameter is 3.23 Å (C—B), 3.34 Å (B—B). The more electronegative carbon atoms attract a greater electron density than do the boron atoms of the cage, imparting measurable dipole moments to the *ortho*- and *meta*-isomers[34] and causing there to be decreasing negative charge on the boron atoms of 1,2-$C_2B_{10}H_{12}$ in the sequence 9,12 > 8,10 > 4,5,7,11 > 3,6 [35]. For the *meta* isomer, 1,7-$C_2B_{10}H_{12}$, decreasing negative charge on the boron atoms is expected in the sequence 9,10 > 5,12 ≥ 4,6,8,11 > 2,3 [33], whereas in the *para*-isomer, all ten boron atoms are equivalent.

4.3.3 Halogenation of the dicarba-*closo*-dodecaboranes

The icosahedral carborane skeleton is sufficiently stable to allow halogenation of substituent hydrogens under vigorous conditions to be used as a guide to the electron distribution in the cage. Indeed, all ten *B*-attached hydrogen atoms of $1,2\text{-}C_2B_{10}H_{12}$ can be substituted by photochemical chlorination using gaseous chlorine[36]. The decachloro derivative, $C_2H_2B_{10}Cl_{10}$, is a dibasic acid of strength comparable to carboxylic acids (pK_A c. 7), forming salts with tertiary amines, from which the perchlorinated derivative can be prepared:

$$C_2H_2B_{10}Cl_{10} \xrightarrow{2Et_3N} (Et_3NH)_2(C_2B_{10}Cl_{10}) \xrightarrow{Cl_2} C_2B_{10}Cl_{12}$$

Electrophilic halogenation in the presence of aluminium halide replaces the boron-attached hydrogens in a sequence consistent with the ground-state charge distribution deduced from molecular orbital calculations[31, 35, 37, 38]. For example, bromination of $1,2\text{-}C_2B_{10}H_{12}$ by $Br_2/AlBr_3$ affords successively the 9,12-dibromo-, 8(or 10),9,12-tribromo-, and 8,9,10,12-tetrabromo- compounds[31], the degree of substitution being limited apparently by the bulk of the substituents. Electrophilic chlorination follows a similar sequence, though affording more fully halogenated end products. For $1,7\text{-}C_2B_{10}H_{12}$, the preferred sequence for electrophilic substitution is $9,10 \gg 4,6,8,11 > 5,12 \gg 2,3$ [39].

Electrophilic iodination (and bromination) of the *ortho-* and *meta-*carboranes has been carried out in aqueous sulphuric acid solution[40] with comparable results to those obtained with aluminium halide catalyst. For example, 9-iodo-*ortho*-carborane, 9-iodo-*meta*-carborane and 9,12-diiodo-*meta*-carborane have been prepared thus using an iodine/iodate iodinating mixture. The substituent iodine can be replaced by chlorine using copper(I) chloride at 250–350 °C [41]; reduction to the parent carborane is effected by sodium in liquid ammonia[42]; and an Ullmann-type coupling reaction occurs with copper[43]:

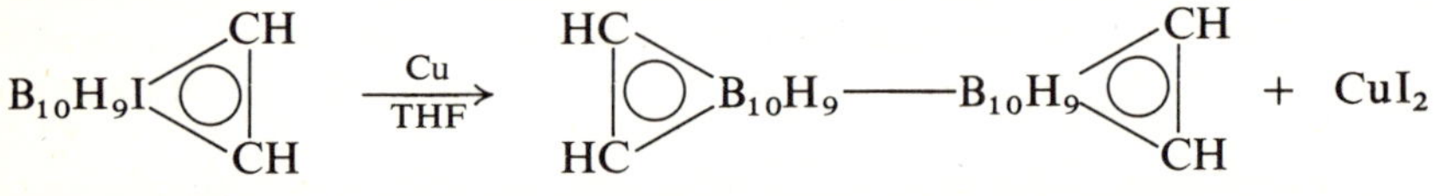

$$B_{10}H_9I\underset{CH}{\overset{CH}{\diamondsuit}} \xrightarrow{Cu \atop THF} \underset{HC}{\overset{HC}{\diamondsuit}}B_{10}H_9\text{—}B_{10}H_9\underset{CH}{\overset{CH}{\diamondsuit}} + CuI_2$$

The electrophilic halogenation of *para*-carborane proceeds much less readily than that of the *ortho-* or *meta-*isomer[25, 44], as had been predicted by Lipscomb on the basis of his charge distribution calculations, from which the maximum negative charges were estimated to be -0.28 e on the 9 and 12 positions of *ortho*-carborane, -0.16 e on the 9 and 10 borons of *meta*-carborane and -0.03 e on the ten equivalent borons of *para*-carborane. The sharply differing reactivities of *meta-* and *para-*carborane allow preferential chlorination of the former to be used to facilitate separation of pure samples of the latter from the mixture obtained by thermal isomerisation of *ortho*-carborane[44].

All three carborane isomers can be fluorinated using fluorine in liquid hydrogen fluoride, the final products being the *B*-deca-fluoro derivatives[45]:

$$C_2H_2B_{10}H_{10} + 10F_2 \xrightarrow{HF; 0\,°C} C_2H_2B_{10}F_{10} + 10HF$$

Though the *C*-attached hydrogens are unaffected under these conditions, they can be fluorinated via the C,C'-dilithio derivative:

$$B_{10}H_{10}C_2Li_2 \xrightarrow{2ClO_3F} B_{10}H_{10}C_2F_2 \xrightarrow{10F_2} B_{10}F_{10}C_2F_2$$

All three B-decafluorocarboranes are hydrolysed by water or on exposure to moist air, in marked contrast to other B-poly-halogenated carboranes, which are hydrolytically stable.

4.3.4 C-organo-substituted dicarba-*closo*-dodecaboranes

The weakly acidic carbon-attached hydrogen atoms of $1,2$-$C_2B_{10}H_{12}$ can be replaced by lithium by reaction with lithium alkyl or amide[2-9]:

$$B_{10}H_{10}C_2H_2 \xrightarrow{n\text{-BuLi}} B_{10}H_{10}C_2HLi \xrightarrow{n\text{-BuLi}} B_{10}H_{10}C_2Li_2$$

The C-lithio derivatives behave like typical organolithium reagents RLi, affording RX + LiX with halogens X_2, RR' with alkyl or aryl halides R'X, RCOR' with R'COCl, RCO_2Li with CO_2, RNO with NOCl, etc. (R throughout is the carboranyl residue). Most of this work is of a vintage to be accessible through the early reviews[2-7]. Among the more interesting products are some cyclic systems derived from the dilithio derivative of $1,2$-$C_2B_{10}H_{12}$. Phosgene, for example, affords a product in which two carborane residues are bridged by two CO groups[46]:

$$2B_{10}H_{10}C_2Li_2 + 2COCl_2 \longrightarrow$$

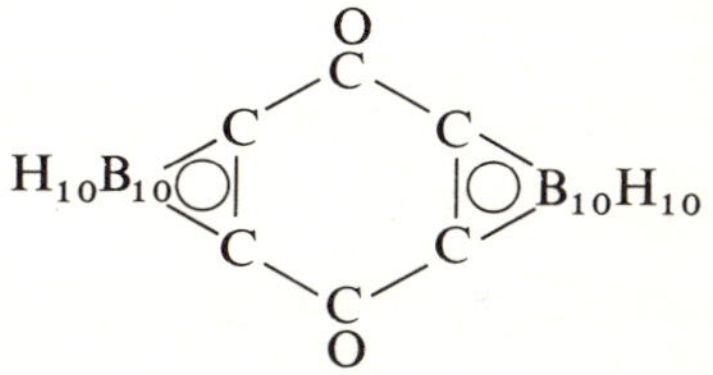

The six carbon atoms of the bridging unit are not quite co-planar, adopting a slight chair configuration.

Benzocarborane, prepared by the reaction sequence[47]

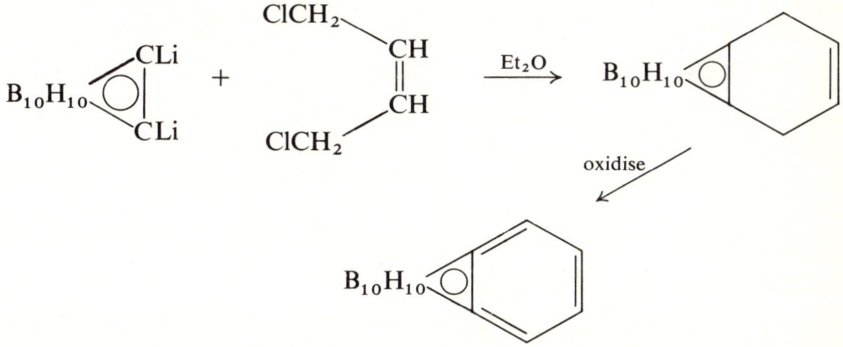

'benzocarborane'

provided evidence of conjugation of the 'aromatic' carborane nucleus with an organic π-system. That its six-membered carbon ring had appreciable aromatic character was indicated by its ^{1}H n.m.r. spectrum, which contained a multiplet centred at 3.63 τ (between typical olefinic and aromatic values), and by its u.v. spectrum, also appropriate for an aromatic system. In the analogous naphthacarborane, prepared by a similar sequence of reactions[48],

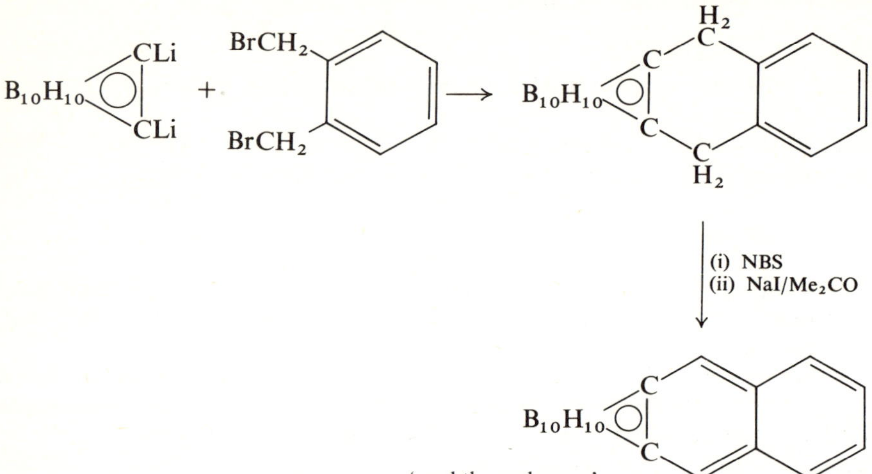

'naphtha carborane'

the inner ring appears to have some aromatic character, even though the double bonds are relatively localised, so that the outer ring is susceptible to attack by dienophiles (the bulk of the carborane residue inhibits attack at the inner ring).

Though most work has been carried out on the substitution of 1,2-$C_2$$B_{10}H_{12}$, and cyclic derivatives of the type just mentioned cannot be prepared from the 1,7-isomer (polymers tend to form instead), the same route into the derivative chemistry of *meta*-carborane is available, i.e., through the lithio derivatives. Carboranyl Grignard reagents have also found some use. These are accessible by various routes, e.g.,

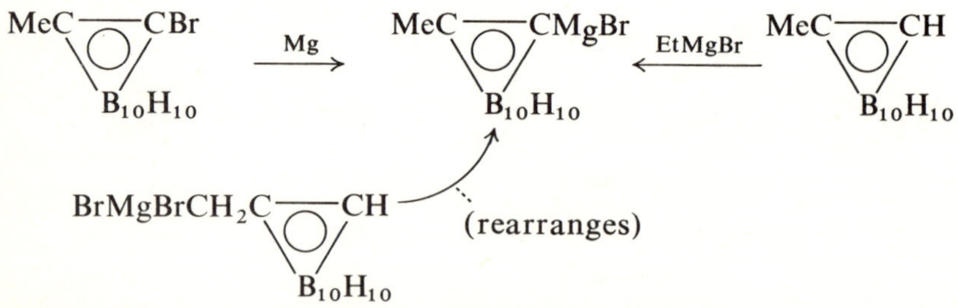

References 49 and 50 provide a lead into their literature.

A major use of carboranylmagnesium and more particularly carboranyl-lithium derivatives has been in the preparation of carboranyl derivatives of many other elements, generally by reaction with a halide of the other element. This use is outlined in the following section.

CARBORANES AND METALLOCARBORANES

Note: Many papers additional to those cited in Section 4.3.4 above have been concerned with carboranyl derivatives containing functional groups. They have dealt with the following classes of derivative: (a) carboranyl haloalkanes[49, 50, 172–174]; (b) alkenes[175, 176]; (c) carboxylic acids[175, 178–183]; (d) alcohols[184–188]; (e) ketones[188–193]; (f) aldehydes[194–199]; (g) amines[200–202]; and (h) hydroxycarboranes[203].

4.3.5   C-carboranyl derivatives of some metals and metalloids[13]

4.3.5.1   General features

Many reactions have been investigated of the type

$$xB_{10}H_{10}C_2RLi + Cl_xMX_y \rightarrow (B_{10}H_{10}C_2R)_xMX_y + xLiCl$$

in which R = H, alkyl, or aryl; M = a metal or metalloid; and X_y are the ligands retained by the metal or metalloid during the reaction. Derivatives of boron, silicon, germanium, tin, phosphorus, arsenic, antimony, sulphur, mercury, cobalt, nickel, copper and gold have been prepared in this way[13]. The bulk of the carboranyl residue, particularly the *ortho*-carboranyl residue with a phenyl group on the other carbon atom, prevents the attachment of many such groups to a small metal or metalloid atom, so that for example incomplete reaction occurs with the Group IV tetrachlorides[51]:

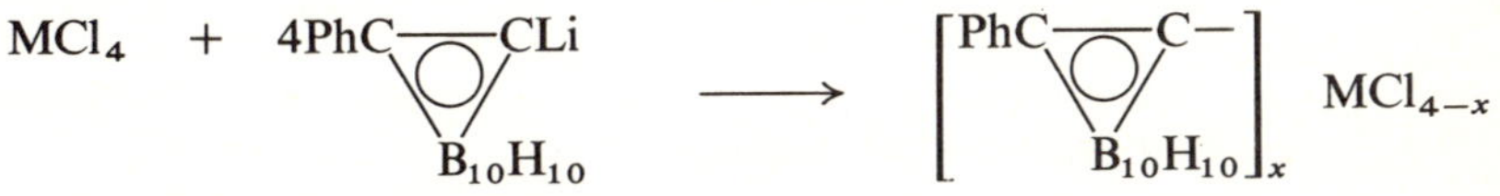

$$MCl_4 + 4PhC\text{—}CLi\underset{B_{10}H_{10}}{\diagup\diagdown} \longrightarrow \left[PhC\text{—}C\text{—}\underset{B_{10}H_{10}}{\diagup\diagdown}\right]_x MCl_{4-x}$$

$x = 2$ when M = Si, Ge; $x = 3$ when M = Sn.

Phosphorus trichloride similarly gives the bis(carboranyl) derivative $(B_{10}H_{10}C_2Ph)_2PCl$, whereas the trichlorides of arsenic and antimony give tris derivatives $(B_{10}H_{10}C_2Ph)_3M$.

Use of the dilithio carborane allows substitution at both carbon atoms, e.g.[51, 52],

$$B_{10}H_{10}C_2Li_2 + 2R_3MCl \rightarrow B_{10}H_{10}C_2(MR_3)_2 \quad (M = Si, Ge, Sn)$$

Cyclic products may results if a dihalide is used:

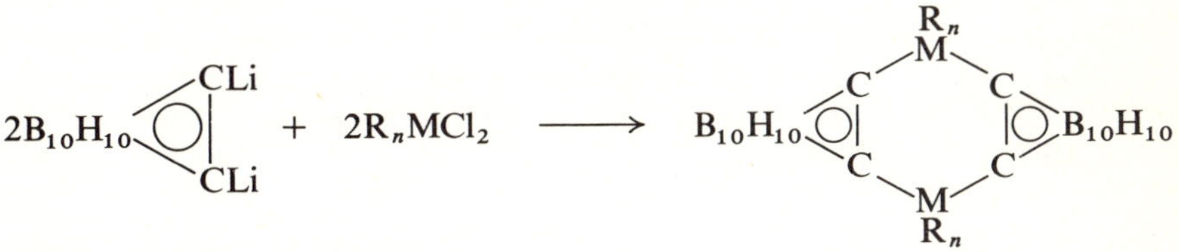

$$2B_{10}H_{10}\underset{CLi}{\overset{CLi}{\diagup\diagdown}} + 2R_nMCl_2 \longrightarrow B_{10}H_{10}\underset{C}{\overset{C\text{—}M\text{—}C}{\diagup\diagdown_{R_n}\diagdown}}\underset{R_n}{\overset{}{B_{10}H_{10}}}$$

$(R_nM = Me_2Si, PhP, PhAs\ [53])$.

4.3.5.2   Some silicon compounds

Derivatives with silicon attached to the carborane residue have attracted particular attention as materials from which to build silicone-type polymers containing the highly thermally stable carborane residue. The subject was reviewed recently[11], but merits brief mention here.

The pronounced tendency for the *ortho*-carborane residue to participate in ring systems of the type already described thwarted efforts to incorporate it in a polysiloxane backbone, e.g.

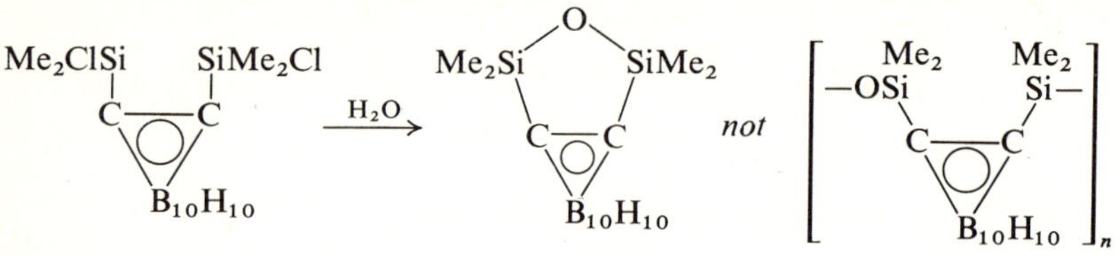

$$Me_2ClSi{-}C{\equiv}C{-}SiMe_2Cl \; (B_{10}H_{10}) \xrightarrow{H_2O} Me_2Si{<}^O_{C{\equiv}C}{>}SiMe_2 \; (B_{10}H_{10}) \; \textit{not} \; \left[\begin{array}{c} Me_2 \quad Me_2 \\ -OSi \quad Si- \\ C{\equiv}C \\ B_{10}H_{10} \end{array} \right]_n$$

However, the positions of the carbon atoms in *meta*-carborane do not allow the formation of such small rings, so polysiloxanes incorporating this isomer can be prepared, e.g.

$$1,7\text{-}B_{10}H_{10}C_2(SiMe_2Cl)_2 + 1,7\text{-}B_{10}H_{10}C_2(SiMe_2OMe)_2 \xrightarrow{-2MeCl} \frac{1}{n}\left[\begin{array}{c} -OSiCB_{10}H_{10}CSi- \\ Me_2 \qquad\qquad Me_2 \end{array} \right]_n$$

Many such polymers have been prepared and some (available under the trade name 'Dexsil') appear to have potential application where elastomeric properties are required above about 550 °C.

Low molecular weight siloxanes incorporating a single carborane unit in the molecule also appear likely to prove useful, as oils of high thermal stability[54]. They can be prepared as follows:

$$B_{10}H_{10}C_2Li_2 + 2Cl{-}\left(\overset{Me_2}{Si\text{-}O}\right)_n{-}\overset{Me_2}{SiCl} \rightarrow B_{10}H_{10}C_2\left[\left(\overset{Me_2}{SiO}\right)_n SiCl\right]_2$$

$$\downarrow 2R_3SiOH$$

$$B_{10}H_{10}C_2\left[\left(\overset{Me_2}{SiO}\right)_{n+1} SiR_3\right]_2$$

4.3.5.3 Carboranyl-sulphur compounds

Bis(mercapto)carboranes $B_{10}H_{10}C_2(SR)_2$ can be prepared from organic disulphides RSSR and dilithiocarboranes (*ortho*- or *meta*-), or by reaction of the dilithio derivative with elemental sulphur followed by treatment of the product with alkyl halide[55]:

$$B_{10}H_{10}C_2Li_2 + 2S \rightarrow B_{10}H_{10}C_2(SLi)_2 \begin{array}{c} \xrightarrow{RX} B_{10}H_{10}C_2(SR)_2 \\ \xrightarrow{H_2O} B_{10}H_{10}C_2(SH)_2 \end{array}$$

The compounds 1,2- or 1,7-$B_{10}H_{10}C_2(SMe)_2$ differ from organic sulphides in not forming adducts with chlorine or bromine, an effect attributed to the electron-withdrawing properties of the carborane residue.

The mercaptide 1,2-$B_{10}H_{10}C_2(SH)_2$ readily acts as a bidentate ligand in which the donor atoms are a little further apart than in the *ortho*-phenylene analogue[55–57]:

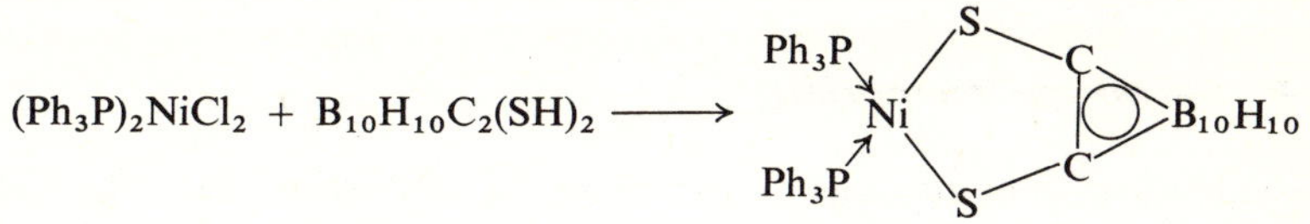

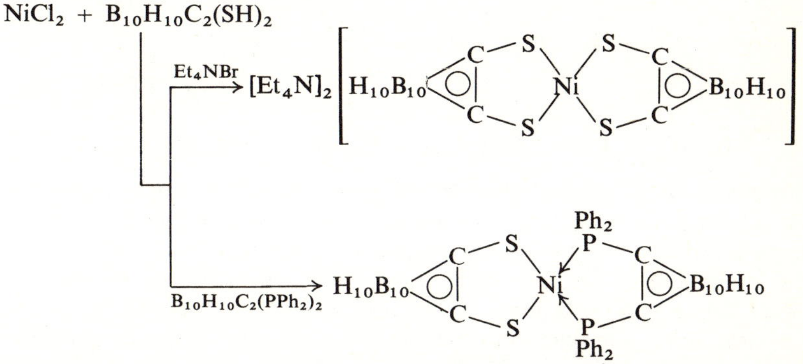

$$(Ph_3P)_2NiCl_2 + B_{10}H_{10}C_2(SH)_2 \longrightarrow \underset{Ph_3P}{\overset{Ph_3P}{}}Ni\underset{S}{\overset{S}{\diagdown}}\underset{C}{\overset{C}{\diagup}}B_{10}H_{10}$$

$NiCl_2 + B_{10}H_{10}C_2(SH)_2$

$$\xrightarrow{Et_4NBr} [Et_4N]_2 \left[H_{10}B_{10}\underset{C-S}{\overset{C-S}{\diagup\diagdown}}Ni\underset{S-C}{\overset{S-C}{\diagdown\diagup}}B_{10}H_{10} \right]$$

$$B_{10}H_{10}C_2(PPh_2)_2 \longrightarrow H_{10}B_{10}\underset{C-S}{\overset{C-S}{\diagup\diagdown}}Ni\underset{P-C}{\overset{Ph_2}{\overset{P-C}{\diagdown\diagup}}}\underset{Ph_2}{B_{10}H_{10}}$$

The electronic spectra of the nickel(II) complexes indicate that the Ni—S bonds are predominantly sigma in character, rather unexpectedly, since the carborane network had been considered likely to allow extensive π-delocalisation, conferring strongly π-acid character on the ligand. No particular enhancement of the π-acid character of phosphine ligands is found either when these are attached to the carborane residue, as in the case of the third nickel complex shown above, or in related *ortho*-carboranylene-diarsine complexes[58]:

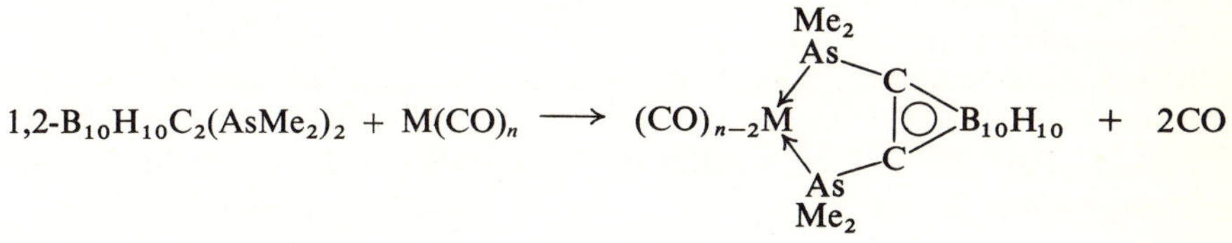

$$1,2\text{-}B_{10}H_{10}C_2(AsMe_2)_2 + M(CO)_n \longrightarrow (CO)_{n-2}M\underset{As}{\overset{As}{\diagup\diagdown}}\underset{Me_2}{\overset{Me_2}{}}B_{10}H_{10} + 2CO$$

M = Ni, n = 4; M = Fe, n = 5; M = Mo, n = 6.

The difunctionality and non-adjacent carbon atoms of the *meta*- and *para*- carborane derivatives 1,7- and 1,12-$B_{10}H_{10}C_2(SCl)_2$ [from $B_{10}H_{10}C_2(SH)_2 + 2Cl_2$] make these derivatives suitable polymer intermediates. They have been converted into polymeric sulphides by condensation with the dilithio derivative 1,7- (or 1,12-) $B_{10}H_{10}C_2Li_2$ [59]:

$$ClSCB_{10}H_{10}CSCl + LiCB_{10}H_{10}CLi \rightarrow \frac{2}{n}(-SCB_{10}H_{10}C-)_n$$

Corresponding disulphides have been prepared from 1,7- and 1,12-$B_{10}H_{10}$

$C_2(SCl)_2$ by heating them with ethanol, and polythiosulphinates by treating them with water. All these polymers are of low to moderate molecular weight, melting in the range 210–310 °C.

4.3.5.4 Carboranylmercurials

The divalency of mercury allows two types of carboranyl derivative to be prepared[60]:

$$B_{10}H_{10}C_2RLi \begin{array}{c} \xrightarrow{MeHgX} B_{10}H_{10}C_2RHgMe \\ \\ \xrightarrow{HgX_2} (B_{10}H_{10}C_2R)_2Hg \end{array}$$

$$B_{10}H_{10}C_2RH + MeHgOH \rightarrow B_{10}H_{10}C_2RHgMe + H_2O$$

The linear C—Hg—C skeletons of the products allow ample room for the bulky substituents, and these compounds are unusually thermally stable for organomercurials; the bis compounds are stable to above 300 °C, and the mixed compounds do not disproportionate on melting (in the range 150–200 °C). In their reactions, carboranyl mercurials resemble pentafluoro- and pentachloro-phenylmercurials, and indeed the electronegativities of the carboranyl group and pentachlorophenyl groups appear to be comparable. Moreover, the carboranyl carbon is well shielded from electrophilic attack, which occurs only under relatively severe conditions:

$$(B_{10}H_{10}C_2Ph)_2Hg + HgCl_2 \xrightarrow[210\,°C/8\,h]{PhNO_2} 2(B_{10}H_{10}C_2Ph)HgCl$$

(20% yield)

Both carboranylmercury halides and bis(carboranyl)-mercurials act as Lewis acids towards such bases as phenanthroline and bipyridyl.

Meta-carboranylmercurials are rather more reactive, and prone to rearrangement reactions, than their *ortho*-carboranyl counterparts, presumably because of both steric factors [in the case of substituted compounds $(B_{10}H_{10}C_2R)HgX$] and the weaker electron-withdrawing properties of the *meta*-carboranyl group[62].

4.3.5.5 Transition metal derivatives

Several σ-carboranyl derivatives of transition metals have been prepared by reactions of metal halides with carboranyl-lithium reagents, e.g.[63, 64],

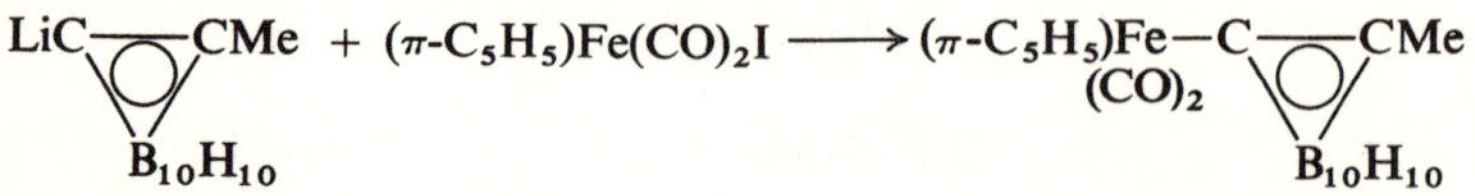

$$\text{LiC}\underset{B_{10}H_{10}}{\diagdown\diagup}\text{CMe} + (\pi\text{-}C_5H_5)Fe(CO)_2I \longrightarrow (\pi\text{-}C_5H_5)Fe-C\underset{B_{10}H_{10}}{\diagdown\diagup}CMe$$
$$(CO)_2$$

$$Ph_3PAuCl + B_{10}H_{10}C_2RLi \rightarrow Ph_3PAuC_2B_{10}H_{10}R$$

CARBORANES AND METALLOCARBORANES

Typically, these gold compounds have higher thermal stability and lower reactivity than alkyl or aryl derivatives of the metal. For example, the Au—C bond is retained during the oxidation[64]

$$Ph_3PAu^IC_2B_{10}H_{10}Ph + Br_2 \rightarrow Ph_3PAu^{III}(C_2B_{10}H_{10}Ph)Br_2$$

Some novel chelated biscarborane transition metal complexes have also been described[65]. These are accessible by the following reaction sequence:

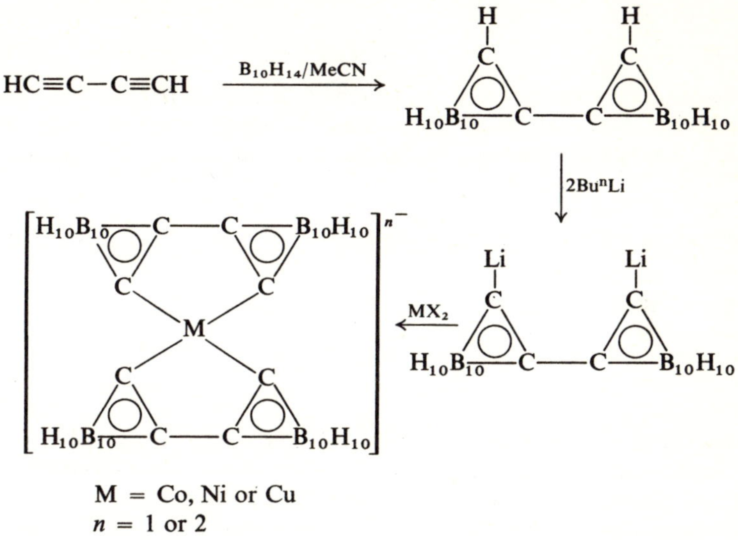

$$HC\equiv C-C\equiv CH \xrightarrow{B_{10}H_{14}/MeCN}$$

M = Co, Ni or Cu
n = 1 or 2

4.4 OTHER DICARBA-BORANES $C_2B_mH_{m+2, 4\ or\ 6}$

This section is concerned with the *closo-* compounds $C_2B_mH_{m+2}$, where m = 3,4,5,6,7,8, or 9; the *nido-* compounds $C_2B_mH_{m+4}$, where m = 3,4, or 9; and the *arachno-* compound $C_2B_7H_{13}$. It is convenient to treat these in the same section, because the *nido-* and *arachno-* compounds are key intermediates in the preparation of the *closo-* species, and because their structures and reactivities may then more readily be compared.

4.4.1 The dicarba-undecaboranes $C_2B_9H_{11}$ and $C_2B_9H_{13}$

These carboranes, containing two carbon and nine boron skeletal atoms, can be prepared by alcoholic alkaline degradation of icosahedral carboranes[29]. Both 1,2- and 1,7-$C_2B_{10}H_{12}$ react with methoxide ion, which attacks the most positively charged boron atom of the cage (in each case the boron in the 3 position) causing its selective removal. The products, the ions (3)-1,2-$C_2B_9H_{12}^-$ and (3)-1,7-$C_2B_9H_{12}^-$ [the prefix '(3)' signifies the vacant cage position in these *nido-* anions], can be isolated as their caesium or potassium salts, or converted in about 90% yield into the neutral 1,2- or 1,7-dicarba-*nido*-undecaborane(13), $C_2B_9H_{13}$, by treatment with anhydrous hydrogen chloride or polyphosphoric acid[66]. These *nido-* carboranes lose hydrogen and collapse to a single *closo-* structure when heated at 110–150 °C in an inert solvent:

$$1,2\text{-}C_2B_{10}H_{12} \xrightarrow{\text{MeO}^-/\text{MeOH}} (3)\text{-}1,2\text{-}C_2B_9H_{12}^- \xrightarrow{\text{acid}} (3)\text{-}1,2\text{-}C_2B_9H_{13}$$

$$1,7\text{-}C_2B_{10}H_{12} \qquad (3)\text{-}1,7\text{-}C_2B_9H_{12}^- \qquad (3)\text{-}1,7\text{-}C_2B_9H_{13} \Bigg] \begin{array}{c} 110\text{-} \\ 150\,°\text{C} \end{array}$$

$$2,3\text{-}C_2B_9H_{11}$$

Dicarba-undecaboranes with *C*-attached substituents can be prepared similarly by degradation of substituted icosahedral carboranes $B_{10}H_{10}C_2R^1R^2$. A plausible mechanism for the degradation involves the following steps[29]:

$$C_2B_{10}H_{12} \xrightarrow{\text{MeO}^-} [C_2B_{10}H_{12}(\text{OMe})]^- \xrightarrow{\text{MeO}^-} [C_2B_{10}H_{12}(\text{OMe})_2]^{2-}$$

$$C_2B_9H_{12}^- + \text{MeO}^- \xleftarrow{\text{MeOH}} C_2B_9H_{11}^{2-} + \text{HB(OMe)}_2 \longleftarrow$$

$$\downarrow \text{MeOH}$$

$$H_2 + B(\text{OMe})_3$$

The structures of the *nido-* anions (3)-1,2- and (3)-1,7-$C_2B_9H_{11}^{2-}$ and of the *closo-* carborane 2,3-$C_2B_9H_{11}$ (established by x-ray crystallographic study of the *C,C*-dimethyl derivative[67]) are shown in Figure 4.5.

The anions (3)-1,2- and (3)-1,7-$C_2B_9H_{12}^-$ are believed to have structures

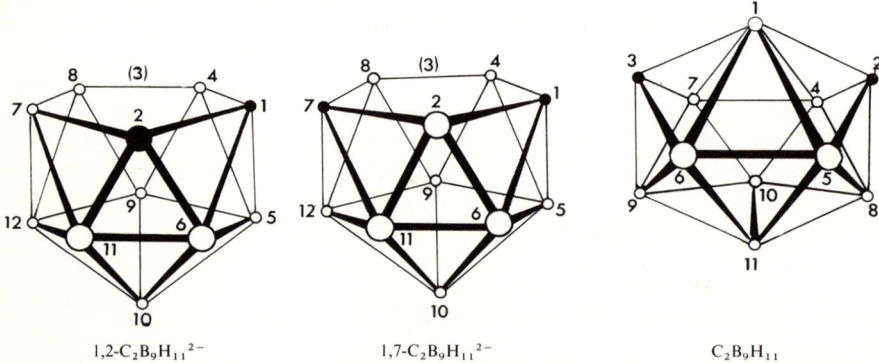

Figure 4.5 The skeletons of the *nido*-anions(3)-1,2- and (3)-1,7-$C_2B_9H_{11}^{2-}$, and of the *closo*-carborane $C_2B_9H_{11}$

derived from those of $C_2B_9H_{11}^{2-}$ by adding a proton either rapidly tautomerising between various B—H—B and C—H—B bridging positions and/or BH_2 positions about the open pentagonal face, or in the icosahedral position previously held by the boron atom in the parent $C_2B_{10}H_{12}$. (The *nido-*carborane $C_2B_9H_{13}$ would have a similar skeleton, and two extra protons).

The presence of this potentially acidic twelfth hydrogen atom in $C_2B_9H_{12}^-$ is well established by such chemical reactions as the following:

$$C_2B_9H_{12}^- \underset{[H^+]}{\overset{NaH/THF}{\rightleftharpoons}} C_2B_9H_{11}^{2-}$$

However, spectroscopically it has proved elusive, though its interaction with a boron atom of the open pentagonal face is believed responsible for fine structure on one of the high field doublets in the 60 MHz ^{11}B n.m.r. spectrum of the salt $K^+(3)$-$1,2$-$C_2B_9H_{12}^-$ [29].

Species $C_2B_9H_{11} \cdot L$, isoelectronic with the anion $C_2B_9H_{12}^-$, are readily formed by dicarba-*closo*-undecaborane with Lewis bases L like EtNC or Ph_3P [66]. These adducts are believed to have *nido* icosahedral-fragment (3)-1,7-$C_2B_9H_{12}^-$ type structures with the Lewis base L coordinated to the boron atom in the pentagonal face linked to both carbon atoms. Isomeric adducts result from the reaction[68]

$$(3)\text{-}1,2\text{-}C_2B_9H_{12}^- + 2FeCl_3 + L \rightarrow (3)\text{-}1,2\text{-}C_2B_9H_{11}L + 2FeCl_2 + HCl + Cl^-$$

(where L = THF, pyridine, Et_2S, or MeCN). In these, the base is believed to be attached to one of the three boron atoms in the open pentagonal face of the icosahedral fragment. Consistent with this, these compounds (3)-1,2-$C_2B_9H_{11}L$ exist in two isomeric forms.

4.4.2 Dicarba-boranes containing six, seven, or eight boron atoms

4.4.2.1 The arachno-carborane $C_2B_7H_{13}$

Dicarba-boranes containing six, seven, or eight boron atoms can be prepared via the *arachno*- carborane $C_2B_7H_{13}$, which is itself the somewhat surprising product of oxidative degradation of $C_2B_9H_{11}$ by chromic acid[66]:

$$C_2B_9H_{11} + 6H_2O \xrightarrow[\text{degradation}]{\text{chromic acid}} C_2B_7H_{13} + 2B(OH)_3 + 4H^+ + 4e^-$$

$$\downarrow \begin{array}{c} 200\text{-}225\,°C \\ Ph_2O \text{ solvent} \end{array}$$

$$C_2B_6H_8 + C_2B_7H_9 + C_2B_8H_{10}$$

C-alkylated derivatives undergo similar reactions. An x-ray crystallographic study of the dimethyl derivative, $C_2B_7H_{11}Me_2$ [69], has confirmed the skeletal arrangement which had been deduced from spectroscopic studies[70] and which is illustrated in Figure 4.6. This skeletal arrangement is that appropriate for a nine-atom cage with twelve pairs of skeletal bonding electrons (regarding $C_2B_7H_{13}$ as $C_2B_7H_9^{4-}$ plus $4H^+$), i.e., it may be regarded as based on the eleven-cornered octadecahedron of $C_2B_9H_{11}$ but with two corners left vacant. It is often referred to as if it were an icosahedral fragment, which is understandable in view of the close correspondence between most of the atom positions of the octadecahedron and icosahedron (see Figure 4.1).

The four 'extra' hydrogen atoms of $C_2B_7H_{13}$ occupy two terminal CH positions and two BHB bridging positions. Interestingly, deuterium tracer studies have shown that the bridging BHB hydrogens exchange readily with one of the two hydrogens of the CH_2 groups, and that CH hydrogen is axially orientated with respect to the chair-shaped six-membered ring which defines the open face of this compound (Figure 4.6)[66].

Apparently the CH (axial) bond is significantly weaker than the CH (equatorial) bond. The axial CH bond is that which formally may be counted

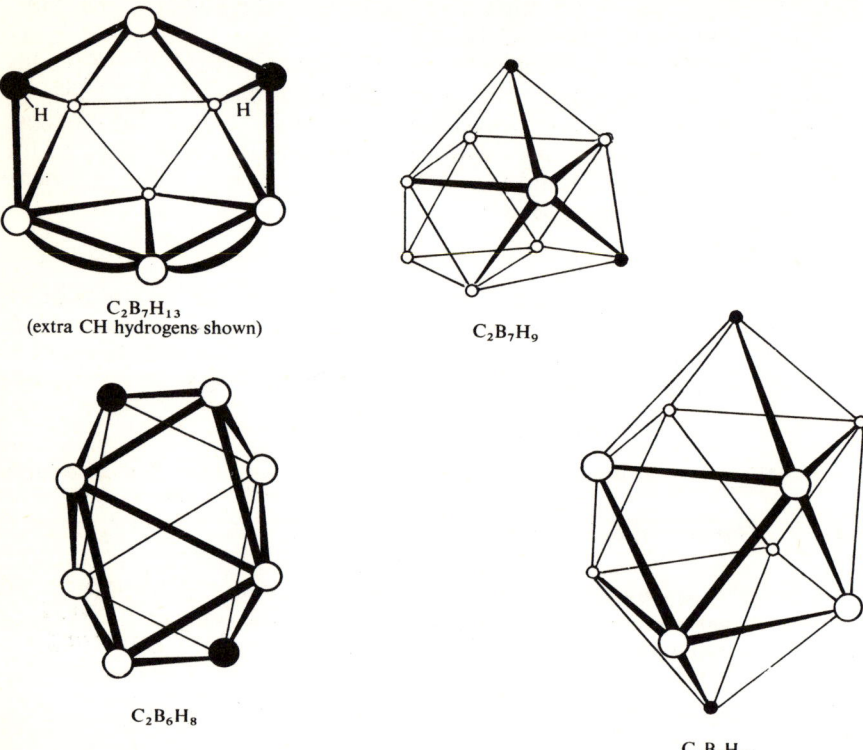

Figure 4.6 The skeletons of the *arachno*- carborane $C_2B_7H_{13}$, and of the *closo*- carboranes $C_2B_6H_8$, $C_2B_7H_9$ and $C_2B_8H_{10}$ formed by its thermal decomposition

among the skeletal bonds, and MO theory applied to $C_2B_7H_{13}$ indicates that the axial protons are substantially more positively charged than the equatorial protons of the CH_2 groups, the relative positive charges decreasing in the sequence

$$\text{axial CH} > \text{bridging BHB} > \text{equatorial CH}$$

4.4.2.2 The preparation of $C_2B_6H_8$, $C_2B_7H_9$, and $C_2B_8H_{10}$

The thermal decomposition of the *arachno*-carborane $C_2B_7H_{13}$ which provides a route to the B_6, B_7, and B_8 *closo*- species apparently involves predominantly disproportionation into the B_6 and B_8 species, which are obtained in about 30% yield, much more than the B_7 compound. When the pyrolysis is carried out in the presence of diborane, the yield of 1,6-$C_2B_8H_{10}$ is improved, and some 1,7-$C_2B_{10}H_{12}$ is also generated[66, 71]. An improved synthesis of 1,7-$C_2B_6H_8$ is the slow low-pressure pyrolysis of $C_2B_7H_{13}$ (or a C-methyl derivative), the other main products being 2,4-$C_2B_5H_7$, 1,7-$C_2B_7H_9$, and 1,6-$C_2B_8H_{10}$ (or their C-methyl derivatives)[72]. The dimethyl derivative $B_6H_6C_2Me_2$ was first prepared from the reaction between hexaborane(10) and dimethylacetylene[73]:

$$B_6H_{10} + \text{excess MeC}\vdots\text{CMe} \xrightarrow[\text{12 min at 35 °C}]{\text{u.v. gas phase}} B_6H_6C_2Me_2$$

The pyrolytic routes to these medium-sized carboranes rarely afford high yields of particular products, which must moreover be separated from several other products. An improved synthesis of 1,6-$C_2B_8H_{10}$ avoids these disadvantages, affording yields greater than 70% with sodium borohydride as the by-product: this involves the reaction between $C_2B_7H_{13}$, $NaC_2B_7H_{12}$ (from $C_2B_7H_{13}$ + NaH), and diborane in tetrahydrofuran[74].

4.4.2.3 The structures of $C_2B_6H_8$, $C_2B_7H_9$, and $C_2B_8H_{10}$

The structures as determined by x-ray crystallographic studies on C-methylated derivatives, are shown in Figure 4.6. The first[75] has a skeleton slightly distorted (towards a square antiprism) from the idealised D_{2d} dodecahedral geometry adopted also by B_8Cl_8 and $B_8H_8^{2-}$. The second, $C_2B_7H_9$ (as represented by $C_2B_7H_7Me_2$)[76], has the tricapped prism geometry (essentially D_{3h}) of $B_9H_9^{2-}$, while the skeleton of $C_2B_8H_{10}$ is a bicapped Archimedean antiprism[77] effectively isostructural with the isoelectronic anion $B_{10}H_{10}^{2-}$. All three structures contain two types of skeletal site: one 5-coordinate, the other 6-coordinate. Significantly, the carbon atoms preferentially occupy the 5-coordinate sites.

The charge distribution within these carborane skeletons has been calculated by molecular orbital methods[75-77], which substantiate the simple predictive rule suggested earlier[78], which is that boron atoms in order of increasing positive charge are those bonded to (a) no C atoms; (b) one C atom; and (c) two C atoms, and that for a given category, (a), (b) or (c), a BH unit bonded to four other BH or CH units is more negative than one bonded to five other such units.

4.4.2.4 Some reactions of $C_2B_7H_9$ and $C_2B_8H_{10}$

These charge distributions are supported by the few substitution reactions that have now been carried out. For example, electrophilic substitution in 1,6-$C_2B_7H_9$ occurs at the 8 (apical) position[79]:

1,6-$C_2B_7H_9$ $\xrightarrow{C_2H_4/AlCl_3}$ 8-$EtC_2B_7H_8$

1,6-$C_2B_7H_9$ $\xrightarrow{Br_2/AlBr_3/CS_2}$ 8-$BrC_2B_7H_8$

Some derivative chemistry of 1,6- and 1,10-$C_2B_8H_{10}$ has been described[80]. The two isomers differ markedly in their degradative stabilities. Whereas 1,6-$C_2B_8H_{10}$ is hydrolysed by both acid and base at 25 °C to the *arachno-* anion $C_2B_7H_{12}^-$ and boric acid, more than 50% of the 1,10-isomer can be recovered after treatment at 300 °C for 12 h with neat piperidine. Presumably this reflects the chemical equivalence of the eight boron atoms of the 1,10-isomer (cf., 1,12-$C_2B_{10}H_{12}$). Also like its icosahedral homologue, 1,10-$C_2B_8H_{10}$ reacts with chlorine in carbon tetrachloride to form the fully B-chlorinated product, 1,10-$C_2B_8H_2Cl_8$. Substitution at carbon is effected through the C-lithio derivative, e.g.[80]:

$$1,10\text{-}C_2B_8H_9R \xrightarrow{Bu^nLi} 1,10\text{-}C_2B_8H_8RLi \begin{cases} \xrightarrow{R'I} 1,10\text{-}C_2B_8H_8RR' \\ \xrightarrow{CO_2} 1,10\text{-}C_2B_8H_8RCO_2Li \\ \xrightarrow{I_2} 1,10\text{-}C_2B_8H_8RI \end{cases}$$

Also[69],

$$1,10\text{-}C_2B_8H_{10} \xrightarrow{2Bu^nLi} 1,10\text{-}C_2B_8H_8Li_2$$

$$\downarrow (\pi\text{-}C_5H_5)Fe(CO)_2I$$

$$(\pi\text{-}C_5H_5)Fe\text{-}CB_8H_8C\text{-}Fe(\pi\text{-}C_5H_5)$$
$$\quad(CO)_2 \qquad\qquad (CO)_2$$

4.4.3 Dicarba-boranes containing three, four, or five boron atoms

The known lower dicarba-boranes include the *closo-* compounds 1,5-$C_2B_3H_5$, 1,2- and 1,6-$C_2B_4H_6$, 2,4-$C_2B_5H_7$ and the *nido-* compounds 1,2-$C_2B_3H_7$ and 2,3-$C_2B_4H_8$ (see Figure 4.7), and many derivatives of these, particularly alkyl derivatives.

4.4.3.1 *Preparation of the lower dicarba-boranes*

The lower carboranes are usually prepared by reactions between the lower boranes and acetylenes[81–86]. Complex mixtures of alkylboranes and/or carboranes and/or alkylcarboranes generally result, the identities and proportions of the components varying with the borane (B_2H_6, B_4H_{10}, or B_5H_9) and the acetylene (C_2H_2, C_2HR, or C_2R_2), and also with the conditions. Under high energy conditions[81,84], e.g., silent electrical discharge or flash reactions or high temperature reactions, *closo-* carborane products predominate; *nido-* carboranes and alkylboranes are the major products of reactions under low energy conditions[82–86], e.g. moderate heat. For example, the reaction between diborane and acetylene gives ethyl-diboranes if carried out at 85 °C [82], and a mixture of the *closo-* carboranes $C_2B_3H_5$, $C_2B_4H_6$, and $C_2B_5H_7$, and assorted methyl derivatives thereof, if the gas mixture diluted with helium is subjected to a.c. discharge[81]. The smallest known *nido-* carborane, 1,2-$C_2B_3H_7$, can be obtained in 3–4% yield from the gas phase reaction between tetraborane(10) and acetylene at 50 °C [86]; above 100 °C, the same reaction affords only *closo-* carboranes of the type obtained from the B_2H_6-C_2H_2 reaction [84].

The best reaction for preparing significant quantities of the lower carboranes is that between pentaborane(9) and acetylene. The *nido-* compound 2,3-$C_2B_4H_8$ is formed in 30–40% yield from pentaborane(9) and acetylene

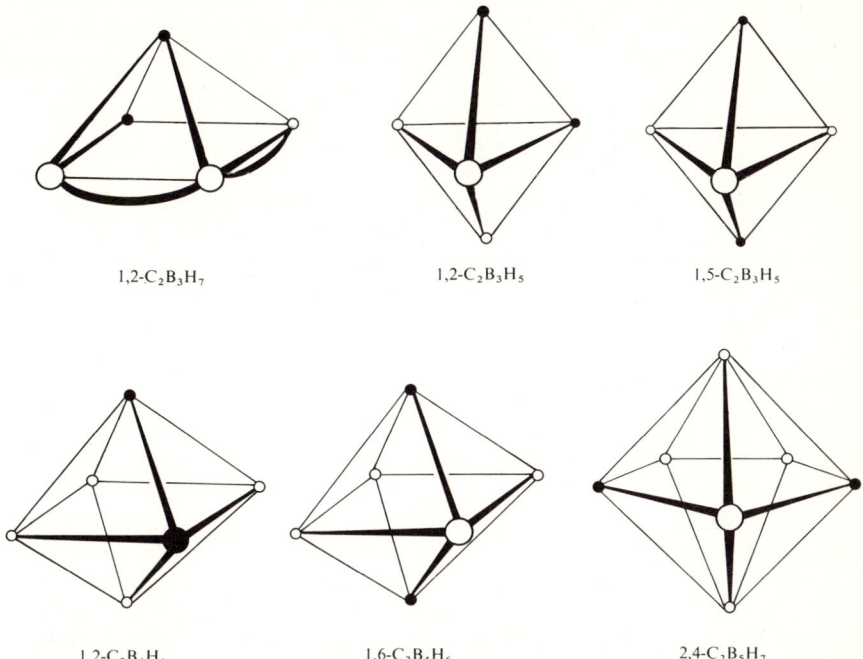

Figure 4.7 Skeletons of the carboranes 1,2-$C_2B_3H_7$, 1,2- and 1,5-$C_2B_3H_5$, 1,2- and 1,6-$C_2B_4H_6$, and 2,4-$C_2B_5H_7$

at 215 °C [85]; brief treatment at a higher temperature, either of the C_2H_2/B_5H_9 mixture [87] or of preformed 2,3-$C_2B_4H_8$ [88], gives mixtures of the three *closo-* carboranes $C_2B_3H_5$, $C_2B_4H_6$ and $C_2B_5H_7$ in as high as 50–60% collective yield at a rate of as much as 400 g/h [87].

$$B_5H_9 + C_2H_2 \xrightarrow{215\,°C} 30\text{–}40\%\ 2,3\text{-}C_2B_4H_8\ [85]$$

$$\Bigg\downarrow \begin{array}{l}500\,°C;\\ 0.5\ s\ [87]\end{array} \qquad\qquad \Bigg\downarrow \begin{array}{l}450\,°C;\ 1\text{–}3\ s\ [88]\\ \text{or u.v. 23 °C 2 h}\end{array}$$

1,5-$C_2B_3H_5$; 1,2- and 1,6-$C_2B_4H_6$; and 2,4-$C_2B_5H_7$

Alternative routes to mixed alkylated carboranes start from organoboranes, e.g. the dehalogenation of alkylboron halides, the pyrolysis of alkyldiboranes, and the action of boron hydrides on alkynylboranes. Such reactions have been discussed in some detail by Köster and Grassberger, whose review[6] concentrates on the lower carboranes.

4.4.3.2   1,2-Dicarba-nido-pentaborane(7), $C_2B_3H_7$

The infrared and ^1H and ^{11}B n.m.r. spectra of this compound are consistent with the structure shown in Figure 4.7, with a pyramidal skeleton analogous to that of the isoelectronic B_5H_9. The two carbon atoms occupy an apical and a basal position and the two basal B---B edges are hydrogen-bridged[86].

4.4.3.3   2,3-Dicarba-nido-hexaborane(8), $C_2B_4H_8$

That this compound is undoubtedly a member of the family of *nido-* compounds isoelectronic with, and structurally analogous to, hexaborane(10) was early established by an x-ray crystallographic study of the C,C'-dimethyl derivative[89], which showed the C—C interatomic distance to be only 1.43 Å, comparable to that in graphite. The n.m.r. spectra of the parent compound can be interpreted in detail on such a model[90], which has two basal BHB bridges (Figure 4.2). One bridging proton is removed by reaction with hydride ion[91]:

$$2,3\text{-}C_2B_4H_8 \xrightarrow{\text{NaH/diglyme}} Na^+C_2B_4H_7^- \xrightarrow{\text{DCl}} C_2B_4H_7D$$

The related dianion, $C_2B_4H_6^{2-}$, is known in the form of transition metal complexes (see below).

Deuteriation of $C_2B_4H_8$ can be effected not only by the deprotonation-protonation sequence just illustrated, but also by equilibration with molecular deuterium in the presence of Cr_2O_3/Al_2O_3, when all the *B*-attached (terminal and bridging) hydrogen atoms exchange. In the $Me_2C_2B_4H_6$-DCl reaction, the terminal but not bridging hydrogens exchange[92]. Halogenation of $C_2B_4H_8$ with X_2/AlX_3 gives products $C_2B_4H_7X$ apparently halogenated at a basal boron adjacent to a carbon atom[93].

4.4.3.4   The dicarba-closo-pentaboranes, $C_2B_3H_5$

From its spectra, the dicarba-*closo*-pentaborane, preparable by the thermal decomposition of $C_2B_4H_8$ or a C_2H_2/B_5H_9 mixture, is the 1,5-isomer (Figure 4.7) with axial carbons and equatorial borons in a trigonal biprismatic skeleton[94, 95]. This is the isomer predicted to be most stable by Hoffmann and Lipscomb[96]. A second isomer, 1,2-$C_2B_3H_5$, is known only in the form of a dimethyl derivative, 2,3-Me_2-1,2-$C_2B_3H_3$, which does *not* isomerise to the methylated 1,5-isomer on heating[97].

1,5-Dicarba-*closo*-pentaborane, $C_2B_3H_5$, is the smallest and most volatile known dicarba-*closo*-borane (b.p. $-3.7\,°C$, m.p. $-126.4\,°C$). Air-stable, it does not react at 20 °C with acetone, trimethylamine, or water (contrast the lower boranes), and is thermally stable to 150 °C.

4.4.3.5 The dicarba-closo-hexaboranes, 1,2- and 1,6- $C_2B_4H_6$

Both possible isomers of this tetragonal-bipyramidal carborane are known, the structures being clearly indicated by their vibrational and n.m.r. spectra[98] and by the microwave spectrum in the case of 1,2-$C_2B_4H_6$ [96]. At 250 °C, the 1,2-isomer rearranges smoothly to the 1,6-isomer, which has been calculated to be some 15 kcal mol^{-1} the more stable isomer[100, 101].

A pentagonal-pyramidal or trigonal-prismatic intermediate in this rearrangement may be envisaged[10].

4.4.3.6 2,4-Dicarba-closo-heptaborane, $C_2B_5H_7$

The pentagonal-bipyramidal structure inferred from the i.r. and n.m.r. spectra[102] of $C_2B_5H_7$ was confirmed by microwave spectroscopy[103]. The 2,4-dicarba species is the only isomer so far characterised and is that indicated by theory[96, 101] to have lowest energy.

As $C_2B_5H_7$ is the *closo*-carborane which is obtained in best yield from B_5H_9-C_2H_2 reactions, it is available in large enough quantities for some of its chemistry to be explored. It can be *C*-lithiated by butyl-lithium (though much more slowly than $C_2B_{10}H_{12}$), and this affords a route to *C*-substituted derivatives[104]:

$$HCB_5H_5CH \xrightarrow{2n\text{-BuLi}} LiCB_5H_5CLi \xrightarrow{2RX} RCB_5H_5CR$$

The monolithio derivative HCB_5H_5CLi apparently polymerises too readily to provide an analogous route to monosubstituted species.

The parent carborane with $Br_2/AlBr_3$ is brominated at the 5-position, that calculated to be the most negatively charged[96, 101]. Electrophilic chlorination involves the same site, though borons 1 and 3 are also chlorinated if no $AlCl_3$ is added[105].

Whereas the icosahedral carboranes $C_2B_{10}H_{12}$ generate *nido*- anions on alkaline degradation, $C_2B_5H_7$ is completely degraded by base at room temperature, apparently because the apical boron atoms (1 and 7), which are those most positively charged[96, 101] are so accessible to attack. Accordingly, base degradation of $C_2B_5H_7$ does not provide the route into $C_2B_4H_6^{2-}$-metallocarborane chemistry that had been hoped for by analogy with the behaviour of icosahedral carboranes.

4.5 CARBORANES WITH OTHER THAN TWO SKELETAL CARBON ATOMS

The chemistry of carboranes containing other than two carbon atoms is much less developed than that of dicarba species, to a considerable extent because the standard methods of synthesising carboranes, starting from acetylenes, furnish carbon atoms in pairs. Moreover, carboranes with odd numbers of skeletal carbon atoms generally lack the stability associated with the *closo*- structures of neutral dicarba-compounds. For example, the neutral monocarba-species CB_mH_{m+2} which have the appropriate numbers of skeletal bonding electrons for *closo-* structures, have one extra hydrogen atom to accommodate additional to their $(m+1)$ terminal hydrogen atoms, and this serves as a source of reactivity. Anions $[CB_mH_{m+1}]^-$ are likely to prove relatively stable species, but have been little studied as yet.

4.5.1 Carboranes containing one skeletal carbon atom

The known monocarba-boranes are confined to two groups; one group, the CB_5 species, are preparable by adding a single skeletal carbon atom to the five skeletal borons of pentaborane(9); the other group, the various CB_{10} species (and some CB_9 and CB_{11} species derivable from these) are preparable by adding a single skeletal carbon atom to the ten skeletal borons of decaborane(14).

4.5.1.1 Monocarba-*closo*-hexaborane(7), CB_5H_7

This was the first carborane with only one skeletal carbon atom to be prepared, as a product of decomposition of 1-methylpentaborane(9) by electrical discharge[106]; it has since been made by pyrolysis of 1-MeB$_5$H$_8$ [107] and by the action of atomic carbon (from a carbon arc) on B_5H_9 [108]. Yields are about 2–3%.

Its ^1H n.m.r. spectrum is consistent with an octahedral arrangement of one terminal CH and five terminal BH groups (of three types, ratio 2:2:1) together with a single bridging hydrogen. In the ^{11}B n.m.r. spectrum, the two upfield doublets (each of area 2) merge to a single doublet (area 4) at 100 °C, apparently as the bridging hydrogen, thought to be located over a B_3 octahedral face in the ground state, moves rapidly from one face to another round the B_{eq}—B_{ax} edges[109].

4.5.1.2 Monocarba-*nido*-hexaborane(9), CB_5H_9

The 2-, 3- and 4-methyl derivatives of 2-carba-*nido*-hexaborane(9) (Figure 4.2) are minor products of the reaction between acetylene and pentaborane(9) at 215 °C. Their spectra indicate them to be further examples of carboranes with a pentagonal pyramidal structure. They have apical boron, and three BHB bridging units in the pentagonal base[110]. The *C*-methyl *B*-pentaethyl derivative results from the action of lithium on ethylboron difluoride in tetrahydrofuran, while the parent compound CB_5H_9 can be prepared by the action of tetramethylammonium borohydride on the dicarba-*closo*-borane, $C_2B_6H_8$, in diglyme at 100 °C [111]. Whereas the removal of boron atoms from carborane cages by both pyrolytic and chemical methods is well known, this last reaction appears to be the first example of the removal of a cage *carbon* atom from a *closo-* carborane.

The 2-ethyl derivative, $EtCB_5H_8$, is the exclusive product of a different preparative procedure which uses methylacetylene as its source of carbon[110]:

$$B_5H_9 + MeC\!:\!CLi \xrightarrow{BF_3/diglyme} 2\text{-EtCB}_5H_8$$

4.5.1.3 The CB_9, CB_{10} and CB_{11} carboranes

Monocarba-*nido*-undecaborane(14), $CB_{10}H_{14}$, is the parent species of this group of monocarba-boranes. A single skeletal carbon atom can be added to the ten skeletal boron atoms of decaborane(14) by reaction of the latter with isocyanides[112]:

$$B_{10}H_{14} + RNC \rightarrow B_{10}H_{12}CNH_2R$$

As some alkyl isocyanides are polymerised by decaborane, a more general route to these derivatives is by alkylation of a cyano-decaborane anion[113]:

$$B_{10}H_{14} \xrightarrow{CN^-;\ -H_2} B_{10}H_{12}CN^- \begin{array}{c} \xrightarrow[(ii)\ H_2O]{(i)\ RI/THF} B_{10}H_{12}CNR_3 \\ \\ \xrightarrow[(ii)\ OH^-]{(i)\ Me_3SiCl} B_{10}H_{12}CNH_3 \end{array}$$

The nitrogen in these C—N compounds is not a skeletal atom (see Figure 4.8); they are effectively zwitterionic derivatives $B_{10}H_{12}\overset{-}{C}\text{-}\overset{+}{N}H_nR_{3-n}$ of the

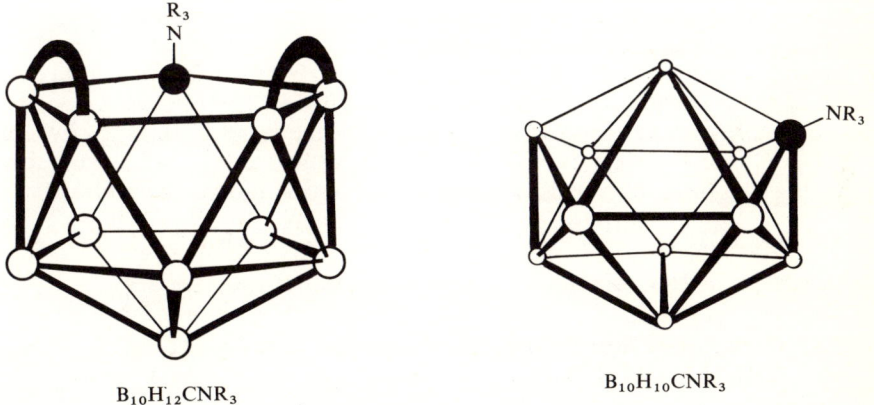

$B_{10}H_{12}CNR_3 \qquad\qquad B_{10}H_{10}CNR_3$

Figure 4.8 Probable structures of $B_{10}H_{12}CNR_3$ and $B_{10}H_{10}CNR_3$

anion $B_{10}H_{12}CH^-$, into which they can be converted by alkali metal or hydride ion, e.g.[114],

$$B_{10}H_{12}CNR_3 \xrightarrow{Na/THF} Na_3B_{10}H_{10}CH,2THF \xrightarrow{H_2O} 80\%\ NaB_{10}H_{12}CH$$

Similarly, $\quad B_{10}H_{12}CNR_3 \xrightarrow{NaH/THF} 63\%\ NaB_{10}H_{12}CH$

All these derivatives are believed to have *nido*- structures based on an icosahedron but with one corner vacant, with the carbon atom one of the five atoms surrounding the open face and the extra hydrogen atoms fulfilling a bridging role between pairs of boron atoms around the open face (Figure 4.8). Oxidative removal of these bridging hydrogens from $B_{10}H_{12}CH^-$ and from $B_{10}H_{12}CNH_nR_{3-n}$ generates the *closo*- species $B_{10}H_{10}CH^-$ and $B_{10}H_{10}CNH_nR_{3-n}$ (Figure 4.8)[114].

$$Na_3B_{10}H_{10}CH \cdot 2THF + I_2 \rightarrow Na^+B_{10}H_{10}CH^- + 2NaI + 2THF$$
$$B_{10}H_{12}CNR_3 + 2NaH + I_2 \rightarrow B_{10}H_{10}CNR_3 + 2NaI + 2H_2$$

Halogenation of these *closo*- compounds by molecular halogen (Cl_2, Br_2 or I_2) apparently occurs at the 4 or 6 position[114].

The presumably bicapped Archimedean antiprismatic anion $B_9H_9CH^-$ and icosahedral $B_{11}H_{11}CH^-$ have been prepared by the thermal degradative disproportionation of $B_{10}H_{12}CH^-$ [115]:

$$2CsB_{10}H_{12}CH \xrightarrow{300-320\,°C} Cs\text{-}1\text{-}B_9H_9CH + CsB_{11}H_{11}CH + 2H_2$$

4.5.2 Carboranes containing three skeletal carbon atoms; 2,3,4-$C_3B_3H_7$

The only tricarba-boranes so far characterised are tricarba-*nido*-hexaborane(7) $C_3B_3H_7$ and some alkyl derivatives thereof, which are among the products of the gas-phase reaction between tetraborane(10) and acetylene at room temperature[83, 84]. When this reaction is carried out at 50 °C using C_2D_2, C-deuteriated $C_3B_3H_7$ derivatives are produced, but the boron-attached hydrogen atoms, whether terminal or bridging, remain undeuteriated, showing that there is negligible C—D(H) bond breaking during the formation of these tricarba-hexaboranes[116].

Spectroscopic studies on these tricarba-hexaboranes indicate that the three skeletal carbon atoms occupy adjacent basal (2,3,4) positions on the pentagonal pyramid expected for a member of the series $C_aB_mH_{m+4}$ (where $a+m = 6$) (see Figure 4.2). The single bridging hydrogen that presumably links the two basal borons is removed by the action of hydride ion[116]:

$$Me_2C_3B_3H_5 \xrightarrow{NaH/THF} Na^+Me_3C_3B_3H_4^- \xrightarrow{DCl} Me_3C_3B_3H_4D$$

In this last reaction, it is the bridging hydrogen that is deuteriated.

4.5.3 Carboranes containing four skeletal carbon atoms; $C_4B_2H_6$, $C_4B_6H_{10}$ and $C_4B_7H_{11}$

The only tetracarba-borane certainly characterised so far is the compound 2,3,4,5-tetracarba-*nido*-hexaborane, the final member of the *nido*-hexaborane series of pentagonal pyramidal carboranes (see Figure 4.2). The parent compound is obtained in low yield from the low-pressure pyrolysis (550 °C) of 1,2-tetramethylenediborane(6), which effectively provides the skeletal atoms already suitably linked[117]:

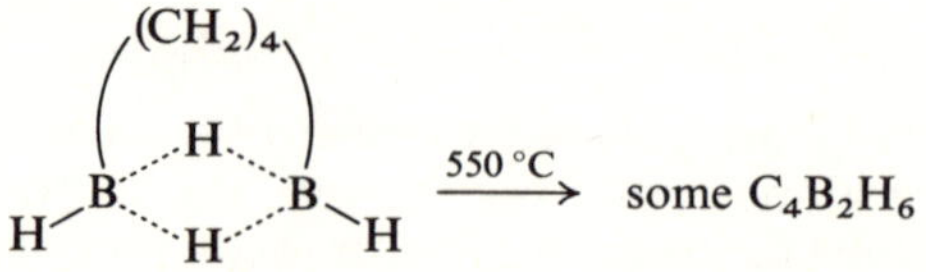

$$\text{(structure with (CH}_2\text{)}_4\text{ bridge between two BH groups with bridging H)} \xrightarrow{550\,°C} \text{some } C_4B_2H_6$$

Several alkyl derivatives had been prepared earlier by reactions such as the following equally elegant synthesis[118]:

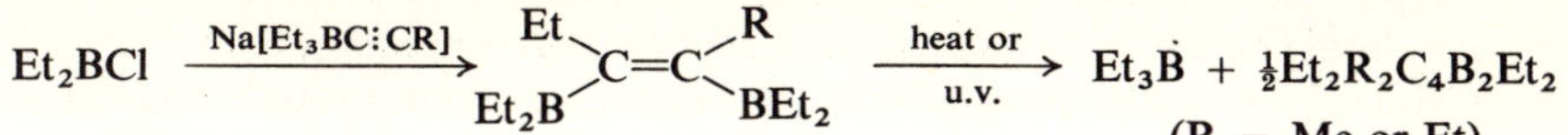

$$Et_2BCl \xrightarrow{Na[Et_3BC:CR]} \underset{Et_2B}{\overset{Et}{>}}C=C\underset{BEt_2}{\overset{R}{<}} \xrightarrow[u.v.]{heat\ or} Et_3B + \tfrac{1}{2}Et_2R_2C_4B_2Et_2$$

(R = Me or Et)

It appears likely that two other tetracarba-boranes have been prepared. An uncharacterised compound $C_{20}B_6H_{42}$ reported in 1965[119] is now believed[7, 16] to be an alkyl derivative of $C_4B_6H_{10}$, with methyl groups on the carbons and ethyl groups on the borons, i.e., $Me_4C_4B_6Et_6$. A *nido*-structure with a skeleton like that of decaborane(10) would be expected for this compound. The second probable tetracarba-borane is the species $C_4B_7H_{11}$; mass spectroscopic evidence has been obtained of its presence among the products of copyrolysis of $C_2B_4H_6$ and B_2H_6 [16]. This should have a *nido*- structure like that of $C_2B_9H_{11}^{2-}$ (see Figure 4.5), the 'dicarbollide' anion known so widely in the form of transition metal complexes, as outlined below.

4.6 METALLOCARBORANES AND RELATED SPECIES[14]

4.6.1 Introduction

This section is concerned with compounds in which a metal or metalloid atom effectively occupies the spare cage site of a *nido*- carborane residue, thus regenerating the closed cage structure from which the *nido*- carborane unit

Table 4.2 The metals and metalloids to which particular carborane anions are known to coordinate

Carborane anion	Metal or metalloid (oxidation state in parentheses)
$C_2B_9H_{11}^{2-}$	Be(II)[120]; Al(III)[121-123]; Ge(II)[124]; Sn(II)[124]; Pb(II)[124]; Cr(0,III)[132, 134]; Mo(0)[132]; W(0)[132]; Mn(I)[148]; Re(I)[148, 149]; Fe(I,II,III)[131, 132, 138, 150, 151]; Co(I,II,III)[132, 139-141, 146, 150]; Ni(II,III,IV)[132, 135, 137, 142, 143, 145, 147, 152]; Pd(II,III,IV)[135]; Cu(II,III)[132, 136, 137, 144]; Au(II,III)[137]
$C_2B_8H_{10}^{4-}$	Co(III)[153-155]
$C_2B_7H_9^{2-}$	Co(III)[156-159]
$C_2B_6H_8^{2-}$	Mn(I)[160, 161]; Co(III)[162]
$C_2B_6H_8^{4-}$	Co(III)[162]
$C_2B_4H_6^{2-}$	Ga(III)[125]; In(III)[126]; Fe(II)[126]
$C_2B_3H_7^{2-}$	Fe(II)[126]
$CB_{10}H_{11}^{3-}$	Ge(IV)[128]; P(III)[127, 130]; As(III)[128-130]; Sb(III)[128, 129]; Cr(II,III)[163, 166]; Mo(II)[166]; W(II)[166]; Mn(I,IV)[163, 164]; Fe(III)[163, 164]; Co(III,IV)[163, 164]; Ni(IV)[163-165]; Cu(III)[164]
$CB_9H_{10}P^{2-}$	Fe(II)[167]; Co(III)[167]
$CB_9H_{10}As^{2-}$	Fe(II)[128, 129]; Co(III)[128, 129]
$C_3B_3H_6^-$	Mn(I)[168, 169]

may be regarded as derived. A few species in which two metal atoms occupy the two spare cage sites of an *arachno-* carborane residue are also described. Both types may formally be regarded as coordination complexes of metal (or metalloid) cations with carborane anions, e.g., the *nido* $[C_2B_mH_{m+2}]^{2-}$ or $[CB_mH_{m+1}]^{3-}$, the *arachno* $[C_2B_mH_{m+2}]^{4-}$ (even though the anions themselves may be unknown in uncoordinated form), and it is in this manner that they have been treated in Table 4.2, which lists the carborane anions known to complex with metals or metalloids, and the metal or metalloid cations to which they are known to coordinate.

Compounds with a metal or metalloid atom terminally attached to a carbon atom of a carborane are not included in Table 4.2. They have been described earlier in this survey, under the derivative chemistry of the carborane in question.

4.6.2 Main group metallocarboranes

These may be classified as formally derived from three *nido-* anions; (1) the dicarba-*nido*-undecahydroundecaborate(2-) ion, $C_2B_9H_{11}^{2-}$, known as the 'dicarbollide' anion (from the Spanish for 'jar'), which is an icosahedral fragment (see Figure 4.5); (2) the anion $C_2B_4H_6^{2-}$, a pentagonal pyramid; and (3) the anion $CB_{10}H_{11}^{3-}$, another icosahedral fragment.

4.6.2.1 Species formally derived from $C_2B_9H_{11}^{2-}$

The *nido-* carborane $C_2B_9H_{13}$ acts as a protic acid towards dimethyl-beryllium, cleaving both methyl groups[120]:

$$C_2B_9H_{13} + Me_2Be(OEt_2)_2 \xrightarrow{PhH} C_2B_9H_{11}Be \cdot OEt_2 \xrightarrow{Me_3N} C_2B_9H_{11}Be \cdot NMe_3$$

The products may be regarded as coordination complexes of the cation Be^{2+} with the anion $C_2B_9H_{11}^{2-}$, with donor species Et_2O or Me_3N completing the coordination sphere about the metal. Alternatively, as $(L \rightarrow Be)^{2+}$ and $(H-B)^{2+}$ are isoelectronic, the products $C_2B_9H_{11}Be \cdot L$, effectively derivatives of $C_2B_9H_{11}BeH^-$, incorporate covalently bound beryllium as the twelfth cage atom (Figure 4.9).

A similar reaction occurs with aluminium trialkyls[121]:

$$R_3Al + (3)\text{-}1,2\text{-}C_2B_9H_{13} \xrightarrow{50\,°C} 1,2\text{-}C_2B_9H_{12}AlR_2 \xrightarrow{80\,°C} 1,2\text{-}C_2B_9H_{11}AlR$$

The end-product, $1,2\text{-}C_2B_9H_{11}AlR$, has the icosahedral structure (confirmed by an x-ray study) expected for a species derivable from *ortho-* carborane by replacing BH by AlR. The intermediate $C_2B_9H_{12}AlR_2$, however, has but one of the two bridging protons of $C_2B_9H_{13}$ replaced by AlR_2, which indeed (by x-ray crystallography) occupies a bridging position between two borons of the pentagonal face[122].

The *closo-* aluminium compounds can also be prepared from organo-aluminium dihalides[123]:

$$(3)\text{-}1,2\text{-}C_2B_9H_{11}^{2-} + RAlCl_2 \rightarrow 1,2\text{-}C_2B_9H_{11}AlR + 2Cl^-$$

Anhydrous hydrochloric acid regenerates $C_2B_9H_{13}$ from $C_2B_9H_{11}AlR$. At 410 °C, 1,2-$C_2B_9H_{11}AlEt$ rearranges to the 1,7-isomer.

Some Group IV derivatives $C_2B_9H_{11}M$ (M = Ge, Sn, or Pb) are known. In these, the metal atom has no terminal substituent (Figure 4.9)—a Group

$C_2B_9H_{11}^{2-}$ X
               |
               M

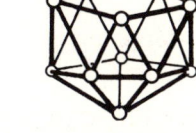

M:— Be[120];   Al[121–123]; Ge, Sn, Pb[124];
X:— Et$_2$O, Me$_3$N; R;
M:— Cr[132], Fe[132, 138], Ni[152]; Cr, Mo, W[132], Mn, Re[148, 149]
X:— π-C$_5$H$_5$; (CO)$_3$

$CB_{10}H_{11}^{3-}$ E   E:—P[127], As[128, 129], Sb[128, 129] GeMe[128]

Figure 4.9 Complexes of the *nido*- anions $C_2B_9H_{11}^{2-}$ and $CB_{10}H_{11}^{3-}$

IV element with a lone pair is isoelectronic with B—H. They are prepared (and react) as follows[124]:

$$Na_2C_2B_9H_{11} + MX_2 \rightarrow C_2B_9H_{11}M \begin{array}{l} \xrightarrow{OMe^-/MeOH} C_2B_9H_{12}^- \\ \xrightarrow{HCl} C_2B_9H_{13} \end{array}$$

The ^{11}B n.m.r. spectrum of the lead compound indicates that the metal may be 'slipped' towards the facial boron atoms.

4.6.2.2 Species formally derived from $C_2B_4H_6^{2-}$

The *nido*- carborane $C_2B_4H_8$ (Figure 4.2) acts as a protic acid towards trimethyl-gallium or -indium, cleaving two methyl groups[125, 126]:

$$2,3\text{-}C_2B_4H_8 + Me_3M \rightarrow C_2B_4H_6MMe + 2MeH$$

The products are believed to be isostructural with the *closo*- carborane $C_2B_5H_7$ (see Figure 4.12)

4.6.2.3 Species formally derived from $CB_{10}H_{11}^{3-}$

The triple negative charge of the *nido*- anion $CB_{10}H_{11}^{3-}$ makes it an appropriate species from which to prepare neutral derivatives $CB_{10}H_{11}E$ (Figure 4.9) by combination with a triply-charged cation with a lone pair (e.g.,

P^{3+}, As^{3+}, or Sb^{3+}) or terminal substituent (e.g., RGe^{3+}):

$$EX_3 \xrightarrow{Na_3CB_{10}H_{11}} 1,2\text{-}CB_{10}H_{11}E \xrightarrow{450-500\,°C} 1,7\text{-}CB_{10}H_{11}E$$
(E = P^{127}, $As^{128, 129}$, $Sb^{128, 129}$ or $MeGe^{128}$)

These *closo*- icosahedral species $CB_{10}H_{11}E$ can be converted into *nido*- anions $CB_9H_{11}E^-$ by heating them with an excess of piperidine. Treatment of the phosphacarba- and arsacarba-ions with methyl iodide gives neutral *nido*- species $CB_9H_{10}EMe$, e.g.,

$$CB_{10}H_{11}P \xrightarrow{\text{piperidine}} CB_9H_{11}P^- \xrightarrow{MeI} CB_9H_{11}PMe$$

The anions $CB_9H_{11}E^-$ form transition metal complexes either by coordinating to a metal which occupies the spare icosahedral site (see below) or by use of the phosphorus or arsenic lone pair as in the reaction[130]:

$$Me_4N^+[7,8\text{-}CB_9H_{11}P]^- + Mo(CO)_6 \xrightarrow{u.v.}$$
$$Me_4N^+[7,8\text{-}CB_9H_{11}P\text{-}Mo(CO)_5]^-$$

4.6.3 Transition metal complexes of the dicarbollide ion, $C_2B_9H_{11}^{2-}$

Many stable transition metal complexes of the dicarbollide anion (3)-1,2-$C_2B_9H_{11}^{2-}$, and a few of the isomeric anion (3)-1,7-$C_2B_9H_{11}^{2-}$, have been prepared. Two types of complex are known: the first, bis(dicarbollyl) complexes $[M(C_2B_9H_{11})_2]^{n-}$, have sandwich structures (cf., metallocenes) with the metal between the two dicarbollide anions; whereas in the second type, mixed ligand complexes $(C_2B_9H_{11})ML_x$, the coordination sphere of the metal contains x other ligands L as well as the carborane anion.

4.6.3.1 Bis(dicarbollyl) complexes $[M(C_2B_9H_{11})_2]^{n-}$

These can be prepared from the disodium salt $Na_2C_2B_9H_{11}$ (from $C_2B_9H_{12}^-$ + NaH) and an appropriate metal halide, commonly in tetrahydrofuran[131-133]:

$$C_2B_9H_{11}^{2-} \xrightarrow[\text{THF}]{FeCl_2} [Fe^{II}(C_2B_9H_{11})_2]^{2-} \xrightleftharpoons[\text{Na/Hg}]{\text{air}} [Fe^{III}(C_2B_9H_{11})_2]^-$$

Subsequent oxidation or reduction may be needed if the complex initially formed contains the metal in other than the desired oxidation state. The hydrolytic stability of the products allows them alternatively to be prepared in aqueous solution, e.g., from the metal halide and the anion $C_2B_9H_{12}^-$ in 40% aqueous sodium hydroxide, when $C_2B_9H_{11}^{2-}$ is formed *in situ*. Metals known to form bis(dicarbollyls) include chromium(III)[134], iron(II) and (III)[131-132], cobalt(II) and (III)[132], nickel(II), (III) and (IV)[132, 135], palladium(II), (III), and (IV)[135], copper(II) and (III)[132, 136] and gold(II) and (III)[137].

Crystallographic studies on a number of bis(dicarbollyl) derivatives have shown that the anion (3)-1,2-$C_2B_9H_{11}^{2-}$ (or C- or B- terminally substituted derivatives thereof) can coordinate in two distinct ways, illustrated in Figure 4.10. Complexes $M[(3)\text{-}1,2\text{-}C_2B_9H_{11}]_2^{n-}$, etc., of metals M with a d^3

(Cr^{III})[134], d^5 (Fe^{III})[138], d^6 (Fe^{II}, Co^{III} [139–141], or Ni^{IV} [142]) or d^7 (Co^{II} or Ni^{III} [135]) electronic configuration have the metal located over the centre of the pentagonal face of the anion $C_2B_9H_{11}^{2-}$. The other type of complex, formed by ions with a d^8 (Ni^{II} [135, 137, 143] or Cu^{III} [137]) or d^9 (Cu^{II} [144]) configuration, has a 'slipped' structure in which the metal is displaced towards the three boron atoms of the pentagonal face (as if the bonding resembled that in π-allyl rather than π-cyclopentadienyl complexes). Preferential bonding to boron also appears to be a feature of the complex $[Ni(1,7-C_2B_9H_{11})_2]^{2-}$ in which the two carbon atoms of each ligand still border the pentagonal face but are no longer adjacent (see Figure 4.5). In this d^8 complex, the metal is displaced only 0.15 Å from a central position[145] (cf., 0.60 Å for the related complex[137] $[Ni(1,2-C_2B_9H_{11})_2]^{2-}$), but the carbon atoms are bent away from the nickel back into the cage, so that the pentagonal face is no longer planar.

The bis(dicarbollyls) are less reactive substances than bis(cyclopentadienyl) complexes, with which their metal coordination spheres may be compared. Thus, salts of the anion $[Cr(1,2-C_2B_9H_{11})_2]^-$ are air- and moisture-stable (contrast chromocinium salts which are readily hydrolysed) and thermally stable to 250 °C [134]. Of the iron(II) and iron(III) complexes $[Fe(1,2-C_2B_9H_{11})_2]^{2-}$ and $[Fe(1,2-C_2B_9H_{11})_2]^-$, the latter is air-stable (the former

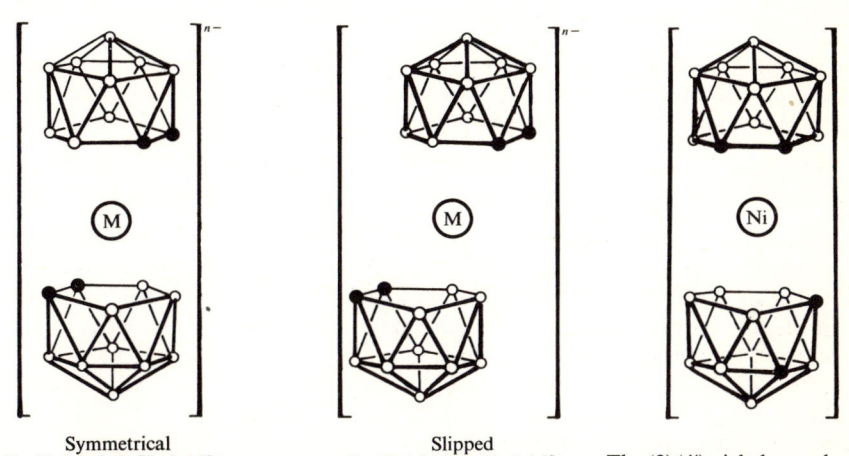

Symmetrical
$[\{\pi\text{-}(3)\text{-}1,2\text{-}C_2B_9H_{11}\}_2M]^{n-}$

Slipped
$[\{\pi\text{-}(3)\text{-}1,2\text{-}C_2B_9H_{11}\}_2M]^{n-}$

The (3),(4') nickel complex

Figure 4.10 The skeletons of symmetrical and slipped bis(π-(3)-1,2-dicarbollide) complexes and of the nickel complex $[\pi\text{-}(3)\text{-}1,2\text{-}B_9H_9C_2Me_2]Ni^{IV}[\pi\text{-}(4)\text{-}1,2\text{-}B_9H_9C_2Me_2]$ [147].

being oxidised slowly as solid, quickly in solution), does not react with concentrated acid, and can be heated unchanged to 300 °C [132]. In the case of the cobalt(II) and cobalt(III) complexes, the higher oxidation state is again the more stable, being generated in the direct synthesis even at the expense of reduction of Co^{2+} to cobalt metal:

$$3CoCl_2 + 4(1,2\text{-}C_2B_9H_{11})^{2-} \rightarrow 2[Co(1,2\text{-}C_2B_9H_{11})_2]^- + 6Cl^- + Co$$

A comparison of the ^{59}Co n.q.r. spectra of the Co^{III} complexes $[Co(1,2\text{-}C_2B_9H_{11})_2]^-$ and $[(\pi\text{-}C_5H_5)_2Co]^+$ indicates a very close similarity between the bonding properties of the dicarbollide and cyclopentadienide ligands[146].

When the complex $K[Co(1,2-C_2B_9H_{11})_2]$ is treated with aluminium chloride in carbon disulphide and hydrogen chloride is subsequently bubbled through the mixture, the product is a neutral species, $Co(C_2B_9H_{10})_2S_2CH$, in which the two staggered dicarbollide residues, sharing the cobalt as a common apex, are linked through an $-S-\overset{+}{C}C-S-$ unit spanning borons 8 and 8'[141]. A hexabromo derivative has been prepared from another substitution reaction[140]:

$$[Co(C_2B_9H_{11})_2]^- \xrightarrow{Br_2/HOAc} [Co(C_2B_9H_8Br_3)_2]^-$$

The three borons substituted on each dicarbollide residue are those furthest from the two carbon atoms, i.e., numbers 8, 9 and 12.

The redox chemistry of bis(dicarbollyl) derivatives of nickel and palladium involves three formal oxidation states of the metal[132, 142, 145]:

$$2(C_2B_9H_{11})^{2-} + Ni^{2+} \xrightarrow[\text{or hot aqueous NaOH}]{\text{THF; air}} [Ni^{III}(C_2B_9H_{11})_2]^-$$
(paramagnetic)

Na/Hg ↙   ↓ Fe^{3+} aq.

$[Ni^{II}(C_2B_9H_{11})_2]^{2-}$ $Ni^{IV}(C_2B_9H_{11})_2$
(very air sensitive)   (diamagnetic)

In the Ni^{IV} complex $Ni(1,2-C_2B_9H_{11})_2$ [142], although the metal as expected occupies the common apex of the two icosahedra, the orientation of the ligands is such as to cause the two pairs of carbon atoms to be as close to each other as is compatible with a staggered arrangement of the cages (cf., $Co(C_2B_9H_{11})_2S_2CH$ [141]).

Further structural variations have emerged from a study of some C,C'-dimethyldicarbollyl complexes of nickel and palladium[135]. In addition to forming the usual type of complex, with the metal in the 3 position, coordinated $1,2-C_2Me_2B_9H_9^{2-}$ can rearrange to give complexes in which the metal is in the 4 position, one of the two carbon atoms moving from the pentagonal face away from the immediate coordination sphere of the metal, as shown in Figure 4.10 for the complex $d,l-[\pi-(3)-1,2-B_9H_9C_2Me_2]Ni^{IV}[\pi-(4)-1,2-B_9H_9C_2Me_2]$ [147]. Complexes in which both ligands appear to have thus rearranged are also known[135].

Further information on bis(dicarbollyl) complexes will be found in Todd's review[14].

4.6.3.2 Mixed ligand dicarbollyl complexes $(C_2B_9H_{11})ML_x$

Attention has been focused principally on two types of mixed ligand complex; dicarbollylmetal cyclopentadienyls and dicarbollylmetal carbonyls (Figure 4.9). Both types are represented by chromium sub-group examples[132]:

$$CrCl_3 + Na_2C_2B_9H_{11} + NaC_5H_5 \xrightarrow{THF} (\pi-C_5H_5)Cr^{III}(C_2B_9H_{11}) + 3NaCl$$

$$C_2B_9H_{11}^{2-} + M(CO)_6 \xrightarrow{u.v.; THF} (C_2B_9H_{11})M(CO)_3^{2-} + 3CO$$
(M = Cr, Mo, or W)

These dicarbollylmetal carbonyl anions react as nucleophiles, undergoing protonation and alkylation to give the hydride $(C_2B_9H_{11})M(CO)_3H^-$ and alkyl [e.g., $(C_2B_9H_{11})M(CO)_3Me^-$], respectively, and giving binuclear metal–metal bonded species with further hexacarbonyl, e.g.,

$$(C_2B_9H_{11})Mo(CO)_3^{2-} + W(CO)_6 \rightarrow (C_2B_9H_{11})(CO)_3MoW(CO)_5^{2-} + CO$$

Related manganese and rhenium anions $(C_2B_9H_{11})M(CO)_3^-$ have been prepared from the carbonyl bromide[148]:

$$Na_2C_2B_9H_{11} + BrM(CO)_5 \xrightarrow{THF} Na(C_2B_9H_{11})M(CO)_3 + NaBr + 2CO$$

Sodium bromide precipitates immediately the solutions are mixed, but carbon monoxide is evolved only when the solution is boiled, indicating that an intermediate $(C_2B_9H_{11})M(CO)_5^-$ may be involved.

An x-ray single crystal study of the rhenium complex $Cs(C_2B_9H_{11})Re(CO)_3$ has confirmed the (3)-1,2- geometry of the dicarbollide group and terminal nature of the three carbonyls[149].

The dicarbollyl-cyclopentadienyl-iron complex $(C_5H_5)Fe^{III}(C_2B_9H_{11})$ (from $Na_2C_2B_9H_{11}$, NaC_5H_5 and $FeCl_2$ in tetrahydrofuran) was the first neutral sublimable complex of the dicarbollide series, and its structure determination provided the first x-ray crystallographic evidence of the mode of coordination of the dicarbollide ligand (see Figure 4.9)[138]. It can be reduced to a related iron(II) complex by sodium amalgam[132]:

$$(C_5H_5)Fe(C_2B_9H_{11}) \xrightarrow{Na/Hg} [(C_5H_5)Fe(C_2B_9H_{11})]^-$$

A binuclear dicarbollyliron carbonyl species results from the action of dicarbollide ion on iron pentacarbonyl[150, 151], whereas dicobalt octacarbonyl

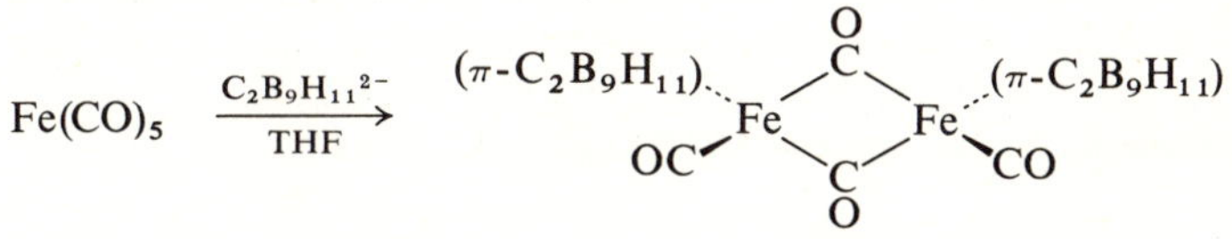

$$Fe(CO)_5 \xrightarrow[THF]{C_2B_9H_{11}^{2-}} (\pi\text{-}C_2B_9H_{11})\text{Fe}\underset{\underset{O}{C}}{\overset{\overset{O}{C}}{\underset{\diagdown}{\diagup}}}\text{Fe}(\pi\text{-}C_2B_9H_{11})$$
(with OC and CO terminal)

gives the mononuclear anionic complex[150] $(\pi\text{-}C_2B_9H_{11})Co(CO)_2^-$.

Cyclic voltammetric studies of the oxidation and reduction potentials of the complex $(\pi\text{-}C_5H_5)Ni^{III}(\pi\text{-}C_2B_9H_{11})$, and of the related species $(\pi\text{-}C_5H_5)_2Ni^+$ and $[Ni(\pi\text{-}C_2B_9H_{11})_2]^-$, have illustrated the relative capacities of the two ligands $C_5H_5^-$ and $C_2B_9H_{11}^{2-}$ to stabilise different formal oxidation states of the metal[152]. The higher charge of the dicarbollyl ligand allows it to be associated with higher formal metal oxidation states than are commonly stable with the cyclopentadienyl ligand. Thus the complex $[(\pi\text{-}C_5H_5)_2Ni^{IV}]^{2+}$, which could only be detected electrochemically using low temperatures, is a very powerful oxidising agent, whereas the neutral complex $(\pi\text{-}C_2B_9H_{11})_2Ni^{IV}$ is a relatively feeble oxidising agent. The cation $[(\pi\text{-}C_5H_5)Ni^{IV}(\pi\text{-}C_2B_9H_{11})]^+$ is intermediate in oxidising power. Conversely, whereas nickelocene $(\pi-C_5H_5)_2Ni^{II}$ is but weakly reducing, both $[(\pi-C_5H_5)Ni^{II}(\pi-C_2B_9H_{11})]^-$ and $[Ni^{II}(\pi-C_2B_9H_{11})_2]^{2-}$ are powerful reducing agents.

4.6.4 Other transition metal-carborane complexes

4.6.4.1 Complexes of the dicarbacanastide ion $(3,6)$-$1,2$-$C_2B_8H_{10}^{4-}$

Structurally related to the above dicarbollides are the dicarbacanastides, complexes (e.g. see Figure 4.11) of the *arachno-* icosahedral fragment $C_2B_8H_{10}^{4-}$, referred to as '-canastides' from the Spanish 'canasta' (= basket). The $(3,6)$-$1,2$-dicarbacanastide(-4) ion is a ten-particle icosahedral fragment; the two vacant sites, numbers 3 and 6, are those next to both carbon atoms (which form the handle of the basket). The binuclear complex anion $[(C_2B_9H_{11})Co^{III}(C_2B_8H_{10})Co^{III}(C_2B_9H_{11})]^{2-}$, shown in Figure 4.11, consists of

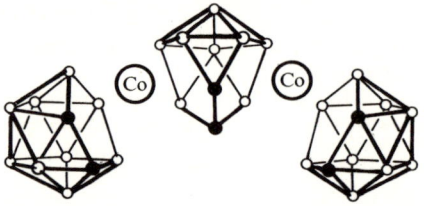

Figure 4.11 The dicarbacanastide complex $[(C_2B_9H_{11})Co^{III}(C_2B_8H_{10})Co^{III}(C_2B_9H_{11})]^{2-}$

three linked icosahedra, the central one of which is defined by the ten dicarbacanastide skeletal atoms and the two cobalt atoms. First obtained as a minor by-product of the synthesis of $[Co^{III}(C_2B_9H_{11})_2]^-$ from $C_2B_9H_{11}^-$ and $Co(OH)_2$ [153], it can be prepared in better yield (*c.* 15%) by reaction of $[Co^{III}(C_2B_9H_{11})_2]^-$ with 30% aqueous sodium hydroxide at 100 °C for several hours (cf., the alkaline degradation of the parent carborane $C_2B_{10}H_{12}$) [154]. Further alkaline degradation leads to the trinuclear species $[(C_2B_9H_{11})Co^{III}(C_2B_8H_{10})Co^{III}(C_2B_8H_{10})Co^{III}(C_2B_9H_{11})]^{3-}$ with two dicarbacanastide residues linked to the central cobalt[155].

4.6.4.2 Complexes of the dicarbazapide ion $C_2B_7H_9^{2-}$

Complexes of the *nido-* anion $C_2B_7H_9^{2-}$ (sometimes referred to as the 'dicarbazapyl' ligand from the Spanish 'zapato' = shoe) can be prepared from the *arachno-* carborane $1,3$-$C_2B_7H_{13}$ by the following type of reaction sequence[156, 157]:

$$1,3\text{-}C_2B_7H_{13} \xrightarrow{2NaH} 1,3\text{-}C_2B_7H_{11}^{2-} \xrightarrow[70\,°C]{CoCl_2} [Co^{III}(C_2B_7H_9)_2]^-$$

(brown complex, 51% yield)

The product has a structure based on a bicapped Archimedean antiprism,

CARBORANES AND METALLOCARBORANES

with the metal occupying a non-apical(2) position (see Figure 4.12), and carbon atoms in positions 1 and 6 [156]. If the reaction is carried out at 25 °C, a red isomer is obtained, the ^{11}B n.m.r. spectrum of which is consistent with a higher symmetry ligand with carbon atoms occupying positions 6 and 7. When heated to 315 °C and 250 °C, respectively, the brown and red isomers rearrange to a third, orange isomer in which the non-apical carbon atom is

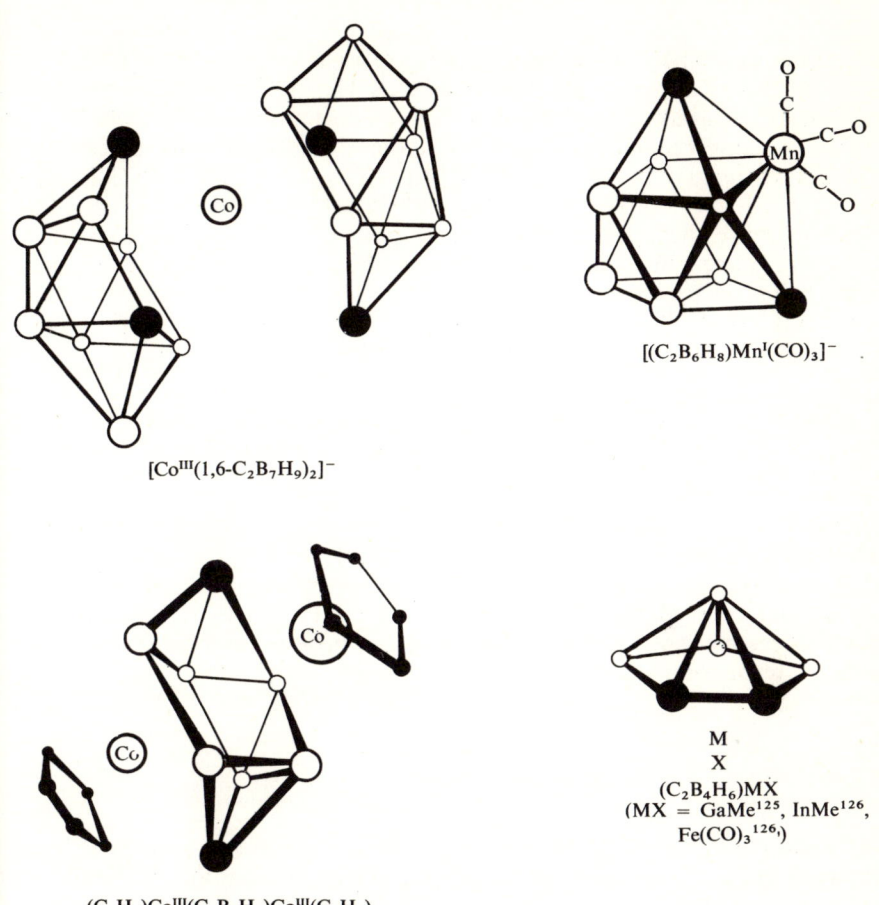

Figure 4.12 Further metallocarboranes

believed to have moved out of the bonding face into the second apical position [157, 158], i.e., this isomer is believed to be $[Co^{III}(\pi\text{-}(2)\text{-}1,10\text{-}C_2B_7H_9)_2]^-$.

The mixed ligand neutral complex $(C_5H_5)Co^{III}(C_2B_7H_9)$, with the carbon atoms of the dicarbazapyl ligand in the 1,6 positions, has been prepared from cobalt(II) chloride and a solution containing both $C_2B_7H_{11}^{2-}$ and $C_5H_5^-$ ions. This rearranges to the 1,10 isomer at 315 °C, with activation enthalpy ΔH^* 34 kcal mol^{-1} and entropy ΔS^* c. 3 kcal mol^{-1}, consistent with a smooth changeover of atom positions without discrete bond-breaking and -forming[157]. Electrophilic substitution of the complex $(C_5H_5)Co^{III}$ $(\pi\text{-}(2)\text{-}1,6\text{-}C_2B_7H_9)$, e.g., using $MeCOCl/AlCl_3$, apparently occurs at the boron atom in the 8-position[159].

4.6.4.3 Complexes of the ions $C_2B_6H_8^{2-}$ and $C_2B_6H_8^{4-}$

The structures expected for these ligands are a *nido-* structure for $C_2B_6H_8^{2-}$ based on the tricapped trigonal prism of $C_2B_7H_9$, and an *arachno-* structure for $C_2B_6H_8^{4-}$ based on the bicapped Archimedean antiprism of $C_2B_8H_{10}$.

Two transition metal complexes of the dianion $C_2B_6H_8^{2-}$ have been described. Treatment of $BrMn(CO)_5$ [160] or $Mn_2(CO)_{10}$ [161] in THF with $C_2B_7H_{11}^{2-}$ ion afforded the complex $[(C_2B_6H_8)Mn^I(CO)_3]^-$, in which the $Mn(CO)_3$ unit apparently occupies a prism corner position between two equivalent apical carbon atoms, the third apical position of the tricapped trigonal prism being occupied by a boron atom (see Figure 4.12). A related cobalt complex $(C_5H_5)Co^{III}(C_2B_6H_8)$, presumably with a similar structure, $Co(C_5H_5)$ replacing $Mn(CO)_3$, is one product of the reaction sequence[162]:

$$1,7\text{-}C_2B_6H_8 \xrightarrow[\text{(ii) Excess } NaC_5H_5 + CoCl_2]{\text{(i) } 2Na^+ \text{naphthalene}^- \text{ in THF}} (C_5H_5)Co^{III}(C_2B_6H_8) + (C_5H_5)Co^{III}(C_2B_6H_8)Co^{III}(C_5H_5)$$

This reaction illustrates the structural effect of adding a pair of electrons to a neutral *closo-* carborane $C_2B_mH_{m+2}$, which is thereby converted into the *nido-* anion $C_2B_mH_{m+2}^{2-}$. Indeed, the second product of this reaction, $(C_5H_5)Co^{III}(C_2B_6H_8)Co^{III}(C_5H_5)$, illustrates the effect of adding *two* pairs of electrons to $C_2B_6H_8$ to convert it into $C_2B_6H_8^{4-}$, the *arachno-* species based on the bicapped Archimedean antiprism believed to form the skeleton of this neutral complex (Figure 4.12) [162].

4.6.4.4 Complexes of the ion $C_2B_4H_6^{2-}$

The *nido-* carborane $C_2B_4H_8$ appears to be a ready-made precursor from which to prepare complexes of the pentagonal pyramidal anion $C_2B_4H_6^{2-}$, but few complexes have been prepared to date. As well as the gallium and indium compounds already described, an iron complex $(\pi\text{-}C_2B_4H_6)Fe(CO)_3$ (Figure 4.12) is known[126] as a product of the reaction between $C_2B_4H_8$ and an excess of iron pentacarbonyl at 240 °C.

4.6.4.5 Complexes of the ion $C_2B_3H_7^{2-}$

The reaction between $C_2B_4H_8$ and $Fe(CO)_5$ just mentioned affords a second product, $(\pi\text{-}C_2B_3H_7)Fe(CO)_3$, which is of particular interest as the first example of a complex of the ion $C_2B_3H_7^{2-}$ believed to have a cyclic structure (cf. the isoelectronic $C_5H_5^-$) with two BHB bridges, the skeleton of which is formally derived from the pentagonal bipyramid of $C_2B_5H_7$ by removal of both apical borons[126].

4.6.4.6 Complexes of the ion $CB_{10}H_{11}^{3-}$

Several transition metal complexes are known containing the ion $CB_{10}H_{11}^{3-}$ or a related ion $[CB_{10}H_{10}NR_xH_{3-x}]^{2-}$, in which the group NR_xH_{3-x} is a terminally attached substituent on the carbon atom. Three general methods of preparation of such complexes have been used:

(a) addition of n-BuLi to a mixture of the metal halide and $B_{10}H_{12}C\cdot NH_3$ or $B_{10}H_{10}CH^-$ in tetrahydrofuran[163],

(b) reaction of $B_{10}H_{12}C\cdot NMe_3$ with sodium in tetrahydrofuran to precipitate $Na_3B_{10}H_{10}CH\cdot 2THF$, followed by addition of the metal halide[164]:

$$B_{10}H_{12}C\cdot NMe_3 \xrightarrow{Na/THF} B_{10}H_{10}CH^{3-}$$
$$\xrightarrow{CoCl_2} Co + NaCl + [Co^{III}(CB_{10}H_{11})_2]^{3-}$$

(c) reaction of a metal halide with $B_{10}H_{12}CH^-$ or $B_{10}H_{12}CNR_xH_{3-x}$ in concentrated aqueous sodium hydroxide[163–165]:

$$NiCl_2 + CsCB_{10}H_{13} \xrightarrow{aq.\ NaOH} Na_2[Ni^{IV}(CB_{10}H_{11})_2]$$

Bis complexes $[M(CB_{10}H_{11})_2]^{n-}$ of several metals M have been prepared. They are probably structurally analogous to the dicarbollide metal sandwich compounds[163, 164], the greater negative charge of the ligand $CB_{10}H_{11}^{3-}$ making it even more effective than $C_2B_9H_{11}^{2-}$ at stabilising high oxidation states. The preferred oxidation state of the metal in complexes $[M(CB_{10}H_{11})_2]^{n-}$ is +3 for chromium, iron and cobalt, and +4 for manganese and nickel.

The amine group of complexes derived from $B_{10}H_{12}C\cdot NH_3$ provides access to other C-substituted species, e.g.[163],

$$Ni^{IV}(B_{10}H_{10}CNH_3)_2 \begin{cases} \xrightarrow{Me_2SO_4} Ni^{IV}(B_{10}H_{10}CNHMe_2)_2 \\ \xrightarrow{HNO_2} Ni^{IV}(B_{10}H_{10}COH)_2 \end{cases}$$

Mixed ligand complexes are accessible by the usual types of reaction, e.g.[164],

$$BrMn(CO)_5 + B_{10}H_{10}CH^{3-} \rightarrow (B_{10}H_{10}CH)Mn(CO)_3^{2-} + Br^- + 2CO$$

Other carbonyl complexes have resulted from incorporation of a carbonyl carbon into the cage of the decaborane anion $B_{10}H_{13}^-$ [166].

$$NaB_{10}H_{13} + M(CO)_6 \xrightarrow{u.v.} Na^+[(B_{10}H_{10}COH)M(CO)_4]^-$$

$$\downarrow NaH \qquad M = Cr, Mo, W$$

$$Na_2[B_{10}H_{10}\underline{COMCO}(CO)_3]$$

An x-ray crystallographic study has shown that in the anion $[B_{10}H_{10}COMoCO(CO)_3]^{2-}$ the molybdenum completes the icosahedral CB_{10} framework. An unusual feature of the structure is that the metal is bound to the carbon of a carboxylate group which is linked through one oxygen to the CB_{10} cage carbon. Atypically for carboranes and metallocarboranes, these last complexes are degraded by base with removal of a *carbon* (rather than a boron) atom from the icosahedral framework, generating *nido*-metalloboranes $[B_{10}H_{12}M(CO)_4]^{2-}$.

4.6.4.7 Complexes of the ions $CB_9H_{10}P^{2-}$ and $CB_9H_{10}As^{2-}$

The *nido*- phosphacarborane anions 1,2- and 1,7-$B_9H_{10}CHP^-$ and the *P*-methylated neutral species $B_9H_{10}CHPMe$, can be deprotonated with sodium hydride to form 'phosphacarbollide' ions $B_9H_9CHP^{2-}$ which form sandwich bis-complexes with transition metals, e.g.[167],

$$1,7\text{-}B_{10}H_{10}CHP \xrightarrow{\text{piperidine}} 1,7\text{-}B_9H_{10}CHP^- \xrightarrow[\text{(ii) FeCl}_2]{\text{(i) NaH}} [(3)\text{-}1,7\text{-}B_9H_9CHP]_2Fe^{2-}$$

$$\downarrow \text{MeI} \qquad\qquad\qquad\qquad \downarrow \text{MeI}$$

$$1,7\text{-}B_9H_{10}CHPMe \xrightarrow[\text{(ii) FeCl}_2]{\text{(i) NaH}} [(3)\text{-}1,7\text{-}B_9H_9CHPMe]_2Fe$$

Iron(II) and cobalt(III) complexes of this type, with both 1,2- and 1,7-$B_9H_9CHP^{2-}$ ligands, have been characterised. The arsacarborane anions 1,2- and 1,7-$B_9H_9CHAs^{2-}$ form similar complexes[128, 129], which however could be only partially methylated to derivatives of the type $[(1,7\text{-}B_9H_9CHAsMe)M(1,7\text{-}B_9H_9CHAs)]^-$. Mixed ligand species such as $(C_5H_5)Co^{III}(1,2\text{-}B_9H_9CHAs)$, are also known:

$$7,8\text{-}B_9H_{10}CHAs^- + C_5H_6 \xrightarrow[\text{CoCl}_2]{\text{Et}_3\text{N/THF}} (C_5H_5)Co^{III}(1,2\text{-}B_9H_9CHAs)$$

An x-ray crystallographic study has shown the arsenic and carbon atoms to be adjacent in the product. Since under the mild conditions used to synthesise the complex, migration of the arsenic and carbon atoms would be unlikely, it is concluded that these two atoms were also adjacent in 1,2-$B_{10}H_{10}CHAs$ and in 7,8-$B_9H_{10}CHAs^-$.

4.6.4.8 Complexes of the ion

The pentagonal pyramidal tricarbahexahydrohexaborate(-1) anion, $C_3B_3H_6^-$, like its analogue $C_2B_4H_6^{2-}$, appears capable of π-bonding to transition metals to form metallocene and dicarbollide analogues. Attempts at the preparation of such complexes from the sodium salts of the methylated anions 2,3- and 2,4-$Me_2C_3B_3H_4^-$ and $FeCl_2$ or $Mn_2(CO)_{10}$ were, however, unsuccessful; complex reactions occurred in which the carborane anions suffered partial decomposition[168]. A complex $(MeC_3B_3H_5)Mn(CO)_3$ of the monomethyl anion was obtained by heating $Mn_2(CO)_{10}$ and $MeC_3B_3H_6$ in 1:2 molar proportion at 175–200 °C [169]. Its spectra are consistent with a pentagonal bipyramidal arrangement of the C_3B_3Mn skeletal atoms.

References

1. Bobinski, J. (1964). *J. Chem. Educ.*, **41,** 500
2. Onak, T. P. (1965). *Advan. Organometal. Chem.*, **3,** 263
3. Stanko, V. I., Chapovskii, Yu. A., Brattsev, V. A. and Zakharkin, L. I. (1965). *Chem. Rev. USSR*, **34,** 424
4. Issleib, K., Linder, R. and Tzschach, A. (1966). *Z. Chem.*, **6,** 1
5. Hawthorne, M. F. (1966). *Endeavour*, **25,** 146
6. Köster, R. and Grassberger, M. A. (1967). *Angew. Chem. Int. Ed. Engl.*, **6,** 218
7. Williams, R. E., p. 37 of ref. 8
8. Brotherton, R. J. and Steinberg, H. (1970). *Progress in Boron Chemistry*, Vol. 2 (Oxford: Pergamon Press)
9. Grimes, R. N. (1971). *Carboranes* (New York: Academic Press)
10. Lipscomb, W. N. (1966). *Science*, **153,** 373
11. Heying, T. L. p. 119 of ref. 8
12. Eaton, G. R. and Lipscomb, W. N. (1969). *N.M.R. Studies of Boron Hydrides and Related Compounds*, (New York: W. A. Benjamin)
13. Bregadze, V. I. and Okhlobystin, O. Yu. (1969). *Organometal. Chem. Rev.*, **4A,** 345
14. Todd, L. J. (1970). *Advan. Organometal. Chem.*, **8,** 87
15. Hawthorne, M. F. (1968). *Accounts Chem. Res.*, **1,** 281
16. Williams, R. E. (1971). *Inorg. Chem.*, **10,** 210
17. (1968). *Inorg. Chem.*, **7,** 1945
18. Lipscomb, W. N. (1963). *Boron Hydrides*, 86, (New York: W. A. Benjamin)
19. Klanberg, F., Eaton, D. R., Guggenberger, L. J. and Muetterties, E. L. (1967). *Inorg. Chem.*, **6,** 1271
20. Stanko, V. I., Gol'tyapin, Yu. V. and Klimova, T. P. (1969). *J. Gen. Chem. USSR*, **39,** 1261
21. Hawthorne, M. F. and Wegner, P. A. (1968). *J. Amer. Chem. Soc.*, **90,** 896
22. Roscoe, J. S., Konypricha, S. and Papetti, S. (1970). *Inorg. Chem.*, **9,** 1561
23. Sieckhaus, J. J., Semenuk, N. S., Knowles, T. A. and Schroeder, H. (1969). *Inorg. Chem.*, **8,** 2452
24. Zakharkin, L. I. and Kalinin, V. N. (1966, 1967). *J. Gen. Chem. USSR*, **36,** 376; **37,** 266
25. Stanko, V. I., Klimova, A. I. and Gol'tyapin, Yu. V. (1969). *J. Gen. Chem. USSR*, **39,** 1765
26. Kaesz, H. D., Bau, R., Beall, H. A. and Lipscomb, W. N. (1967). *J. Amer. Chem. Soc.*, **89,** 4218
27. Hart, H. V. and Lipscomb, W. N. (1969). *J. Amer. Chem. Soc.*, **91,** 771

28. Stanko, V. I., Brattsev, V. A. and Gol'tyapin, Yu. V. (1969). *J. Gen. Chem. USSR*, **39**, 1142, 2623
29. Hawthorne, M. F., Young, D. C., Garrett, P. M., Owen, D. A., Schwenn, S. G., Tebbe, F. N. and Wegner, P. A. (1968). *J. Amer. Chem. Soc.*, **90**, 862
30. Zakharkin, L. I. and Kalinin, V. N. (1965). *J. Gen. Chem. USSR*, **35**, 1693
31. Potenza, J. A. and Lipscomb, W. N. (1966). *Inorg. Chem.*, **5**, 1471, 1478, 1483
32. Wunderlick, J. A. and Lipscomb, W. N. (1960). *J. Amer. Chem. Soc.*, **82**, 4427
33. Beall, H. and Lipscomb, W. N. (1967). *Inorg. Chem.*, **6**, 874
34. Laubengayer, A. W. and Rysz, W. R. (1965). *Inorg. Chem.*, **4**, 1513
35. Potenza, J. A., Lipscomb, W. N., Vickers, G. D. and Schroeder, H. (1966). *J. Amer. Chem. Soc.*, **88**, 628
36. Zakharkin, L. I. and Ogarodinikova, N. A. (1968). *J. Organometal. Chem.*, **12**, 13
37. Zakharkin, L. I. and Kalinin, V. N. (1967). *J. Gen. Chem. USSR*, **37**, 889, 914
38. Stanko, V. I. and Klimova, A. I. (1968). *J. Gen. Chem. USSR*, **38**, 1147
39. Stanko, V. I. Klimova, A. I. and Titova, N. S. (1968). *J. Gen. Chem. USSR*, **38**, 2718
40. Stanko, V. I., Brattsev, V. A., Vostrikova, T. N. and Danilova, G. N. (1968). *J. Gen. Chem. USSR*, **38**, 1300
41. Zakharkin, L. I. and Kalinin, V. N. (1967). *Bull. Acad. Sci. USSR*, 2460
42. Zakharkin, L. I., Kalinin, V. N. and Snyakin, A. P. (1967). *Bull. Acad. Sci. USSR*, 1809
43. Stanko, V. I., Gol'tyapin, Yu. V. and Brattsev, V. A. (1967). *J. Gen. Chem. USSR*, **37**, 2247
44. Stanko, V. I. and Gol'tyapin, Yu. V. (1969, 1970). *J. Gen. Chem. USSR*, **39**, 676; **40**, 115
45. Kongpricha, S. and Schroeder, H. (1969). *Inorg. Chem.*, **8**, 2449
46. Rudolph, R. W., Pflug, J. L., Bock, C. M. and Hodgson, M. (1970). *Inorg. Chem.*, **9**, 2274.
47. Hota, N. K. and Matteson, D. S. (1968). *J. Amer. Chem. Soc.*, **90**, 3570
48. Matteson, D. S. and Davis, R. A. (1970). *Chem. Commun.*, 669
49. Stanko, V. I., Anorova, G. A. and Klimova, T. V. (1969). *J. Gen. Chem. USSR*, **39**, 1541
50. Stanko, V. I. and Anorova, G. A. (1969). *J. Gen. Chem. USSR*, **39**, 1550
51. Zakharkin, L. I., Bregadze, V. I. and Okhlobystin, O. Yu. (1965). *J. Organometal. Chem.*, **4**, 211
52. Schroeder, H., Papetti, S., Alexander, R. P., Sieckhaus, J. F. and Heying, T. L. (1969). *Inorg. Chem.*, **8**, 2444
53. Smith, H. D. (1969). *Inorg. Chem.*, **8**, 676
54. Scott, R. N., Papetti, S. and Schroeder, H. A. (1970). *Inorg. Chem.*, **9**, 2597
55. Smith, H. D., Obenland, C. O. and Papetti, S. (1966). *Inorg. Chem.*, **5**, 1013
56. Smith, H. D., Robinson, M. A. and Papetti, S. (1967). *Inorg. Chem.*, **6**, 1014
57. Zakharkin, L. I. and Zhigareva, G. G. (1967). *J. Gen. Chem. USSR*, **37**, 2646
58. Zaborowski, R. and Cohn, K. (1969). *Inorg. Chem.*, **8**, 678
59. Semenuk, N. S., Papetti, S. and Schroeder, H. (1969). *Inorg. Chem.*, **8**, 2441
60. Zakharkin, L. I., Bregadze, V. I. and Okhlobystin, O. Yu. (1966). *J. Organometal. Chem.*, **6**, 228.
61. Zakharkin, L. I., Kalinin, V. N. and Podvisotskaya, L. S. (1968). *Bull. Acad. Sci. USSR*, 677
62. Zakharkin, L. I. and Podvisotskaya, L. S. (1967). *J. Organometal. Chem.*, **7**, 385
63. Smart, J. C., Garrett, P. M. and Hawthorne, M. F. (1969). *J. Amer. Chem. Soc.*, **91**, 1031
64. Mitchell, C. M. and Stone, F. G. A. (1970). *Chem. Commun.*, 1263
65. Owen, D. A. and Hawthorne, M. F. (1970). *J. Amer. Chem. Soc.*, **92**, 3194
66. Tebbe, F. N., Garrett, P. M. and Hawthorne, M. F. (1968). *J. Amer. Chem. Soc.*, **90**, 869
67. Tsai, C. and Streib, W. E. (1966). *J. Amer. Chem. Soc.*, **88**, 4513
68. Young, D. C., Howe, D. V. and Hawthorne, M. F. (1969). *J. Amer. Chem. Soc.*, **91**, 859
69. Voet, D. and Lipscomb, W. N. (1967). *Inorg. Chem.*, **6**, 113
70. Tebbe, F. N., Garrett, P. M. and Hawthorne, M. F. (1966). *J. Amer. Chem. Soc.*, **88**, 607
71. Garrett, P. M., Smart, J. C., Ditta, G. S. and Hawthorne, M. F. (1969). *Inorg. Chem.*, **8**, 1907
72. Dunks, G. B. and Hawthorne, M. F. (1968). *Inorg. Chem.*, **7**, 1038
73. Williams, R. E. and Gerhart, F. J. (1965). *J. Amer. Chem. Soc.*, **87**, 3513
74. Garrett, P. M., Ditta, G. S. and Hawthorne, M. F. (1970). *Inorg. Chem.*, **9**, 1947
75. Hart, H. and Lipscomb, W. N. (1968). *Inorg. Chem.*, **7**, 1070
76. Koetzle, T. F., Scarbrough, F. E. and Lipscomb, W. N. (1968). *Inorg. Chem.*, **7**, 1076

77. Koetzle, T. F. and Lipscomb, W. N. (1970). *Inorg. Chem.*, **9,** 2279
78. Boer, F. P., Potenza, J. A. and Lipscomb, W. N. (1966). *Inorg. Chem.*, **5,** 1301
79. Dunks, G. B. and Hawthorne, M. F. (1970). *Inorg. Chem.*, **9,** 893
80. Garrett, P. M., Smart, J. C. and Hawthorne, M. F. (1969). *J. Amer. Chem. Soc.*, **91,** 4707
81. Grimes, R. N. (1966). *J. Amer. Chem. Soc.*, **88,** 1895
82. Lindner, H. H. and Onak, T. (1966). *J. Amer. Chem. Soc.*, **88,** 1886
83. Grimes, R. N. and Bramlett, C. L. (1967). *J. Amer. Chem. Soc.*, **89,** 2557
84. Grimes, R. N., Bramlett, C. L. and Vance, R. L. (1968, 1969). *Inorg. Chem.*, **7,** 1066; **8,** 55
85. Onak, T. P., Dunks, G. B., Spielman, J. R., Gerhart, F. J. and Williams, R. E. (1966). *J. Amer. Chem. Soc.*, **88,** 2061
86. Franz, D. A. and Grimes, R. N. (1970). *J. Amer. Chem. Soc.*, **92,** 1438
87. Ditter, J. F., Klusmann, E. B., Oakes, J. D. and Williams, R. E. (1970). *Inorg. Chem.*, **9,** 889
88. Ditter, J. F. (1968). *Inorg. Chem.*, **7,** 1748
89. Boer, F. P., Streib, W. E. and Lipscomb, W. N. (1964). *Inorg. Chem.*, **3,** 1666
90. Onak, T., Marynick, D., Mattschei, P. and Dunks, G. (1968). *Inorg. Chem.*, **7,** 1754
91. Onak, T. and Dunks, G. B. (1966). *Inorg. Chem.*, **5,** 439
92. Spielman, J. R., Warren, R., Dunks, G. B., Scott, J. E. and Onak, T. (1968). *Inorg. Chem.*, **7,** 216
93. Spielman, J. R., Dunks, G. B. and Warren, R. (1969). *Inorg. Chem.*, **8,** 2172
94. Shapiro, I., Good, C. D. and Williams, R. E. (1962). *J. Amer. Chem. Soc.*, **84,** 3837
95. Köster, R. and Rotermund, G. W. (1964). *Tetrahedron Lett.*, 1667
96. Hoffmann, R. and Lipscomb, W. N. (1962). *J. Chem. Phys.*, **36,** 3489
97. Grimes, R. N. (1967). *J. Organometal Chem.*, **8,** 45
98. Shapiro, I., Keilin, B., Williams, R. E. and Good, C. D. (1963). *J. Amer. Chem. Soc.*, **85,** 3167
99. Beaudet, R. A. and Poynter, R. L. (1970). *J. Chem. Phys.*, **53,** 1899
100. Epstein, I. R., Koetzle, T. F., Stevens, R. M. and Lipscomb, W. N. (1970). *J. Amer. Chem. Soc.*, **92,** 7019
101. Koetzle, T. F. and Lipscomb, W. N. (1970). *Inorg. Chem.*, **9,** 2743
102. Onak, T. P., Gerhart, F. J. and Williams, R. E. (1963). *J. Amer. Chem. Soc.*, **85,** 3378
103. Onak, T., Dunks, G. B., Beaudet, R. A. and Poynter, R. L. (1966). *J. Amer. Chem. Soc.*, **88,** 4622
104. Olsen, R. R. and Grimes, R. N. (1970). *J. Amer. Chem. Soc.*, **92,** 5072
105. Warren, R., Paquin, D., Onak, T., Dunks, G. and Spielman, J. R. (1970). *Inorg. Chem.*, **9,** 2285
106. Onak, T., Drake, R. and Dunks, G. (1965). *J. Amer. Chem. Soc.*, **87,** 2505
107. Onak, T., Mattschei, P. and Groszek, E. (1969). *J. Chem. Soc. A*, 1990
108. Prince, S. R. and Schaeffer, R. (1968). *Chem. Commun.*, 451
109. Onak, T. and Leach, J. B. (1971). *Chem. Commun.*, 76
110. Onak, T. P., Dunks, G. B., Spielman, J. R., Gerhart, F. J. and Williams, R. E. (1966). *J. Amer. Chem. Soc.*, **87,** 2061
111. Dunks, G. B. and Hawthorne, M. F. (1969). *Inorg. Chem.*, **8,** 2667
112. Hyatt, D. E., Owen, D. A. and Todd, L. J. (1966). *Inorg. Chem.*, **5,** 1749
113. Scholer, F. R. and Todd, L. J. (1968). *J. Organometal. Chem.*, **14,** 261
114. Hyatt, D. E., Scholer, F. R., Todd, L. J. and Warner, J. L. (1967). *Inorg. Chem.*, **6,** 2229
115. Knoth, W. H. (1967). *J. Amer. Chem. Soc.*, **89,** 1274
116. Franz, D. A., Howard, J. W. and Grimes, R. N. (1969). *J. Amer. Chem. Soc.*, **91,** 4010
117. Onak, T. P. and Wong, G. T. F. (1970). *J. Amer. Chem. Soc.*, **92,** 5226
118. Binger, P. (1966). *Tetrahedron Lett.*, 2675
119. Köster, R. and Rotermund, G. W. (1965). *Tetrahedron Lett.*, 777
120. Popp, G. and Hawthorne, M. F. (1968). *J. Amer. Chem. Soc.*, **90,** 6553
121. Young, D. A. T., Willey, G. R., Hawthorne, M. F., Churchill, M. R. and Reis, A. H. (1970). *J. Amer. Chem. Soc.*, **92,** 6663
122. Churchill, M. R., Reis, A. H., Young, D. A. T., Willey, G. R. and Hawthorne, M. F. (1971). *Chem. Commun.*, 298
123. Mikhailov, B. M. and Potapova, T. V. (1968). *Bull. Acad. Sci. USSR*, **5,** 1153
124. Rudolph, R. W., Voorhees, R. L. and Cochoy, R. E. (1970). *J. Amer. Chem. Soc.*, **92,** 3351

125. Grimes, R. N. and Rademaker, W. J. (1969). *J. Amer. Chem. Soc.,* **91,** 6498
126. Grimes, R. N. (1971). *J. Amer. Chem. Soc.,* **93,** 261
127. Todd, L. J., Little, J. L. and Silverstein, H. T. (1969). *Inorg. Chem.,* **8,** 1698
128. Todd, L. J., Burke, A. R., Silverstein, H. T., Little, L. J. and Wikholm, G. S. (1969). *J. Amer. Chem. Soc.,* **91,** 3376
129. Todd, L. J., Burke, A. R., Garber, A. R., Silverstein, H. T. and Storhoff, B. N. (1970). *Inorg. Chem.,* **9,** 2175
130. Silverstein, H. T., Bier, D. C. and Todd, J. L. (1969). *J. Organometal. Chem.,* **21,** 139
131. Hawthorne, M. F., Young, D. C. and Wegner, P. A. (1965). *J. Amer. Chem. Soc.,* **87,** 1818
132. Hawthorne, M. F., Young, D. C., Andrews, J. D., Howe, D. V., Pilling, R. L., Pitts, A. D., Reintjes, M., Warren, L. F. and Wegner, P. A. (1968). *J. Amer. Chem. Soc.,* **90,** 879
133. Hawthorne, M. F., Andrews, T. D., Garrett, P. M., Olsen, F. P., Reintjes, M., Tebbe, F. N., Warren, L. F., Wegner, P. A. and Young, D. C. (1967). *Inorg. Syn.,* **10,** 111
134. Rühle, H. W. and Hawthorne, M. F. (1968). *Inorg. Chem.,* **7,** 2279
135. Warren, L. F. and Hawthorne, M. F. (1970). *J. Amer. Chem. Soc.,* **92,** 1157
136. Brattsev, V. A. and Stanko, V. I. (1968). *J. Gen. Chem. USSR,* **38,** 2721
137. Wing, R. M. (1968). *J. Amer. Chem. Soc.,* **90,** 4828
138. Zalkin, A., Templeton, D. H. and Hopkins, T. E. (1965). *J. Amer. Chem. Soc.,* **87,** 3987
139. Zalkin, A., Hopkins, T. E. and Templeton, D. H. (1966). *Inorg. Chem.,* **5,** 1189
140. DeBoer, B. G., Zalkin, A. and Templeton, D. H. (1968). *Inorg. Chem.,* **7,** 2288
141. Churchill, M. R., Gold, K., Francis, J. N. and Hawthorne, M. F. (1969). *J. Amer. Chem. Soc.,* **91,** 1222
142. St. Clair, D., Zalkin, A. and Templeton, D. H. (1970). *J. Amer. Chem. Soc.,* **92,** 1173
143. Warren, L. F. and Hawthorne, M. F. (1968). *J. Amer. Chem. Soc.,* **90,** 4823
144. Wing, R. M. (1967). *J. Amer. Chem. Soc.,* **89,** 5599
145. Wing, R. M. (1970). *J. Amer. Chem. Soc.,* **92,** 1187
146. Harris, C. B. (1968). *Inorg. Chem.,* **7,** 1517
147. Churchill, M. R. and Gold, K. (1970). *J. Amer. Chem. Soc.,* **92,** 1180
148. Hawthorne, M. F. and Andrews, T. D. (1965). *J. Amer. Chem. Soc.,* **87,** 2496
149. Zalkin, A., Hopkins, T. E. and Templeton, D. H. (1966). *Inorg. Chem.,* **5,** 1189
150. Hawthorne, M. F. and Rühle, H. W. (1969). *Inorg. Chem.,* **8,** 176
151. Greene, P. T. and Bryan, R. F. (1970). *Inorg. Chem.,* **9,** 1464
152. Wilson, R. J., Warren, L. F. and Hawthorne, M. F. (1969). *J. Amer. Chem. Soc.,* **91,** 758
153. Francis, J. N. and Hawthorne, M. F. (1968). *J. Amer. Chem. Soc.,* **90,** 1663
154. St. Clair, D., Zalkin, A. and Templeton, D. H. (1969). *Inorg. Chem.,* **8,** 2080
155. Churchill, M. R., Reis, A. H., Francis, J. N. and Hawthorne, M. F. (1970). *J. Amer. Chem. Soc.,* **92,** 4993
156. Hawthorne, M. F. and George, T. A. (1967). *J. Amer. Chem. Soc.,* **89,** 7114
157. George, T. A. and Hawthorne, M. F. (1969). *J. Amer. Chem. Soc.,* **91,** 5475
158. Reeke, G. N., Vincent, R. L. and Lipscomb, W. N. (1968). *J. Amer. Chem. Soc.,* **90,** 1661
159. Graybill, B. M. and Hawthorne, M. F. (1969). *Inorg. Chem.,* **8,** 1799
160. Hawthorne, M. F. and Pitts, A. D. (1967). *J. Amer. Chem. Soc.,* **89,** 7115
161. George, T. A. and Hawthorne, M. F. (1969). *Inorg. Chem.,* **8,** 1801
162. Dunks, G. B. and Hawthorne, M. F. (1970). *J. Amer. Chem. Soc.,* **92,** 7213
163. Knoth, W. H. (1967). *J. Amer. Chem. Soc.,* **89,** 3343
164. Hyatt, D. E., Little, J. L., Moran, J. T., Scholer, F. R. and Todd, L. J. (1967). *J. Amer. Chem. Soc.,* **89,** 3342
165. Knoth, W. H., Little, J. L. and Todd, L. J. (1968). *Inorg. Syn.,* **11,** 41
166. Wegner, P. A., Guggenberger, L. J. and Muetterties, E. L. (1970). *J. Amer. Chem. Soc..* **92,** 3473
167. Todd, L. J., Paul, I. C., Little, J. L., Welcker, P. S. and Peterson, C. R. (1968). *J. Amer. Chem. Soc.,* **90,** 4489
168. Franz, D. A., Howard, J. W. and Grimes, R. N. (1969). *J. Amer. Chem. Soc.,* **91,** 4010
169. Howard, J. W. and Grimes, R. N. (1969). *J. Amer. Chem. Soc.,* **91,** 6499
170. Odom, J. D. and Schaeffer, R. (1970). p. 141 of ref. 8
171. Schmid, G. (1970). *Angew. Chem. Int. Ed. Engl.,* **9,** 819
172. Stanko, V. I., Anorova, G. A. and Klimova, T. V. (1968). *J. Gen. Chem. USSR,* **38,** 1146
173. Zakharkin, L. I., Grebennikov, A. V., Vinogradova, L. E. and Leites, L. A. (1968). *J. Gen. Chem. USSR,* **38,** 1008

174. Stanko, V. I. and Anorova, G. A. (1969). *J. Gen. Chem. USSR*, **39**, 204
175. Grafstein, D., Bobinski, J., Dvorak, T., Smith, H., Schwartz, N., Cohen, M. S. and Fein, M. M. (1963). *Inorg. Chem.*, **2**, 1120
176. Fein, M. M., Bobinski, J., Mayes, N., Schwartz, N. and Cohen, M. S. (1963). *Inorg. Chem.*, **2**, 1111
177. Hill, W. E. (1968). *Inorg. Chem.*, **7**, 222
178. Stanko, V. I. (1965). *J. Gen. Chem. USSR*, **35**, 1140
179. Zakharkin, L. I., Chapovskii, Yu. A., Brattsev, V. A. and Stanko, V. I. (1966). *J. Gen. Chem. USSR*, **36**, 892
180. Stanko, V. I. and Klimova, A. I. (1966). *J. Gen. Chem. USSR*, **36**, 165
181. Stanko, V. I. and Anorova, G. A. (1966). *J. Gen. Chem. USSR*, **36**, 961
182. Stanko, V. I., Klimova, A. I. and Kashin, A. N. (1969). *J. Gen. Chem. USSR*, **39**, 1857
183. Heying, T. L., Ager, J. W., Clark, S. L., Alexander, R. P., Papetti, S., Reid, J. A. and Trotz, S. I. (1963). *Inorg. Chem.*, **2**, 1097
184. Zakharkin, L. I., Brattsev, V. A. and Stanko, V. I. (1966). *J. Gen. Chem. USSR*, **36**, 899
185. Zakharkin, L. I. and Kazantsev, A. V. (1966). *J. Gen. Chem. USSR*, **36**, 960
186. Obenland, C. and Papetti, S. (1966). *J. Org. Chem.*, **31**, 3868.
187. Zakharkin, L. I. and L'vov, A. I. (1967). *J. Gen. Chem. USSR*, **37**, 1154
188. Zakharkin, L. I. and Kazantsev, A. V. (1967). *J. Gen. Chem. USSR*, **37**, 519
189. Zakharkin, L. I., Grebennikov, A. V. and Savira, L. A. (1968). *Bull. Acad. Sci. USSR*, 1076
190. Zakharkin, L. I. and Kazantsev, A. V. (1966). *J. Gen. Chem. USSR*, **36**, 958
191. Stanko, V. I., Klimova, A. I., Chapovskii, Yu. A and Klimova, T. P. (1966). *J. Gen. Chem. USSR*, **36**, 1773
192. L'vov, A. I. and Zakharkin, L. I. (1967). *Bull. Acad. Sci. USSR*, 2527
193. Zakharkin, L. I., L'vov, A. I., Soshka, S. A. and Shepilov, C. I. P. (1968). *J. Gen. Chem. USSR.*, **38**, 258
194. Stanko, V. I., Brattsev, V. A., Al'perovich, N. E. and Titova, N. S. (1966). *J. Gen. Chem. USSR*, **36**, 1856
195. Zakharkin, L. I. and L'vov, A. I. (1967). *J. Gen. Chem. USSR*, **37**, 696
196. Zakharkin, L. I. and L'vov, A. I. (1966). *J. Gen. Chem. USSR*, **36**, 777
197. Stanko, V. I., Brattsev, V. A., Al'perovich, N. E. and Titova, N. S. (1968). *J. Gen. Chem. USSR*, **38**, 1015
198. Brattsev, V. A., Al'perovich, N. E. and Stanko, V. I. (1970). *J. Gen. Chem. USSR*, **40**, 1317
199. Brattsev, V. A. and Stanko, V. I. (1969). *J. Gen. Chem. USSR*, **39**, 1143
200. Zakharkin, L. I. and Kalinin, V. N. (1965). *J. Gen. Chem. USSR*, **35**, 1878
201. Zakharkin, L. I., Kalinin, V. N. and Gedymir, V. V. (1969). *J. Organometal. Chem.*, **16**, 371
202. Zakharkin, L. I., Kalinin, V. N., Gedymir, V. V. and Dzarasova, G. S. (1970). *J. Organometal. Chem.*, **23**, 303
203. Zakharkin, L. I. and Zhigareva, G. G. (1969). *J. Gen. Chem. USSR*, **39**, 1856

5
Other Aspects of Boron Chemistry

R. H. CRAGG
University of Kent

5.1	INTRODUCTION *(which includes the main general sources of reference to boron compounds published over the last ten years)*	186
5.2	BORON: CHEMICAL AND PHYSICAL PROPERTIES OF	186
5.3	BORON COMPOUNDS CONTAINING BORON–HALOGEN BONDS	187
	5.3.1 *Introduction (which includes the main sources of reference)*	187
	5.3.2 *Chemical and physical properties of boron halides*	187
	5.3.3 *Chemical and physical properties of diboron tetrahalides*	191
	5.3.4 *Coordination compounds of boron trifluoride*	191
	5.3.5 *Coordination compounds of boron trichloride*	194
	5.3.6 *Coordination compounds of boron tribromide*	195
	5.3.7 *Coordination compounds of boron halides*	195
5.4	BORON COMPOUNDS CONTAINING BORON–NITROGEN BONDS	196
	5.4.1 *Introduction (which includes the main sources of reference)*	196
	5.4.2 *Preparation and chemistry of compounds containing boron–nitrogen bonds*	197
	5.4.3 *Structural and physical properties of compounds containing boron–nitrogen bonds*	200
	5.4.4 *Borazines*	201
5.5	COMPOUNDS CONTAINING BORON–OXYGEN BONDS	203
	5.5.1 *Introduction (which includes the main sources of reference)*	203
	5.5.2 *Preparation and chemistry of compounds containing boron–oxygen bonds*	204
	5.5.3 *Physical properties of compounds containing boron–oxygen bonds*	207

5.6	BORON COMPOUNDS CONTAINING BORON–SULPHUR BONDS	210
	5.6.1 Introduction (which includes the main sources of reference)	210
	5.6.2 Preparation and chemistry of compounds containing boron–sulphur bonds	210
	5.6.3 Physical properties of compounds containing boron–sulphur bonds	213

5.1 INTRODUCTION

The growth of publications in the field of 'The Chemistry of Boron and its Compounds' over the last decade has been remarkable. One of the first comprehensive reviews of boron compounds[1] appeared in 1956, and since that time almost all aspects of boron chemistry have been surveyed. Two general texts on boron compounds have been published[2,3], and also two conference proceedings—one concerned with inorganic polymers[4] and the other with bonding in boron compounds[5]. General reviews have been concerned with redistribution reactions[6], nuclear magnetic resonance of boron compounds[7], insertion reactions[8], pseudohalides[9], and organic compounds of boron[10]. Two volumes of a comprehensive review of organoboron chemistry have appeared[11], as well as the first three volumes in a series dealing with specialist articles in boron chemistry[12]. There is also a book on the organic chemistry of boron[13]. Finally, reviews on all aspects of boron chemistry are to be found in *Organometallic Chemistry Reviews*, *Annual Reviews of Organometallic Chemistry*, and *Annual Reports of the Chemical Society*.

The purpose of this contribution is to provide a comprehensive survey of publications on boron chemistry which have appeared in the last 2 years. Since there are other chapters relating to boron hydrides, carboranes, and organometallic compounds, references concerned with these areas are generally excluded from this review. However, some comments are made to contributions concerned with organoboranes where the reactions discussed do not specifically involve the chemistry of the boron–carbon or boron–hydrogen bond.

5.2 BORON: CHEMICAL AND PHYSICAL PROPERTIES OF

There has been a recent review of the preparation, properties, and applications of elemental boron[14]. A continuous process for the production of boron, by the reduction of vaporised liquid trichloroborane by hydrogen, has been patented[15]. The thermal conductivity[16], as well as the optical and photoelectric properties[17], of β-rhombohedral boron have been investigated. Calculations on the hyperfine-splitting constant for boron have been carried out using the V.H.F. and G.F. methods[18], and the electric quadrupole moment for ^{11}B determined (Q = 0.037 barns). The boron 1s binding energies have been measured for 25 boron compounds[19]; the results were found to correlate linearly with the calculated atomic charges. Two methods for the determination of boron have been reported: proton excitation was used for the x-ray spectrochemical detection of boron as an impurity, at the level of 0.0014% in nickel alloys[20], while using a boron–curcumin colour complex and thin layer chromatography, boron at a level of 0.5% was detected in caviar samples[21]. The γ-radiolysis of liquid ammonia, in the presence of boron compounds, has been studied[22], the results suggest that Lewis acids or their ammonia adducts induce stability against radiation decomposition of the hydrazine formed, thereby considerably increasing its yield from the radiolysis of ammonia. Calculations on a series of boron compounds indicated that the B—X bond order increases in the series[23]:

$$H < C \leqq I < S \leqq Br < O < Cl < F \leqq N.$$

Investigations into seed germination of a polycarp under greenhouse conditions showed that boron had no effect on the root/top ratio[24].

5.3 BORON COMPOUNDS CONTAINING BORON–HALOGEN BONDS

5.3.1 Introduction

One of the first reviews of compounds containing boron–halogen bonds dealt with coordination compounds of boron trichloride[25]; this was shortly followed by one on the coordination compounds of boron tribromide and boron tri-iodide[26]. The quartet was completed with a review on the coordination compounds of boron trifluoride[27]. A later article discussed the reactions of boron trichloride with organic compounds[28], and recently the addition reactions of boron tribromide, diboron tetrachloride, and other boron halides to unsaturated compounds have been surveyed[29]. The large number of papers published each year concerned with coordination compounds of boron halides appears to be disproportionate to those on other areas of boron chemistry.

5.3.2 Chemical and physical properties of boron halides

Boron halides have been used as catalysts in the polymerisation of α-olefins to polyolefins[30]. Boron trifluoride is an effective catalyst for the cyclisation of 4-arylolefins[31]. Calculations on the electronic structure of boron trifluoride, by variation of the boron–fluorine bond length to find a minimum energy situation, have resulted in the determination of the energies of π-localisation and reorganisation for the molecule[32]. The electronic structure of boron trifluoride has been calculated by a self-consistent field approach[33]: the results give a value of 34 kcal mol^{-1} for the energy of the process, BF$_3$ (planar) $\rightarrow$ BF$_3$ (pyramidal), and 50 kcal mol^{-1} for the corresponding π energy change; this suggests a regain of σ energy on reorganisation. *Ab initio* studies of the electronic structures of a series of fluoroboranes, including boron trifluoride, have been made[34]. The high-resolution photoelectron spectrum of boron trifluoride has been assigned; the highest occupied orbital is of π and not σ type[35]. The energy of separation of the π type orbitals, for all four boron halides, is reported: it was suggested that though boron trifluoride has the greatest π-energy stabilisation, the π-back donation is the smallest. Dioxygenyltetrafluoroborate has been prepared and its infrared spectrum reported[36].

$$BF_3 + O_2F_2 \rightarrow O_2BF_4$$

X-ray powder determination shows that the compound is orthorhombic and isomorphous with the nitrosyl analogue.

The gases, produced by heating sulphuric acid and fluorospar, were dissolved in boric acid to give fluoroboric acid and pure silica[37]. A series of tin tetrafluoroborates were obtained from the reaction of the etherate of boron trifluoride and the corresponding tin fluoride[38]. The bis(chlorosulphur)nitrogen cation, characterised by mass spectrometry and x-ray analysis, has been obtained from the reaction of boron trichloride and sulphur nitride trifluoride[39]. The complex readily decomposed:

$$3BCl_3 + 2NSF_3 \rightarrow [NS_2Cl_2]^+[BCl_4]^- + \tfrac{1}{2}N_2 + \tfrac{3}{2}Cl_2 + 2BF_3$$

The reaction of trithiazylchloride with chlorine and boron trichloride afforded the same complex.

The preparation of compounds containing boron–phosphorus bonds has been reported[40]. Boron trichloride reacts readily with lithium phosphides:

$$BCl_3 + 3LiPEt_2 \rightarrow B(PEt_2)_3 \xrightarrow{LiPEt_2} LiB(PEt_2)_4 \xrightarrow{BCl_3} [B(PEt_2)_2]_2$$

Boron tribromide has been used in the quantitative preparation of a series of actinide bromides and their complexes[41].

$$PaCl_5 + BBr_3 \rightarrow PaBr_5$$
$$UCl_5Ph_3PO + BBr_3 \rightarrow UBr_5Ph_3PO$$

Monobromo- and dibromo-alkylselenoboranes have been obtained from the reaction of alkylselenols and boron tribromide[42].

$$BBr_3 + EtSeH \rightarrow EtSeBBr_2 + HBr$$

Studies on the cleavage of methyl ethers with boron tribromide suggested that the mechanism of the reaction is more complex than that for the corresponding reaction with boron trichloride and that the length of the aliphatic chain had some effect on the reaction[43]. Di-iodophenylborane reacted with tropylium iodide or cycloheptatriene to give tropylium tri-iodophenylborate[44].

$$2\,C_7H_8 + 3PhBI_2 \longrightarrow 2\,C_7H_7^+ [PhBI_3]^- + PhBH_2$$

The reaction with dimethyl disulphide was studied.

$$C_7H_7^+ + [PhBI_3]^- + MeSSMe \rightarrow PhB(SMe)_2 + C_7H_7^+I_3^-$$

Mono, di, and tri-thienylboranes were obtained from the interaction of boron tri-iodide and 2-iodothiophenes[45].

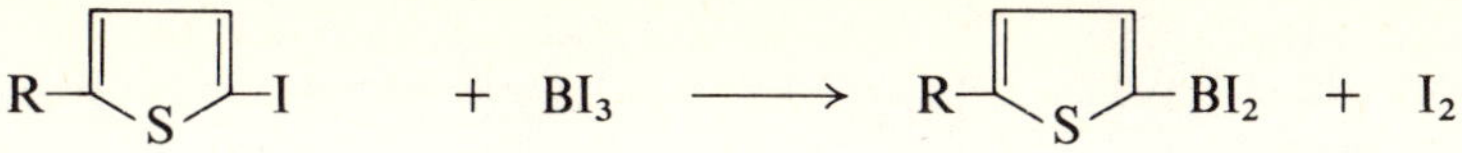

$$R-\underset{S}{\underset{|}{\bigcirc}}-I + BI_3 \longrightarrow R-\underset{S}{\underset{|}{\bigcirc}}-BI_2 + I_2$$

Aryldi-iodoboranes were obtained from the reaction of boron tri-iodide and aryl iodides[46]

$$RI + BI_3 \to RBI_2 + I_2$$

Di-iodophenylborane has been obtained from the reaction of boron tri-iodide and tetraphenyltin[47]. Three methods have been reported for the preparation of a series of 2-halo-1,3,2-diazaboracycloalkanes which were characterised spectroscopically: interaction of diamines with a trialkylamine boron trihalide complex, displacement of a dimethylamino group (by boron trihalide) from the corresponding heterocycle, or halogen exchange using titanium tetrafluoride[48].

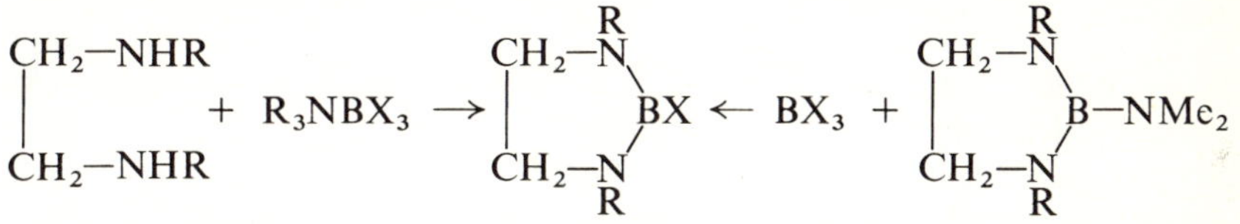

$$\begin{array}{c}CH_2-NHR\\|\\CH_2-NHR\end{array} + R_3NBX_3 \to \begin{array}{c}CH_2-N{\scriptstyle R}\\|\quad\quad\diagdown\\\quad\quad\quad BX\\|\quad\quad\diagup\\CH_2-N\\{\scriptstyle R}\end{array} \leftarrow BX_3 + \begin{array}{c}CH_2-N{\scriptstyle R}\\|\quad\quad\diagdown\\\quad\quad\quad B-NMe_2\\|\quad\quad\diagup\\CH_2-N\\{\scriptstyle R}\end{array}$$

Aliphatic selenols and selenphenol react with boron halides to give the tris(organoseleno)boranes[49]:

$$BX_3 + PhSeH \to B(SePh)_3 + 3HX$$
$$(X = Cl, Br, I)$$

The reactions of selenoboranes with nucleophiles and electrophiles have been investigated. Diphenylketimido-boron dihalides have been synthesised as follows[50]:

$$Ph_2C = NSiMe_3 + BX_3 \to Ph_2C = NBX_2 + Me_3SiX$$

The reaction of boron trihalides with $Ph_2C = NLi$ has been used in the preparation of diphenylketimidoboron dihalides[51]. The difluoro compound $(Ph_2C = NBF_2)_n$ was difficult to separate from lithium fluoride and was obtained from the redistribution reaction:

$$\tfrac{n}{3}(Ph_2CN)_3B + \tfrac{2n}{3}BF_3 \cdot Et_2O \to [(Ph_2CN)BF_2]_n$$

Cryoscopic, infrared, and mass spectrometric data suggest dimeric structures when X = I and Br, and possibly Cl. The preparation of aldimidoboron dichlorides proved unsuccessful. The stereochemistry of the reactions of boron trihalides and a number of active alkoxy- and amino-silanes have been investigated[52].

$$R_3SiOR + BX_3 \to R_3SiX + ROBX_2$$

A four-centre mechanism has been suggested for these reactions which, with the exception of the reaction of the silanol and boron trichloride, generally proceed with an inversion of configuration at silicon. The reactions of boron tribromide and trichloride with hexafluoroisopropylidenimine have been studied[53].

$$(CF_3)_2C = NH + BCl_3 \to (CF_3)_2-C(Cl)-NHBCl_2 \xrightarrow{H_2O} (CF_3)_2C(Cl)NH_2$$

In contrast, the reaction with boron trifluoride gave a complex.

Boron trihalides (F, Cl, Br) readily react with methoxytrifluorosilane to give the corresponding dihalomethoxyborane[54].

$$MeOSiF_3 + BBr_3 \rightarrow MeOBBr_2 + BrSiF_3$$

The infrared spectra of all the binary mixed halides of boron have been observed and assigned[55]. The frequency of the fundamental vibrational modes above 300 cm^{-1} for di-iodofluoroborane and iodofluoroborane have been observed and assigned. The reaction between phosphorus trioxide and boron trifluoride etherate has been found to be complex due to scrambling amongst the molecular species in the mixture[56]. Three fluorophosphines were identified as products of the reaction:

$$P_4O_6 + BF_3Et_2O \rightarrow PF_3 + F_2POPF_2 + F_2POEt$$

The infrared and Raman spectra of difluoromethylborane and difluoroperdeuteromethylborane have been measured in the gas and liquid phases respectively[57]. On the basis of a molecular model in which there was free rotation of the methyl group about the boron carbon bond, a complete vibrational assignment was determined. The anharmonic coefficients and rotational vibration interaction constant for boron trifluoride have been calculated[58]. The heat of evaporation of boron trifluoride etherate has a calculated value of 14 kcal mol^{-1} [59]. The BF_2 radical has been detected and its structure determined (F–B̂–F 112 degrees) from results obtained from the electron spin resonance studies of γ irradiated boron trifluoride trapped in a xenon matrix at 4.0 K[60]; analysis of the hyperfine structure gives the orbital distribution of the unpaired electron. By examination of negative ion formation in the gas phase from boron trifluoride, the heat of formation of the fluoroborate ion has been determined as $\leqslant 15.5$ eV[61]. The molecular ions BX_2Y^+ and $BXYZ^+$ (where X,Y, and Z = F, Cl, Br, I, NMe$_2$, OMe, OPr^n, SMe, and Et), which were obtained by redistribution reactions, have been characterised by mass spectrometry and their first ionisation potentials have been measured[62]: the results suggest that the interaction between a given p_π orbital on a ligand and the p_π orbital on the boron atom decreases in the order B—N > B—O $\simeq$ B—S > B—Hal. The mass spectra of boron trichloride and diboron tetrachloride have been measured and the ion yield curves for the molecular ions obtained[63]; the bond dissociation energy for the boron–boron bond was 87.6 kcal mol^{-1}. The ^{35}Cl nuclear quadrupole resonance spectra of boron trichloride and of some amine and nitrile complexes have been measured[64]. It was observed that the ^{35}Cl n.q.r. frequencies of B-trichloroborazine were temperature dependent; also the ^{11}B quadrupole coupling constants in solid boron halides have been measured[66]. Nuclear magnetic resonance studies, (^{11}B of BF$_3$ and ^{19}F and ^{10}BF$_3$) at a series of temperatures have been made on liquid boron trifluoride[67]. There is a linear relationship between the ^{11}B—^{19}F coupling constant and the ^{19}F chemical shift for boron trifluoride and related compounds[65]. The anisotropy of rotational reorientation in liquid boron trichloride has been determined from results of ^{11}B and ^{35}Cl relaxation measurements[68].

5.3.3 Chemical and physical properties of diboron tetrahalides

The boron subhalides and related compounds containing boron–boron bonds have been reviewed[69]. Diboron tetrachloride, in greater than 80% yield, was obtained at room temperature by halogen exchange between boron trichloride and diboron tetrafluoride[70]; by contrast, dimethylfluoroborane and diboron tetrachloride react in a reverse sense:

$$4Me_2BF + B_2Cl_4 \rightarrow B_2F_4 + 4Me_2BCl$$

A very much improved synthesis of tetraboron tetrachloride has been published[71]. By passing diboron tetrachloride through a mercury discharge tube, up to 10 mg of tetraboron tetrachloride per hour were obtained. The pyrolysis of the volatile decomposition products of diboron tetrachloride have been studied and thermally-stable, yellow-orange boron sub-chloride B_9Cl_9 has been identified[72]. The infrared spectra of diboron tetrachloride and tetrafluoride, isolated in a matrix of solid argon at liquid hydrogen temperature, have been reported[73]. The Raman spectra, in the liquid phase, of both compounds confirm that their structures have the staggered configuration. An electron diffraction study of diboron tetrachloride, based on a D_{2d} symmetry shows B—Cl 1.750 Å and B—B 1.702 Å[74]. The barrier to rotation, based on results obtained at five temperatures is 1.85 kcal mol^{-1}

5.3.4 Coordination compounds of boron trifluoride

Primary amine boron trifluoride adducts, or salts of fluoroboric acid, are readily dehydrofluorinated by the adducts of hindered tertiary amine complexes of boron trifluoride[75]. By this method fluoroborazines have been obtained. The thermodynamic aspects of the systems were discussed in order to explain the difference in relative ease of dehydrohalogenation of primary amine adducts of boron trifluoride and boron trichloride. In the presence of boron trifluoride, the esters $(MeO)_3P{=}X$ (where X = S or Se) undergo isomerisation to form $(MeO)_2MeXP{=}OBF_3$, as shown by calorimetric and n.m.r. results[76]; isomerisations were 30–40% complete in six hours. A stable 1:1 complex resulted from the exothermic addition of a dialkylphosphonate and gaseous boron trifluoride[77]; its heat of formation was determined, and complex formation was accompanied by a large increase in J_{P-H}. The complex between triphenylmethylamine and 1,2-bisdifluoroborylethane has been shown to be non-chelated in non polar solvents[78] and the existence of the oxotrifluoroborate ion $[BF_3O]^{2-}$ has been demonstrated[79]. Amine complexes of nickel(III) have been prepared using nitrosyl tetrafluoroborate[80]. X-ray diffraction studies on a single crystal of trifluorophosphine-tris(difluoroboryl)borane show the molecule to have C_{3v} symmetry with a central boron atom tetrahedrally bonded to three BF_2 groups and a PF_3 group[81]. The first example of a tin Group III complex has been reported[82].

$$BF_3 \cdot Et_2O + Cp_2Sn \xrightarrow{THF} Cp_2Sn \cdot BF_3$$

The reaction of boron trifluoride with $SnCl_3^-$ and $GeCl_3^-$ has been reinvestigated and the results show that there is no evidence for the formation of complexes such as $(Cl_3SnBF_3)^-$ as had originally been thought[83]. The reaction is now regarded as a Cl^- transfer.

$$SnCl_3^- + BF_3 \rightarrow SnCl_2 + BF_3Cl^-$$

The dipole moments, refractive indices, and densities of twelve equimolar complexes of boron trifluoride and ethers have been recorded[84]. The dipole moments and heats of formation of the complexes between boron trifluoride and dimethylaniline and methylphenylaniline have been determined[85]; the results indicate that the charge transfer in both complexes is similar. Boron trifluoride forms a conducting solution with trifluoromethane[86]; upon addition of trimethylamine, to the solution, $Me_3N \cdot BF_3$ and $Me_3NH \cdot BF_4$ were obtained. The boron trifluoride complex with sulphur tetrafluoride, in liquid hydrogen fluoride, was found to have a conductivity near to that of potassium fluoroborate indicating that the complex was fully ionised[87].

$$SF_4 \cdot BF_3 \xrightarrow{HF} SF_3^+ BF_4^-$$

The ultraviolet and ^1H n.m.r. spectra of a series of phenols dissolved in $BF_3 \cdot$sulpholan and $HF \cdot BF_3 \cdot$sulpholan show that the boron trifluoride complex formation always occurs via the oxygen atom but that there may be tautomerism in the adduct[88]. A study has been made of the basicity of organophosphorus compounds in their Lewis complexes with boron trifluoride, using ^1H, ^{19}F, ^{11}B, and ^{31}P n.m.r. spectra[89]. Although the boron resonance is insensitive to the strength of the complex, the fluorine resonance indicated the possibility of various types of σ-complexes with either the phosphorus or another hetero atom of the ligand molecule. The presence of molecular motion in the solid state in a series of complexes of boron trifluoride with amine donors has been observed using broad line n.m.r. spectroscopy[90]. In a study of the ^1H n.m.r. spectra of the 1 : 1 complex formed between diethyl ketone and boron trifluoride, in $CHCl_2F$ solution at temperatures below $-120\,°C$, separate signals from the two ethyl groups were observed[91], suggesting cis-trans-isomerism about the carbon–oxygen bond. ^1H n.m.r. spectroscopy has been used in a study of the effect of complex formation of boron trifluoride with aldehydes[92]. The energy of rotation about the C—C bond has been calculated. Protonation and complex formation at the oxygen atom would stabilise resonance forms of the type shown below. The increase in π-bond character of the C—C bond was investigated by a study of the chemical shift of the formyl hydrogen in the complex compared to that in the free aldehyde.

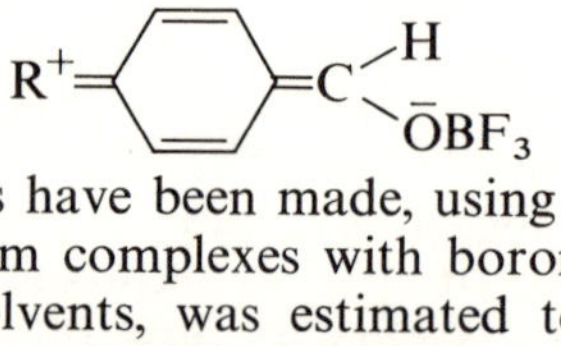

$$R^+ = \underset{}{\underset{}{\bigcirc}} = C \underset{\overline{O}BF_3}{\overset{H}{\diagup}}$$

Qualitative measurements have been made, using ^1H n.m.r., on the ability of various solvents to form complexes with boron trifluoride[93]. The complexing ability, of the solvents, was estimated to decrease in the order tetramethylene sulphoxide $>$ DMF $\sim$ DMSO $>$ tetramethylurea $>>$ THF $>>$ Et_2O $>$ Me_2CO $>$ tetramethylene sulphone. The magneto-optical properties of the addition compounds of mono- and di-alkylfluoroboranes with tertiary amines have been measured[94]. The vibrational spectra of the ammonia-boron trifluoride adduct, and the five isotopic species, have been determined using a C_{3v} symmetry model[95]. The force constant for the boron-nitrogen bond has been calculated at 3.97 mdyn/Å. The infrared spectrum of the pyridine-boron trifluoride complex has been reported[96]. The lowering of the asymmetric BF_3 stretching frequency of the acceptor gave an estimate of the dipolar moment in the py···BF_3 complex. The infrared and Raman spectra, recorded at $-196\,°C$, for the five isotopic species of the boron trifluoride-acetonitrile complex have been reported[97] and a full set of assignments for the fundamental vibrations for the complex were recorded. The Raman spectrum of the 2:1 adduct of ClF with boron trifluoride shows that the cation has the asymmetric Cl_2^+F structure[98]. The complex formed between selenium tetrafluoride and boron trifluoride has been shown, by Raman spectroscopy, to have the structure $Se^+F_3BF_4^-$ [99]; in nitrobenzene the complex was only slightly dissociated. The infrared spectra of the acetone-boron trifluoride complex, and the isotopic complexes, have been reported[100] as well as the infrared and Raman spectra of the boron trifluoride-dimethyl sulphoxide complex[101]. Thermogravimetric studies show that boron trifluoride and potassium superoxide react at liquid nitrogen temperature to form an unstable complex $3KO_2BF_3$; at $160\,°C$, boron trifluoride vapour gives KBF_4 and oxygen[102]. The enthalpies of the reaction of boron trifluoride with tertiary alkyl- and aryl-phosphines have been measured in benzene by direct thermochemical measurements[103]; the donor-acceptor bonds in the phosphine-boron trifluoride adducts were found to be much weaker (by about 20 kcal mol^{-1}) than in the corresponding borane adducts. Kinetic studies have been made to ascertain the effect of boron trifluoride etherate as a catalyst in the polymerisation of epichlorohydrin[104]. The polymerisation rate in the presence of $Et_3O^+BF_4^-$ was found to proceed at a measurable rate only when the temperature of the reaction was greater than $35\,°C$. The kinetics of the tetrahydrofuran polymerisation of the boron trifluoride-THF complex have been studied, in the presence of glycidol nitrate, in 1,2-dichloroethane[105]. Boron trifluoride complexes with amines (eighteen complexes reported) have been used as epoxy-curing agents[106]; however, no definite relationship between the curing activity of the complexes and the basicity of the substituent in the complex was established. The uniform curing of paper has been obtained by impregnation of the paper with a boron trifluoride-amine complex[107]. This treatment improves the sizing and flame resistance of the paper. Boron difluoride chelates, of the type shown below, have been found to be very efficient catalysts for the polymerisation of epoxy resins[108].

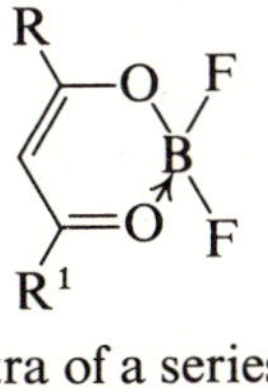

The ^1H, ^{11}B, and ^{19}F n.m.r. spectra of a series of boron difluoride complexes of β-di and $\beta\beta'$-tri-carbonyl compounds have been measured[109]. The reactions of chelates of β-dicarboxylic acids with boron trifluoride, have been studied[110]. The preparation of the dimethyl and diethyl ether complexes of boron trifluoride have been patented[111]. The cleavage of amides by methanolic boron trifluoride has been reported[112]; the products of the reaction, volatile methyl esters, were found to be suitable for study by gas chromatography.

5.3.5 Coordination compounds of boron trichloride

Adducts of boron trichloride with substituted anilines have been shown by spectroscopic methods to be σ-complexes with the amino protons retaining their identity[113]. The complexes in methyl cyanide undergo an exchange with the elimination of hydrogen chloride and this reaction was followed by ^{1}H n.m.r. spectroscopy. The reaction of disulphur dinitride with boron trichloride in methylene chloride has been studied; $S_4N_4BCl_3$, $S_2N_2(BCl_3)_2$, and a polymer $(S_2N_2BCl_3)_n$, were obtained as products, each of which could be made the major product depending upon the reaction conditions[114]. The properties of $S_2N_2BCl_3$ and $S_2N_2(BCl_3)_2$ indicate that the S_2N_2 ring remains intact in these compounds. Antimony pentachloride displaces boron trichloride from these complexes, but the polymer is inert to reaction. At 0 °C, $S_2N_2(BCl_3)_2$ loses boron trichloride to form the adduct $S_2N_2BCl_3$. With boron trifluoride only the complex $S_4N_4BF_3$ was obtained.

$$SbCl_5 + S_2N_2BCl_3 \rightarrow BCl_3 + S_2N_2(SbCl_5)_2$$

$$S_2N_2(BCl_3)_2 \underset{BCl_3}{\overset{0°C}{\rightleftharpoons}} S_2N_2BCl_3 + BCl_3$$
$$-78°C$$

Boron trichloride forms a 1:1 complex with tellurium tetrachloride[115]. Molar conductance measurements in a series of solvents, such as nitrobenzene, indicated that the complex was fully ionised $TeCl_3^+ BCl_4^-$. The 1:1 nature of the complex was demonstrated by conductimetric titration with amines. The complexes of boron trichloride and of boron tribromide with diacetyl- and dibenzoyl-hydrazines have been studied[116]. Molar conductance measurements and infrared spectral studies were used to indicate the structure shown below for the complexes.

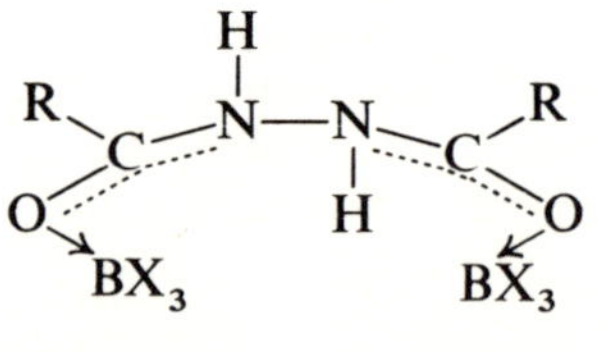

The complexes of di-π-cyclopentadienyltitanium chloride with Group III halides have been shown to have the structure below[117].

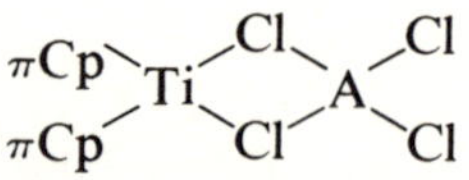

[A = B, Al, Ga, In]

E.P.R. hyperfine interaction of the unpaired electron of the Ti^{III} with the magnetic nuclei of the Group III elements was observed and from isotopic coupling the unpaired spin density in the relevant s-orbitals of the Group III elements was estimated. The results showed the following Lewis acid trends:

$$GaCl_3 > BCl_3 > AlCl_3 > InCl_3$$

Methane- and ethane-sulphenyl chloride form complexes with boron trichloride of the structure $(RSCl)_2 \cdot BCl_3$, but with disulphides complexes of the type $R_2S_2 \cdot BCl_3$ were obtained[118]. The trigonal boron cation

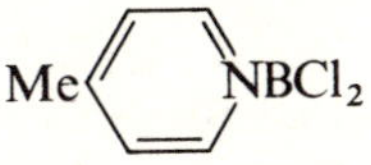

$$Me\text{-}C_5H_3N\text{-}BCl_2$$

has been obtained from a mixture of 4-methylpyridine, boron trichloride, and aluminium chloride[119]; ^1H and ^{11}B n.m.r. studies and conductance measurements supported the proposed structure. Boron trichloride complexes with triethylphosphine have been prepared[120] and the thermal decomposition of acylazide boron trihalide adducts studied[121].

5.3.6 Coordination compounds of boron tribromide

The molar conductance of the complex of boron tribromide with cyanoacetamide in nitrobenzene shows it to be non-ionic[122]. The infrared spectrum indicated that the bonding was through the carbonyl oxygen. Benzil formed an unstable 1:2 addition complex with boron tribromide[123]; the heat of the reaction was measured.

5.3.7 Coordination compounds of boron halides

The preparation and properties of a series of complexes between dialkyl sulphides and dihaloboranes RBX_2, where X = Cl, Br, I, have been published and the Lewis acidity of the dihaloboranes discussed[124]. Wide line n.m.r. studies of trimethylamine complexes of the boron compounds BX_3, where X = F, Cl, Br, and H, show that rotation of the methyl groups and rotation about the boron–nitrogen axis occur in the solid state[125]. The complex between ammonia and boron trifluoride was also studied. The activation energies for the rotation of the boron–nitrogen bond range from 1.6 kcal mol^{-1} for the boron trifluoride complex to 9.3 kcal mol^{-1} for the boron trichloride complex. A ^1H n.m.r. study of solutions of boron trihalides BX_3 and trimethylamine–boron trihalide complexes, $L \cdot BY_3$, showed peaks due to the adduct and also the adduct formed by displacement[126]. The crystal structures of the acetonitrile complexes of boron trifluoride and boron trichloride have been determined[127]. The boron–nitrogen bond in the BCl_3 complex is shorter than that in the BF_3 complex. The results were rationalised in terms of a model for donor-boron halide interactions where the acceptor strength of the boron halide increases with increasing distortion of the BX_3 group. The ^1H n.m.r. spectra of trimethylamine complexes of the boron halides show the chemical shifts in the order $BI_3 > BBr_3 > BCl_3 > BF_3$[128]. Amide hydrohalides and methylimide hydrohalides are readily obtained from the hydrolysis and methanolysis of nitrile solutions of boron trihalides[129].

$$\text{MeCN·BX}_3 + \text{H}_2\text{O} \xrightarrow{\text{MeCN}} \text{jellies} \rightarrow y\text{B(OH)}_3 + 1\text{-}y\text{MeCN·BX}_3$$

$$\text{MeCN·BX}_3 + 4\text{H}_2\text{O} \xrightarrow{\text{MeCN}} \text{B(OH)}_3 + \text{AcNH}_3^+\text{Cl}^- + 2\text{HCl} (X = \text{Cl})$$

Methanolysis gives the methylimide hydrohalide

$$\text{MeCNBX}_3 + 4\text{MeOH} \xrightarrow{\text{MeCN}} \text{AcMeC} = \text{NH}_2^+\text{Cl}^- + \text{B(OMe)}_3 + 2\text{HX}$$

The ultraviolet spectra of a series of pyridine complexes of boron trihalides (F, Cl, Br) show a bathochromic shift of the longest wave π—π^* transition which is dependent upon the acceptor strength of the boron halide[130]. The observed shifts range from 59 nm for the 3Cl—py·BF$_3$ complex to 151 nm for the 3Cl—py·BBr$_3$ complex. Ligand exchange reactions, by ^{1}H n.m.r. studies, have been made on a series of complexes of the boron trihalides (Cl, Br, I) with donors R$_2$E (E = S, Se, Te) and Ph$_3$PE (E = S and Se)[131]. The heat of reaction of boron trichloride and boron tribromide with a series of anthrones have been measured[132]; the results were compared with the values for the corresponding aluminium complexes, and show the following order of acceptor strengths: BBr$_3$ > BCl$_3$ > AlBr$_3$ > AlCl$_3$. Complexes of MeBX$_2$ and Me$_2$BX (X is halogen) and haloacetonitriles have been characterised by infrared spectroscopy[133]. The adducts between boron trifluoride and boron trichloride with trimethylsilylmethylenedimethyl sulphurane have been prepared[134]. The boron trichloride complex readily decomposed to give trimethylchlorosilane.

The molar conductances of the triethylammonium salts of phenyltrichloroborate, tetraboromoborate, tetraiodoborate, and of some pyridine complexes have been measured in acetonitrile[135]; an ionisation mechanism was proposed to explain the high limiting molar conductance of tetraethylammonium tetraiodoborate.

$$\text{Et}_4\text{NBI}_{4s} \rightarrow \text{Et}_4\text{N}^+ + \text{BI}_2(\text{MeCN})_n^+ + 2\text{I}^-$$

The nature of the solute species in acetonitrile for a series of boron trihalides and phenyl-substituted boron halides and their complexes with acetonitrile and pyridine have been studied by conductance, infrared and n.m.r. spectroscopic techniques[136]. The difference in the behaviour of the boron halides was rationalised in terms of the relative acceptor strengths of the boron halides, and donor strengths of pyridine and the solvent molecules. In acetonitrile, boron trichloride and tribromide are weak electrolytes. Boron tri-iodide, on the other hand, was found to be a strong 1 : 1 electrolyte with $S_2BI_2^+$ and I^- as the ionic species.

5.4 BORON COMPOUNDS CONTAINING BORON–NITROGEN BONDS

5.4.1 Introduction

Over the past ten years there has been considerable activity in the chemistry and structure of compounds containing boron–nitrogen bonds. Most of the effort has been centred around the nature of the boron–nitrogen bond and the resulting properties of compounds containing such bonds. A wealth of information appeared in 1964 when thirty-two papers given at a conference on boron–nitrogen chemistry were published in one volume[137]. A year later a book on *Boron-Nitrogen Compounds*[138] was published and this included references to most of the published work on aminoboranes; a further review on aminoboranes has recently appeared[139]. In addition to the chapters on boron–nitrogen compounds which can be found in the general texts mentioned in the introduction to this review, a further review is included in a volume on the *Developments in Inorganic Nitrogen Chemistry*[140]. Two reviews dealing specifically with heterocyclic systems containing boron–nitrogen bonds have been published[141, 142] as well as two general reviews on heterocyclic organoboron compounds[143, 144]. A review has been published concerning the preparation and properties of borazides[145]. Four reviews on the preparation, substitution reactions, and properties of borazines have been published[146–149]. As a guide to those who require to synthesise aminoboranes, a review on *Boron-Nitrogen Compounds* can be found in *Preparative Inorganic Reactions*[150].

5.4.2 Preparation and chemistry of compounds containing boron–nitrogen bonds

The preparation of boron nitride, from boric acid and sodamide at a temperature of 900 °C, has been patented[151], and the reactivity of boron nitride has been compared to the nitrides of aluminium and silicon[152]. The reaction between boric acid and ammonium thiocyanate gave hexagonal boron nitride[153]. It was found that the normal process of annealing boron nitride to a temperature of 2350 °C produced an increase in density of the compound but had very little effect on the preferred orientation of the crystals[154]. Boron nitride prepared in this way was found to be soft and easily cleaned. A further industrial use of boron nitride has been proposed[155]. The addition of boron nitride to hard metal powders, prior to sintering, produces hard metallic materials which have self-lubricating properties. Bor(di)imine, $B_2(NH)_2$, is readily obtained by the deoxygenation of a quaternary amine salt, prepared by the reaction of boric oxide and an organic amine such as ethylenediamine, in an ammonia atmosphere[156]. From bor(di)imine, high purity boron nitride was obtained. A study of the reactions of metal alkyls with tris(dimethylamino)borane showed the order of reactivity to be $R_3Al > R_3B \sim R_2Zn > R_2Cd \gg R_4Sn$ [157]. Although the reaction with trimethylalane was complete within a few hours, the reactions with other metal alkyls took several days. The preparation of a series of 2-amino-1,3,2-diazaboracycloalkanes, from the reaction between trisdimethylaminoborane and an N,N'-disubstituted diamine, has been reported[158].

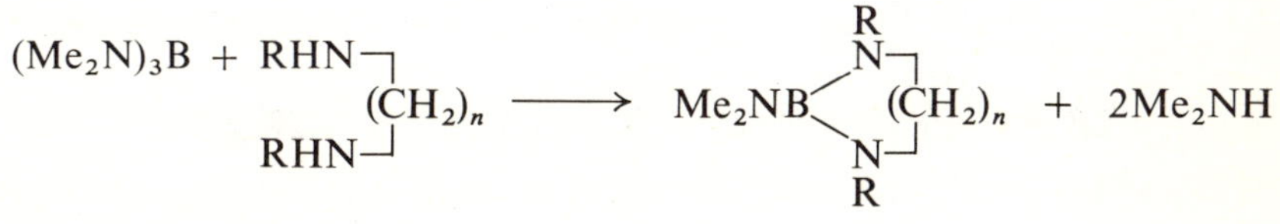

$$(Me_2N)_3B + \begin{matrix} RHN\!\!-\!\! \\ (CH_2)_n \\ RHN\!\!-\!\! \end{matrix} \longrightarrow Me_2NB\begin{matrix} R \\ N\!\!-\!\! \\ (CH_2)_n \\ N\!\!-\!\! \\ R \end{matrix} + 2Me_2NH$$

However, with ethylene diamine a polymer was obtained:

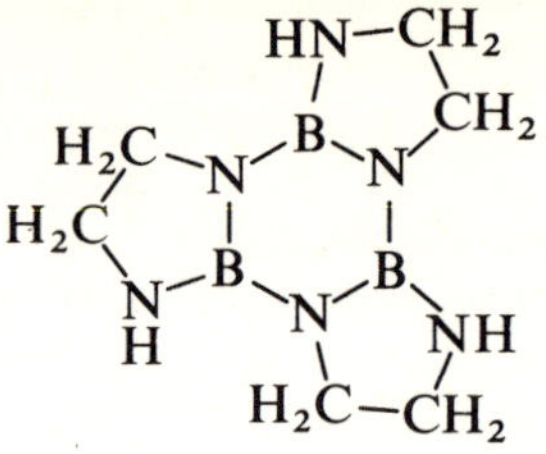

$$\begin{array}{c}
\text{HN–CH}_2 \\
\text{H}_2\text{C–N}^{\text{B}}\text{N–CH}_2 \\
\text{H}_2\text{C} \quad | \quad | \\
\text{N–B–N–B} \\
\text{H} \quad \text{N} \quad \text{NH} \\
\text{H}_2\text{C–CH}_2
\end{array}$$

The syntheses and structures of a series of monomeric and dimeric hexahydro-1,2,4,5,3,6-tetraborin derivatives have been reported[159].

$$2RB(NR_2')_2 + 2R''NHNHR'' \longrightarrow RB\begin{array}{c}R''\ R''\\N-N\\N-N\\R''\ R''\end{array}BR + 4R_2'NH$$

Mass spectrometric evidence supported the tetrameric formulation for (HBNMeNMe)$_4$. The reactions of hydrazinoboranes, with amines, boron trifluoride, boron trichloride, and hydrogen chloride have been studied[160]. Hydrazine was found to react with phenylboroxine to give the heterocycle (1)[161] shown below. However, the melting point of the heterocycle was 90 °C lower than that reported for the compound in the literature. It was suggested that the lower melting point was due to the inclusion of polymer(2), which had also been prepared by an alternative route.

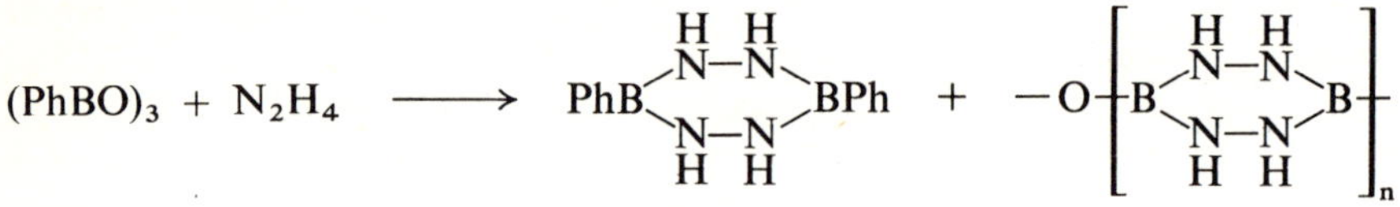

$$(PhBO)_3 + N_2H_4 \longrightarrow PhB\begin{array}{c}H\ H\\N-N\\N-N\\H\ H\end{array}BPh + \left[-O-B\begin{array}{c}H\ H\\N-N\\N-N\\H\ H\end{array}B-\right]_n$$

(1) \qquad\qquad (2)

Monomeric heterocyclic boron compounds, of the type shown below, which are relatively resistant to oxidation and thermal decomposition, have been obtained from the reaction of the *N*-lithioderivative of bismethylaminomethylborane and a series of chlorosilanes[162].

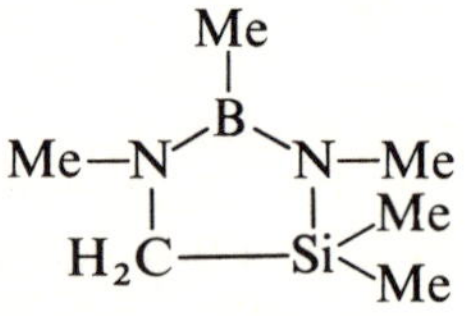

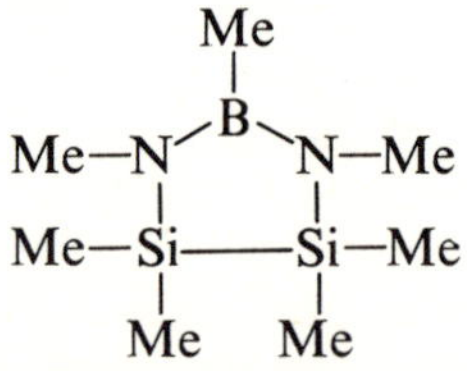

$$\begin{array}{c}
\text{Me} \\
| \\
\text{Me–N}^{\text{B}}\text{N–Me} \\
| \quad | \quad \text{Me} \\
\text{H}_2\text{C—Si}^{\text{Me}}
\end{array} \qquad \begin{array}{c}
\text{Me} \\
| \\
\text{Me–N}^{\text{B}}\text{N–Me} \\
| \quad | \\
\text{Me–Si—Si–Me} \\
| \quad | \\
\text{Me} \quad \text{Me}
\end{array}$$

The ^{11}B n.m.r. results were consistent with there being considerable p_π–p_π bonding in the boron–nitrogen bond in these heterocycles. Three methods have been outlined for the preparation of diborylamines of the type $(R_2B)_2NH$ and $(R_2B)_2NR$ [163], e.g.,

$$2R_2BCl + R_3'SiNRSiR_3' \rightarrow R_2B\text{—}NR\text{—}BR_2 + 2R_3'SiCl$$

Diborylamines are thermodynamically unstable and decompose to give trialkylboranes and borazines. A series of amino- and alkoxy-azidoboranes have been obtained from the reaction of an organosilicon azide and a chloroborane[164].

$$(MeO)_2BCl + Bu_3SiN_3 \rightarrow (MeO)_2BN_3 + Bu_3SiCl$$

Thermal studies on the azidoboranes in cyclohexene have been investigated. The facile reaction of compounds containing silicon–nitrogen bonds with boron halides has been used by several workers for the preparation of a variety of boron–nitrogen compounds. Bis(diphenylboryl)di-imine was obtained from the reaction of chlorodiphenylborane and bis(trimethylsilyl)di-imine[165].

$$Me_3Si-N=N-SiMe_3 + 2Ph_2BCl \rightarrow 2Me_3SiCl + Ph_2B-N=N-BPh_2$$
$$+ \text{polymer}$$

The reaction of mono-, di-, and tri-haloboranes with heptamethyldisilazane leads to a stepwise replacement of the halogen by the Me_3SiNMe group[166].

$$Me_3Si-N(Me)-SiMe_3 + {>}B-Cl \longrightarrow Me_3SiCl + {>}B-N(Me)-SiMe_3$$

Dimethylaminopolyboranes, of the type $B_n(NR_2)_{n+2}$, have been obtained by the dehalogenation of chlorodimethylaminoboranes by a liquid sodium/potassium alloy[167]. In these compounds the boron atoms form chains and the chain length is related to the number of dimethylamino groups substituted on the boron atom. The reaction of dimethylaminopolyboranes and molecular oxygen gives an aminoboroxine.

$$2(Me_2N)_2BCl + Me_2NBCl_2 + 4K \rightarrow 4KCl + (Me_2N)_2B-B(NMe_2)-B(NMe_2)_2$$
$$\downarrow O_2$$
$$\tfrac{2}{3}(Me_2NBO)_3 + (Me_2N)_3B\cdot$$

Deeply coloured boron–nitrogen radicals, of the type R_2BL (L = py), have been observed in the dehalogenation of pyridine-chlorodialkylboranes with lithium in tetrahydrofuran[168]. Salt-like species containing cationic radicals, e.g. $Et_4B_2(py)_4^+$, have been obtained. The reaction of enaminoboranes with nitriles has been investigated and compounds of the type (3) have been synthesised[169]. With aqueous $HCl/CHCl_3$ solution (3) reacts to give the heterocycle (4).

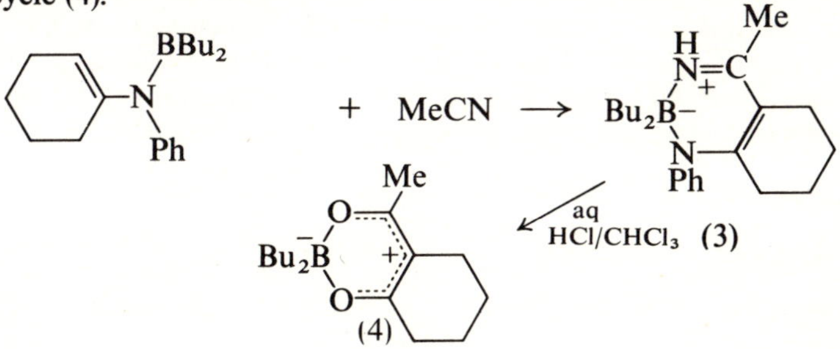

There has been much interest in the structure and chemistry of aldimino- and ketimino-boranes. The reactions between boron halides and diphenylketiminotrimethylsilane have been fully investigated[170].

$$2Ph_2C{=}NSiMe_3 + PhBCl_2 \rightarrow (Ph_2C{=}N)_2BPh + 2Me_3SiCl$$

The preparation of aldiminoboranes and their cyclic dimers[171] and diphenylketimidoboron dihalides[51] has been published. Mass spectrometric, cryoscopic, and infrared evidence supports the dimeric structures for the diphenylketimidoboron dibromide and di-iodide. The preparation of a series of monomeric aryl- and diaryl- methyleneaminoboranes by five methods has been reported[172]; e.g.,

$$R_2C{=}NLi + R'_2BX \rightarrow R_2C{=}NBR'_2 + LiX$$

The infrared spectra of the monomeric compounds have a characteristic absorption in the range 1762–1820 cm^{-1} which has been assigned to the asymmetric stretching (C$=$N$\rightleftharpoons$B) vibration. In associated compounds, bands in the region 1610–1645 cm^{-1}, characteristic of bridging imino groups, were found.

5.4.3 Structural and physical properties of compounds containing boron–nitrogen bonds

Over the past 2 years interest has been sustained in the structural aspects of compounds containing boron–nitrogen bonds. The molecular structure of tris-dimethylaminoborane has been determined by electron diffraction in the vapour phase[173]. The boron-nitrogen bond length, 1.431 Å, is much shorter than the corresponding bond length in tetrahedral compounds. The results show that although there is rotation of the dimethylamino group out of the BN$_3$ plane, there is still sufficient overlap between the lone pair on each nitrogen atom and the vacant p-orbital of the boron atom to effect delocalisation resulting in π-bonding. The crystal structure of dimethyldimethyl aminoborane at −95 °C has been determined and the boron–nitrogen bond length found to be 1.42 Å[174]. Due to the planarity of the molecule, π-bonding is suggested between the boron and nitrogen atoms, and molecular orbital calculations have given a value of 1.6 for the boron–nitrogen bond order. The crystal and molecular structure of the trimethylamine–boron trichloride complex has been studied by x-ray analysis[175], and found to be monomeric with a boron–nitrogen bond length of 1.575 Å; the structure of the compound was determined in order to provide a comparison between an amineborane and a borazine. The crystal and molecular structure of 1,3,5-trisdimethylamino-1,3,5-triboracyclohexane has been determined[176] and the boron–nitrogen bond length found to be 1.40 Å. As the boron–nitrogen bond length differs so little from that in borazines, it was suggested that the bond order in the two classes of compound is similar. The crystal and molecular structure of hexakis(trimethylsilyl)-2,4-dimino-1,3,2,4-diazaboretidene has been determined[177]. The molecule is planar with a boron–nitrogen bond length of 1.54 Å; that the exocyclic boron–nitrogen bond is so short is surprising since the substituents on the boron and nitrogen atoms are arranged perpendicular to one another preventing the formation of a classical π-bond. The molecular structure in the gas phase of dimethylcyclotetrazenoborane shows a short boron–nitrogen bond length, 1.413 Å, and this suggests the presence of extensive delocalisation of the π electrons in the N_4B ring[178]. The crystal structure of 1,8,10,9-triazaboradecalin shows the structure to be that of a distorted naphthalene compound[179]; the N_3B group is planar with a boron–nitrogen bond length of 1.421 Å.

The barriers due to rotation about the boron–nitrogen bond in a series of aminoboranes have been found, by ^1H n.m.r., to be about 12 kcal mol^{-1} [180]. The activation parameters, for those compounds where the barriers were high enough to measure, were reported. ^1H n.m.r. has been used to measure the substituent dependence of the rotation barriers around the boron–nitrogen bond in a series of N,N'-dialkylaminoboranes[181]; the results show that the substituents have a large effect and in some cases large changes of rotation occur when alkyl groups on the boron atom are replaced by a chlorine atom or a second amino group. The lowering of the rotation barrier by about 10 kcal mol^{-1} in bisaminoboranes compared with that in monoaminoboranes was rationalised in terms of the lowering of the double bond character or each boron–nitrogen bond. Of the fifteen compounds measured, $\Delta G^{\ddagger}$ varied from 23.7 to 10.2 kcal mol^{-1}. It should be noted that in some cases the spectra were recorded using carbon disulphide as a solvent. ^1H n.m.r. and ^{11}B n.m.r. studies have been made of a series of aminoboranes in which restricted rotation about the boron–nitrogen bond was expected[182]. Long range spin-spin coupling between hydrogen and boron has been observed in the structural arrangement HCNB[183]. The coupling could only be observed for tetrahedral boron compounds coordinated with nitrogen, chlorine, or bromine. The coupling was destroyed when the substituents on boron were hydrogen or fluorine. The first and second order quadrupole effects of the ^{11}B n.m.r. have been measured in polycrystalline samples of dimethyldimethylaminoborane and dimethylaminodihaloboranes (F, Cl, and Br)[184]. The nuclear quadrupole coupling constants and asymmetric parameters of ^{11}B n.m.r. have been determined. The vibrational spectra and force constants of some disilylaminohaloboranes have been reported[185]. The results indicate that N-silylation weakens the boron–nitrogen bond. In bis(trimethylsilyl)-aminodichloroborane, the boron–nitrogen stretching force constant had a calculated value of 6 mdyn Å^{-1}. A mass spectral study on a series of heterocyclic five membered boranes showed that the cyclic boronium ions were highly unstable[186]. This suggests that for boronium ions a linear structure is more thermodynamically stable than for its cyclic analogue. The mass spectral determination of the heat of atomisation for BCN and BNC has been reported[187], and the mass spectra fragmentation of cyclic amineboranes and their spirocatechol derivatives have been discussed[188].

5.4.4 Borazines

There has recently been a renewal of interest in the chemistry and structures of borazine and its derivatives, especially with regard to the effects of substitution on the boron atoms on the delocalised π electrons in the borazine ring. From the photolysis of a borazine–methyl bromide mixture in the gas phase, B-monobromoborazine has been obtained[189]. B-chloroborazine was obtained similarly using hydrogen chloride, methyl chloride, carbon tetrachloride, or chloroform. The infrared spectra and mass spectra of the two compounds and their deuterated analogues were recorded. Pyridine forms a complex with B-trichloroborazine which on reaction with lithium borohydride yields B-monochloroborazine as the major product[190].

$$\text{ClB}\underset{\underset{\text{Cl}}{\text{B}}}{\overset{\overset{\text{H}}{\text{N}}}{\diagup\diagdown}}\text{BCl} \cdot \text{py} \xrightarrow{\text{Li BH}_4}{\text{Et}_2\text{O}} \text{HB}\underset{\underset{\text{H}}{\text{B}}}{\overset{\overset{\text{H}}{\text{N}}}{\diagup\diagdown}}\text{BCl} + \text{ClB}\underset{\underset{\text{H}}{\text{B}}}{\overset{\overset{\text{H}}{\text{N}}}{\diagup\diagdown}}\text{BCl} + \text{HB}\underset{\underset{\text{H}}{\text{B}}}{\overset{\overset{\text{H}}{\text{N}}}{\diagup\diagdown}}\text{BH}$$

$$\qquad\qquad\qquad\qquad\qquad\quad 50\% \qquad\quad 6\% \qquad\quad 17\%$$

By contrast the analogous bromo complex, on reduction, gave mainly borazine. ^{1}H n.m.r. studies support the idea that chlorine atoms are weak electron-releasing substituents. The preparation and properties of unsymmetrical B-fluoroborazines have been reported[191]. Double silicon–nitrogen bond cleavage of compounds of the type $\text{RN}(\text{BR}'\text{—NR—SiR}_3)_2$, by their reaction with dihaloboranes, has been used as a preparative method for the synthesis of symmetric and asymmetric borazines[192].

$$\text{RN}(\text{BR}'\text{—NR—SiR}_3)_2 + \text{R}''\text{BCl}_2 \longrightarrow 2\text{R}_3\text{SiCl} + \text{R}'\text{N}\underset{\underset{\text{R}'}{\text{B—N}}}{\overset{\overset{\text{R}'}{\text{B—N}}}{\diagup\diagdown}}\text{BR}''$$

Reactions with analogous dihalides of silicon, tin, and phosphorus were not successful. The reaction with sulphur dichloride gave a thiaborazine which could not be separated from the B-triphenyl-N-trimethylborazine also formed in the reaction.

$$\text{MeN}\underset{\underset{\underset{\text{Ph Me}}{|\ \ |}}{\text{B—N—SiR}_3}}{\overset{\overset{\overset{\text{Ph Me}}{|\ \ |}}{\text{B—N—SiR}_3}}{\diagup\diagdown}} + \text{SCl}_2 \longrightarrow 2\text{R}_3\text{SiCl} + \underset{\underset{\text{Me}}{\text{PhB}\diagdown_{\text{N}}\diagup\text{S}}}{\overset{\overset{\text{Ph}}{\text{B}}}{\text{MeN}\diagup\diagdown\text{NMe}}}$$

Studies on the thermal and chemical stability of the compounds formed by the polycondensation of various B-diaminoborazines with diisocyanates, have been made[193]. Non-empirical LCAO—MO—SCF calculations on borazine found that the planar (D_{3h}) model was energetically preferred to the non-planar (C_2) form by 40.5 kcal mol^{-1} [194], and ab initio calculations showed the barrier for rotation in borazine to be 3.064 kcal mol^{-1} [195]. A normal coordinate analysis, based on the reassignment of the borazine fundamental vibrations, showed that the force constant for the boron–nitrogen bond was around 5.5 mdyn/Å, which is smaller than the previously reported value[196]. An ultraviolet study of 2,4,6-trisethylamino-1,3,5-triethylborazine, in hexane has been carried out and the π—π^* transition assigned[197]. The results indicate that the substitution of an ethylamino group on the borazine ring does not cause any decrease in its pseudoaromatic character. The infrared spectra of isotopically substituted N-trimethylborazines have been reported[198]. On the basis of theoretical calculations, the strongest band observed in the infrared spectra of the various borazines at about 1400 cm^{-1} was assigned to the high frequency boron–nitrogen stretching E' mode which is also coupled with the symmetric N-methyl deformation vibrations. The vibrational frequencies of borazine and its symmetrically deuterated derivatives have been reported[199], and the results used in the calculations of a modified valence force field for these compounds. The molecular structure of borazine has been redetermined by electron diffraction and the boron–nitrogen bond length determined as 1.4355 Å [200]. The molecular structure of B-monoaminoborazine in the gas phase has been investigated by electron diffraction[201]. The best description of the molecule is that it is essentially planar with the exception of the amine hydrogens. The bond lengths calculated were boron–nitrogen$_{ring}$, 1.418 Å, and boron–nitrogen$_{amine}$, 1.498 Å.

Germylborazines have been obtained from the reaction of a chloroborazine and H$_3$GeK or Ph$_3$GeK[202].

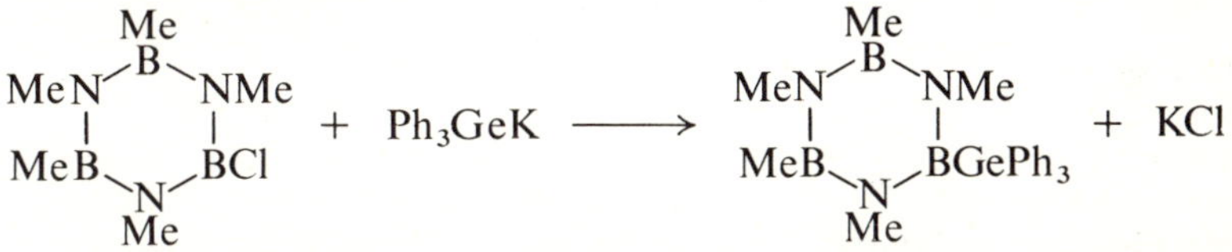

$$\text{MeN} \overset{\text{Me}}{\underset{}{\text{B}}} \text{NMe} \quad + \text{ Ph}_3\text{GeK} \longrightarrow \text{MeN} \overset{\text{Me}}{\underset{}{\text{B}}} \text{NMe} \quad + \text{ KCl}$$
(MeB—N(Me)—BCl → MeB—N(Me)—BGePh$_3$)

The infrared spectra show that the substitution of both the triphenylgermyl and trihydrogermyl groups on the boron atom causes a decrease in the p_π–p_π bonding between the boron and nitrogen atoms in the ring compared to that of hydrogen or a methyl group. The following order of decreasing p_π–p_π bonding was suggested: H ~ Me > GeH$_3$ > Cl ≫ Ph$_3$Ge ≫ Ph$_3$Sn. A study of the effects of substituents in a series of B-monosubstituted borazines of the type H$_2$XBN$_3$H$_3$ (where X = Me, Me$_2$N, MeO, F, Cl, Br) has been made, and the infrared, ^{1}H n.m.r., ^{11}B n.m.r., and mass spectra of the compounds reported[203]. The ^{1}H n.m.r. and ^{11}B chemical shift results supported the view that the π-electrons of borazine are at least partially delocalised and that the substituents interact with this π system. It was rather surprising, however, that no general trends could be observed from the infrared and mass spectral studies.

5.5 COMPOUNDS CONTAINING BORON–OXYGEN BONDS

5.5.1 Introduction

Over the past two years there has been considerable activity in the chemistry of compounds containing boron–oxygen bonds. A number of reviews have been published, in addition to those included in the standard texts, as mentioned in the introduction. A comprehensive review of the use of borax as a primary standard has appeared[203]. Unsaturated organoboron heterocycles containing boron–oxygen bonds have been reviewed with the emphasis on compounds containing boron–chelate bonds[204]. The preparation, thermal stability, and crystal structure of the oxyhalides of Group III elements have been reviewed[205], and a correlation made between the bond dissociation energies with reactivities of the oxyhalides. The preparation, properties and reactions of esters of boric acid have been surveyed in order to obtain information as to their importance as plasticisers for poly(vinyl chloride)[206]. Finally, a review has appeared on the effect of boric ester neighbouring groups in organic systems[207].

5.5.2 Preparation and chemistry of compounds containing boron–oxygen bonds

High purity boric oxide has been obtained from the reaction of a metal borate and sulphuric acid[208]; the reaction mixture was heated to give two molten phases from which boric oxide was separated.

$$Na_2B_4O_7 \cdot 5H_2O + H_2SO_4 \to B_2O_3 \text{ (98.3\% pure)} + NaSO_4$$

However, the reaction between an alkali metal borate and sulphur trioxide gave boric acid[209], and the preparation of boric acid from boric oxide has been patented[210]. Resins of high hydrolytic and thermal stability have been obtained by condensing boric acid, or oxide, with polyhydric phenols or aromatic polyamines[211]. Boric acid has been found to be a good catalyst in the preparation of melamine from ammonia and hydrogen isocyanate[212]. The reaction of boric acid and sodium carbonate at 720 °C gave a good yield of sodium borate[213]. In air flame studies of hydrocarbon slurries containing boron, BO, BO_2, and HBO_2 were identified as the boron-containing products[214]. At high temperatures, boric acid reacted with ammonia to give crystalline boron nitride[215]. Methyl borate has been obtained from the reaction of boric acid and dimethyl phthalate in the presence of sulphuric acid[216]. When using butyl borate instead of boric acid, methyl borate was obtained in quantitative yields. Boron esters have been obtained in high yields from the reaction of a trialkylborane and an amine-N-oxide[217]. The well-known reaction of boric acid with an alcohol has been used to obtain butyl and isoamyl borates[218]. The reaction of the bis(trifluoromethyl)nitroxide radical with boron halides gives the corresponding orthoborate $B[ON(CF_3)_2]_3$ [219]; the reaction with boron tribromide was rapid, but in the case of boron trichloride, iodine was used as a catalyst. Arylboronic acids have been obtained from the ultraviolet irradiation, followed by hydrolysis, of a mixture of boron tribromide (or tri-iodide) and benzene or toluene[220]; it was suggested that initially arylboron dihalides were formed by photolysis of the boron halide, followed by the interaction of the dihalogenoboryl radical with the aromatic compound. Dimethoxy-t-butylperoxyborane (5) has been synthesised and its thermal decomposition investigated[221].

$$(MeO)_2BCl + Me_3COOH \to (MeO)_2BOOCMe_3 + HCl$$
$$(5)$$

A series of organoperoxyboranes have been synthesised analogously[222]. The infrared and n.m.r. spectra were recorded and compound (5) was found to be 30% associated. Aliphatic and aromatic diboronic acids, of the type

$(HO)_2BRB(OH)_2$, have been obtained from the reaction of methyl borate with a difunctional Grignard reagent[223]. In a similar way, the reaction of Grignard reagents with diboronic acids gave the corresponding diborinic acids[224]. A convenient preparation and full characterisation of dimethylborinic anhydride has been reported[225]. The dehydration of hydroxyethylboronophthalide gave a dimeric boronic ester (below) which was relatively stable to hydrolysis[226].

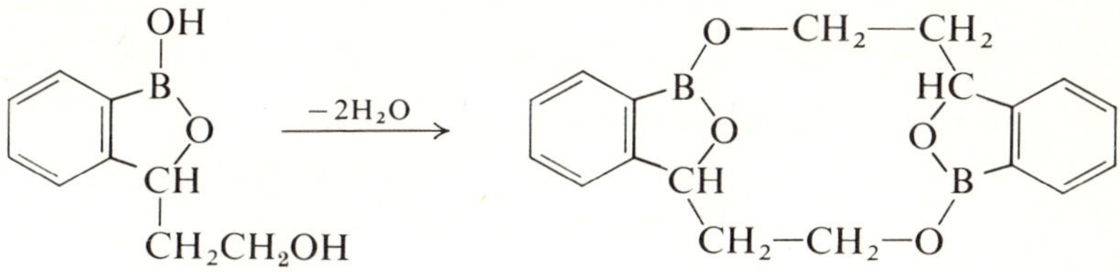

The reaction between boric acid and pinacone hydrate was found to yield a 1:1 complex which was formulated as the heterocycle shown below[227].

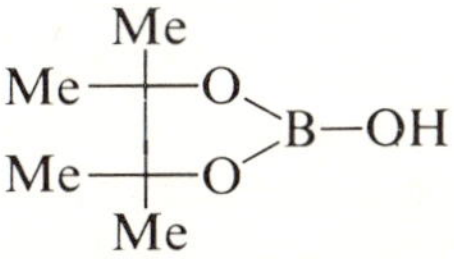

The synthesis and reactions of various heterocyclic compounds have been reported. The sodium salt of 2-mercaptopyridine-N-oxide and oxybis-diphenylborane were refluxed in methanol and (6), which was characterised by infrared and ultraviolet spectroscopy, was obtained[228]. The formation of chelates between (6) and Ni^{2+}, Cu^{2+} and Cd^{2+} were studied.

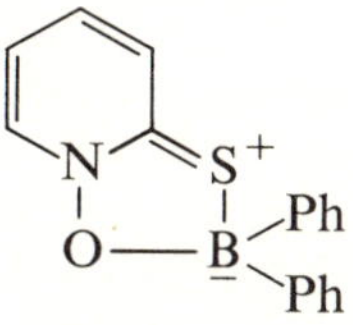

(6)

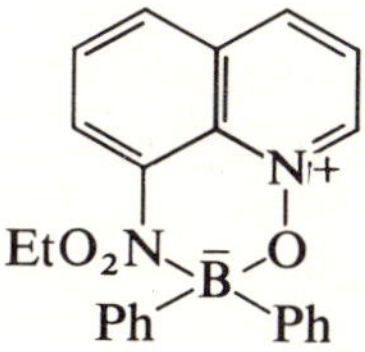

(7) 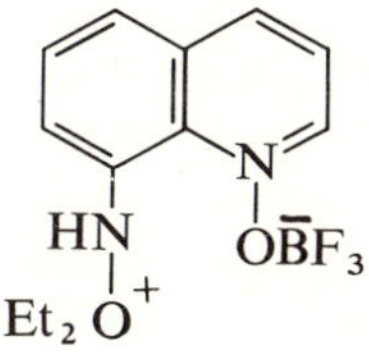   (8)

The preparation of the chelate (7) and its reaction with boron trifluoride to give (8) have been reported[229]. The following structures give some idea of the various types of chelates of boron compounds that have been prepared[229, 230].

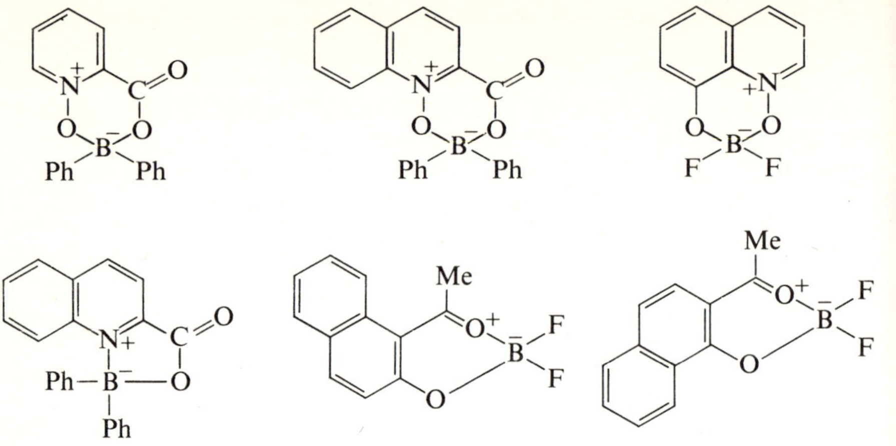

The reactions of boric acid with chlorotrimethylsilane and dichlorodimethylsilane have been used to prepare siloxanes and borosiloxanes, respectively[231].

$$[Me_2SiCl]_2O + polymer \xleftarrow{Me_2SiCl_2} H_3BO_3 \xrightarrow{XSMe_3SiCl} (Me_3SiO)_3B$$

The reactions of tri(bistrifluoromethylnitroxido)borane with amines such as pyridine and trimethylamine have been studied and the resultant 1:1 complexes have been characterised by ^{19}F n.m.r. spectra[232].

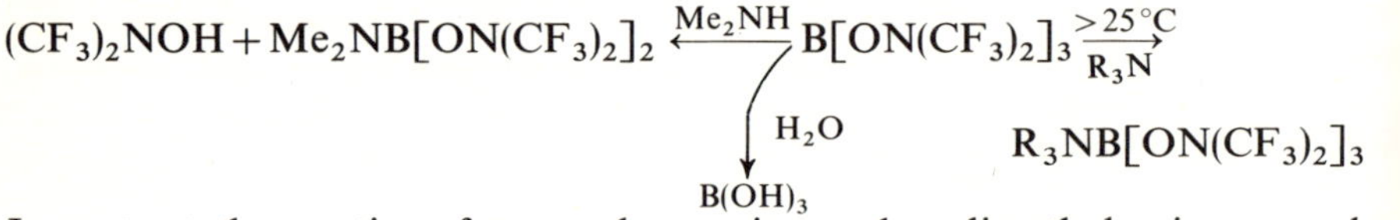

$$(CF_3)_2NOH + Me_2NB[ON(CF_3)_2]_2 \xleftrightarrow{Me_2NH} B[ON(CF_3)_2]_3 \xrightarrow[R_3N]{>25°C} R_3NB[ON(CF_3)_2]_3$$

$$\downarrow H_2O$$

$$B(OH)_3$$

In contrast, the reaction of a secondary amine, such as dimethylamine, caused a displacement reaction. No complex was formed with phosphorus trichloride or tribromide.

The reactions of phosphites and orthoborates have been studied[233].

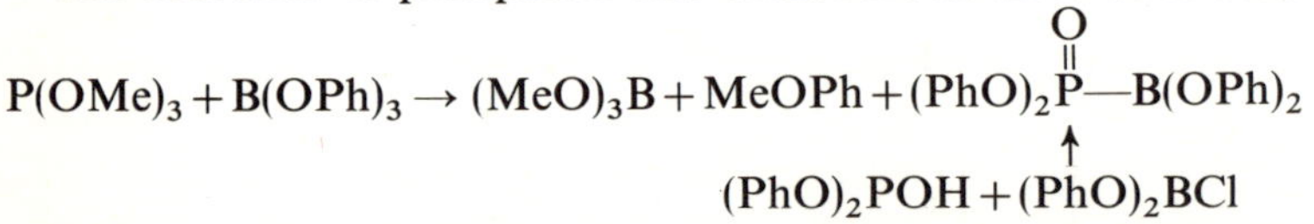

$$P(OMe)_3 + B(OPh)_3 \rightarrow (MeO)_3B + MeOPh + (PhO)_2\overset{O}{\underset{\|}{P}}-B(OPh)_2$$

$$\uparrow$$

$$(PhO)_2POH + (PhO)_2BCl$$

However, phenyl phosphite and phenyl borate interacted to form a 1:1 complex. Boron bis(trifluoromethyl)phosphinites have been obtained by the cleavage of $(CF_3)_2P-O-P(CF_3)_2$ with haloboranes[234]. In this way $[(CF_3)_2PO]_3B$, $MeB[OP(CF_3)_2]_2$, and $Me_2B[OP(CF_3)_2]$ were obtained, and their infrared and n.m.r. spectra recorded. None of the compounds showed any tendencies towards Arbuzov type rearrangements to form boron–phosphorus bonds. The ultraviolet induced reaction of benzyl borate in bromobenzene produced on hydrolysis, dibenzyl ether, benzyl bromide, benzaldehyde, and benzene[235]; the reaction of 1-phenylethyl borate was also studied and the use of this method for the preparation of ethers was discussed. Phenyl borate was obtained from the reaction of methyl borate with phenol[236], and its reactions with carboxylic acids and acetates were described.

Methyl and ethyl borates were found to form complexes with sulphur trioxide, of stoichiometry $(MeO)_3B \cdot 1.5SO_3$ and $(EtO)_3B \cdot SO_3$[237]. Confirmation of these complexes was provided by infrared spectra and chemical analysis. Trimethylborane, in 50% yield, was obtained from the reaction between methyl borate and trimethylalane at low temperatures[238]. Trimethoxyboroxine forms a series of 1:1 complexes with transition metal halides[239]. The kinetics of the thermal decomposition of tris(t-butylperoxy)-borane have been studied in nonane and cumene[240]; in nonane, boron esters were identified as one of the products of decomposition. The effect of the boric acid concentration on the reactions with anions of tartaric, citric, malonic, and lactic acids in ion exchange reactions has been studied[241] and complexes were investigated. Potentiometric titrations and ion exchange measurements were used to show that in the systems boric acid—lactic acid and boric acid—malonic acid complex formation occurs and leads to the formation of two types of complexes in which there is either one or two molecules of acid to one molecule of boric acid[242]. The complex acids were stronger than the individual constituents. The existence of a weakly dissociated 1:1 complex between boric acid and ethylene glycol was demonstrated from graphs depicting deviations from property additivity as a function of concentration[243]. The formation of complexes between boric acid and salicylic acid has been observed potentiometrically using dilute solutions of sodium hydroxide[244]. When a 1:2 mixture of boric acid and citric acid was evaporated to dryness a syrup was obtained which, on drying above sulphuric acid, yielded the compound $H_3BO_3 \cdot 2C_6H_8O_7$[246]; x-ray diffraction studies and thermogravimetric analysis showed that the complex formed was an individual compound and not a mixture. Thermogravimetric analysis, infrared studies, and x-ray analysis were used to demonstrate that pinacol and boric acid form a pinacol borate–anhydride equilibrium, the former being converted to the latter at 120–60 °C[246]. Several patents have been registered concerned with the properties of boron-oxygen containing compounds: these include the preparation of borate esters[247] and their properties as antioxidants for polymers and lubricants[248]. Borates of N-(hydroxyalkyl)-nitrogen heterocyclic saturated compounds have been used as stabilisers for plastics and hydrocarbon distillates[249]. Borates, boric acid, boric oxide, and boroxines, when used in conjunction with phosphoramides, were effective stabilisers for poly(phenylene ether)[250]. The mechanisms of the interaction of acyloxyalkoxyboranes with amines have been investigated[251].

5.5.3 Physical properties of compounds containing boron–oxygen bonds

P.P.P. and V.E.S.C.F. molecular orbital calculations on phenylboronic acid and phenylboroxine showed that the red shift of the second band (in going from the acid to the anhydride) was due to the conjugation in the boroxine ring[252]. Molecular orbital studies have also been carried out on $H_2B_2O_3$ [253] and the borate anion[254]. The ultraviolet spectra of phenylboroxine and the p-bromo-derivative have been measured[255]. A comparison of the spectra with those of the corresponding boronic acids suggested a stronger interaction between the phenyl groups and the boroxine ring. In a study of the ultraviolet spectra of arylboronic acids it was concluded that the $B(OH)_2$ group was responsible for the bathochromic shifts[256]. The structure of boroxine in the gas phase[256] shows the molecule to have a planar six-membered ring with D_{3h} symmetry and a boron–oxygen bond length of 1.375 Å. Infrared and nuclear magnetic resonance studies have been used to show that boroxine and azo compounds form 1 : 1 addition compounds[258]. An x-ray diffraction analysis of tribenotalarene showed it to contain a nearly planar six-membered boron–oxygen ring[259]. The structure of boric oxide is made up of a random three-dimensional network of BO_2 triangles with a relatively high proportion of six-membered boroxine rings[260]; the stabilisation energy of the $(B—O)_3$ ring was estimated as about 5–10 kcal g^{-1} formula weight. The infrared spectra of the borates $(RO)_3B$ [where $R = C_2D_5$, $CD(CD_3)_2$, CH_2CF_3, and $CH(CF_3)_2$] have been recorded and a comparison made with ethyl and isopropyl borates[261]; in the spectra of the fully deuterated derivatives, the splitting of about 50 cm^{-1} due to the $^{10}BO_3$ and $^{11}BO_3$ asymmetric stretching frequencies was larger than in the non-deuterated case. Infrared spectroscopy has been used to show that three types of surface—BH groups are obtained from the pyrolysis of B—OMe groups on boron-silica surfaces[262]. The thermal collapse of the chemisorbed alkoxy groups is therefore a suitable method for the production of hydride species. The infrared spectra of some natural and synthetic borates have been investigated[263] and the infrared spectra of the pyroborates $Mg_2B_2O_5$, $Mn_2B_2O_5$, and $Fe_2B_2O_5$ studied[264]. Vibrational assignments for $H_2B_2O_3$ have been made assuming the molecule to have C_{2v} symmetry[265]. An infrared study of alkali-metal derivatives of dipinacol hydrate showed that the observed band at 964–968 cm^{-1} was indicative of the boron being tetracoordinate in the complexes[266]. The magnetic properties of a series of boroxines $(XBO)_3$ (where $X = Et$, Me, Bu, C_6H_{13}, MeO, or EtO) have been measured[267]. Diamagnetism, which if high, has been suggested as evidence of aromaticity, was not as high as that of borazines. The hydrogen-bonding and association of dimethylborinic acid in inert solvents has been studied by n.m.r. spectroscopy[268]. The ^{11}B n.m.r. spectra of three-coordinate boron in vitreous and crystalline boric oxide have been measured[269]; a planar structure with three-fold symmetry was proposed for the BO_3 group in both cases and a shorter relaxation time and a higher degree of thermal motion in boric oxide than for borates supported the structural interpretation. Fourier transforms have been found useful in the analysis of powder patterns perturbed by second order nuclear quadrupole interactions: this technique has been applied to the spectra of vitreous and crystalline boric oxide[270]; the shape of the spectra indicated an almost threefold symmetry configuration of the BO_3 group in these substances. The ^{11}B n.m.r. chemical shifts of aqueous solutions (50 g/l) of sodium tetraborate have been measured over the pH 2–12 range. The results showed that the chemical shift at high pH corresponded to the tetrahedral $B(OH)_4^-$ ion and at low pH to that of boric acid. Only one ^{11}B n.m.r. resonance was observed, thus indicating a rapidly interconverting equilibrium. The ^{11}B n.m.r. spectra of some cyclic boron esters show that the chemical shifts of the neat esters, relative to methyl borate, appear to be diagnostic for ring size[272]; when solutions of phenylboronates in coordinating solvents were studied, it was found that ^{11}B resonance lines were narrowed considerably with increasing dilution. The ^{11}B—H coupling constant in methyl borate has been determined from proton spin-echo n.m.r. spectroscopy[273]. Infrared and ^{1}H n.m.r. spectra were used to assign the structure of the asymmetric borate salt $[Me_2NH_2]^+$ $[(PhCO_2)_3BOMe]^-$ [274].

The BOCN molecule has been observed mass spectrometrically during the initial stages of the vaporisation of boron nitride from a titanium coated graphite (Knudsen) effusion cell[275]. Rearrangement under electron impact appears to be a general property of phenylboronates of diols[276]; the products of the rearrangement were hydrocarbon ions containing 7, 8, 9, or 10 carbon atoms depending upon the structure of the esters. The fragmentation pattern of 2-phenyl-1,3,2-dioxoborolane indicated an unexpected stability of the parent molecule[277]; a major re-arrangement process leading to the formation of a species, assigned as the tropylium ion, was observed. From the mass spectra of boric acid, the electron affinites for BO_2 and BO were determined[278]. The mass spectra of a series of equimolar mixtures of borates have been reported and the ion abundances used to distinguish fragments which could only come from the mixed borates[279]; further evidence for mixed borates was obtained by metastable ion analysis and supported the redistribution reaction:

$$B(OR)_3 + B(OR')_3 \rightleftharpoons B(OR)_2OR' + B(OR)(OR')_2$$

The mass spectra of a series of boroxines show that, apart from triphenylboroxine, the molecular ion is in low abundance due to the fragmentation of the exocyclic groups attached to boron[280]; the gas chromatographic behaviour of the boroxines was also discussed.

The formation of complexes between trivalent boron compounds and substituted o-diphenols has been studied by potentiometric and spectrophotometric methods[281]; correlations were made between stability constants and the acidities of the diphenols. Borate complexes of fructose and D-ribose, but not of L-cysteine migrate electrophoretically as the borate complexes[282]. Complexes between boric acid and benzoylacetone, formed in concentrated sulphuric acid after the addition of ether to form ices at 77 K, were studied phosphorimetrically[283]; the complexes shown below were identified.

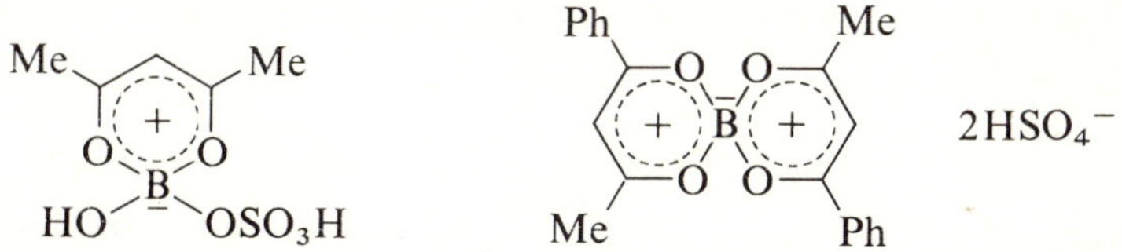

The integral heats of solution of boric acid solutions of calcium nitrate have been measured and the ternary enthalphy concentration diagram constructed[284]; for the calcium nitrate—boric acid—water system, the enthalphy depends very little upon the concentration of calcium nitrate.

Peroxyborates, when carefully heated, can be modified in such a way as to release large quantities of oxygen when they are dissolved in water[285]; the modified peroxyborates have unusually high paramagnetic susceptibilities. Solution thermochemistry has been used to determine the ring strain in B-phenyl and B-butyl-dioxaboron compounds[286]; the values obtained ranged from 4 kcal mol^{-1} for $\overline{PhBO(CH_2)_3O}$ to 13 kcal mol^{-1} for $\overline{Bu^nBO(CH_2)_4O}$.

5.6 BORON COMPOUNDS CONTAINING BORON–SULPHUR BONDS

5.6.1 Introduction

In comparison with other areas of boron chemistry, that of compounds containing boron–sulphur bonds has been relatively little investigated. The first review of all aspects of boron-sulphur compounds was published in 1966 [287] and was followed 3 years later by a comprehensive survey of boron-sulphur compounds for the period 1950–1967 [288]. This has since been brought up to date by two further articles[289, 290]. The chemistry of the borthiins has also been reviewed[291]. As well as reviews which were included in the general texts referred to earlier, others have recently appeared on sulphur-containing organic compounds of boron[292], and on cyclic boron–sulphur compounds[293]. Finally, the preparation, structures, and main chemical and physical characteristic properties or boron sulphides and ternary boron–S–metal compounds have been surveyed[294].

5.6.2 Preparation and chemistry of compounds containing boron–sulphur bonds

The main advance in new methods of preparation of thioboranes has come from a study of redox reactions of iodoboranes. Diphenyliodoborane reacts readily with dialkyl and diaryl disulphides to give high yields of the corresponding thioborane[295].

$$Ph_2BI + MeSSMe \rightarrow 2Ph_2BSMe + I_2$$

Although di-iodophenylborane reacted analogously, the main product of the reaction with di-t-butyl disulphide was the heterocycle (9) [296].

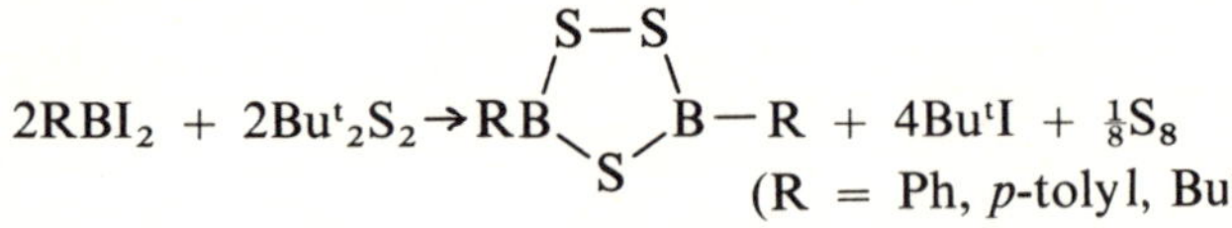

$$2RBI_2 + 2Bu^t{}_2S_2 \rightarrow RB\underset{S}{\overset{S-S}{\diagup\diagdown}}B-R + 4Bu^tI + \tfrac{1}{8}S_8$$
$$(R = Ph, p\text{-tolyl}, Bu^n)$$
$$(9)$$

Reactions were also carried out using dichlorophenylborane. The heterocycle (9) was also formed from the reaction of an aryldi-iodoborane and sulphur. The reactivity of iodoboranes is such that diiodomesitylborane reacted readily with dialkyl sulphides to give the corresponding thioborane[297].

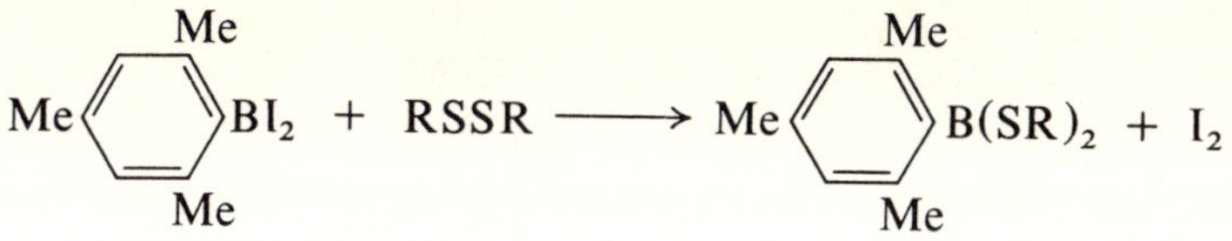

$$\text{Me-}\underset{\text{Me}}{\overset{\text{Me}}{\bigcirc}}\text{-BI}_2 + \text{RSSR} \longrightarrow \text{Me-}\underset{\text{Me}}{\overset{\text{Me}}{\bigcirc}}\text{-B(SR)}_2 + \text{I}_2$$

Thioboranes $RSBX_2$ $(RS)_2BX$, and $(RS)_3B$ have been obtained in high yields in the redox reactions between tri-iodoborane and dialkyl disulphides[298]; their thermal and hydrolytic stabilities and their propensities towards association were examined. It was not possible to prepare methanethiodifluoroborane. The reactions of dialkyliodoboranes with sulphur, or dialkylchloroborane with H_2S_2 or Na_2S_2, leads to bis(dialkylboryl)-disulphides which have high thermal but low hydrolytic stability[299].

$$R_2BI + \tfrac{1}{4}S_8 \rightarrow R_2B-S-S-BR_2 + I_2$$

These compounds react with triphenylphosphine to give the corresponding sulphide.

$$R_2B-S-S-BR_2 + Ph_3P \rightarrow R_2B-S-BR_2 + Ph_3PS$$

The chemistry of the trithiadiborolane system has been investigated. The preparations and properties of nine compounds of type (10) are documented[300].

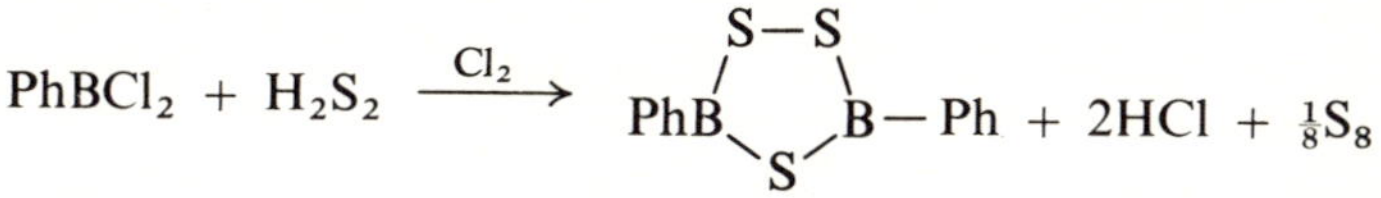

$$PhBCl_2 + H_2S_2 \xrightarrow{Cl_2} PhB\underset{S}{\overset{S-S}{\diagup\diagdown}}B-Ph + 2HCl + \tfrac{1}{8}S_8$$

(10)

Orthothioborates have been obtained from the reaction between a mercaptan and sulphuretted sodium borohydride[301].

$$NaBH_2S_3 + RSH \rightarrow (RS)_3B + H_2 + RSSR + NaS_3H$$

The thioborates readily react with ketones to give quantitative yields of thioacetals.

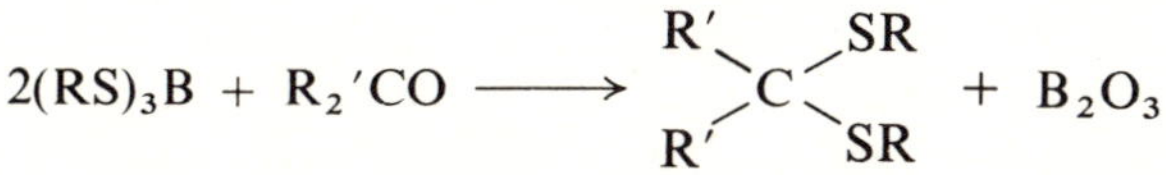

$$2(RS)_3B + R_2'CO \longrightarrow \underset{R'}{\overset{R'}{\diagdown}}C\underset{SR}{\overset{SR}{\diagup}} + B_2O_3$$

Trisalkylthioboranes, which have been used as polythene stabilisers, were obtained from the reaction of mercaptans and BOCl [302]. The formation of compounds containing boron–sulphur bonds has been reported in the reactions of boron halides or aminodihaloboranes with sodium dimethyldithiocarbamate[303].

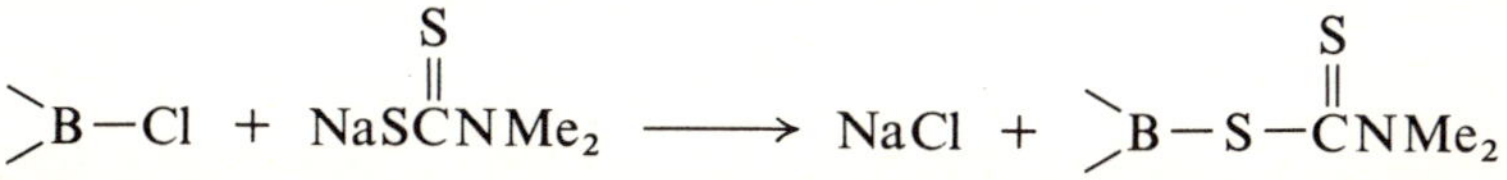

$$>\!B-Cl + NaS\overset{S}{\overset{\|}{C}}NMe_2 \longrightarrow NaCl + >\!B-S-\overset{S}{\overset{\|}{C}}NMe_2$$

By contrast o-phenylenechloroboronate and the sodium salt gave the corresponding aminoborane.

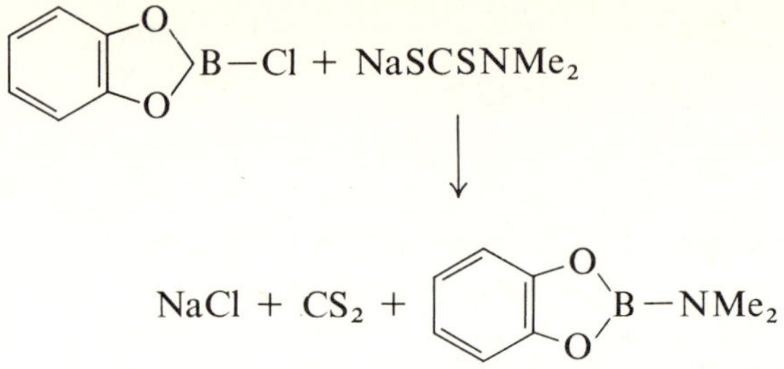

$$\text{catechol-B-Cl} + \text{NaSCSNMe}_2$$
$$\downarrow$$
$$\text{NaCl} + \text{CS}_2 + \text{catechol-B-NMe}_2$$

Boron sulphide was obtained from the reaction of boron trichloride and disilthiane[304].

$$(\text{H}_3\text{Si})_2\text{S} + \text{BCl}_3 \rightarrow \text{H}_3\text{SiCl} + \text{B}_2\text{S}_3$$

Thioboranes are reactive intermediates in the preparation of organosulphur compounds. High yields of thioacetals were obtained at ambient temperatures from the reaction of an aldehyde and a thioborane[305].

$$\text{PhCH}=\text{CHCHO} + (\text{EtS})_3\text{B} \rightarrow \text{PhCH}=\text{CHCH}(\text{SEt})_2 + \text{B}_2\text{O}_3$$

The reactions of thioboranes with sulphenic esters yield unsymmetrical disulphides[306].

$$\text{PhSOMe} + (\text{MeS})_3\text{B} \rightarrow \text{PhSSMe} + (\text{MeO})_3\text{B}$$

Boron sulphide is effective for the conversion of non-enolisable ketones, 2-pyrones, and 4-pyrones into the corresponding thiones[307].

Perhalogenacetones inserted into boron–sulphur bonds to give the corresponding boron esters[308]: e.g.,

$$\text{B(SMe)}_3 + 3(\text{CF}_3)_2\text{CO} \rightarrow \text{B}[\text{OC}(\text{CF}_3)_2\text{SMe}]_3$$

Similar reactions were carried out with PhB(SMe)$_2$ and Ph$_2$BSMe.

In contrast, the reaction of trisalkylthioboranes with carboxylic acids is complex[309]. The interaction of molar portions of a series of carboxylic acids and tris(ethylthio)borane, in a refluxing solvent, gave high yields of the corresponding thioester; two well-defined stages in the reaction were identified and the reactions provide a method for the activation of carboxylic acids.

Boron–sulphur substituted iminoboranes were obtained from the thioboration of nitriles[310]. Iminoboranes, having two thio substituents on each carbon and boron atom, were obtained from the reaction of an organothiocyanate and a tris(alkylthio)borane.

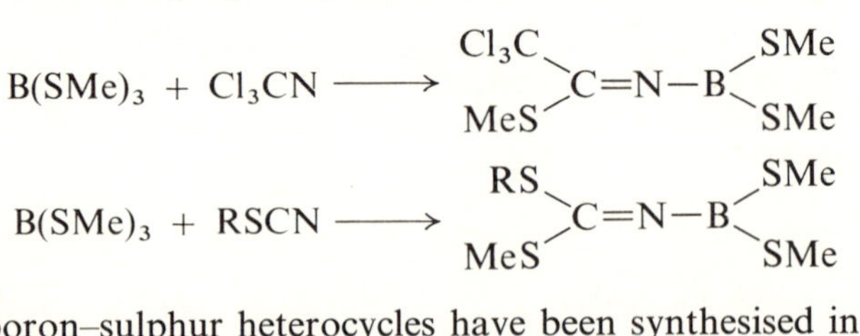

$$\text{B(SMe)}_3 + \text{Cl}_3\text{CN} \longrightarrow \underset{\text{MeS}}{\overset{\text{Cl}_3\text{C}}{>}}\text{C}=\text{N}-\text{B}\underset{\text{SMe}}{\overset{\text{SMe}}{<}}$$

$$\text{B(SMe)}_3 + \text{RSCN} \longrightarrow \underset{\text{MeS}}{\overset{\text{RS}}{>}}\text{C}=\text{N}-\text{B}\underset{\text{SMe}}{\overset{\text{SMe}}{<}}$$

Organoboron–sulphur heterocycles have been synthesised in high yields by the room temperature reaction of a thioborane and a difunctional organic compound, such as mercaptoethanol[311]:

OTHER ASPECTS OF BORON CHEMISTRY 213

PhB(SEt)$_2$ + $\begin{array}{l}\text{CH}_2\text{—OH}\\ |\\ \text{CH}_2\text{—SH}\end{array}$ ⟶ $\begin{array}{l}\text{H}_2\text{C—O}\\ \phantom{\text{H}_2\text{C—}}\diagdown\\ \phantom{\text{H}_2\text{C—O}}\text{B—Ph} + 2\text{EtSH}\\ \phantom{\text{H}_2\text{C—}}\diagup\\ \text{H}_2\text{C—S}\end{array}$

Unsymmetrical borazines have been obtained from the reactions of B-alkylthioborazines with mercuric chloride or a Grignard reagent[312].

$\underset{\underset{H}{N}}{\overset{\overset{SBu}{|}}{\underset{HN}{B}}\diagup\diagdown{NH}}$
$\underset{BuSB}{|}\underset{BSBu}{|}$
$\xrightarrow{\text{MeMgBr}}$
$\underset{\underset{H}{N}}{\overset{\overset{SBu}{|}}{\underset{HN}{B}}\diagup\diagdown{NH}}$
$\underset{MeB}{|}\underset{BSBu}{|}$
$\xrightarrow{\text{HgCl}_2}$
$\underset{\underset{H}{N}}{\overset{\overset{Cl}{|}}{\underset{HN}{B}}\diagup\diagdown{NH}}$
$\underset{MeB}{|}\underset{BCl}{|}$

5.6.3 Physical properties of compounds containing boron–sulphur bonds

The presence of BS$_2$ and BS has been observed in the thermal dissociation of B$_2$S$_3$ [313]. The vaporisation of crystalline BS$_2$ has been studied over the temperature range 550–110 °C and thermodynamic analysis showed that the major constituents of the saturated vapour at >550 °C are BS$_2$, (BS$_2$)$_2$, and (BS$_2$)$_4$ [314]. The vaporisation of boron-sulphur compounds has been investigated mass spectrometrically for a glassy sulphur-rich B$_2$S$_3$ system[315]; the volatile species from a sample of B$_2$Se$_2$ were identified by the same technique[316]. From mass spectral data, the dissociation energy of the boron–sulphur bond has been determined as 143 ± 6 kcal mol^{-1} and also the heat of atomisation for BS$_2$ (264.4 ± 8 kcal mol^{-1})[317]. The mass spectra of B$_2$(S$_2$C$_2$H$_4$)$_2$, and of related nitrogen and oxygen derivatives, have been reported and results interpreted (absence of cyclic boronium ions) on the basis of structure (11) rather than (12)[318]; (see also ref. 319).

$\begin{array}{c}\text{H}_2\text{C—CH}_2\\ \diagup\diagdown\\ \text{S}\text{S}\\ \diagdown\diagup\\ \text{B—B}\\ \diagup\diagdown\\ \text{S}\text{S}\\ \diagdown\diagup\\ \text{H}_2\text{C—CH}_2\end{array}$
(11)

$\begin{array}{c}\text{H}_2\text{C}\diagdown_\text{S}\diagdown\diagup^\text{S}\diagup\text{CH}_2\\ |\text{B—B}|\\ \text{H}_2\text{C}\diagup^\text{S}\diagup\diagdown_\text{S}\diagdown\text{CH}_2\end{array}$
(12)

The mass spectra of phenyldithiaborolane and phenyloxathioborolane have been reported; the electron impact-induced rearrangement to form the tropylium ion from the former and not the latter compound was rationalised in terms of the relative bond energies between the atoms in the heterocyclic ring[320].

The infrared spectra of sodium and potassium metathioborates indicated that they were isostructural with alkali metaborates[321]. Five active in-plane vibrations of the planar B$_3$S$_6^{3-}$ ion were assigned for both salts.

The n.m.r., magnetic susceptibilities, and rotary power have been measured for the dithiaboracycloalkanes (13) (X = Ph or Cl) and (14)[322].

$$XB\begin{smallmatrix}S-CH_2\\|\\S-CH_2\end{smallmatrix} \qquad ClB\begin{smallmatrix}S-CH_2\\CH_2\\S-CH_2\end{smallmatrix}$$

(13)   (14)

Compound (14) was found to exist in an associated form through B → S coordination. The triethylamine addition compounds, for both heterocyclic boranes, were obtained.

The preparation of B_2S_3 single crystals from vitreous B_2S_3 has been reported, along with the infrared spectra and x-ray patterns for the crystals grown[323]; the unit cell has been determined by x-ray diffraction[324].

References

1. Lappert, M. F. (1956). *Chem. Rev.,* **56,** 959
2. Adams, R. M. (1964). *Boron, Metallo-Boron Compounds and Boranes.* (New York: Interscience)
3. Muetterties, E. L. (1967). *The Chemistry of Boron and its Compounds.* (New York: Wiley)
4. Lappert, M. F. and Leigh, G. J. (1962). *Developments in Inorganic Polymer Chemistry.* (New York: Elsevier)
5. Centre National de la Recherche Scientifique (1970). *La Nature et les Propriétés des Liasons de Coordination.* (Paris)
6. Lockhart, J. C. (1965). *Chem. Rev.,* **65,** 131
7. Eaton, G. R. (1969). *J. Chem. Educ.,* **46,** 547
8. Lappert, M. F. and Prokai, B. (1967). *Advan. Organomet. Chem.,* **5,** 225. (New York: Academic Press)
9. Lappert, M. F. and Pyszora, H. (1966). *Advan. Inorg. Chem. Radiochem.,* **9,** 133. (London: Academic Press)
10. Mikhailov, B. M. (1959). *Usp. Khim.,* **28,** 1450
11. Steinberg, H. and Brotherton, R. J. (1964). **1** (1966) **2,** *Organoboron Chemistry.* (New York: Interscience)
12. Steinberg, H. and McCloskey, A. L. (1964). *Progress in Boron Chemistry.* (Oxford: Pergammon Press) (Brotherton, R. J. and Steinberg, H. (1970). **2,** 3)
13. Gerrard, W. (1961). *Organic Chemistry of Boron.* (London: Academic Press)
14. Bower, J. G. (1970). *Progress in Boron Chemistry,* **2,** 231. (Oxford: Pergammon Press)
15. Kuehl, D. K. (1970). *Off. Gaz.,* **3,** 488, 152
16. Petrov, A. V., Germaidze, M. S., Golikova, O. A., Kiskachi, A. Yu. and Malveev, V. N. (1969). *Fiz. Tverd. Tela,* **11,** 907
17. Jaumann, J. and Werheit, H. (1969). *Phys. Status Solidi.,* **33,** 587
18. Goddard, W. A. (1969). *Phys. Rev.,* **182,** 48
19. Hendrickson, D. N., Hollander, J. M. and Jolly, W. L. (1970). *Inorg. Chem.,* **9,** 612
20. Kamada, H., Inoue, R., Teraswa, M., Gohshi, Y., Kamei, H. and Fugii, I. (1969). *Anal. Chim. Acta,* **46,** 107
21. Brunstad, J. W. (1969). *J. Ass. Offic. Anal. Chem.,* **52,** 487
22. Stoecklin, G., Plott, G. and Heckel, E. (1969). *Int. J. Appl. Radiat. Isotop.,* **20,** 399
23. Maijs, L. (1969). *Latv. PSR Zinat. Akad. Vestis, Kim. Ser.,* 247
24. Chatterton, J. N., McKell, C. M., Goodwin, J. R. and Bingham, F. T. (1969). *J. Agron. J.,* **61,** 451
25. Martin, D. R. (1944). *Chem. Rev.,* **34,** 462
26. Martin, D. R. (1948). *Chem. Rev.,* **42,** 582
27. Greenwood, N. N. and Martin, R. L. (1954). *Quart. Rev. Chem. Soc.,* **8,** 1
28. Gerrard, W. and Lappert, M. F. (1958). *Chem. Rev.,* **58,** 1081

29. Polivka, Z. and Ferles, M. (1968). *Chem. Listy.*, **62**, 869
30. Fodor, L. M. (1969). *Off. Gaz.*, **2**, 444, 153
31. Canonne, P. and Regnault, A. (1969). *Can. J. Chem.*, **47**, 2837
32. Armstrong, D. R. and Perkins, P. G. (1969). *Theor. Chim. Acta*, **15**, 413
33. Armstrong, D. R. and Perkins, P. G. (1969). *Chem. Commun.*, 856
34. Schwartz, M. E. and Allen, L. C. (1970). *J. Amer. Chem. Soc.*, **92**, 1466
35. Bassett, P. J. and Lloyd, D. R. (1970). *Chem. Commun.*, 36
36. Goetschel, C. T., Campanile, V. A., Wagner, C. D. and Wilson, J. N. (1969). *J. Amer. Chem. Soc.*, 1969, **91**, 4702
37. Singh, G., Chakrabarti, H. K. and Banergee, T. (1969). *Indian Pat.*, 110, 834
38. Lorberth, J. (1969). *J. Organometal. Chem.*, **17**, 151
39. Glemser, O., Krebs, B., Wegener, J. and Kindler, E. (1969). *Angew. Chem. Int. Ed. Engl.*, **8**, 598
40. Fritz, G. and Pfannerer, F. (1970). *Z. Anorg. Allg. Chem.*, **373**, 30
41. Brown, D., Hill, J. and Rickard, C. E. F. (1970). *J. Chem. Soc. A*, 476
42. Schmidt, M. and Block, H. D. (1970). *Z. Anorg. Allg. Chem.*, **377**, 305
43. Stehle, V., Brini, M. and Pousse, A. (1969). *Bull. Soc. Chim. Fr.*, 2171
44. Siebert, W. (1970). *Angew. Chem. Int. Ed. Engl.*, **9**, 734
45. Siebert, W. (1970). *Chem. Ber.*, **103**, 2308
46. Siebert, W., Rittig, F. R. and Schmidt, M. (1970). *J. Organometal. Chem.*, **25**, 305
47. Nölh, H. and Petz, W. (1969). *J. Organometal. Chem.*, **20**, 35
48. Wang, T., Busse, P. J. and Niedenzu, K. (1970). *Inorg. Chem.*, **9**, 2150
49. Schmidt, M. and Block, H. D. (1970). *J. Organometal. Chem.*, **25**, 17
50. Summerfield, C. and Wade, K. (1969). *J. Chem. Soc. A*, 1487
51. Jennings, J. R., Pattison, I. and Wade, K. (1969). *J. Chem. Soc. A*, 565
52. Sommer, L. H., Citron, J. D. and Parker, G. A. (1969). *J. Amer. Chem. Soc.*, **91**, 4729
53. Niedenzu, K., Blick, K. E. and Miller, C. D. (1970). *Inorg. Chem.*, **9**, 975
54. Airey, W. and Sheldrick, G. M. (1970). *J. Inorg. Nucl. Chem.*, **32**, 1827
55. Wolfe, D. F. and Humphrey, G. L. (1969). *J. Mol. Struct.*, **3**, 293
56. Riess, J. G. (1969). *Rec. Chim. Miner*, **6**, 643
57. Reed, P. R. and Lovejoy, R. W. (1970). *Spectrochim. Acta*, **26A**, 1087
58. Ginn, S. G. W., Reichman, S. and Overend, J. (1970). *Spectrochim. Acta*, **26A** 291
59. Utyanskaya, E. Z. and Vinnik, M. I. (1969). *Zh. Fiz. Khim.*, **43**, 2863
60. Nelson, W. and Gordy, W. (1969). *J. Chem. Phys.*, **51**, 4710
61. MacNeil, K. A. G. and Thynne, J. C. J. (1969). *Inorg. Nucl. Chem. Lett.*, **5**, 1009
62. Lappert, M. F., Litzow, M. R., Pedley, J. B., Riley, P. N. K., Spalding, T. R. and Tweedale, A. (1970). *J. Chem. Soc. A*, 2320
63. Dibeler, V. H. and Walker, J. A. (1969). *Inorg. Chem.*, **8**, 50
64. Kaplansky, M. and Whitehead, M. A. (1970). *Can. J. Chem.*, **48**, 697
65. Fieldhouse, S. A. and Peat, I. R. (1969). *J. Phys. Chem.*, **73**, 275
66. Casabella, P. A. and Oga, T. (1969). *J. Chem. Phys.*, **50**, 4814
67. Bacon, J., Gillespie, R. J., Hartman, J. S. and Rao, U. R. K. (1970). *Mol. Phys.*, **18**, 561
68. Allerhand, A. (1970). *J. Chem. Phys.*, **52**, 3596
69. Holliday, A. K. and Massey, A. G. (1962). *Chem. Rev.*, **62**, 303
70. Ritter, J. J. and Coyle, T. D. (1970). *J. Chem. Soc. A*, 1303
71. Kane, J. and Massey, A. G. (1970). *Chem. Commun.*, 378
72. Lanthier, G. F. and Massey, A. G. (1970). *J. Inorg. Nucl. Chem.*, **32**, 1807
73. Nimon, L. A., Seshadri, K. S., White, D. and Taylor, R. C. (1969). *U.S. Govt. Res. Develop. Rep.*, **69**, 88
74. Ryan, R. R. and Hedberg, K. (1969). *J. Chem. Phys.*, **50**, 4986
75. Harris, J. J. and Ruder, B. (1969). *Inorg. Chem.*, **8**, 1258
76. Elegant, L., Gal. J. F. and Azzaro, M. (1969). *Bull. Soc. Chim. Fr.*, 4273
77. Elegant, L., Wolf, R. and Azzaro, M. (1969). *Bull. Soc. Chim. Fr.*, 4269
78. Biallas, M. J. (1969). *J. Amer. Chem. Soc.*, **91**, 7290
79. Clark, M. J. R. and Lynton, H. (1969). *Can. J. Chem.*, **47**, 2943
80. Barefield, E. K. and Busch, D. H. (1970). *Chem. Commun.*, 522
81. DeBoer, B. G., Zalkin, A. and Templeton, D. H. (1969). *Inorg. Chem.*, **8**, 836
82. Harrison, P. G. and Zuckermann, J. J. (1970). *J. Amer. Chem. Soc.*, **92**, 2577
83. Wharf, I. and Shriver, D. F. (1970). *J. Inorg. Nucl. Chem.*, **32**, 1831
84. Topchii, V. A. and Borovikov, Yu. Ya. (1969). *Zh. Obshch. Khim.*, **39**, 2689

85. Sevast'yanova, T. G., Romm, I. P., Gur'yanova, E. N. and Kolli, I. D. (1969). *Zh. Obshch. Khim.*, **39**, 1182
86. Jander, J. and Boerner, D. (1969). *Justus Liebigs. Ann. Chem.*, **726**, 13
87. Azeem, M., Brownstein, M. and Gillespie, R. J. (1969). *Can. J. Chem.*, **47**, 4519
88. Alder, R. W. and Taylor, F. J. (1970). *J. Chem. Soc. B*, 845
89. Elegant, L., Azzaro, M., M-Favelier, R., Marvel, G. and Morize, N. (1969). *Org. Mag. Res.*, **1**, 471
90. Dunell, B. A., Fyfe, C. A., McDowell, C. A. and Ripmeester, J. (1969). *Trans. Faraday Soc.*, **65**, 1153
91. Henriksson, U. and Forsen, S. (1970). *Chem. Commun.*, 1229
92. Greenvald, A. and Rabinovitz, M. (1969). *Chem. Commun.*, 642
93. Fratiello, A. and Schuster, R. E. (1969). *Inorg. Chem.*, **8**, 480
94. Tuchagues, J. P., Laurent, J. P. and Gallais, F. (1969). *C.R.H. Acad. Sci., Ser. C*, **268**, 2125
95. Taylor, R. C., Gabelnick, H. S., Aida, K. and Amster, R. L. (1969). *Inorg. Chem.*, **8**, 605
96. Taillandier, M. and Taillandier, E. (1969). *Spectrochim. Acta*, **25A**, 1807
97. Swanson, B. and Shriver, D. F. (1970). *Inorg. Chem.*, **9**, 1406
98. Gillespie, R. J. and Morton, M. J. (1970). *Inorg. Chem.*, **9**, 811
99. Gillespie, R. J. and Whitla, A. (1970). *Can. J. Chem.*, **48**, 657
100. Forel, M. T., Fouassier, M. and Tranquille, M. (1970). *Spectrochim. Acta*, **26A**, 1716
101. Forel, M. T., Tranquille, M. and Fouassier, M. (1970). *Spectrochim. Acta*, **26A**, 1777
102. Sauka, J., Martinsons, V. and Bruners, V. (1969). *Latv. PSR Zinat. Akad. Vestis, Kim. Ser.*, 157
103. McAllister, T. and Mackle, H. (1969). *Trans. Faraday Soc.*, **65**, 1734
104. Estrin, Ya. I. and Entelis, S. G. (1969). *Vyosokomol. Soedin., Ser. A*, **11**, 1133
105. Kuzaev, A. I., Komratov, G. N., Korovina, G. V. and Entelis, S. G. (1969). *Vysokomol. Soedin., Ser. A*, **11**, 989
106. Zubkova, Z. A., Moshinski, L. Ya. and Romantsevich, M. K. (1969). *Plast. Massy.*, **3**, 24
107. Kenaga, D. L. (1969). *Off. Gaz.*, **3**, 428, 484
108. Chandler, J. F., Dobinson, D., Johnston, E., Jones, M. E. D., Martin, R. J. and Stark, B. P. (1969). *Brit. Polym. J.*, **1**, 208
109. Shapet'ko, N. N., Kurkovskaya, L. N., Medvedeva, V. G., Skoldinov, A. P. and Vasyanina, L. K. (1969). *Zh. Oshch. Khim.*, **39**, 936
110. Medvedeva, V. G., Skoldinov, A. P. and Shapet'ko, N. N. (1969). *Zh. Obshch. Khim.*, **39**, 460
111. Rother, E. (1969). *German Pat.*, 1, 300, 103
112. Hamilton, D. J. and Price, M. J. (1969). *Chem. Commun.*, 414
113. Blackborow, J. R. and Lockhart, J. C. (1969). *J. Chem. Soc. A*, 816
114. Patton, R. L. and Jolly, W. L. (1969). *Inorg. Chem.*, **8**, 1392
115. Paul, R. C. and Paul, K. K. (1969). *Aust. J. Chem.*, **22**, 847
116. Paul, R. C. and Chadha, S. L. (1970). *J. Inorg. Nucl. Chem.*, **32**, 1038
117. H-Olivé, G. and Olivé, S. (1969). *J. Organometal. Chem.*, **17**, 83
118. Peach, M. E. (1969). *Can. J. Chem.*, **47**, 1675
119. Ryschkewitsch, G. E. and Wiggins, J. W. (1970). *J. Amer. Chem. Soc.*, **92**, 1790
120. Laussac, J. P., Jugie, G. and Laurent, J. P. (1969). *C.R.H. Acad. Sci. Ser. C*, **269**, 698
121. Fahr, E. and Neumann, L. (1969). *Annal.*, **721**, 14
122. Paul, R. C. and Chandra, S. L. (1969). *Aust. J. Chem.*, **22**, 1381
123. Paul, R. C., Gupta, S. G., Ahleuvalia, S. C. and Parkash, R. (1969). *Indian J. Chem.*, **7**, 631
124. Schmidt, M. and Rittig, F. R. (1970). *Chem. Ber.*, **103**, 3343
125. Yim, C. T. and Gilson, D. F. R. (1970). *Can. J. Chem.*, **48**, 515
126. Hartman, J. S. and Miller, J. M. (1969). *Inorg. Nucl. Chem. Lett.*, **5**, 831
127. Swanson, B., Shriver, D. F. and Ibers, J. A. (1969). *Inorg. Chem.*, **8**, 2182
128. Clippard, P. H. and Taylor, R. C. (1969). *Inorg. Chem.*, **8**, 2802
129. Blackborow, J. R. (1969). *J. Chem. Soc. C*, 739
130. Hensen, K., Messer, K. P. and Pickel, P. (1970). *Chem. Ber.*, **103**, 2091
131. Schmidt, M. and Block, H. D. (1970). *Chem. Ber.*, **103**, 3705
132. Paul, R. C., Parkash, R. and Subhash, S. C. (1969). *J. Indian Chem. Soc.*, **46**, 525
133. Meller, A. and Osska, A. (1969). *Monatsh. Chem.*, **100**, 1187

134. McMullen, J. C. and Miller, N. E. (1970). *Inorg. Chem.*, **9,** 2291
135. Ahmed, I. Y. and Schmulbach, C. D. (1969). *Inorg. Chem.*, **8,** 1411
136. Schmulbach, C. D. and Ahmed, I. Y. (1969). *Inorg. Chem.*, **8,** 1414
137. Boron-Nitrogen Chemistry. (1964). *Advan. Chem. Set.*, *No. 42, Amer. Chem. Soc.*
138. Niedenzu, K. and Dawson, J. W. (1965). *Boron-Nitrogen Compounds.* (Springer Verlag)
139. Nöth, H. (1964). *Progress in Boron Chemistry*, **3,** 211. (London: Pergammon Press)
140. Ruff, J. K. (1966). *Develop. Inorg. Nitrogen Chem.*, **1,** 470. (Amsterdam: Elsevier)
141. Finch, A., Leach, J. B. and Morris, J. H. (1969). *Organometal. Chem. Rev.*, **4,** 1
142. Payne, D. A. and Eads, E. A. (1964). *J. Chem. Educ.*, **41,** 334
143. Gerrard, W. (1966). *Chem. Ind. (London)*, 832
144. Maitlis, P. M. (1962). *Chem. Rev.*, **62,** 223
145. Paetzold, P. I. (1967). *Fortschr. Chem. Forsch.*, **8,** 437
146. Gutmann, V. and Meller, A. (1965). *Osterr, Chem. Ztg.*, **66,** 324
147. Mikhailov, B. M. (1960). *Russ. Chem. Rev.*, 459
148. Sheldon, J. C. and Smith, B. C. (1960). *Quart. Rev. Chem. Soc.*, **14,** 200
149. Mellon, E. K. and Lagowski, J. J. (1963). *Advan. Inorg. Chem. Radiochem. Chem.*, **5,** 259. (London: Academic Press)
150. Geanangel, R. A. and Shore, S. G. (1966). *Prep. Inorg. React.*, **3,** 123. (London: Interscience)
151. Horiguchi, G. (1969). *Jap. Pat.*, **6,** 901, 699
152. Coles, N. G., Glasson, D. R. and Jayaweera, S. A. A. (1969). *J. Appl. Chem.*, **19,** 178
153. Saito, H. and Ushio, M. (1969). *Yogyo Kyokai Shi.*, **77,** 151
154. Moore, A. W. (1969). *Nature (London)*, **221,** 1133
155. Schneider, R., Wasels, H., Schneider, V. and Wasels, K. P. (1969). *Ger. Pat.*, **1,** 286, 805
156. Grulke, C. A. (1970). *Off. Gaz.*, **3,** 495, 955
157. Abeler, G., Bayrhuber, H. and Nöth, H. (1969). *Chem. Ber.*, **102,** 2249
158. Niedenzu, K., Busse, P. J. and Miller, C. D. (1970). *Inorg. Chem.*, **9,** 977
159. Nöth, H. and Regnet, W. (1969). *Chem. Ber.*, **102,** 167
160. Nöth, H. and Regnet, W. (1969). *Chem. Ber.*, **102,** 2241
161. Ungurenasu, C., Cihoclarn, St. and Popescu, I. (1969). *Tetrahedron Lett.*, **18,** 1435
162. Geisler, I. and Nöth, H. (1969). *Chem. Commun.*, 775
163. Nöth, H. and Vahrenkamp, H. (1969). *J. Organometal. Chem.*, **16,** 357
164. Paetzold, P. I. and Maier, G. (1970). *Chem. Ber.*, **103,** 281
165. Wiberg, N. and Schwenk, G. (1969). *Angew. Chem. Int. Ed. Engl.*, **8,** 755
166. Nöth, H. and Sprague, M. J. (1970). *J. Organometal. Chem.*, **22,** 11
167. Hermannsdörfer, K. H., Matejcikova, E. and Nöth, H. (1970). *Chem. Ber.*, **103,** 516.
168. Köster, R., Bellut, H., Benedikt, G. and Ziegler, E. (1969). *Annal.*, **724,** 34
169. Dorokhov, V. A. and Mikhailov, B. M. (1969). *Dokl. Akad. Nauk SSSR*, **187,** 1300
170. Summerfield, C. and Wade, K. (1969). *J. Chem. Soc. A*, 1487
171. Dorokhov, V. A. and Lappert, M. F. (1969). *J. Chem. Soc. A*, 433
172. Summerfield, C. and Wade, K. (1970). *J. Chem. Soc. A*, 2010
173. Clark, A. H. and Anderson, G. A. (1969). *Chem. Commun.*, 1082
174. Bullen, G. J. and Clark, N. H. (1969). *J. Chem. Soc. A*, 992
175. Hess, H. (1969). *Acta. Crystallogr.*, **25B,** 2338
176. Hess, H. (1969). *Acta. Crystallogr.*, **25B,** 2334
177. Hess, H. (1969). *Acta. Crystallogr.*, **25B,** 2342
178. Chang, C. H., Porter, R. F. and Bauer, S. H. (1969). *Inorg. Chem.*, **8,** 1677
179. Bullen, G. J. and Clarke, N. H. (1969). *J. Chem. Soc. A*, 404
180. Dewar, M. J. S. and Rona, P. *J. Amer. Chem. Soc.*, **91,** 2259
181. Imbery, D., Jaeschke, A. and Friebolin, H. (1970). *Org. Mag. Res.*, **2,** 271
182. Scott, K. N. and Brey, W. S. (1969). *Inorg. Chem.*, **8,** 1703
183. Ryschkewitsch, G. E. and Rademaker, W. J. (1969). *J. Mag. Res.*, **1,** 584
184. Wiedermann, K. and Voitlaender, J. (1969). *Z. Naturforsch. B.*, **24A,** 566
185. Bürger, H. and Hoefler, F. (1970). *Spectrochim. Acta*, **26A,** 31
186. Kotz, J. C., Zanden, R. J. V. and Cooks, R. G. (1970). *Chem. Commun.*, 923
187. Gingerich, K. A. (1969). *Chem. Commun.*, 764
188. Catlin, J. C. and Snyder, H. R. (1969). *J. Org. Chem.*, **34,** 1664
189. Oertel, M. and Porter, R. F. (1970). *Inorg. Chem.*, **9,** 904
190. Beachley, O. T. (1969). *Inorg. Chem.*, **8,** 2665
191. Meller, A., Wojnowska, M. and Marecek, H. (1969). *Monatsh. Chem.*, **100,** 175
192. Nöth, H. and Sprague, M. J. (1970). *J. Organometal. Chem.*, **23,** 323

193. Clement, R. and Proux, Y. (1969). *Bull. Soc. Chim. Fr.*, 558
194. Armstrong, D. R. and Clark, D. T. (1970). *Chem. Commun.*, 99
195. Veillard, A. (1969). *Chem. Phys. Lett.*, **3**, 128
196. Blick, K. E., Dawson, J. W. and Niedenzu, K. (1970). *Inorg. Chem.*, **9**, 1416
197. Graffeuil, M. and Labarre, J. F. (1969). *J. Chim. Phys. Physicochim. Biol.*, **66**, 177
198. Totani, T., Watanabe, H. and Kubo, M. (1969). *Spectrochim. Acta*, **25A**, 585
199. Huler, E., Silberman, E. and Jones, E. A. (1970). *Spectrochim. Acta*, **26A**, 2241
200. Harshbarger, W., Lee, G. H., Porter, R. F. and Bauer, S. H. (1969). *Inorg. Chem.*, **8**, 1683
201. Harshbarger, W., Lee, G. H., Porter, R. F. and Bauer, S. H. (1969). *J. Amer. Chem. Soc.*, **91**, 551
202. Amberger, E. and Stoeger, W. (1969). *J. Organometal. Chem.*, **17**, 187
203. Rokoz, A. and Madej, A. (1969). *Chem. Anal. (Warsaw)*, **14**, 995
204. Bally, I. and Balaban, A. T. (1969). *Stud. Cercet. Chim.*, **17**, 431
205. Segel, B. (1968). *Inorg. Chim. Acta Rev.*, **2**, 137
206. Potnis, S. P. and Khanolkar, A. G. (1969). *Pop. Plast.*, **14**, 21
207. Matteson, D. S. (1970). *Accounts Chem. Res.*, **3**, 186
208. Schumacher, J. C., Rado, T. A. and Fairchild, J. L. (1969). *Off. Gaz.*, 3, 450, 497
209. Morley, H. B., Skrzec, A. R. and Shiloff, J. C. (1969). *Off. Gaz.*, 3, 454, 359
210. Cromwell, T. M. and Sprague, R. W. (1969). *Ger. Pat.*, 1, 803, 912
211. Peterson, K. (1969). *Off. Gaz.*, 3, 445, 430
212. Schmidt, A., Weinrotter, F. and Mueller, W. (1969). *Ger. Pat.*, 1, 294, 386
213. Panchout, S. (1969). *Chim. Anal. (Paris)*, **51**, 490
214. Morrison, M. E. and Scheller, K. (1969). *Combust. Flame*, **13**, 93
215. Nakamura, S. (1969). *Jap. Pat.*, 01, 697
216. Fiksel, L. I. and Zaripov, R. K. (1969). *Tr. Khim. Met. Inst. Akad. Nauk. Kaz. S.S.R.*, **8**, 39
217. Köster, R. (1969). *Ger. Pat.*, 1, 294, 964
218. Nencetti, G. and Tartarelli, R. (1969). *Ann. Chim. (Rome)*, **59**, 727
219. Eméleus, H. J., Spaziante, P. M. and Williamson, S. M. (1970). *J. Inorg. Nucl. Chem.*, **32**, 3219
220. Bowie, R. A. and Musgrave, O. C. (1970). *J. Chem. Soc. C*, 485
221. Gerbert, G. R. and Maslennikov, V. P. (1920). *Zh. Obshch. Khim.*, **40**, 1105
222. Maslennikov, V. P., Gerbert, G. P. and Khodalev, G. F. (1969). *Zh. Obshch. Khim.*, **38**, 1893
223. Coutts, I. G. C., Goldschmid, H. R. and Musgrave, O. C. (1970). *J. Chem. Soc. C*, 488
224. Coutts, I. G. C. and Musgrave, O. C. (1970). *J. Chem. Soc. C*, 2225
225. Lanthier, G. F. and Graham, W. A. G. (1969). *Can. J. Chem.*, **47**, 569
226. Cummings, W. M., Cox, C. H. and Snyder, H. R. (1969). *J. Org. Chem.*, **34**, 1669
227. Kulacheva, V. G., Ben'kovskii, V. G. and Svarcs, E. (1969). *Latv. PSR Zinat. Akad. Vestis, Khim. Ser.*, 152
228. Hohaus, E. and Umland, F. (1969). *Naturwissenschaften*, **56**, 636
229. Hohaus, E. and Umland, F. (1969). *Chem. Ber.*, **102**, 4025
230. Reynolds, G. A. and VanAllan, J. A. (1969). *J. Heterocycl. Chem.*, **6**, 375
231. Shapatin, A. S., Popkov, K. K., Nudel'man, Z. N., Golubtsev, S. A. and Markina, R. F. (1969). *Zh. Obshch. Khim.*, **39**, 554
232. Anderson, S. I. and Shreeve, J. M. (1970). *Inorg. Nucl. Chem. Lett.*, **6**, 1
233. Gertsev, V. V., Vladimirova, L. A. and Karyakin, A. V. (1969). *Zh. Obshch. Khim.*, **39**, 1558
234. Burg, A. B. and Bais, J. S. (1969). *J. Amer. Chem. Soc.*, **91**, 1937
235. Ogata, Y. and Ukigai, T. (1969). *J. Chem. Soc. C*, 2413
236. Gertsev, V. V. (1969). *Zh. Vses. Khim. Obschest.*, **14**, 352
237. Paul, R. C., Narula, S. P. and Makhni, H. S. (1970). *J. Inorg. Nucl. Chem.*, **32**, 3125
238. Casanova, J., Kiefer, H. R. and Williams, R. E. (1969). *Org. Prep. Proceedings*, **1**, 57
239. Wade, R. C. (1970). *Off. Gaz.*, 3, 502, 703
240. Gerbert, G. P., Maslennikov, V. P. and Shushunov, V. A. (1970). *Zh. Obshch. Khim.*, **40**, 131
241. Grekovich, A. L. and Materova, A. E. (1970). *Zh. Neorg. Khim.*, **15**, 187
242. Grekovich, A. L. and Materova, A. E. (1969). *Vestn. Leningrad Univ., Fiz. Khim.*, 127
243. Kalacheva, V. G., Ben'Kovskii, V. G. and Svarcs, G. E. (1969). *Latv. PSR Zinat. Akad. Vestis, Khim. Ser.*, 149
244. Havel, J., Havelkova, L. and Bartusek, M. (1969). *Chem. Zvesti*, **23**, 582

245. Kalacheva, V. G., Gabdzhanov, G. Z., Ni, E. A. and Karazhanov, N. A. (1969). *Zh. Neorg. Khim.,* **14,** 683
246. Kalacheva, V. G., Svarcs, E., Ben'kovskii, V. G. and Guseva, G. P. (1969). *Latv. PSR, Zinat. Akad. Vestis, Kim. Ser.,* 656
247. Russel, J. L. (1969). *Ger. Pat.,* 1, 908, 837
248. Cyba, H. A. (1969). *Off. Gaz.,* 3, 445, 498
249. Cyba, H. A. (1969). *Off. Gaz.,* 3, 446, 808
250. Holoch, K. E., Katchman, A. and Schufelt, R. A. (1969). *Off. Gaz.,* 3, 450, 670
251. Pelter, A. and Levitt, T. E. (1970). *Tetrahedron,* **26,** 1545
252. Bossa, M. and Maraschini, F. (1970). *J. Chem. Soc. A,* 1416
253. Coulson, C. A. (1969). *Acta Crystallogr.,* **25B,** 807
254. Medvinskii, A. A. and Bulgakov, N. N. (1969). *Zh. Strukt. Khim.,* **10,** 304
255. Santucci, L. and Triboulet, C. (1969). *J. Chem. Soc. A,* 392
256. Dang, Q. Q. (1969). *C.R.H. Acad. Sci., Ser. C,* **269,** 1562
257. Chang, C-H., Porter, R. F. and Bauer, S. H. (1969). *Inorg. Chem.,* **8,** 1689
258. Kaldor, A., Pines, I. and Porter, R. F. (1969). *Inorg. Chem.,* **8,** 1418
259. Fleisher, E. D. and Dewar, R. (1970). *Tetrahedron Lett.,* 363
260. Krogh-Moe, J. (1969). *J. Non Cryst. Solids,* **1,** 269
261. Meller, A. and Wojnowska, M. (1969). *Monatsh. Chem.,* **100,** 1489
262. Morterra, C. and Low, M. J. D. (1969). *Chem. Commun.,* 862
263. Valyashko, M. G. and Vlasova, E. V. (1969). *Jena Rev.,* **14,** 3
264. Suknev, V. S. (1970). *Zh. Prikl. Spektrosk.,* **12,** 76
265. Grimm, F. A. and Porter, R. F. (1969). *Inorg. Chem.,* **8,** 731
266. Kalacheva, G. V., Svarcs, E., Ben'kovskii, V. G. and Lesnov, I. D. (1970). *Zh. Neorg. Khim.,* **15,** 401
267. Laurent, J. P., Cros, G. and Pasdeloup, M. (1970). *Bull. Soc. Chim. Fr.,* 836
268. Lanthier, G. F. and Graham, W. A. G. (1969). *Inorg. Chem.,* **8,** 172
269. Svanson, S. E. and Johansson, R. (1969). *Acta. Chem. Scand.,* **23,** 635
270. Svanson, S. E. and Johansson, R. (1969). *Acta Chem. Scand.,* **23,** 628
271. How, M. J., Kennedy, G. R. and Mooney, E. F. (1969). *Chem. Commun.,* 267
272. Cragg, R. H. and Lockhart, J. C. (1969). *J. Inorg. Nucl. Chem.,* **31,** 2282
273. Allerhand, A. and Moll, R. E. (1969). *J. Mag. Res.,* **1,** 488
274. Pelter, A. and Levitt, T. E. (1969). *Chem. Commun.,* 1027
275. Gingerich, K. A. (1970). *Chem. Commun.,* 441
276. McKinley, I. R. and Weigel, H. (1970). *Chem. Commun.,* 1022
277. Cragg, R. H. and Todd, J. F. J. (1970). *Chem. Commun.,* 386
278. Jensen, D. E. (1970). *J. Chem. Phys.,* **52,** 3305
279. Fallon, P. J. and Lockhart, J. C. (1969). *Int. J. Mass Spectrom. Ion Phys.,* **2,** 247
280. Brooks, C. J. W., Harvey, D. J. and Middleditch, B. S. (1970). *Org. Mass Spectrom.,* **3,** 231
281. Meilleur, R. and Benoit, R. L. (1969). *Can. J. Chem.* **47,** 2569
282. Lerch, B. and Stegemann, H. (1969). *Anal. Biochem.,* **29,** 76
283. Marcantonatos, M., Gamba, G. and Monnier, D. (1969). *Helv. Chim. Acta,* **52,** 2183
284. Scholle, S. and Kniza, L. (1969). *Chem. Prum.,* **19,** 203
285. Edwards, J. O., Griscom, D. L., Jones, R. B., Walters, K. L. and Weeks, R. A. (1969). *J. Amer. Chem. Soc.,* **91,** 1095
286. Finch, A., Gardner, P. J., McNamara, P. M. and Wellum, G. R. (1970). *J. Chem. Soc. A,* 3339
287. Cragg, R. H. and Lappert, M. F. (1966). *Organometal. Chem. Rev.,* **1,** 43
288. Cragg, R. H. (1968). *Quart. Rep. Sulfur Chem.,* **3,** 1
289. Cragg, R. H. (1968). *Mech. of Reactions of Sulphur Cpds.,* **2,** 127
290. Cragg, R. H. (1969). *Mech. of Reactions of Sulphur Cpds.,* **3,** 29
291. Wiberg, E. and Sturm, W. (1955). *Angew. Chem.,* **67,** 483
292. Mikhailov, B. M. (1968). *Usp. Khim.,* **37,** 2121
293. Horn, H. G. (1970). *Chem. Ztg.,* **94,** 467
294. Rosso, J. C. and Dubusc, M. (1969). *Chim. Ind. Genie Chim.,* **102,** 405
295. Siebert, W., Schmidt, M. and Gast, E. (1969). *J. Organometal. Chem.,* **20,** 29
296. Schmidt, M. and Rittig, F. R. (1970). *Angew Chem. Int. Ed: Engl.,* **9,** 738
297. Siebert, W., Schaper, K. J. and Schmidt, M. (1970). *J. Organometal. Chem.,* **25,** 315
298. Siebert, W., Rittig, F. R. and Schmidt, M. (1970). *J. Organometal. Chem.,* **22,** 511

299. Siebert, W., Gast, E. and Schmidt, M. (1970). *J. Organometal. Chem.*, **23,** 329
300. Schmidt, M. and Sibert, W. (1969). *Chem. Ber.,* **102,** 2752
301. Lalancette, J. M. and Lanchance, A. (1969). *Can. J. Chem.,* **47,** 859
302. Young, D. W. (1969). *Off. Gaz.,* 3, 423, 359
303. Nöth, H. and Schweizer, P. (1969). *Chem. Ber.,* **102,** 161
304. Girdwell, C. (1969). *J. Inorg. Nucl. Chem.,* **31,** 1030
305. Cragg, R. H. and Husband, J. P. N. (1970). *Inorg. Nucl. Chem. Lett.,* **6,** 773
306. Cragg, R. H., Husband, J. P. N. and Weston, A. F. (1970). *Chem. Commun.,* 1701
307. Dean, F. M., Goodchild, J. and Hill, A. W. (1969). *J. Chem. Soc.,* 2192
308. Abel, E. W., Walker, D. J. and Wingfield, J. N. (1969). *Inorg. Nucl. Chem. Lett.,* **5,** 139
309. Pelter, A., Levitt, T. and Smith, K. (1969). *Chem. Commun.,* 435
310. Meller, A. and Ossko, A. (1970). *Monatsh. Chem.,* **101,** 1104
311. Cragg, R. H. (1969). *Chem. Commun.,* 832
312. Mikhailov, B. M. and Galkin, A. F. (1969). *Izv. Akad. Nauk, SSSR, Ser. Khim.,* 604
313. Grinberg, Ya. Kh., Zhukov, E. G. and Koryazhkin, V. A. (1969). *Dokl. Akad. Nauk, SSSR,* **184,** 847
314. Grinberg, Ya. Kh., Zhukov, E. G. and Koryazhkin, V. A. (1970). *Dokl. Akad. Nauk, SSSR,* **190,** 589
315. Chen, H-Y. and Gilles, P. W. (1970). *J. Amer. Chem. Soc.,* **92,** 2309
316. Melucci, R. C. and Wahlbeck, P. G. (1970). *Inorg. Chem.,* **9,** 1065
317. Gingerich, K. A. (1970). *Chem. Commun.,* 580
318. Brubaber, G. L. and Shore, S. G. (1969). *Inorg. Chem.,* **8,** 2804
319. Kotz, J. C., Zanden, R. J. V. and Cooks, R. G. (1970). *Chem. Commun.,* 923
320. Cragg, R. H., Gallagher, D. A., Husband, J. P. N., Lawson, G. and Todd, J. F. J. (1970). *Chem. Commun.,* 1562
321. Chopin, F. and Turrell, G. (1969). *J. Mol. Struct.,* **3,** 57
322. Bonnet, J. P., Jongler, C. and Laurent, J. P. (1970). *Bull. Soc. Chim. Fr.,* **6,** 2089
323. Zhukov, E. G. and Grinberg, Ya. Kh. (1969). *Izv. Akad. Nauk. SSSR, Neorg. Mater.,* **5,** 1646
324. Chen. H-Y., Conrad, B. R. and Gilles, P. W. (1970). *Inorg. Chem.,* **9,** 1776

6
Carbon and Silicon

J. SIMPSON
University of Otago, Dunedin

6.1	CARBON		223
	6.1.1	Introduction	223
	6.1.2	The element	223
		6.1.2.1 Diamond and graphite	223
		6.1.2.2 Newer forms of carbon	224
		(a) Pyrolytic carbon	224
		(b) Vitreous carbon	224
		(c) Carbon whiskers and fibres	224
		(d) Lamellar compounds of graphite	225
		6.1.2.3 Reactions of carbon vapour	225
	6.1.3	Molecular carbon compounds	225
		6.1.3.1 Carbon-nitrogen compounds	225
		(a) Cyanogen	225
		(b) Cyanogen halides	226
		(c) The cyanotrihydroborate anion	226
		6.1.3.2 Compounds of carbon with the Group VI elements	
		(a) Carbon dioxide and carbon disulphide	227
		(b) Carbon trioxide	228
		(c) Carbon suboxide and carbon subsulphide	228
		6.1.3.3 Carbon halides	229
		(a) Carbon tetrachloride and carbon tetrabromide	229
		(b) Carbon complexes	229
6.2	SILICON		230
	6.2.1	Introduction	230
	6.2.2	Bonding considerations	230
		6.2.2.1 Catenation in silicon compounds	230
		6.2.2.2 Bond energies in silicon compounds	230
		6.2.2.3 π-Bonding in silicon chemistry	231
		(a) $(p\text{-}p)$ π-bonding	231
		(b) $(p\text{-}d)$ π-bonding	231
		(c) $(d\text{-}d)$ π-bonding	232

		6.2.2.4	Stereochemistry of silicon compounds	233

        6.2.2.4   *Stereochemistry of silicon compounds*   233
                (a) *2- and 3-coordinate silicon*   233
                (b) *5- and 6-coordinate silicon*   233
   6.2.3   *The element*   234
        6.2.3.1   *Pure silicon*   234
        6.2.3.2   *Metallurgical silicon*   235
   6.2.4   *Silicon hydrides*   235
        6.2.4.1   *Preparative aspects*   235
        6.2.4.2   *Pyrolysis of the silanes*   236
        6.2.4.3   *Base-catalysed condensation reactions*   236
   6.2.5   *Compounds of silicon with the Group V elements*   236
        6.2.5.1   *Binary silicon–nitrogen compounds*   236
        6.2.5.2   *The silylamines*   237
                (a) *Preparative aspects*   237
                (b) *Structures*   237
                (c) *Function as a Lewis base*   238
        6.2.5.3   *Compounds with silicon–phosphorus, silicon–arsenic, and silicon–antimony bonds*   239
                (a) *Preparative aspects*   239
                (b) *Structures*   239
                (c) *Function as a Lewis base*   240
   6.2.6   *Compounds of silicon with the Group VI elements*   240
        6.2.6.1   *Binary compounds*   240
        6.2.6.2   *Hydride and halogen compounds*   241
                (a) *Preparative aspects*   241
                (b) *Structures*   241
                (c) *Reactions*   242
   6.2.7   *Halides and halogenosilanes*   242
        6.2.7.1   *Halides*   242
        6.2.7.2   *Halogenosilanes*   244
        6.2.7.3   *Complex compounds*   244
   6.2.8   *Compounds with bonds between silicon and a metal or metalloid*   245
        6.2.8.1   *Non-transition metal compounds*   245
                (a) *Alkali metal derivatives*   245
                (b) *Silicon–Group III compounds*   245
                (c) *Silicon–Group IV compounds*   246
        6.2.8.2   *Silicon–transition metal compounds*   246
                (a) *Preparative aspects*   246
                (b) *Reactions*   247

6.3   TABULAR SURVEY   248
   Table 6.1   New forms of carbon   249
   Table 6.2   Reactions of carbon vapour   250
   Table 6.3   Preparations and reactions of carbon compounds   251
   Table 6.4   Preparations and reactions of the silanes   254
   Table 6.5   Preparations and reactions of silicon–nitrogen compounds   256
   Table 6.6   Preparations and reactions of compounds containing

	silicon–phosphorus, silicon–arsenic, and silicon–antimony bonds	260
Table 6.7	Preparations and reactions of silicon–oxygen compounds	262
Table 6.8	Preparations and reactions of compounds of silicon with S, Se, Te	266
Table 6.9	Preparations and reactions of silicon halides and halogenosilanes	268
Table 6.10	Preparations and reactions of compounds of silicon with metals and metalloids	272
Table 6.11	Infrared and Raman studies on carbon and silicon compounds	278
Table 6.12	Nuclear magnetic resonance studies on silicon compounds	279
Table 6.13	Electronic spectra of carbon and silicon compounds	279
Table 6.14	Crystal structures of carbon and silicon compounds	280
Table 6.15	Mass and photoelectron spectra of carbon and silicon compounds	280
Table 6.16	Thermodynamic data for carbon and silicon compounds	281
Table 6.17	Molecular orbital calculations	281
Table 6.18	Miscellaneous studies	281
Table 6.19	Electron diffraction studies	282
Table 6.20	Supplementary references	283

ACKNOWLEDGEMENTS 283

6.1 CARBON

6.1.1 Introduction

The element carbon is notable among the Group IV elements in its almost limitless ability for catenation. This fact, together with the ready formation of bonds between carbon and most other elements of the Periodic Table, gives rise to the great multiplicity of organic and organometallic compounds. In this Section however, we shall be concerned with the Inorganic Chemistry of carbon, excluding such topics as organometallic chemistry and the carbides which will be covered elsewhere in the Review.

6.1.2 The element

6.1.2.1 Diamond and graphite

The best-known forms of elemental carbon, diamond and graphite, are notable for their commercial utility[1-6]. In recent years, a great deal of research effort has been directed to the production of synthetic diamonds from graphite, principally for industrial use. The processes involve extremely high temperatures and pressures and in some cases, the use of transition metal catalysts[7–14].

The use of graphite as a moderator in nuclear reactors has led to the production of numerous 'nuclear graphites'[15] differing in pore distribution and resistance to radiation damage and oxidation. Another fruitful area of graphite research has been the development of procedures to grow single graphite crystals[16–18]. This has enabled workers to collect a considerable volume of reliable physical data on this important material[19–27].

6.1.2.2 Newer forms of carbon

In addition to synthetic diamond and graphite, the demands of space technology, nuclear science, and related fields have led to the production of a number of new carbonaceous materials which have earned, for carbon, the accolade 'an old but new material'[28–31]. Of these recently-developed forms the most important are pyrolytic carbon, vitreous carbon, carbon whiskers and fibres, and the lamellar compounds of graphite.

(a) *Pyrolytic carbon* — Pyrolytic carbon, often referred to as pyrolytic graphite, is one of the two major types of graphitising carbon. It is prepared either by streaming a gaseous hydrocarbon over a heated substrate, or by suspending the substrate in a fluidised bed[32–34]. The second type of graphitising carbon, formed as a coke by solidification from the liquid or plastic state, has also been reviewed recently[35].

The structure of pyrolytic carbon differs from that of crystalline graphite in that crystals of the former are composed of imperfect hexagonally-bonded sheets of carbon atoms, with a greater spacing between the adjacent parallel planes than is found in graphite itself[26, 29, 36]. One of the major applications of pyrolytic carbon is in the production of dispersed nuclear fuel[37, 38], due to its high resistance to radiation damage and low porosity, which prevents passage of fission products. Pyrolytic carbon is also used as a material for rocket nozzles and nose cones, as it distributes heat evenly over the whole surface coating, at the same time insulating the less-refractory substrate material[29, 39]. Furthermore, its high resistance to attack by body fluids has led to its use for surgical implants[40].

(b) *Vitreous carbon* — Vitreous carbon is prepared by the pyrolysis of organic polymers, such as phenol formaldehyde resin, in which there is some degree of cross-linking[41–46]. The vitreous form has a low density and permeability compared to normal graphites, coupled with a higher elastic modulus and greater strength. Vitreous carbon is inert to a wide range of chemical agents, including hydrofluoric acid, and most strong acids and alkalis, and work is in progress to explore its potential for the manufacture of laboratory 'glassware' and vacuum apparatus[29].

(c) *Carbon whiskers and fibres* — The inherent brittleness of normal forms of graphite makes it impossible to use them under tensile forces. This lack of tensile strength has been attributed to the presence of microcracks in normal graphite[47, 48], since in the absence of such dislocations, the rupture of graphite materials should theoretically be dependent on the force required to break the strong covalent carbon–carbon bond.

Attempts to improve the tensile strength of carbon materials have led to the production of graphitic whiskers and carbon fibres. The former are prepared in a d.c. arc between carbon electrodes at high pressure[49], whilst fibres are obtained by the pyrolysis of organic polymers, such as rayon or polyacrylonitrile, in a dry inert atmosphere[50-60]. For both these materials, the absence of mobile dislocations[61, 62] raises the elastic moduli to values at least approaching that of the carbon–carbon bond.

The difficulty in production of carbon whiskers seems likely to preclude their large-scale commercial utilisation, but the production of carbon fibre-reinforcement laminates is already possible on an industrial scale, and the products have found numerous applications[63-66].

(d) *Lamellar compounds of graphite* — The ability of graphite to absorb metal atoms or radical species between the carbon sheets, giving so-called lamellar compounds, has been the subject of a number of reviews[67-72]. The absorbed species can be either electron donors or acceptors, and the resulting compounds either retain the electrical conductivity of graphite, often to an enhanced degree — intercalation compounds — or become non-conducting. Typical intercalating agents are the alkali metals and a number of metal oxides and halides, while non-conducting lamellar graphite is found only with graphite oxide and fluoride.

The use of intercalation compounds, in particular those of sodium and potassium, as polymerisation and isomerisation catalysts has excited considerable interest, and early problems of catalyst poisoning have now been overcome[71, 73, 74]. The phenomenon of intercalation has also been used to improve the bonding between carbon fibres and resin in a number of fibre-reinforced plastics[29].

6.1.2.3 Reactions of carbon vapour

In addition to significant advances in the understanding of the thermodynamic and kinetic aspects of carbon vaporisation[75], the use of carbon vapour in chemical synthesis has led to the production of a number of new chemical species[75-78]. For example, the co-condensation of carbon vapour, from a carbon arc, with diboron tetrachloride afforded a mixture of $C(BCl_2)_4$, $ClC(BCl_2)_3$ and $Cl_2C(BCl_2)_2$.

6.1.3 Molecular carbon compounds

6.1.3.1 *Carbon–nitrogen compounds*

The most active area of research involving carbon–nitrogen compounds is undoubtedly the use of cyanide, cyanate, and thiocyanate ions as ligands to transition metals, but coverage of this aspect is beyond the scope of this article. However, the search for improved polymeric materials and the requirements of synthetic organic chemistry have stimulated interest in several other aspects of carbon–nitrogen chemistry.

(a) *Cyanogen* — The chemistry of cyanogen and of its polymeric form, paracyanogen $(CN)_x$, has been reviewed[79-82]. The polymerisation of cyanogen has been shown to be extremely dependent on the purity of the gaseous dimer, no appreciable polymerisation occurring between 300–600 °C in the absence of impurities[83].

Phosphine derivatives of nickel(0), palladium(0), and platinum(0) are readily oxidised by treatment with cyanogen (Equation (6.1))[84].

$$(Ph_3P)_4Pt + (CN)_2 = (Ph_3P)_2Pt(CN)_2 + 2Ph_3P \qquad (6.1)$$

This is a rare example of an oxidative addition reaction in which a carbon–carbon bond is broken.

The spectral properties of the cyanogen radical have been widely studied[85]; a recent development in this field has been the use of the radical as a laser source in the gas phase[86].

(b) *Cyanogen halides* — The preparation of cyanogen fluoride by thermal degradation of cyanuric fluoride $(FCN)_3$ [87], completed the series of cyanogen halides[88]. Numerous industrial processes for the large-scale preparation of cyanogen chloride have been developed[89–94], as this is the precursor of the trimeric cyanuric chloride $(ClCN)_3$ which is of considerable importance in the manufacture of dyestuffs, insecticides, plastics, and synthetic rubbers[95].

(c) *The cyanotrihydroborate anion* — The lithium salt of the cyanotrihydroborate anion, $LiBH_3CN$, was first isolated by Wittig[96] in 1951. More recent work has shown it to be an extremely versatile reducing agent in organic chemistry[97], as it is stable in acid solution up to pH 3 and yet will reduce acid-sensitive functional groups quite readily at pH 7.

X-ray data[98] on the unsolvated lithium salt suggest a dimeric structure (Figure 6.1), in which the lithium ions are equidistant from, and coplanar

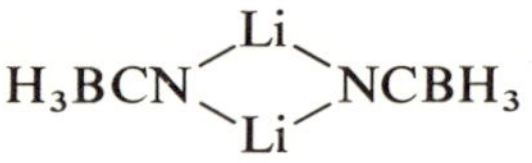

$$H_3BCN\underset{Li}{\overset{Li}{\diamondsuit}}NCBH_3$$

Figure 6.1 Structure of $LiBH_3CN$

with, the nitrogen atoms. Infrared and n.m.r. data on a series of cyanotrihydroborate salts and their deuterio derivatives have been investigated[99], and the results compared with data for other BH_3X compounds.

Surprisingly, the first report of the use of the BH_3CN^- ion as a ligand to transition metals did not appear until 1970 (Equation (6.2))[100].

$$(Ph_3P)_3AgCl + NaBH_3CN = (Ph_3P)_3Ag(NCBH_3) + NaCl \qquad (6.2)$$

Although the transition metals used in this investigation were known to form stable borohydride complexes involving hydrogen-bridged structures, infrared, conductivity, and magnetic data suggest that bonding to the transition metal in these compounds is through the nitrogen atom, giving a $M-NCBH_3$-type structure.

6.1.3.2 Compounds of carbon with the Group VI elements

Compounds of carbon are known with all the Group VI elements, but oxygen and sulphur derivatives are the most widely studied. Indeed the reaction between carbon and oxygen represents one of the most fundamental chemical processes known to man but is still a subject of considerable research interest[101-105]. In this Section, the chemistry of carbon monoxide and its role as a ligand in transition metal chemistry will not be considered as it will be dealt with elsewhere in this series.

(a) *Carbon dioxide and carbon disulphide* — The manufacture and uses of carbon dioxide have been the subject of a number of reviews[106-108]. The availability of carbon dioxide in both solid and liquid forms has greatly increased the efficiency of refrigerated transportation, while gaseous CO_2 is used to an increasing extent in the manufacture of rubber and plastic foams, and in large scale chemical production. The use of carbon disulphide as a solvent in industrial processes has declined markedly, due mainly to the hazards of working with a highly-flammable and toxic material. However, CS_2 is still produced on a large scale, chiefly by the catalytic reaction between natural gas and sulphur, for use in the manufacture of viscose rayon and cellophane, and in the preparation of carbon tetrachloride[109-111].

The ability of carbon dioxide, carbon disulphide, and the mixed species carbonyl sulphide (COS) to insert into a variety of element–element bonds has been widely investigated[112]. For example, the use of CS_2-insertion in the preparation of transition metal dithiocarbamate complexes is found to have considerable advantages over the more obvious preparative route from dithiocarbamate salts[113,114].

Treatment of tris-(triphenylphosphine)platinum(0) with carbon disulphide equation (6.3) was found by Wilkinson[115] to give an oxidative-addition reaction resulting in a platinum complex with CS_2 as a π-bonding ligand.

$$Pt(PPh_3)_3 + CS_2 = (Ph_3P)_2PtCS_2 + Ph_3P \qquad (6.3)$$

Reactions of this type have now been extended to include a number of other transition metal systems[116-120a], and a recent investigation[121] of the crystal structure of the original $(Ph_3P)_2PtCS_2$ shows that the coordinated CS_2 molecule is bent with an S—C—S bond angle of 136.2 degrees. This conformation closely parallels that expected for the free molecule in its first excited state, an observation which is explained in terms of Mason's theory[122,123] on the geometry of coordinated ligands. As well as forming π-complexes, the CS_2 molecule can act as a simple donor ligand through sulphur[117,118], and in the molecule $RhCl(CS_2)_2(PPh_3)_2$ both types of coordination are displayed[124].

Attempts to prepare the analogous π-CO_2 complexes have been less successful[125-129]. Thus, no reaction occurs between CO_2 and $(Ph_3P)_3Pt$ in the absence of oxygen, while admission of oxygen to the system leads to the formation of a carbonate complex $(Ph_3P)_2PtCO_3$ [125,126]. Iwashita and Hayata[129] however, report the reaction of a rhodium complex $Rh_2(CO)_4(PPh_3)_4$ with molecular oxygen to give a π-CO_2 complex, so assigned on the basis of infrared and analytical data.

A further advance in the field of carbon–sulphur chemistry came with the preparation of the first thiocarbonyl compound, $[RhCl(CS)(PPh_3)_2]$, in which the CS moiety replaces the well-known carbonyl group as a ligand. Since then, a number of other thiocarbonyl complexes have been reported[117,118,131-137], and in each case, carbon disulphide, or one of its derivatives, was used as a source of CS. The transient nature of carbon monosulphide[138] seems likely to preclude more direct routes to thiocarbonyl complexes, although reactions of trapped CS and its selenium analogue are known[139]. A molecular orbital calculation[140] of the energy of orbitals available to the CS molecule for bonding to a transition metal predicts that thiocarbonyl complexes should be more stable than their carbonyl analogues, but as yet there is little chemical evidence to support this suggestion.

(b) *Carbon trioxide* — The molecule carbon trioxide was first postulated in 1962 as an unstable intermediate to account for the rapid exchange between carbon dioxide and oxygen[141], and in 1966 the species was isolated in a solid CO_2 matrix and identified by infrared spectroscopy[142]. Preparation is effected by one of three processes. In the first, solid CO_2 at 77 K is photolysed, the second uses the photolysis of an ozone–CO_2 matrix at 50–60 K, while in the third process gaseous CO_2 is subjected to a radio-frequency discharge and the product is collected by a 'sweep-trapping' technique[143].

The structure of the carbon trioxide molecule has been the subject of speculation and a number of possibilities were suggested (Figure 6.2). Infrared

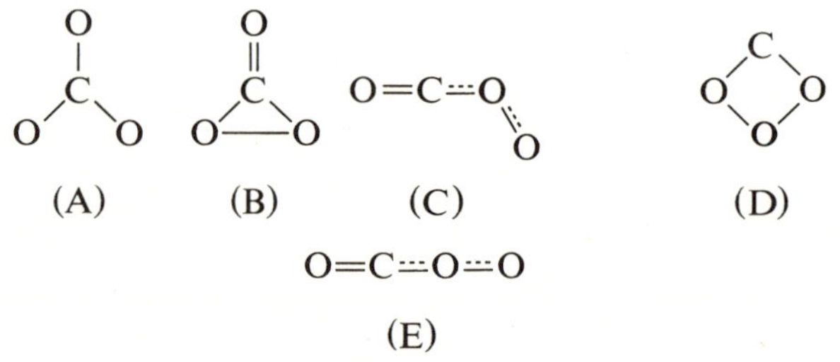

Figure 6.2 Possible structures for the CO_3 molecule

evidence favours the C_{2v} structure (B) and this is supported by *ab initio* calculations, which show that B is the lowest energy form[144, 145].

(c) *Carbon suboxide and carbon subsulphide* — The molecular structure of carbon suboxide[146–148] has been the subject of considerable discussion in the recent literature[149–153]. Although it is now generally accepted that the mole-

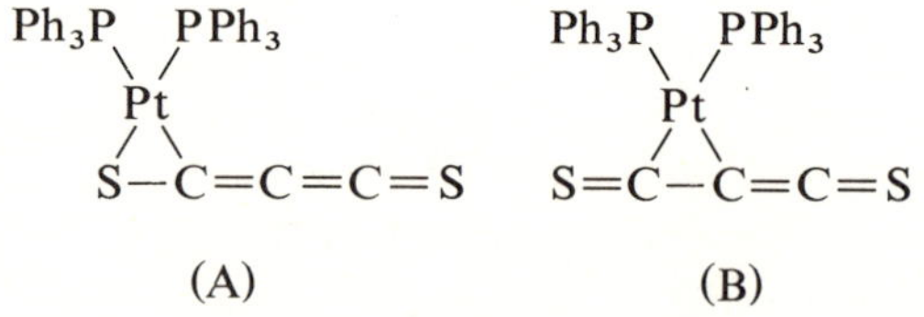

Figure 6.3 Possible structures for the $Pt(PPh_3)_2(C_3S_2)$ molecule

cule is linear, considerable disagreement exists as to the complete interpretation of its vibrational spectrum. In particular, the CCC-bending potential function is still the subject of lively debate[154–159]. A linear structure is also accepted for carbon subsulphide, but here again there is disagreement over the assignment of vibrational frequencies[160, 161].

Also of recent interest in the chemistry of these molecules has been the investigation of their potential as ligands. Blues and Bryce-Smith[162] have discussed the preparation and properties of silver complexes of carbon suboxide, while Ginsberg and Silverthorn[163, 164] report a number of oxidative-addition reactions involving carbon subsulphide. Typical of the latter is that between C_3S_2 and $Pt(PPh_3)_3$ (Equation (6.4)).

$$Pt(PPh_3)_3 + C_3S_2 \rightarrow Pt(PPh_3)_2(C_3S_2) + PPh_3 \qquad (6.4)$$

There are two possible structures for the resulting platinum complex (Figure 6.3) and structure (A), in which the bonding to platinum is similar to that in the π-carbon disulphide complexes, is favoured on the basis of infrared data. No information is as yet available regarding the conformation, linear or otherwise, of the C_3S_2 molecule in these complexes.

6.1.3.3 Carbon halides

(a) *Carbon tetrachloride and carbon tetrabromide* — The technology of carbon tetrachloride[165, 166] and carbon tetrabromide[167] has been reviewed. CCl_4 has been largely superseded as a solvent by perchloroethylene and other less hazardous chlorinated hydrocarbons. Its principal industrial use is now in the preparation of fluorinated chlorocarbons for use as refrigerants and propellants. The high density of CBr_4 makes it particularly suitable for use in mineral processing[167].

(b) *Carbon complexes* — An interesting extension to the inorganic chemistry of carbon has been the preparation of carbon complexes of the type $[H_2C\ diars]^{2+}[X^-]_2$ and $[C(diars)_2]^{4+}[X^-]_4$, where 'diars' is the bidentate ligand o-phenylene-bis-(dimethylarsine) and X = Br, I, and ClO_4 [168]. These compounds were isolated as white crystalline solids following the direct reaction of methylene halide or carbon tetrahalide with the di(tertiary arsine). Conductivity data were consistent with the ionic formulation of the complex species. Another cationic carbon complex is formed[169] by the reaction of carbon tetrachloride with the tetrachloroantimony azide dimer (Equation 6.5). An infrared spectrum of the complex indicates

$$3SbCl_4N_3 + CCl_4 = [C(N_3)_3]^+[SbCl_6]^- + 2SbCl_5 \qquad (6.5)$$

a cation with C_{3v} symmetry, having a coplanar arrangement of α-nitrogen atoms around the carbon.

The structures of a number of carbanion species have been investigated[170–174], including the planar tricyano-methanide ion $[C(CN)_3]^-$ [170] and the nitroform anion $[C(NO_2)_3]^-$, which is found to adopt a configuration dependent on the associated cation[173, 174]. Evidence has also been presented for the existence of the anion CCl_5^- [175]; and a number of reports of the carbon tetrahalides behaving as Lewis acids, with both anionic and neutral donors, have appeared[176–179].

6.2 SILICON

6.2.1 Introduction

The last decade has seen a tremendous upsurge of interest in the chemistry of silicon, spurred on by increased commercial utilisation of this abundant element and the intriguing problems posed by a multitude of new silicon compounds. The inorganic chemistry of silicon covers a very wide range of research interests and this article will, of necessity, exclude major topics such as solid-state silicon chemistry, the silicates, silicides, and the rapidly expanding field of organosilicon chemistry, all of which will be covered elsewhere in this series. In this context, an organometallic compound will be defined as one having a silicon–carbon bond, and will therefore, embrace compounds such as Me_3SiH and $EtSiCl_3$ which have both inorganic and organic ligands.

6.2.2 Bonding considerations

Because the position of silicon in the second row of the Periodic Table, there are very great differences between the chemistry of carbon and silicon. In particular, the availability of low lying d orbitals to silicon and the possibility of their involvement in bond formation has been used to explain some of these differences. For example, the facile formation of 5- and 6-coordinate silicon complexes, and the unexpected stereochemistry of a number of silicon–nitrogen and silicon–oxygen derivatives, is generally rationalised in terms of d orbital participation in the bonding. It is, however, possible to explain many of these observations using only the occupied atomic orbitals of the silicon atom[180]. These problems have been the *raison d'etre* of a large body of research in silicon chemistry for a number of years, and will doubtless remain so until a theory emerges which is acceptable to all protagonists in this challenging field.

6.2.2.1 Catenation in silicon compounds

The ability to form stable catenated compounds has long been regarded as a unique characteristic of the element carbon. More recent studies[181-193] have shown that the other Group IV elements, and silicon in particular, are capable of forming very stable chain and ring compounds, provided the substituents on silicon are carefully chosen. Thus, although the catenated silicon hydrides react readily with air and moisture due to the labile nature of the Si—H bond, polysilanes with the heavier halogens or organic groups on silicon are much more stable.

6.2.2.2 Bond energies in silicon compounds

Bond dissociation energies are extremely important to any discussion of chemical bonding as they provide the most obvious criterion of the strength of a bond between two elements. Unfortunately, the bond dissociation energy data available for silicon compounds are often contradictory and not infrequently in direct conflict with observed chemical behaviour[190, 192, 194], an obvious reflection of the variety of techniques employed by different workers. For example, for the Si—Si bond, values ranging from 205 to 360 kJ mol^{-1} are to be found in the literature for hexamethyldisilane, Me$_3$SiSiMe$_3$, while for disilane, Si$_2$H$_6$, itself the range is 213 to 351 kJ mol^{-1} [198-200].

However, in 1967 an elegant kinetic determination of the Si—Si bond dissociation energy in Me$_3$SiSiMe$_3$ [201, 202] yielded a value of 280±12.5 kJ mol^{-1}, and this has since been supported by mass spectrometric determinations[203, 204]. This seemingly consistent value has been used to determine dissociation energies for a number of Si—X bonds in organometallic systems[202, 205], but considerable work remains to be done before the bond energies of inorganic silicon compounds can be quoted with any degree of confidence.

6.2.2.3 π-Bonding in silicon chemistry

(a) *(p–p) π-bonding* — The apparent inability of silicon and its congeners in Group IV to form stable derivatives containing double and triple bonds analogous to those found in carbon compounds has been the subject of continued speculation. Although the existence of monomeric species such as SiO [206] and SiNH[207] has been established under what may be described as 'forcing' conditions, and (p–p) π-bonded intermediates have been postulated in a number of reactions[208-211], no stable silicon analogues of the olefins, aldehydes, etc. have been isolated.

The most common reasons advanced for the failure of silicon to form a (p–p) π-bond are that π-overlap with the more diffuse 3p orbitals is likely to be poor[212], and that repulsions between the inner electron shells precludes the close approach of atoms required to achieve a multiple bond[213]. Both of these explanations have been criticised[190, 194, 214-216], and there is a growing tendency to assign the lack of (p–p) π-bonding in silicon chemistry to the inherent strength of σ-bonds formed by silicon[217], and to the thermodynamic and kinetic instability of the potentially π-bonded monomer species relative to σ-bonded polymer or oligomer units.

(b) *(p–d) π-bonding* — The concept of d orbital involvement in bonding for the second row elements is by no means a new one[218, 219], and its application to silicon chemistry has been extensively reviewed[181, 190, 194, 216, 210-225]. The most quoted example of (p–d) π-bonding in inorganic silicon chemistry is undoubtedly for the molecule trisilylamine, (SiH$_3$)$_3$N, which was found by Hedburg[226] to have a planar heavy-atom skeleton in contrast to trimethylamine (CH$_3$)$_3$N which is pyramidal. This structure was rationalised by assuming that the nitrogen atom forms σ-bonds in a trigonal plane leaving the lone pair of electrons in a pure p orbital at right angles to this plane. The energy of this p orbital may be lowered by donating electron density from it into the correct symmetry vacant d orbitals of the adjacent silicon atom. The weak donor ability of the nitrogen atom in (SiH$_3$)$_3$N [227] offered evidence of the involvement of this lone pair in bonding to silicon.

An alteration in the bond angle at atoms adjacent to silicon, relative to the analogous carbon compound, is generally explained in terms of (p–d) π-bonding, as is any shortening of the Si—X bond compared with the expected single bond length[194]. In recent years, direct methods of structural determination, such as electron diffraction, have been used increasingly to obtain these types of data, since a number of structural assignments based on vibrational spectroscopy have been found to be in error[228–231a].

Since Ebsworth[232] has pointed out that there may be considerable overlap between a d orbital on silicon and a lone pair of electrons on an adjacent atom which occupies a tetrahedral site with respect to a central Si atom, it would be rash to conclude that drastic structural changes invariably accompany (p–d) π-bonding. Hence the possibility of π-character in the Si—P bond of trisilylphosphine $(SiH_3)_3P$ cannot be ruled out simply because it does not have a planar heavy-atom skeleton. In fact, the angle at phosphorus in this molecule is considerably less than the tetrahedral angle[231], which effectively precludes a (p–d) π-interaction as the lone-pair electrons must reside in what is essentially a pure s orbital. However some type of d orbital involvement is still to be inferred from the poor donor ability of the P atom, and the possibility of an (s–d) σ-interaction with the d_z^2 orbital on silicon has been discussed[233].

Apart from structural investigations, there are numerous other techniques which have been used to obtain evidence both for and against the (p–d) π-bond. In an excellent review of the subject, Ebsworth[194] examines manifestations of the (p–d) π-bond through the agency of vibrational spectroscopy, n.m.r., e.s.r., n.q.r., u.v. and visible spectra, dipole moments, and investigations of the acid–base properties of atoms and groups bound to silicon, in addition to the structural determinations already referred to. From this wealth of often conflicting evidence he concludes that (p–d) π-bonding is most likely to be important between silicon and the first row elements N, O, and F, while the position regarding heavier substituent atoms is uncertain. He also warns that a number of observations cannot be rationalised in terms of (p–d) π-bonding, although the majority can. Certainly these problems, with others yet unknown, will provide stimulus for further efforts in this area of silicon chemistry.

(c) *(d–d) π-bonding* — A rapid expansion in the chemistry of silicon bound to transition metals has occurred recently (see Section 6.2.8) and investigations of the physical and chemical properties of compounds containing a Si—M bond have led to the belief that silicon substituents have marked π-acceptor properties[234, 235]. Multiple-bond character in the Si—M bond of transition metal derivatives is thought to derive from overlap of the filled d orbitals of the transition metal with the vacant silicon d orbitals. Considerable support for this (d–d) π-bond theory has come from infrared studies, particularly of the effect on carbonyl stretching frequencies of competition between SiX_3 and carbonyl ligands for the d electrons of the transition metal[236, 238], although disagreements exist as to the degree of emphasis to be placed on (d–d) π-bonding in interpreting the results[239].

Structural determinations have been completed on a number of these compounds[240], and shortening of the Si—M bond relative to the theoretical single bond length has been taken as indicative of a (d–d) π-interaction. The major difficulty with this approach has been obtaining the initial estimate of the single bond length—derived from half the sum of the appropriate atomic radii—as no one set of values for the radii of the transition metals is generally accepted[234, 240].

Recent molecular orbital calculations on silylcobalt tetracarbonyls $[X_3SiCo(CO)_4; X = F,Cl]$ [240] are consistent with some degree of (d–d) π-character in the Si—Co bond. Two other novel observations result from this study. Evidence is presented for a second type of π-interaction (d→σ*) in the Si—Co bond, involving the cobalt $3d_{xy,yz}$ orbitals and the silicon $p_{x,y}$ orbitals. This type of bonding had been suggested previously to account for the stability of perfluoroalkyl transition metal derivatives[241, 242]. Furthermore, the calculation predicts a large bonding interaction between the silicon substituent and the carbon atoms of the equatorial carbonyl groups. This, together with a weak bonding interaction between the equatorial and apical carbon atoms, produces a cage-bonding effect over the molecule as a whole. Support for these intramolecular interactions comes from the fact that the equatorial carbonyl groups in $X_3SiCo(CO)_4$ are in fact bent towards the silicon substituents[243, 244], and also from a subsequent interpretation of their mass-spectral fragmentation patterns[245]. No obvious means is yet available to distinguish between (d–d) π- or (d–σ*) π-bonding and these intramolecular interactions. It may be therefore that effects such as bond shortening, previously attributed solely to (d–d) π-bonding, are due to a combination of these bonding phenomena.

6.2.2.4 *Stereochemistry of silicon compounds*

The most usual stereochemistry of simple molecular silicon compounds is tetrahedral, in keeping with the position of silicon in the Periodic Table[245, 247]. Even when the groups attached to silicon vary widely in size and complexity, departure from this tetrahedral arrangement is in general only slight. Other coordination numbers for the silicon atom are characterised by particular stereochemistries, and the determination of structures, especially of silicon complexes, has commanded considerable recent attention.

(a) *2- and 3-coordinate silicon*—Divalent silicon compounds do not exist as stable monomers under normal conditions of temperature and pressure. Despite this instability, a considerable amount of information has been collected on these species both in the gas phase and in solid matrices[248, 249]. The molecules SiX_2 (X = halogen, hydrogen) are found to have a bent ground-state configuration, with bond angles at silicon considerably less than the tetrahedral angle[249].

Trivalent silicon compounds have been identified as highly reactive radicals, and an e.s.r. study[250] of $SiH_3\cdot$, trapped in a krypton matrix, indicates a pyramidal structure in contrast to the planar methyl radical. Although evidence has been presented for the existence of a 3-coordinate anion $SiCl_3^-$ [251], no structural information is yet available.

(b) *5- and 6-coordinate silicon*—Considerable advances have been made in the last few years in the stereochemistry of 5-coordinate silicon compounds[247, 252–254]. The situation with regard to these compounds is often complicated by the fact that molecules such as $SiH_3I·NMe_3$ and $SiF_4·NMe_3$, which from stoichiometry are potentially 5-coordinate, are found either to be ionic $[SiH_3NMe_3]^+I^-$ [255] or to have a halogen-bridged structure[256].

In a review of the acceptor properties of the Group IV elements published in 1963, Beattie[257] remarked that of the numerous 1:1 adducts of 4-coordinate compounds with monodentate donor ligands only one ($Me_3SnCl·pyridine$) had a known stereochemistry. More recently however, examples of anionic (SiF_5^-)[258], neutral ($SiCl_4NMe_3$)[259], and cationic ($SiH_3·$ 2-pyridine)$^+$ [260] silicon complexes have been assigned trigonal-bipyramidal structures. There have also been numerous examples of 5-coordinate silicon species postulated as intermediates in the reactions of silicon compounds[183, 194, 247].

The position of octahedral 6-coordinate silicon compounds is somewhat better established[247], and again examples of anionic, neutral, and cationic species are known. Of particular recent interest has been the use of full x-ray examinations of 6-coordinate species to distinguish between *cis* and *trans* isomers[261, 262], often a difficult exercise by more conventional spectroscopic techniques[263, 264].

The role of d orbitals in σ-bond formation for 5- and 6-coordinate silicon compounds has been reviewed[194, 265, 266]. The observed stereochemistries can generally be explained on models favouring either complete d orbital involvement or the total neglect of d orbital contributions. A more realistic view[194] would seem to be that the true situation lies somewhere between these two extremes, and since an acceptable method of estimating the degree of d orbital participation in bonding has not yet been devised, this question too must be added to the list of imponderables regarding the part played by d orbitals in silicon chemistry.

6.2.3 The element

The technology of elemental silicon can be divided into two distinct parts: that of pure silicon, used for semiconductors; and metallurgical silicon, for use in silicon-based alloys, silicides, and as a starting material in the direct synthesis of organosilicon compounds[267-269].

6.2.3.1 Pure silicon

The level of impurities acceptable in silicon to be used for semiconductor devices is extremely low, being measured in parts-per-billion[267]. Silicon of this purity cannot be produced by normal metallurgical procedures, such as zone refining, and is generally obtained by the reduction of highly purified samples of trichlorosilane or silicon tetrachloride[270, 271]. The thermal decomposition of monosilane or silicon tetraiodide has also been used to prepare high-purity silicon, but technical difficulties have so far prevented the adoption of these processes on a commercial scale[267]. Thermal decomposition of SiH_4 is used in the deposition of thin silicon films, often with the addition of gaseous compounds such as phosphine, PH_3, or diborane, B_2H_6, to provide the doping elements, which alter the resistivity of the resulting semiconductor[270, 272].

6.2.3.2 Metallurgical silicon

Metallurgical silicon of 99% purity or less is produced by the reduction of silicon dioxide, in the form of quartzite rock, with coke or charcoal in an electric furnace. It is used extensively in the production of superior non-ferrous metal alloys[267–269, 273, 274].

The direct synthesis of organosilicon compounds, precursors of the extremely versatile silicone polymers, has been extensively reviewed[275–281]. The reaction between organic halides and silicon generally takes place at a temperature of 300 °C in the presence of a copper catalyst, but the use of 'active' silicon, prepared by the reaction of calcium silicide with chlorine[283], reduces the operating temperatures and in many cases obviates the use of a catalyst[279]. Recent improvements in the direct synthesis of alkoxysilanes, from silicon and alcohols, have come with the use of silicone oil suspensions of the silicon–copper mixture as reaction media[284]. The process appears to be dependent on the acidity of the organic species as, to date, successful conversion to alkoxysilanes has been achieved only with primary alcohols[284].

6.2.4 Silicon hydrides

The study of hydrides as a general compound class, and of the silicon hydrides in particular, are rapidly expanding research fields, as evidenced by the number of recent books and review articles devoted to the subject[285–294]. Work in this field has been encouraged by more general use of the vacuum techniques pioneered by Stock at the turn of the century[295], and by the fact that the simplicity of the hydrides and their derivatives render them particularly suitable for analysis by modern spectroscopic techniques[296, 297]. In this section, we shall consider the parent silanes, Si_nH_{2n+2}, while the chemistry of their Group V–VII and metal derivatives will be examined subsequently in the appropriate sections.

6.2.4.1 Preparative aspects

In a recent review of the methods of synthesis of the Group IV hydrides, Jolly and Norman[298] recommend reduction of silicon tetrachloride with lithium aluminium hydride as the most reliable method of preparing monosilane. Disilane[299] and trisilane[300, 301] may be obtained similarly from the appropriate chlorosilane, although great care must be exercised in maintaining the correct experimental conditions to prevent undue cleavage of the Si—Si bond[302, 303]. Higher hydrides of silicon can be prepared either by the acid hydroslysis of metal silicides[304–306], or by the use of electric-discharge reactions[307, 309]. In both cases, identification of the isomeric species from

Si_4H_{10} upwards is considerably aided by the use of gas chromatographic separation of the hydride mixtures[309-311].

6.2.4.2 Pyrolysis of the silanes

The pyrolysis of monosilane and its deuteriated analogue has been examined in detail. Purnell and Walsh[312] favoured a mechanism in which the formation of silene radicals (SiH_2) plays an important part. In a more recent study Ring et al.[313] dispute this view, arguing that SiH_2 radicals derive from secondary decomposition of the product disilane, and propose an alternative free-radical chain mechanism involving the initial decomposition of SiH_4 into silyl (SiH_3) radicals and hydrogen atoms. In contrast to these studies, the photochemical and electric-discharge decomposition of monosilane is thought to involve molecular hydrogen elimination with the formation of silene radicals[314].

The pyrolysis of disilane to give trisilane, and of trisilane to give tetrasilane[315], has been explained in terms of SiH_2-formation followed by silene insertion into an Si—H bond. Support for this mechanism for Si_2H_6 pyrolysis comes from co-pyrolysis studies. For example, in the co-pyrolysis of Si_2H_6 and Me_2SiH_2, silene insertion into a Si—H bond of dimethylsilane produces $Me_2SiHSiH_3$ as well as the Si_2H_6 pyrolysis products[316, 317].

6.2.4.3 Base-catalysed condensation reactions

The failure to obtain reasonable yields of disilane from the coupling reaction between potassium silyl and monobromosilane was attributed to the subsequent potassium halide-catalysed condensation of the Si_2H_6 product[318]. Condensation of Si_2H_6 to monosilane and a polymeric species $(SiH_2)_x$ has since been found to occur in the presence of a number of alkali metal halides and hydrides[319, 320]. In the hydride-catalysed reactions, metal silyls, $MSiH_3$, are formed concurrently[319], and a conductivity study[320] confirms the presence of conducting species in the system. The mechanism of these reactions was originally thought to involve silene radical species[319], but the failure to observe silene insertion into the Si—H bonds of organosilanes present in the reaction mixtures has cast serious doubts on this postulate[321]. A mechanism, similar to that proposed for the amine-catalysed condensation of hexachlorodisilane, has been suggested as an alternative, but this must rely on salt-catalysed hydrogen–deuterium exchange to explain the range of deuteriated monosilanes observed when Si_2H_6/Si_2D_6 mixtures react with LiCl[319, 321].

6.2.5 Compounds of silicon with the group V elements

6.2.5.1 Binary silicon–nitrogen compounds

Silicon nitride, Si_3N_4, belongs to the diamond-like class of nitrides characterised by predominantly covalent bonding[322, 323]. Interest in the use of silicon nitride films as insulators in the semiconductor field has developed during the last decade[324]. The films are generally deposited on the semiconductor substrate by decomposition of monosilane and nitrogen or ammonia either pyrolytically, or using electric discharge techniques[325-327]. Pyrolysis of tri-imidodisilane, $Si_2(NH)_3$, in the presence of ammonium chloride gave a new, moisture-sensitive nitride, Si_2N_2, thought to contain Si—Si bonds. Further pyrolysis of this material produced either SiN or Si_3N_4 depending on the conditions[328, 329].

6.2.5.2 The silylamines

This area of silicon–nitrogen chemistry has been the subject of a number of reviews[330-340], in most of which hydride and halide derivatives are considered in conjunction with the numerous organosilicon compounds with Si—N linkages. The inorganic cyclosilazanes are discussed in a comprehensive book by Haiduc[340] on inorganic and organometallic ring systems.

(a) *Preparative aspects*[341] – The reaction between monohalosilanes, SiH_3X, and ammonia to give trisilylamine, $(SiH_3)_3N$, was thought to proceed via the intermediate formation of mono- and di-silyl substituted amines, which then either condensed or reacted with excess SiH_3X to give the final product (Equations (6.6)–(6.8))[285, 287].

$$2SiH_3NH_2 \to (SiH_3)_2NH + NH_3 \qquad (6.6)$$
$$3(SiH_3)_2NH \to 2(SiH_3)_3N + NH_3 \qquad (6.7)$$
$$(SiH_3)_2NH + SiH_3X + NH_3 \to (SiH_3)_3N + NH_4X \qquad (6.8)$$

Support for mechanisms of this type came with the isolation of disilazane, $(SiH_3)_2NH$ (Equation (6.9))[342].

$$2Ph_2NSiH_3 + NH_3 \to 2Ph_2NH + (SiH_3)_2NH \qquad (6.9)$$

Subsequent experiments have shown that disilazane disproportionates readily at 0 °C according to (Equation (6.7)), and that it reacts rapidly with SiH_3I to give $(SiH_3)_3N$ as predicted in (Equation (6.8))[343].

The reaction of halosilanes with ammonia is further complicated by the fact that the product $(SiH_3)_3N$ undergoes a facile base-catalysed condensation reaction in the presence of ammonia. The products of this ancillary reaction are monosilane and polymeric Si—N compounds. An extensive study of this reaction[344] showed that condensation occurs in the liquid rather than the gas phase, and proceeds via an intermolecular mechanism. Disilazane condenses similarly in the presence of ammonia[343], and these reactions appear to be general for silylamines with SiH bonds[292], extending also to silylhydrazine derivatives[345].

The preparation[346, 347] of a lithium salt of hexachlorodisilazane can be expected to prelude a wide derivative chemistry similar to that of hexamethyldisilazane[333, 336]. However, the expected products are not always obtained from the reactions of $LiN(SiCl_3)_2$ with compounds containing more than one Si—X bond, due to decomposition of the lithium salt.

(b) *Structures* – The effects of d orbital participation in the bonding on the structure of $(SiH_3)_3N$ has already been mentioned (Section 6.2.2.3(b)). The structures of several other silylamines have been examined recently by gas phase electron diffraction[349-354] (see Table 6.19) and in all cases, the observed Si—N—Si angles (120–128 degrees), a planar arrangement at the nitrogen atom, and a 'short' Si—N bond have been interpreted in terms of (p-d)π-character in the Si—N bond. The linear SiNCO and SiNCS skeletons of the silyl isocyanate[355, 356] and isothiocyanate[357-359] molecules have been rationalised similarly, but it is interesting to note that a recent electron-diffraction study[360] of the molecule $SiCl_3NCO$ shows a non-linear skeleton similar to that found for the methyl derivative[361] and in silyl azide, SiH_3N_3[355], contrary to the indications from analysis of the infrared spectrum[362].

(c) *Function as a Lewis base* – The reduction in the base strength of $(SiH_3)_3N$ relative to $(CH_3)_3N$ was discussed previously (Section 6.2.2.3(b)). Examination of the donor–acceptor reactions of a number of silylamines[292] leads to the conclusion that the base strength of the nitrogen atom, with respect to both boron and silicon compounds as reference acids, decreases with progressive silyl substitution. Similarly, the base strength of the silyl-amine increases with increasing nitrogen substitution on silicon. These observations may both be explained in terms of (p-d)π-bonding in the Si—N bond.

The donor ability of silylamines towards halides as Lewis acids is difficult to assess due to the irreversible cleavage of the Si—H bond[363] which generally accompanies adduct formation (Equation (6.10)):

$$(SiH_3)_3N + BCl_3 \rightarrow (SiH_3)_3NBCl_3 \rightarrow SiH_3Cl + (SiH_3)_2NBCl_2 \quad {}^{227} \quad (6.10)$$

The mechanism of these cleavage reactions has been a subject of speculation[287]. One obvious pathway involves the formation of a 4-centre transition state (Figure 6.4) following coordination of the boron halide[333], but a recent

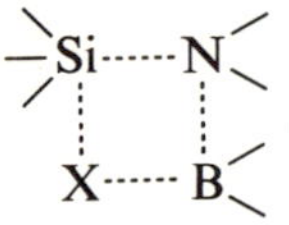

Figure 6.4 4-centre transition state for the reaction of silylamines with the boron halides

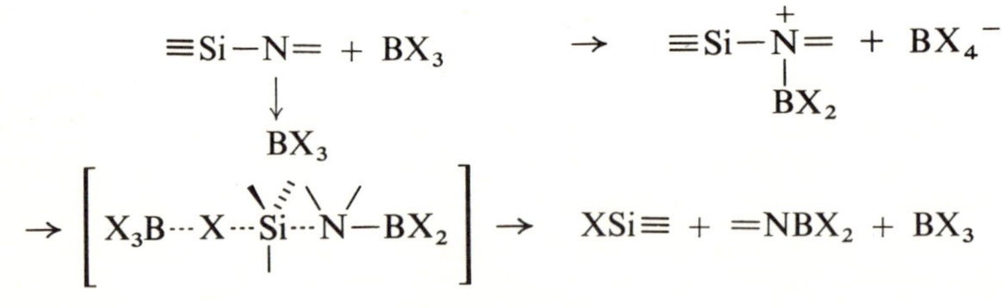

$$\equiv Si-N= \ + \ BX_3 \quad \rightarrow \quad \equiv Si-\overset{+}{N}= \ + \ BX_4^-$$
$$\downarrow \qquad\qquad\qquad\qquad\quad |$$
$$BX_3 \qquad\qquad\qquad\quad BX_2$$

$$\rightarrow \left[X_3B\cdots X\cdots Si\cdots N-BX_2 \right] \rightarrow XSi\equiv \ + \ =NBX_2 \ + \ BX_3$$

Figure 6.5

investigation by Sommer *et al.*[364], using optically-active silylamines and siloxanes, shows that retention of configuration for the product halide, required by the 4-centre mechanism, is observed only in a minority of the reactions studied. They propose an alternative mechanism involving a second mole of boron halide (Figure 6.5) in which retention of configuration is not observed.

6.2.5.3 Compounds with silicon–phosphorus, silicon–arsenic, and silicon–antimony bonds

The inorganic chemistry of silicon bound to the heavier elements of Group V is predominately that of the hydrides[339, 365–368], although halide and hydride–halide derivatives have been reported recently[369–372].

(a) *Preparative aspects* — The preparative chemistry of the silicon–Group V hydrides has undergone a remarkable transformation in the last decade[339]. The original preparative methods for the monosilyl compounds—pyrolysis[373] and electric discharge[374]—gave low yields and required tedious vacuum fractionation to obtain samples of reasonable purity. An initial advance came from a careful examination[375] of the reaction of halosilanes with alkali metal phosphine and arsine compounds, KMH_2, originally believed to produce only the trisilyl derivatives $(SiH_3)_3M$ in moderate yields[376]. By using an excess of the alkali metal compound, however, solid derivatives $KMHSiH_3$ and $KM(SiH_3)_2$ are produced. These on treatment with hydrogen sulphide gave good yields of the mono and disilyl species[375]. The reaction between the readily available halosilanes and lithium tetraphosphine aluminate, $LiAl(PH_2)_4$ [377–380], or its arsenic analogue[375, 381], proved even more successful. Moreover, silyl-substituted alkylphosphines could also be prepared in this way using the appropriate alkylphosphine reagent[382].

Re-distribution or exchange of substituents on silicon[383], or between silicon and the substituents of other Group IV elements[384], are well known in the field of organometallic chemistry. When applied to the chemistry of Group IV–Group V hydrides the exchange reaction has proved to be extremely useful in synthetic and spectroscopic work[339]. In particular, exchange reactions have been used in the preparation of disilanyl[385] and halogen substituted compounds[370], to obtain germyl compounds from the more readily available silyl derivatives[370, 385–387] (Equation (6.11))

$$(SiH_3)_3P + 3GeH_3Br \rightarrow (GeH_3)_3P + 3SiH_3Br \qquad (6.11)$$

and for the synthesis of preferentially deuterated species as an aid to vibrational analysis[388–390].

(b) *Structures* — The absence of a polarised Raman line assignable to the MSi_3 out-of-plane deformation mode in the spectra of $(SiH_3)_3P$ [391] and $(SiH_3)_3As$ [231a] suggested that planar selection rules were obeyed, while the vibrational spectra of $(SiH_3)_3Sb$ [231a] allowed no firm prediction of the molecular geometry. Also, an extended Hückel molecular-orbital treatment indicated that the most stable configuration for $(SiH_3)_3P$ should be planar[392]. On this evidence it was concluded that, by analogy with $(SiH_3)_3N$, (p-d) π-interactions between silicon and the elements of the second and third rows of the Periodic Table might be of stereochemical importance. Subsequent gas-phase electron-diffraction studies[231, 393, 394] have shown that in fact these molecules are pyramidal, with bond angles at the Group V element which preclude any possibility of a (p-d)π-bond. Furthermore, the coupling constant, $J_{PH'}$ obtained from the proton n.m.r. spectra of silyl- and disilanyl-phosphines[388, 395] is little different from that found in the pyramidal parent molecule PH_3. This suggests that no marked change in the s character of the

P—H bond, and hence in the geometry of the molecule, occurs on bonding to silicon.

(c) *Function as a Lewis base* — The failure of $(SiH_3)_3P$ to form complexes with boron trifluoride[398], halosilanes, or pentacarbonyl iron[233] is indicative of weak nucleophilic character, and the proposed explanation of this effect in terms of $(s \rightarrow d)\sigma$ bonding has already been mentioned (Section 6.2.2.3(b)). The molecules SiH_3PH_2, $(SiH_3)_2PH$, and $Si_2H_5PH_2$ give borane adducts when sealed with diborane under high pressures[395–397]. Under these conditions, decomposition of the adduct leads to formation of the appropriate silane and a polymeric material (Equation (6.12)), but if the pressure is released dissociation occurs readily at room temperature.

$$SiH_3PH_2BH_3 \rightarrow SiH_4 + \tfrac{1}{x}(PH_2BH_2)_x \quad ^{397} \qquad (6.12)$$

The corresponding reaction with $(SiH_3)_3P$ has not been reported, so that a direct comparison of the effect of silyl substitution on the donor ability of phosphorus in these compounds is not yet possible.

A novel feature of the reactions of B_2H_6 and BF_3 with certain Si—P and Si—As derivatives is the apparent ability of the Lewis acid species to catalyse self-condensation reactions (Equation (6.13))[395, 398, 399] which are normally found to be extremely slow[400].

$$5SiH_3AsH_2 \xrightarrow[\text{sealed tube}]{B_2H_6} (SiH_3)_2AsH + (SiH_3)_3As + 3AsH_3 \qquad (6.13)$$

Co-condensation of SiH_3PH_2 and $Si_2H_5PH_2$ in the presence of BF_3 has been used to prepare a 'mixed' silyl-disilanyl-phosphine[298] (Equation (6.14)).

$$2SiH_3PH_2 + Si_2H_5PH_2 \xrightarrow{BF_3} (SiH_3)_2PSi_2H_5 + 2PH_3 \qquad (6.14)$$

6.2.6 Compounds of silicon with the group VI elements

The marked tendency of silicon to bond to oxygen is evidenced by the extensive chemistries of silica[401, 404], the silicates[405], and the silicones[282, 406–409], which, as mentioned previously, are beyond the scope of this article. The wide field of silicon–oxygen heterocyclic chemistry, embracing the cyclosiloxanes, has also been reviewed[340, 410].

6.2.6.1 Binary compounds

Of the several silicon suboxides[411] and sulphides[412, 413] the existence of silicon monoxide has already been mentioned (Section 6.2.2.3(a)). A material generally referred to as silicon monoxide has been widely used in microelectronics as a dielectric in thin film capacitors and as a passivating layer[324]. In a recent review of evidence for the existence of a discrete SiO species in the solid phase, Benyon[414] concludes that if SiO does exist in the so-called monoxide films then it can only be in the form of crystallites too small to be identified by x-ray techniques. The weight of evidence points to

CARBON AND SILICON

disproportionation of SiO below 1200 °C into Si and SiO_2, so that the films produced are likely to be a mixture of these two. A similar disproportionation is thought to occur when gaseous silicon monosulphide is condensed[412].

6.2.6.2 Hydride and halogen compounds

(a) *Preparative aspects* — The simplest member of the family of Si—O hydrides, SiH_3OH has never been isolated[339], although it has been named as a possible intermediate in the hydrolysis of numerous silyl compounds. Similarly, hydrolysis of di- and tri-halosilanes gives polymeric species rather than the silanol derivatives[295, 415]. The corresponding silyl derivatives of the heavier Group VI elements are obtained in a redistribution reaction between the disilyl compound and the appropriate Group VI hydride (Equation (6.15))[416, 417], while the more stable trihalosilanethiols have been known for some time[412].

$$(SiH_3)_2E + H_2E \rightleftharpoons 2SiH_3EH \quad (E = S, Se, Te,) \tag{6.15}$$

Disiloxane, $(SiH_3)_2O$, is obtained by the hydrolysis of a wide variety of silyl compounds[339], and reaction of $(SiH_3)_2O$ with phosphorus pentafluoride gives the partially-fluorinated derivatives $FSiH_2OSiH_3$ and $(FSiH_2)_2O$[418]. The chemistry of halosiloxanes and the related siloxene compounds has been reviewed[419]. A recent preparative route to $(SiH_3)_2S$ and $(SiH_3)_2Se$ involves the reaction of $(SiH_3)_3N$ with the appropriate Group VI hydride[420], while $(SiH_3)_2Te$ results from the reaction of monobromosilane with lithium telluride at low temperature[417]. Alkyl- and aryl-siloxanes have been prepared by the reaction of alcohols with disilyl compounds[421–423] or amine adducts of halosilanes[424, 425], or by treating halosilanes with alkali metal derivatives of the thiols[422]. Co-condensation of SiH_3SH and GeH_3SH or the analogous selenium derivatives gives the mixed species SiH_3EGeH_3 (E = S, Se)[426].

The direct synthesis of alkoxysilanes from silicon and alcohols has already been discussed (Section 6.2.3.2). The preparations and industrial applications of these compounds have been reviewed[427, 428], while the analogous sulphur compounds feature in a survey of the organosulphur derivatives of the Group IV elements[429].

(b) *Structures* — Recent investigations of the structures of methyl[430] and phenyl[431] silyl ethers have shown the skeletal Si—O—C angle in these compounds to be ~120 degrees. These results were surprising in view of the very wide Si—O—Si angles found in disiloxane (144 degrees)[432] and perfluorodisilane (156 degrees)[433] which indicated a marked amount of (p-d) π-bonding in the Si—O bond. As with the Group V hydrides, the Si—E—Si angles in the sulphur and selenium disilyls were close to 90 degrees, indicating a rapid decline in the (p-d)π-interaction between silicon and the second- and third-row elements[194].

The observation of long range H—H' coupling in a variety of sulphur[436, 437] and selenium[437, 438] derivatives of the Group IV elements, H_3M—E—MH'_3 (M = C, Si, Ge; E = S, Se), appears at first sight to be another candidate for explanation in terms of (p-d)π-effects. However, the coupling interaction also occurs in the simple methyl derivatives and is absent from the spectra of oxygen analogues, so that the most obvious explanation would appear to be transmission via the d orbitals of the Group VI element, possibly through a hyperconjugative interaction with the Si—H bonds[287].

(c) *Reactions* — Base strength studies of the siloxanes are consistent with appreciable (p-d)π-bonding in the Si—O linkage[424, 439]. Substitution of a silyl group by a methyl group increases the base strength of the resulting ether[440], in keeping with the observed effect on the bond angle at oxygen. There is evidence that the donor ability of $(SiH_3)_2S$ is even less than that of $(SiH_3)_2O$ [441], but this is not altogether unexpected since the observed structure of $(SiH_3)_2S$ [434] requires the Si—S bonds to have considerable p character, with a resulting increase in the s character of the lone-pair electrons. This in turn would reduce their availability to an external acceptor: furthermore, the possibility of an (s→d)σ-interaction cannot be ruled out.

The Si—O and Si—S bonds of the disilyl compounds are cleaved by a variety of covalent halides[339, 416]. Comparative studies have shown that in $(SiH_3)_2O$ and $(Si_2H_5)_2O$, the Si—O bond in the latter is more readily cleaved by BCl_3. Similarly, in the mixed compound $SiH_3OSi_2H_5$ preferential cleavage of the Si_2H_5—O bond is again observed[442]. These results are rationalised in terms of stabilisation of the transition state by more effective delocalisation of the negative charge from the chlorine atom over the two silicon atoms of the disilanyl group[440].

The base-catalysed condensation of $(SiH_3)_2O$ to SiH_4 and siloxane polymers has been shown to proceed via the intermediate formation of siloxysilanes, $(SiH_3O)_{4-x}SiH_x$ ($x = 2, 3$), (Equations (6.16–6.18))[443]. Methoxysilane reacted similarly[443], while the condensation of

$$2(SiH_3)_2O \rightarrow SiH_4 + (SiH_3O)_2SiH_2 \qquad (6.16)$$

$$(SiH_3O)_2SiH_2 + (SiH_3)_2O \rightarrow SiH_4 + (SiH_3O)_3SiH \qquad (6.17)$$

$$3(SiH_3O)_2SiH_2 \rightarrow SiH_4 + 2(SiH_3O)_3SiH \qquad (6.18)$$

$(SiH_3)_2S$ in the presence of ammonia is thought to involve the initial formation of a 1:2 adduct[444].

6.2.7 Halides and halogenosilanes

The outstanding reactivity of the silicon–halogen bond lends particular importance to the silicon halides as starting materials in the synthesis of countless silicon compounds[419, 445–447]. This reactivity has been assigned to the ability of the silicon atom to expand its coordination number from four to six, thus providing low-energy transition states for reactions with nucleophilic species[419]. The halogenosilanes are also useful in synthetic work, being precursors of the majority of silyl and disilanyl compounds discussed in the other Sections[285, 292, 293].

6.2.7.1 Halides

Of the many silicon halides[419, 448–450] and pseudohalides[451, 452] that are now known, only the tetrachloride is used widely on an industrial scale, mainly in the preparation of fumed silica[402] or high-purity silicon[267, 271]. The preparation of all types of silicon halides has been reviewed[448], as have the redistribution reactions leading to the formation of halides containing two or more different halogens[454].

One of the most notable recent advances in the chemistry of the silicon halides has been the development of experimental procedures which enable the divalent halide radicals SiF_2 and $SiCl_2$ to be used in chemical synthesis[248, 249, 455–459]. These reactive analogues of the organic carbene compounds are generally prepared by a high-temperature reaction between the tetrahalide and elemental silicon under high-vacuum conditions[458]. The radical species are then condensed at $-196\,°C$. There is less flexibility in design of apparatus used to prepare the $SiCl_2$ radical because of its greater reactivity away from the hot reaction zone.

For synthetic purposes, the compounds to be reacted with the silicon dihalides are co-condensed with the radicals at $-196\,°C$. Reaction is observed on warming the condensate to room temperature. $SiCl_2$ inserts readily into H—Cl bonds (Equation (6.19)), whereas the corresponding reaction with SiF_2 gives compounds containing more than one silicon atom as the major products (Equation (6.20)).

$$SiCl_2 + BCl_3 \rightarrow SiCl_3BCl_2 \qquad (6.19)\,^{460}$$

$$5\,SiF_2 + 2\,BF_3 \rightarrow SiF_3SiF_2BF_2 + SiF_3(SiF_2)_2BF_2 \qquad (6.20)\,^{461}$$

This observation led to the belief that, on condensation, SiF_2 formed diradical chains ·SiF_2SiF_2· and ·$SiF_2SiF_2SiF_2$·[457, 458, 461]. This postulate was further supported by analysis of the products formed in the reactions of SiF_2 with unsaturated organic molecules[462, 463]. However, more recent work[464, 468] has cast considerable doubt on this diradical mechanism, and an alternative pathway, involving the monomeric SiF_2 radical as the initial reactive species, has been put forward to explain the observed reactions.

Considerable effort has been directed into the search for more specific routes to catenated halosilanes than are provided by pyrolysis[419, 448], electric-discharge reactions[419, 448, 469–471], or the halogenation of metal silicides[419, 448]. The work of Urry and his collaborators[472] on the amine-catalysed disproportionation of perchlorodi- and tri-silane (Equation (6.21)), and the use of perchlorosilylmercurials (Equation (6.22)),

$$4\,Si_2Cl_6 \xrightarrow{Me_3N} Si_5Cl_{12} + 3\,SiCl_4 \qquad (6.21)\,^{473}$$

$$(Si_2Cl_5)_2Hg \xrightarrow{h\nu} n\text{-}Si_4Cl_{10} + Hg \qquad (6.22)\,^{474}$$

in the synthesis of perchloropolysilanes has gone a long way towards achieving these objectives. It is now confidently predicted that a complete system for the specific synthesis of these halides should be possible in the near future.

As well as these relatively low molecular weight compounds, two distinct types of polymeric perhalopolysilanes are known[419, 475]. The polymerisation or dehalogenation of low molecular weight silicon halides leads to random polymers, characterised by variable stoichiometric compositions. The second polymer species are formed from the halogenation of compounds such as calcium silicide having a preferred polymeric skeleton of silicon atoms. This gives products with a silicon layer structure in which free silicon valences occur giving a 'resonance-stabilised polyradical'[419].

6.2.7.2 Halogenosilanes

Trichlorosilane is unique among the halogenosilanes in being manufactured in commercial quantities[453], primarily for conversion to silicone monomers[191, 282, 476, 477]. This compound has also assumed considerable importance in the fields of synthetic organic and organometallic chemistry as a silylating and reducing agent[478–484], when used in conjunction with a tertiary amine, possibly through the intermediacy of the $SiCl_3^-$ anion[251].

Of the remaining halogenosilanes, the monosilyl compounds SiH_3X are undoubtedly of the greatest synthetic utility at the present time. The development of a synthetic route to these halosilanes by cleavage of the Si—C bond in phenylsilanes (Equation (6.23)) is particularly important, as in this way it is possible to prepare large quantities of halosilanes without using monosilane as a starting material[448].

$$p\text{-}ClC_6H_4SiH_3 + HI \rightarrow SiH_3I + C_6H_5Cl \qquad (6.23)[485]$$

Fluorides cannot be prepared by this method due to disproportionation of the product in the presence of HF. Halides can generally be fluorinated by passage over anhydrous metal fluorides, but a more efficient method involves the reaction of Si—N compounds with BF_3[448] (Equation (6.24)).

$$(Si_2H_5)_3N + BF_3 \rightarrow Si_2H_5F + (Si_2H_5)_2NBF_2 \qquad (6.24)[486]$$

Several new methods of halogenating Si—H bonds have been reported recently. The reaction of BBr_3 with SiH_4 yields bromosilanes[457], but no chlorinated products arise from the SiH_4–BCl_3 reaction under similar conditions[488]. The reactions of Si_2H_6 with silver chloride[489, 490] or BX_3[487, 488, 491, 492] (X = Cl, Br) provide useful methods of obtaining halodisilanes $Si_2H_{6-n}X_n$ (n = 1–4), since no cleavage of the Si—Si bond is observed in either reaction.

The halogenosilanes have been a subject of investigation by a variety of spectroscopic techniques[292]. Vibrational spectra have been assigned, and force constants calculated, for the majority of the halosilanes, and a number of correlations between the Si—H absorption frequencies and the nature of the other substituents on silicon have emerged[292, 419]. Similar rationalisation has been sought for n.m.r. parameters, but although an empirical relationship has been found between the coupling constant $J(^{29}Si\text{—}H)$ for directly-bound protons and an 'effective electronegativity' for the other substituents in both silanes[493] and disilanes[492], many problems remain to be solved before all the factors influencing chemical shifts and coupling constants can be adequately accounted for[194, 287, 292].

6.2.7.3 Complex compounds

A notable feature of the chemistry of the silicon halides and halogenosilanes is their ability to act as electron acceptors[247, 257, 494, 495]. The structures and bonding in silicon complexes have already been discussed (Section 6.2.2.4), and it was noted that recent years have seen profound developments, particularly in the field of 5-coordinate silicon complexes[247]. The anion SiF_5^- was discovered by Clark et al.[258] as the result of an attempt to prepare a tetrafluorethylene derivative of platinum[496]!. Since then, a number of SiF_5^- derivatives have been characterised, and their properties are consistent with a trigonal-bipyramidal structure for the anion[497, 498]. The analogous chloro compound appears to be considerably less stable, for despite evidence from Raman spectra for the existence of this species in solution, no salts of the $SiCl_5^-$ ion could be isolated[499].

Examination of the complex chemistry of the silicon halides and halogenosilanes[247, 500] suggests that silicon tends to behave as a typical 'class (a)'[501] acceptor, forming more stable coordinate links with donor atoms from the first row of the Periodic Table. An apparent exception to this classification was found in a recent study of the amine and phosphine adducts of the tetrahalides[502, 503]. Extremely stable 1:2 adducts were formed between SiX_4 (X = Cl, Br) and trimethylphosphine, whereas the corresponding trimethylamine adducts could not be prepared. Although steric factors are bound to influence the formation of these complexes, it was considered unlikely that they would cause such a pronounced difference in acceptor properties. Furthermore, the donor sequence $Me_3P > Me_3N$ was also found with $SiCl_3^+$ as an electron acceptor[502, 503]. It would therefore seem that the silicon halides act as 'class (b)'[501] acceptors in these compounds, an anomaly similar to that found for BH_3 [504].

6.2.8 Compounds with bonds between silicon and a metal or metalloid

6.2.8.1 Non-transition metal compounds

(a) *Alkali metal derivatives* — Alkali metal silyls, $MSiH_3$, were first prepared in 1961 by the reactions of monosilane with potassium metal, or of disilane with the metal or potassium hydride in 1,2-dimethoxyethane[505]. More recently, hexamethylphosphoramide has been recommended as a solvent for this preparation provided the product is required to be used *in situ* for further reaction[506].

Unlike the corresponding methyl derivatives, alkali metal silyls are found to have a NaCl-type structure at room temperature[505, 507], with an effective radius for the SiH_3^- anion of 2.2 Å [507]. Coupling reactions of metal silyls are often fraught with difficulties[292], particularly if the products are liable to disproportionate (Section 6.2.4.3). However, these compounds have been used to good effect in the preparation of a number of novel silyl derivatives[292, 293].

(b) *Silicon–Group III compounds* — Boron[508] and aluminium[509] silyls can be prepared by the reaction of silyl potassium with the appropriate Group III halide (Equation (6.25)).

$$SiH_3K + ClB(NMe_2)_2 \rightarrow SiH_3B(NMe_2)_2 + KCl \qquad (6.25)$$

For the boron compounds, the presence of a B—N bond stabilises the resulting silylborane. The stability of Si—B compounds is also increased on substituting halogen for hydrogen at the Si atom, and the use of silicon dihalides in the preparation of this type of Si—B compound has already been mentioned. (Section 6.2.7.1).

The reaction of a lithium salt of the octahydropentaborate(-1) anion, $B_5H_8^-$, with monochlorosilane gave a stable compound μ-silylpentaborane(9) (Equation (6.26))[510].

$$LiB_5H_8 + SiH_3Cl \rightarrow \mu\text{-}SiH_3B_5H_8 + LiCl \qquad (6.26)\,[510]$$

By analogy with the structure of the organosilicon compound 1-Br-μ-Me$_3$SiB$_5$H$_7$[511], the silicon atom is μ-SiH$_3$B$_5$H$_8$ is thought to occupy a bridging position between two of the boron atoms in the base of the pentaborane pyramid, a structure that involves a three-centre, two-electron B—Si—B bond[510]. The silicon-bridged compound isomerises readily to give 2-SiH$_3$B$_5$H$_8$ in the presence of weak Lewis bases[510], and silicon-substituted halides of both the bridged and terminal isomers are readily prepared by reaction of the silyl compounds with BX$_3$ or HX—AlX$_3$ (X = Cl, Br)[512].

(c) *Silicon–Group IV compounds* — In addition to the vast number of organosilicon compounds, the chemistry of silicon bound to one of the heavier Group IV elements is growing steadily[192]. Silicon–germanium hydrides can be prepared by coupling[513, 514], electric discharge[310, 515, 516], pyrolysis[517], and alloy hydrolysis reactions[517, 518]. The use of gas chromatographic procedures has led to the separation of the hydride mixtures obtained by the latter three methods[519, 520]. The Si—Sn compound H$_3$SiSnH$_3$ has been prepared but is extremely unstable, decomposing in solution at $-80\,°C$. This lack of stability is thought to reflect the weakness of the Sn—H rather than the Si—Sn bond[192] since silyl derivatives of the alkylstannanes are found to be more stable[522]. Halogeno compounds are less common, but fluorides with Si—Ge bonds can be prepared from reactions with SiF$_2$ radicals.

6.2.8.2 Silicon–transition metal compounds

The synthesis of the compound Me$_3$SiFe(CO)$_2$(π-Cp) in 1956[524] heralded the start of investigations in what has recently been described as 'one of the most active areas for research in inorganic chemistry'[525]. Compounds containing bonds between a Group IV and a transition element have been discussed in a number of reviews[234, 235, 525–532]. Although perhaps the majority of these silicon compounds contain organo substituents at the silicon atom, derivatives of the silicon hydrides, halides and alkoxides are well-represented, and it is these that we shall consider in this section.

(a) *Preparative aspects* — Of the general methods of preparation of silicon–transition metal compounds[525, 530], three are commonly used for the preparation of derivatives with inorganic substituents on silicon. The salt elimination or coupling reaction between a silyl halide and an anionic transition metal species (Equation (6.27))

$$SiH_3I + NaCo(CO)_4 \rightarrow H_3SiCo(CO)_4 + NaI \qquad (6.27)\,[533]$$

is used extensively in the preparation of silyl compounds. Reaction of tri-halosilanes with carbonyl hydrides or binuclear metal carbonyls leads also to the formation of a Si—M bond with elimination of molecular hydrogen (Equations (6.28), (6.29)).

$$F_3SiH + HCo(CO)_4 \rightarrow F_3SiCo(CO)_4 + H_2 \qquad (6.28)\,^{534}$$
$$2Cl_3SiH + Co_2(CO)_8 \rightarrow 2Cl_3SiCo(CO)_4 + H_2 \qquad (6.29)\,^{535}$$

Finally, oxidative addition reactions (Equation (6.30)) are used particularly in

$$HIr(PPh_3)_3(CO) + (EtO)_3SiH \rightarrow (EtO)_3SiIrH_2(PPh_3)_2(CO) + PPh_3 \qquad (6.30)\,^{536}$$

the preparation of chloro- and alkoxy-silane derivatives.

An interesting development in this field has been the reported stabilisation of silicon dihalide molecules by coordination to a transition metal[537]. Silene derivatives of platinum are obtained by the dehalogenation of bis-(trihalo-silyl)platinum compounds (Equation (6.31)).

$$(Ph_3P)_2Pt(SiCl_3)_2 + Ph_3P \rightarrow (Ph_3P)_2Pt(SiCl_2)_2 + Ph_3PCl_2 \quad (6.31)\,^{537}$$

Stabilisation of the SiX_2 molecule is thought to derive from considerable transfer of electron density from the non-bonding orbitals of the platinum and halogen atoms to silicon.

(b) *Reactions* – An obvious similarity exists between silyl and methyl derivatives of the transition elements. Comparison of the two types of compound indicates that the silyl derivatives are generally more stable than their methyl counterparts, and it was observations such as this which led

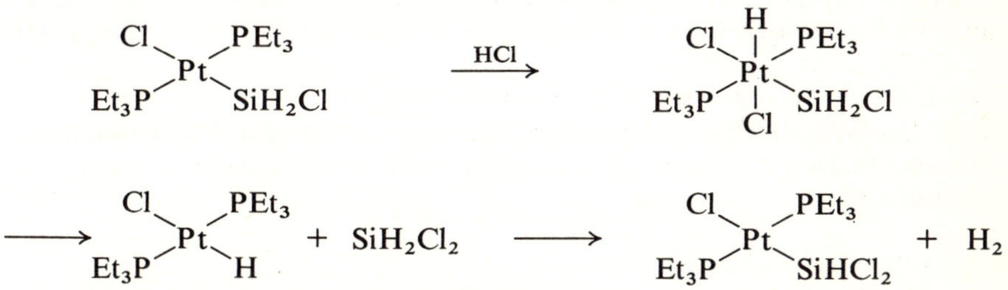

Figure 6.6

to the postulate of multiple bond character in the Si—M bond (Section 6.2.2.3(c))[292, 293, 525]. Cleavage of the Si—X bond (X = N,P,As,S, etc.) by protic reagents is common in numerous silicon compounds. Similar reactions are found with some silyl–transition metal compounds (Equation (6.32))

$$H_3SiCo(CO)_4 + HCl \rightarrow SiH_3Cl + HCo(CO)_4 \qquad (6.32)\,^{538}$$

but with others preferential cleavage of the Si—H bond occurs under forcing conditions (Equation (6.33)).

$$H_3SiMn(CO)_5 + xHCl \rightarrow H_{3-x}Cl_xSiMn(CO)_5 + xH_2 \;(x = 1\text{–}3) \quad (6.33)$$

The reaction between halosilylplatinum compounds and HCl appears to proceed in a similar fashion (Equation (6.34))[540]. However, there is evidence

$$trans\text{-}ClPt(PEt_3)_2SiH_2Cl + HCl \rightarrow ClPt(PEt_3)_2SiHCl_2 + H_2 \;(6.34)\,^{540}$$

that this reaction involves initial cleavage of the Pt—Si bond which is then reformed in a further reaction of the cleavage products (Figure 6.6)[540].

Reaction of $SiH_3Co(CO)_4$ or $SiH_3Mn(CO)_5$ with tertiary amines yields 1:2 addition compounds (Equation (6.35))[260].

$$H_3SiCo(CO)_4 + 2Me_3N \rightarrow [SiH_3 \cdot 2Me_3N]^+[Co(CO)_4]^- \quad (6.35)\ [260]$$

Infrared evidence suggests an ionic structure for the adducts, involving a 5-coordinate silicon cation and the corresponding carbonyl metallate anion. With ammonia, however, cleavage of the Si—M bond occurs to give disilazane and a carbonyl hydride (Equation (6.36))[260].

$$2H_3SiMn(CO)_5 + NH_3 \rightarrow (SiH_3)_2NH + 2HMn(CO)_5 \quad (6.36)\ [260]$$

6.3 TABULAR SURVEY

The following series of tables review data, published in the years 1969 and 1970, on compounds within the scope of this article. Data for 1970 are as comprehensive as the receipt of journals up to the time (March 1970) of completion of the manuscript would allow. Papers which have come to the attention of the author since that time are listed in Table 6.20.

Tables 6.1–6.10 deal with preparations and reactions of carbon and silicon compounds and, as far as possible, the Tables are arranged in the order in which topics are considered in the review section. Standard abbreviations are used for organic groups throughout the text, i.e., Me for methyl, Et for ethyl, Pr for propyl, Bu for butyl, Ph for phenyl, Cp for cyclopentadienyl, and Ac for acetyl.

In Tables 6.11–6.18, papers devoted solely to physical measurements are listed. Because of the considerable interest in structural determinations, molecular parameters, derived from electron-diffraction studies, are set out in Table 6.19.

Key to the 'Physical Data' column in Tables 1–10.
a melting point, boiling point, or vapour pressure
b infrared or Raman spectra
c nuclear magnetic resonance spectra
d electronic spectra
e x-ray data
f mass spectra
g dipole moment
h fluorescence spectra
i structure postulated
j mechanism postulated
k kinetic study
l conductivity study
m other thermodynamic data
n magnetic properties

Table 6.1 New forms of carbon

Carbon type	Source or investigation	References
Pyrolytic carbon	Aliphatic and aromatic hydrocarbons	541
	Methane	542
	Propane	534, 544
	Fluoroethylene	545
	Carbon tetrachloride	542
	Kerosine oils	546
	Sugar	546
	Boron strengthened	547
	Deposition	548–552
	Densification of graphite	553
	In manufacture of tubes	554
	Structure	555
Carbon whiskers	Growth	556
	Use in strengthening metals	557
	Structure	558
Carbon fibres	Cellulose	559–562
	Lignin	563, 564
	Petroleum	565
	Phenol–hexamine polymers	566, 567
	Polyphenylenes	568, 569
	Polyacrylonitrile	570–598
	Polyamides	599, 600
	Polybenzimidazole	601
Carbon fibres	Polyvinyl alcohol	600, 602, 603
	Polyvinyl chloride pitch	604
	Poly[2,2'-(m-phenylene)-5,5'-bibenzimidazole]	605
	Rayon	606–608
	Reinforcement by	609–628
	Metal coated	629, 630
	Strengthening	631
	Structure	632–637
Carbon-coated nuclear fuels	Preparation	638–647
	Irradiation performance	648
Vitreous carbon	Preparation	649
	Oxidation kinetics	650
	Alkali metal intercalation compounds	651
Intercalation compounds of graphite	Alkali metal—preparation	652–655
	As polymerisation catalysts	656
	As ammonia synthesis catalyst	657
	Oxidation	658
	Thermogravimetric analysis	659
	Structure	660
	Electronic structure	661
Intercalation compounds of graphite	Alkali metal, tetrahydrofuran—preparation	662
	Boron – preparation	663
	Chlorine – attempted preparation	664
	Bromine – preparation	665
	Iron halide – preparation	666, 667
	structure	668
	bonding	669

250

Table 6.2 Reactions of carbon vapour

Reactants	Products	Experimental conditions	Physical data	References
C,BCl$_3$	ClC(BCl$_2$)$_3$,Cl$_2$C(BCl$_2$)$_2$	warm reactants from $-196°$C	f	76
	Cl(Cl$_2$B)C=C(BCl$_2$)Cl	warm reactants from $-196°$C		
C,MeBCl$_2$	Me$_2$C(BCl$_2$)$_2$,CH≡CH	warm reactants from $-196°$C	f	76
C,BMe$_3$	Me$_4$B$_2$H$_2$,CH$_2$=CH$_2$,	warm reactants from $-196°$C	f	76
	CH≡CH			
C,B$_2$F$_4$	C(BF$_2$)$_4$,(F$_2$B)$_2$C=C(BF$_2$)$_2$	warm reactants from $-196°$C	f	76
	(BF$_2$)$_2$C=C=C(BF$_2$)$_2$			
C,B$_2$Cl$_4$	C(BCl$_2$)$_4$,ClC(BCl$_2$)$_3$	warm reactants from $-196°$C	f	76
	Cl$_2$C(BCl$_2$)$_2$			
C,SiH$_4$	CH$_3$SiH$_3$,Si$_2$H$_6$,	warm reactants from $-196°$C		78
	CH≡CH,PhH			
C,SiCl$_4$	Cl$_3$SiC≡CCl	warm reactants from $-196°$C	abf	78
C,GeCl$_4$	Cl$_3$GeCCl$_3$,(Cl$_3$Ge)$_2$CCl$_2$,	warm reactants from $-196°$C	abfj	77
	Cl$_3$GeCCl=CCl$_2$			
C,PCl$_3$	Cl$_2$PCCl$_3$,(Cl$_2$P)$_2$CCl$_2$	warm reactants from $-196°$C	afj	77
C,S$_2$Cl$_2$	CSCl$_2$,ClS$_2$CCl$_3$,S$_8$	warm reactants from $-196°$C	afj	77

Table 6.3 Preparations and reactions of carbon compounds

Reactants	Products	Experimental conditions	Yield %	Physical data	References
C,N_2	$(CN)_2$	plasma arc	60		670, 671
$AgC(CN)_3,ClCN$	$C(CN)_4,AgCl$			b	672, 673
$C(CN)_3,M(CN)_4MX$ ($M = Li, X = Cl, M = K, Cs, X = F$)	$MC(CN)_3,XCN$				672, 674
$C(CN)_4,H^+$ or OH^-	$C(CN)_3^-$				672
$CaCN_2,CO_2,H_2SO_4$	H_2CN_2		70–80		675
$(CN)_2,M(PPh_3)_4$ ($M = Pd, Pt$)	$M(CN)_2(PPh_3)_2,PPh_3$	100°C, PhH	40–60	ab	84
$(CN)_2,Ni(PPh_2n\text{-}Pr)_4$	$Ni(CN)_2(PPh_2n\text{-}Pr)_2, PPh_2n\text{-}Pr$	25°C, PhH	61	ab	84
$LiBH_3CN,KF\cdot 2H_2O$	KBH_3CN	0°C, H_2O	75	bc	99
KBH_3CN,DCl,D_2O	KBD_3CN	25°C, D_2O	50	bcjk	99, 676
KBH_3CN,H^+	HCN,H_3BO_3,H_2	25°C		jk	99, 676
$NaBH_3CN,R_3N,HCl$ ($R_3 = Me_3, 4\text{-}MeC_5H_4, O(CH_2)_4H$)	R_3NBH_2CN	0–25°C	30	abc	677
$NaBH_3CN,(CH_2Me_2N)_2,HCl$	$(CH_2Me_2N)_2(BH_2CN)_2$	0–25°C		a	677
$NaBH_3CN,(Ph_3P)_3MCl$ ($M = Cu, Ag$)	$(Ph_3P)_3M(NCBH_3)$	25°C, $CHCl_3/EtOH$		abl	100
$NaBH_3CN,Ni(en)_2Cl_2$	$(en)_2Ni(NCBH_3)_2,THF$	Reflux THF		abl	100
$AgNCO,CNCl$	$NCNCO,(C_2N_2O)_x,AgCl$	25°C, sealed tube		bf	678, 679
CN^-,ClO_2	CNO^-,ClO_2^-	25°C, alkaline solution			680
CO_2,KNH_2,NH_3	$KCO_2NK_2,KCO_2NHK KCO_2NH_2$	Liquid NH_3			681
HCN,Cl_2	$CNCl,HCl$				89–94
HCN,Cl_2	$CNCl,(CNCl)_3$	350°C, carbon catalyst			682
NH_2COCl,HCN,HCl	$CNCl,(CNCl)_3$	Metal chloride catalyst			94
$AgCN,Cl_2O$	$CNCl$	25°C	95	bf	683

Table 6.3 Preparations and reactions of carbon compounds *(continued)*

Reactants	Products	Experimental conditions	Yield %	Physical data	References
CO, O_2, H_2	CO, CO_2			j	684
$CE_2, R_3SnNR'_2$	$R_3SnECENR'_2$	$-78\,°C$	60–100	abc	685
$(E_2 = O_2, OS, S_2; R = Me, Ph; R' = \text{alkyl aryl})$		Light petroleum	Light petroleum		
$CE_2, R_2Sn(NR'_2)_2$	$R_2Sn(ECENR'_2)$	Light petroleum	80–100	a	685
$(E = O, S; R = Me, Ph; R' = \text{alkyl})$					
$CS_2, PhSn(NMe_2)_3$	$PhSn(SCSNMe_2)_3$	Light petroleum	96	a	685
$CO_2, Sn(NMe_2)_4$	$Me_3SnNMeCO_2SnMe_3$	Light petroleum	96	a	685
$CS_2, Sn(NMe_2)_4$	$Sn(SCSNMe_2)_4$	Light petroleum	100	a	685
$CO_2(Me_3Sn)_2NMe$	$Me_3SnNMeCO_2SnMe$	25 °C Light petroleum	98		685
$CE_2, (Me_3Sn)_2NMe$	$(Me_3Sn)_2E, MeNCE'$	$-78\,°C$			
$(E_2 = OS \text{ or } S_2)$	$(E' = O \text{ or } S)$	Light petroleum		a	685
$CS_2, (Me_3Sn)_3N$	$(Me_3Sn)_2S, Me_3SnNCS$	6 °C, PhH	67	a	685
$CO_2(Ph_3P)_2PtOCO_3$	$[(Ph_3P)_2PtOCO_3]PhH$	25 °C, PhH	80–100	abi	127, 128
$CO_2O_2, (Ph_3P)_4M$	$[(Ph_3P)_2MOCO_3]PhH$	25 °C, PhH	80–100	abi	127, 128
$(M = Pd, Pt)$					
$CS_2, (Ph_3P)_2P_2PtO_2$	$[(Ph_3P)_2PtO_2CS_2]$	$0\,°C, PhH/Et_2O$	90		127, 128
$AgOAc, Ac_2O$	AgC_3O_2	Reflux, Ar		bn	162
CS_2, Se	$CSSe$	750 °C, quartz tube	10–15	a	686
CS_2, OH^-	CS_2OH^-, CS_2O^-			k	687
$CS_2, OR^- (R = \text{alkyl})$	CS_2OR^-			k	687
CS_2, SH^-	$CS_3H^-, CS_3^{2-}, S^{2-}$			k	687
CS_2, O_3	COS, CO_2CO, SO_2, O_2	30 °C		k	688
$CS_2, PbMe_4$	PbS	PhH, autoclave			689
$CS_2, PbEt_4$	$PbS, Et_3PbS, PbS_2CEt, (Et_3Pb)_2SO_4$	PhH, autoclave			689
CS_2, Et_3PbOH	$PbS, Et_4Pb, (Et_3Pb)_2SO_4$	PhH, autoclave		abcd	689
$CS_2, M(NR_2)_4$	$M(S_2CNR_2)_4$			eln	113
$(M = Sn, Ti, Zr, V, Nb; R = \text{alkyl})$					

$CS_2,M(CO)_5$			130 °C, high pressure	abcli	690, 691
(R = alkyl,aryl;M = Mn,Re)					
CS_2,NiL_xCl_y	$=[Ni\underset{S}{\overset{S}{\diagup}}C-L]_n$			bdj	114
(L = mono or bidentate amine ligand)					
$CS_2,(R_3P)_2Ni(CO)_2$	$(R_3PCS_2)_nNi$		Reflux	a	692, 693
(R = alkyl, aryl)	(n = 1,2)				
$CS_2,Na[\pi-CpFe(CO)_2CS_2]$	$[\pi-CpFe(CO)_2CS_2]^-$		25 °C, THF	b	135
$[\pi-CpFe(CO)_2CS_2]^-,$ CH_3I,HCl	$[\pi-CpFe(CO)_2CS]^+$	16	25 °C, PhH	b	135
$[\pi-CpFe(CO)_2CSOMe],$ HCl	$[\pi-CpFe(CO)_2(CS)]^+$	45	25 °C, PhH	b	135
$CS_2 \cdot \pi CpMn(CO)_2C_8H_{14}$	$\pi-CpMn(CO)_2(CS), S$		25 °C, PhH	ab	137
$CS_2,RhH(CO)(PPh_3)_3$	$[Rh(CS_2)(CO)(PPh_3)_2]$ $(CS)]S$	60		b	120
$CS_2,Rh(R)(PPh_3)_3$ (R = Me,Ph)	$Rh(CS_2)(CSSR)(PPh_3)_3$		25 °C	ab	120
$CS_2,RhHCl_2(PPh_2Et)_2$	$RhCl(CS_2)(PPh_2Et)_2$		Reflux	ab	120
$CS_2,RhI_2(Me)(PPh_3)_2$ C_6H_6	$RhI_2(CS_2Me)(PPh_3)_2$		Reflux	abc	120
$CS_2,IrH(CO)_2(PPh_3)_2$	$Ir(CO)(CS_2)_2(PPh_3)_2$		Reflux	ab	120
$C_3S_2,trans\text{-}IrCl(CO)$ $(PPh_3)_2$	$IrCl(CO)(PPh_3)_2(C_3S_2)$		5 °C, PhH	b	163, 164
$C_3S_2,trans\text{-}IrCl(CO)$ $(PEtPh_2)_2$	$[IrCl(CO)(PEtPh_2)_2$ $(C_3S_2)_3]_n$		25 °C, PhH	b	163, 164
$C_3S_2,[Ir(Ph_2PCH_2CH_2$ $PPh_2)_2]Cl \cdot xH_2O$	$[Ir(Ph_2PCH_2CH_2PPh_2)_2$ $(C_3S_2)]Cl \cdot xH_2O$		−10 °C, CH_2Cl_2	b	163, 164
$C_3S_2,[Ir(Ph_2PCH_2CH_2$ $PPh_2)_2]Cl \cdot yH_2O$	$\{[Ir(Ph_2PCH_2CH_2PPh_2)_2$ $(C_3S_2)_2]Cl\}_n$		Reflux CH_2Cl_2	b	163, 164
$C_3S_2,Pt(PPh_3)_4$	$Pt(C_3S_2)(PPh_3)_2$		−15 °C, Et_2O	b	163, 164
$C_3S_2,Pt(C_3S_2)(PPh_3)_2$	$[Pt(C_3S_2)_2(PPh_3)_2] \cdot$ $CHCl_3]_n$		Reflux $CHCl_3$	b	163, 164
$CHCl_3,Pd(PPh_3)_4$	$CHCl_2PdCl(PPh_3)_2$		25 °C	ab	694
$CHCl_3,PdCl_2(PPh_3)_2$	$CHCl_2PdCl(PPh_3)_2,$ HCl,CCl_4			ab	694

Table 6.4 Preparation and reactions of the silanes

Reactants	Products	Experimental conditions	Yield %	Physical data	References
$SiCl_4, NaAlH_4, NaH$	$SiH_4, NaCl, AlCl_3$	Reflux THF	100		695
Cl_3SiX, Na_3AlH_6	$SiH_4, NaCl:NaAlCl_4$	THF	low	j	696,697
SiO_2, Na_3AlH_6	SiH_4	700–1300°C			697
SiH_4, H_2	Si, H_2	SiO_2, quartz Al_2O_3			698
SiH_4, CO_2	SiO_2, H_2O	700–1100°C H_2, Ar carrier			699,699a
SiH_4, O_2	SiO_2	>200°C		bm	699
$Si(a)SiH_4, PH_3$	$^{31}SiH_4, ^{31}SiH_3SiH_3$	fast neutron		j	700,701
(b)Si_2H_6, PH_3	$^{31}SiSi_2H_8$	irradiation			
(c)SiH_4, Si_2H_6, PH_3					
SiH_4	$Si_2H_6, H_2 (SiH_2)_n$	(a) 400–600°C		jkm	702,703
		(b) $^{60}Co, \gamma$-radiolysis		j	704
		(a) 460°C, flow system		j	313
SiH_4, SiD_4	$Si_2D_{6-x}H_x, H_2$	(b) 328°C, static system		j	313
	$D_2, HD(x=0\rightarrow 6)$	(c) photolysis		bfj	314
		(d) electric discharge		bfj	314
	Si_2SiH, SiH_2, SiH_3	u.v. photolysis argon matrix		b	705
$SiH_{4-x}D_x$					
$(x = 0\rightarrow 4)$					
SiH_4, C	$CH_3SiH_3, Si_2H_6, CH\equiv CH,$ PhH	warm reactants from $-196°C$		f	76

SiH_4,CH_3		SiH_3,CH_4	29–213 °C	k	706,707
$Si_2H_{6-x}Cl_x,LiAlD_4$ ($x=0\rightarrow6$)		$Si_2H_{6-x}D_x$	u.v.Me_2N_2	f	708
$Si_3Cl_8,LiAlH_4$		Si_3H_8	0 °C,n-Bu_2O	f	301
Si_2H_6	60	Si_3H_8,SiH_4,H_2	370–410 °C	fj	315
$Si_2^iD_6,MeSiH_3$		$SiD_4,MeSiH_2SiD_2H$	flow system 375 °C	bcfj	316,317
Si_2H_6,Me_2SiH_2		$Si_3H_8,Me_2SiHSiH_3$	flow system	fj	316 317
Si_2H_6,Me_3SiH		Me_3SiSiH_3	,,	fj	316,317
$Si_2H_6,HSiCl_3$		Si_3H_8,SiH_4,SiH_2Cl_2	400 °C		316
Si_3H_8		$Si_4H_{10},Si_2H_6,SiH_4,H_2$	flow system 340–370 °C	fj	315
Si_2H_6,KH		SiH_3K,SiH_4	flow system −45 °C,$MeO(CH_2)_2OMe$ conductivity cell	jl	320
$Si_2H_6,LiCl$		$SiH_4,(SiH_2)_x$	25 °C,$MeO(CH_2)_2OMe$	j	321
$(SiBr)_m,LiAlH_4$		$(SiH)_x$	excess $MeSiH_3$ or Me_3SiH	b	709

Table 6.5 Preparations and reactions of silicon-nitrogen compounds

Reactants	Products	Experimental conditions	Yield %	Physical data	References
Si_2N_2	(a) SiN (b) α-Si_3N_4	(a) > 300 °C (b) > 1250 °C			328, 329
$Si_2(NH)_3$	Si_2N_2	60 °C, high vacuum NH_4Cl		i	328, 329
Si,N_2	Si_3N_4	Furnace			710
SiH_4,N_2	Si_3N_4	25–500 °C r.f. glow discharge		b	327
SiH_4,NH_3	Si_3N_4	(a) High temp. N_2 carrier (b) 670–970 °C <10 torr			325 326
Si_3N_4,Li_3N	$LiSi_2N_3$	1200–1900 °C			711
Si_3N_4,Sr_3N	$SrSiN_2,Sr_4SiN_4$	N_2, atm		e	712
$[Si(NH_2)_2]_x,H_2O,NH_3$	Si_2N_2O	960 °C			713
Si_2N_2O	Si_3N_4	1000 °C			713
Si_2N_2O,O_2	$Si_8N_2O_{13}$	1200–1300 °C		e	713
Si_3N_4,O_2	(a) $Si_2(\alpha$-cristobalite) (b) tridymite (c) amorphous SiO_2	1200 °C >1125 °C 1067 °C			714 715 715
SiH_3Cp,Me_3N	SiH_3NMe_2	25 °C			716
$SiH_3Br,LiAl(NR_2)_4$ $(R_2 = MeH,Me_2,Et_2,$ $C_4H_8,C_5H_{10})$	$SiH_3NR_2,LiBr,AlBr_3$	–45 °C monoglyme or diglyme		bc	717
$SiH_3Br,LiAl(NR_2)_4$ $(R_2 = H_2,C_3H_6,C_2H_4)$	SiH_4, polymer	–45 °C monoglyme or diglyme			717
SiH_3Br,KNR_2 $[(R_2 = C_4H_8,C_6H_4(CO)_2,$ $(CH_2CO)_2]$	SiH_3NR_2,KBr	–64–25 °C, TMS Et_2O, cyclohexane		bc	717
SiH_3Br,C_5H_6NH	$SiH_3NC_3H_6,Br(CH_2)_3NH_3Br$	25 °C		bc	717
SiH_3I,Ph_2NH	SiH_3NPh_2,Ph_2NH_2I	–96–25 °C		ab	718
SiH_3NR_2,CE_2 $(R_2 = Me_2,Et_2,C_4H_8,$ $C_5H_{10}; E_2 = O_2,OS,S_2)$	SiH_3ECENR_2	25 °C, sealed tube		c	717

Compound	Conditions	Yield (%)	Notes	Ref
SiH$_3$NMe$_2$ClX (X = NO,CN)	25 °C, sealed tube			717
SiH$_3$NHPh	90 °C		bj	719
SiH$_3$NPh$_2$	310–450 °C		j	718
SiH$_3$NPh$_2$,RNH$_2$ (R = H, Me)	25 °C		j	718
SiH$_3$NPh$_2$,BMe$_3$	No reaction			718
SiH$_2$I$_2$,Ph$_2$NH	$-96 \to -39$ °C		ab	718
SiH$_2$(NPh$_2$)$_2$	0 °C, light petroleum		j	718
(SiH$_2$(NPh$_2$)$_2$,PhNH$_2$	25 °C, 8 weeks			718
SiH$_2$(NPh$_2$)$_2$, BMe$_3$	No reaction			718
SiHCl$_3$,Me$_2$NH	$-96 \to -45$ °C			720
(a) SiHCl$_2$(NMe$_2$)	-70 °C, Et$_2$O 1:2	74	abc	
(b) SiHCl(NMe$_2$)$_2$	-70 °C, Et$_2$O 1:4	76	abc	
(c) SiH(NMe$_2$)$_3$	-70 °C, Et$_2$O 1:6	82	abc	
[Si(NH$_2$)$_2$]$_n$,BCl$_3$		75	abfj	713
SiH$_3$NPh$_2$,NH$_3$	[(SiNH)$_3$(BCl$_2$)$_2$]$_n$,SiCl$_4$ SiH$_3$NPh$_2$			343
SiH$_3$I,PhNH$_2$	(SiH$_3$)$_2$NH,Ph$_2$NH			
	< -46 °C toluene excess			
	SiH$_3$NPh$_2$,			
	25 °C, gas phase or light petroleum	32–91	abf	719
(SiH$_3$)$_2$NH	(SiH$_2$NSiH$_3$)$_n$H$_2$			
(SiH$_3$)$_2$NH	(SiH$_3$)$_3$N,NH$_3$			
(SiH$_3$)$_2$NH,NH$_3$	-80 °C			343
(SiH$_3$)$_2$NH,SiH$_2$NH$_n$	0 °C			343
(SiH$_3$)$_2$NH,SiH$_3$I	(SiH$_3$)$_3$N,NH$_4$I			343
(SiH$_3$)$_2$NMe,ClX (X = NO,CN)	(a) (SiH$_3$)$_2$O,N$_2$, MeCl(X = NO)			343
	-130–0 °C			
	-96–25 °C			
	25 °C, sealed tube			717
	(b) No reaction (X = CN)			
(SiH$_3$)$_2$NPh	SiH$_4$,(SiH$_2$NPh)$_n$		j	719
(SiH$_3$)$_2$NPh,HCl	SiH$_3$NHPh,SiH$_3$Cl		bf	719
	325–380 °C			
	$-96 \to -46$ °C			
	1:1			
(SiH$_3$)$_2$NPh,SiH$_3$I	SiH$_4$, polymer			719
(SiH$_3$)$_2$NPh,BMe$_3$	No reaction			719
SiH$_3$)$_3$N,ClX (X = NO,CN)	No reaction			717
	25 °C, sealed tube			
	-85–25 °C, sealed tube			
	25 °C, sealed tube			
(SiCl$_3$)$_2$NLi,SiSiH$_3$I	(SiCl$_3$)$_2$NSiH$_3$,LiI		abc	348
(SiCl$_3$)$_2$NLi, SiF$_4$	(SiCl$_3$)$_2$NSiF$_3$,Li$_2$SiF$_6$		abc	348
	-78 °C, petrol	91		
	-78 °C, petrol	44		
(SiCl$_3$)$_2$NSiF$_3$	(SiCl$_3$)$_{3-n}$N(SiF$_3$)$_n$ ($n = 0-3$)(SiCl$_3$) (SiF$_2$Cl)(SiCl$_3$) (SiCl$_2$F)$_3$N			348

Table 6.5 Preparations and reactions of silicon–nitrogen compounds (continued)

Reactants	Products	Experimental conditions	Yield %	Physical data	References
$(SiCl_3)_2NLi, SiX_4$ (X = Cl, Br)	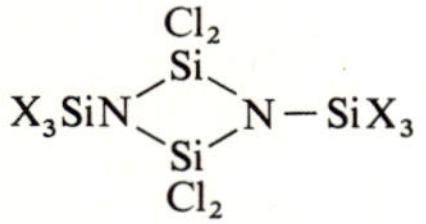 $X_3SiN\begin{smallmatrix}Cl_2\\Si\\\\Si\\Cl_2\end{smallmatrix}N-SiX_3$	50 °C, petrol		a	348
$(SiCl_3)_2NH, Me_2NH$	$[(Me_2N)_3Si]_2NH$, Me_2NH_2Cl	25 °C, petrol	75	abc	721
$[(Me_2N)_3Si]_2NH, NaNH_2$	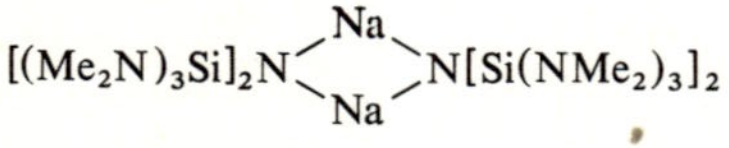 $[(Me_2N)_3Si]_2N\begin{smallmatrix}Na\\\\Na\end{smallmatrix}N[Si(NMe_2)_3]_2$	25 °C, PhH	75		721
$(SiCl_3)_2NH, AgNCO$ $[(OCN)_3Si]_2NH$	$[(OCN)_3Si]_2NH, AgCl$ $Si(NCO)_4, [Si(NCO)_2NH]_n$	25 °C, PhH Heat	77	abc j	721 721
$\begin{smallmatrix}Cl_2\\Si\\HN\quad NH\\\|\quad\quad\|\\Cl_2Si-N-SiCl_2\\H\end{smallmatrix}$, AgNCO (HCCT)	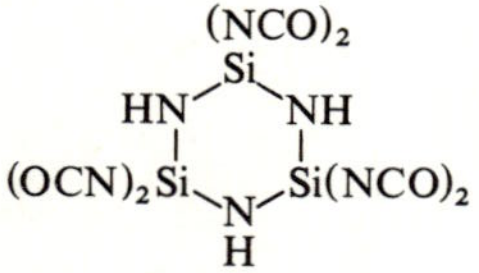 $\begin{smallmatrix}(NCO)_2\\Si\\HN\quad NH\\\|\quad\quad\|\\(OCN)_2Si-N-Si(NCO)_2\\H\end{smallmatrix}$, AgCl	25 °C, Et$_2$O	82		721
HCCT, Me_2NH	$\begin{smallmatrix}(NMe_2)_2\\Si\\HN\quad NH\\\|\quad\quad\|\\(Me_2N)_2Si-N-Si(NMe_2)_2\\H\end{smallmatrix}$	Reflux, petrol	89	a	721

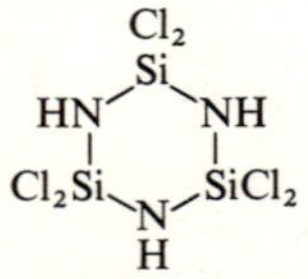

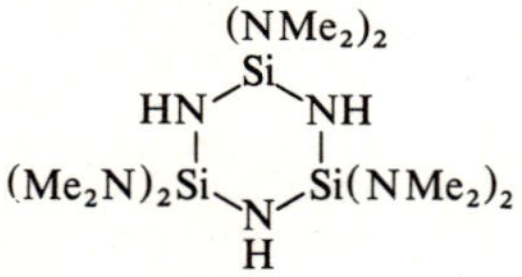

Reactants	Product	Conditions	Yield (%)	Notes	Ref.
HCCT, EtOH, Et$_3$N	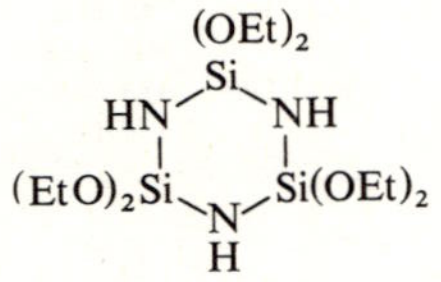 (EtO)$_2$Si[NH-Si(OEt)$_2$-NH]$_2$ (6-membered ring)	$-78\,°$C, petrol	83	a	721
HCCT, LiAlH$_4$	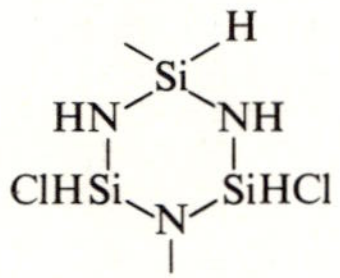 ClHSi[NH-SiH-NH-SiHCl] ring	$-78\,°$C, Et$_2$O		i	721
Si$_2$Ph$_6$, NH$_3$, KNH$_2$	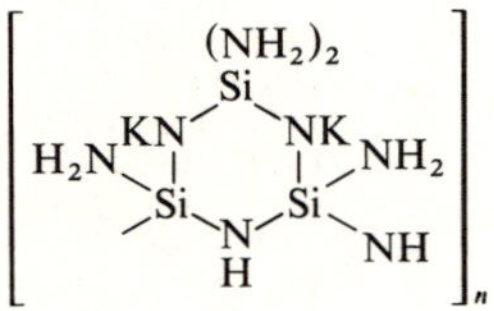 [H$_2$N-Si(NH$_2$)$_2$-NK-Si-NH-Si-NH]$_n$	-75 to $-20\,°$C			722
SiCl$_4$, Ph$_2$C=NLi	(Ph$_2$C=N)$_4$Si, LiCl	90 °C, sealed tube	50	abc	723
SiCl$_4$, C$_4$H$_8$NH	(C$_4$H$_8$N)$_4$Si, C$_4$H$_8$NH$_2$Cl	10 °C, toluene	55	abc	724
(Ph$_3$SiO)$_n$SiCl$_{4-n}$, PhNH$_2$, Me$_3$N	(Ph$_3$SiO)$_n$Si(NHPh)$_{4-n}$				725
SiH$_{4-n}$X$_n$,(CF$_3$)$_2$NO (X = Br, I; n = 1,2)	[(CF$_3$)$_2$NO]$_4$Si	Sealed tube			726, 727
(n−BuO)$_3$SiNH$_2$L (L = SnCl$_4$, TiCl$_4$, AsCl$_3$, Et$_2$PCl, EtPCl$_2$, n-BuPCl$_2$)	(n−BuO)$_3$SiNH$_2$L	70 °C, CCl$_4$		bil	728

Table 6.6 Preparations and reactions of compounds containing silicon–phosphorus, silicon–arsenic, and silicon–antimony bonds

Reactants	Products	Experimental conditions	Yield %	Physical data	References
SiH_3Br, KPH_2, H_2S	$SiH_3PH_2,(SiH_3)_2PH$	H_2S added to solid from SiH_3Br/KPH_2		bcj	375
$SiH_3Br, LiAl(PH_2)_4$	SiH_3PH_2	−45 °C, diglyme	82	j	379
$SiH_3PHK, MeBr$	SiH_3PHMe, KBr	−96 °C, Me_2O		bcj	382
$SiH_3Br, LiAlH(PHMe)_3$	SiH_3PHMe	−45 °C, diglyme		bc	379
$SiH_3Br, LiPMe_2$	SiH_3PMe_2	−130 °C, Me_2O		bc	382
$SiH_3Br, LiAlH_2(PMe_2)_2$	SiH_3PMe_2	−45 °C, diglyme		bc	382
$SiH_3Br, LiPEt_2$	SiH_3PEt_2	−40 °C, Et_2O	42	c	371
SiH_3PH_2, SiD_3Cl	SiD_3PH_2, SiH_3Cl	sealed tube		c	370
SiH_2ClPEt_2	$SiH_2PEt_2, SiCl_2PEt_2$			c	372
SiH_3PH_2, SiH_2Cl_2	SiH_2ClPH_2, SiH_3Cl	sealed tube		c	370
$SiCl_2(PEt_2)_2, SiH_2(PEt_2)_2$	$SiH_2ClPEt_2, SiCl(PEt_2)_3$			c	372
$SiH_3PEt_2, SiHCl_3$	$SiHCl_2PEt_2, SiH_3Cl$			c	372
SiH_3PH_2, GeH_2Cl_2	GeH_2ClPH_2, SiH_3Cl	−60 °C, sealed tube		c	370
$SiH_3PH_2, MeGeH_2Cl$	$MeGeH_2PH_2, SiH_3Cl$	sealed tube		c	370
$SiH_3PH_2, MeGeHCl_2$	$MeGeHClPH_2, SiH_3Cl$	sealed tube		c	370
SiH_3PHMe, HI	$SiH_3I, MePH_2I$			c	382
$(SiH_3)_nPMe_{3-n}$ $(n = 1, 2)$	SiH_3I, Me_nPH_{4-n}, I	25 °C	25	c	382
$SiH_2Br_2, LiAl(PH_2)_4$	$SiH_2(PH_2)_2$	−30 °C, diglyme	41	bcf	379
$SiH_2Br_2, LiPEt_2$	$SiH_2(PEt_2)_2$	−40 °C, Et_2O	37	c	371
$SiH_2PEt_2, LiPEt_2$	$SiH_2(PEt_2)_2, LiH$	Et_2O			729
SiH_3PEt_2	$SiH_2(PEt_2)_2, SiH_4$	25 °C, several months			372
$SiH(PEt_2)_2$	$SiH_2(PEt_2)_2, SiPEt_2)_4$			c	372
$SiHCl(PEt_2)_2$	$SiH_2(PEt_2)_2, SiCl_2(PEt_2)_2$				372
$SiH_2(PH_2)_2, HX$ $(X = Cl, Br)$	SiH_2X_2, PH_3	25 °C			379
$SiH_2(PEt_2)_2, LiPEt_2$	$LiHSi(PEt_2)_2, Et_2PH$				729
$SiHBr_3, LiAl(PH_2)_4$	$SiH(PH_2)_3$	−30 °C, triglyme	18	bcf	379
$SiCl_4, LiAlH_4, LiAl(PH_2)_4$	$SiH(PH_2)_3, SiH_2(PH_2)_2$ SiH_3PH_2	0 °C, triglyme			379
$SiHCl_3, LiPEt_2$	$SiH(PEt_2)_3, SiHCl(PEt_2)_2$	−40 °C, Et_2O	52	c	371
$SiH_3PEt_2, LiPEt_2$	$SiH(PEt_2)_3, LiH$				729
$SiH_3PEt_2, SiHCl(PEt_2)_2$	$SiH(PEt_2)_3, SiH_3Cl$				372
$SiH(PH_2)_3, HX (X = Cl, Br)$	$SiHX_3, PH_3$	25 °C			379
$SiH(PEt_2)_3, LiPEt_2$	$LiSi(PEt_2)_3, Et_2PH$	Et_2O			729

SiH(PEt$_2$)$_3$,LiMe	Me$_4$Si,Me$_3$SiH,Me$_3$SiPEt$_2$			729
SiH(PEt$_2$)$_3$,Lin-Bu	SiH(n-Bu)$_3$,LiPEt$_2$		j	729
SiCl$_4$,LiPEt$_2$	SiPEt$_2$)$_4$,ClSi(PEt$_2$)$_3$, Cl$_2$Si(PEt$_2$)$_2$,Cl$_3$SiPEt$_2$	−70 °C, Et$_2$O		371
(SiH$_3$)$_2$PK,MeBr	(SiH$_3$)$_2$PMe	−96 °C, Me$_2$O	bcj	382
(SiH$_3$)$_2$PH,B$_2$H$_6$	(SiH$_3$)$_2$PHBH$_3$	< −40 °C, sealed tube	c	395
(SiH$_3$)$_2$PHBH$_3$	SiH$_{4+i}$PHBH$_i$	> −40 °C, sealed tube		395
SiH$_3$SiH$_2$Br,LiAl(PH$_2$)$_4$	SiH$_3$SiH$_2$PH$_2$	−45 °C, diglyme		379
SiH$_3$PH$_2$,SiH$_3$SiH$_2$Cl	SiH$_3$SiH$_2$PH$_2$,SiH$_3$Cl	sealed tube	c	370
SiH$_3$SiH$_2$PH$_2$	(Si$_2$H$_5$)$_3$P,PH$_3$	sealed tube	c	395
		B$_2$H$_6$ catalyst		
SiH$_3$SiH$_2$PH$_2$,B$_2$H$_6$	SiH$_3$SiH$_2$PH$_2$BH$_3$	< −40 °C, sealed tube	c	395
Si$_2$H$_6$,(PH$_2$BH$_2$)$_n$	SiH$_3$SiH$_2$PH$_2$BH$_3$	> −40 °C, sealed tube	c	395
SiH$_3$SiH$_2$PH$_2$,B$_2$D$_6$	SiH$_3$SHDPH$_2^1$BD$_2$H	< −40 °C, sealed tube	c	395
(Si$_2$H$_5$)$_3$P	Si$_2$H$_6$,PH$_3$,H$_2$	25 °C, sealed tube		395
SiH$_3$PH$_2$,SiH$_2$ClSiH$_2$Cl	SiH$_3$PH$_2$SiH$_2$PH$_2$...	25 °C, sealed tube		370
SiH$_3$PH$_2$,SiHCl$_2$SiH$_3$Cl	SiH$_2$ClSiH$_2$PH$_2$SiH$_3$Cl	25 °C, sealed tube	c	370
SiH$_3$PH$_2$,SiH$_3$SiHCl$_2$	SiHCl$_2$SiH$_2$PH$_2$SiH$_3$Cl	25 °C, sealed tube	c	370
SiH$_3$Br,LiSi(PEt$_2$)$_3$	SiH$_3$Si(PEt$_2$)$_3$,LiX	−50 °C, Et$_2$O	c	729
(SiH$_3$)$_3$P,HCl	SiH$_3$Cl,PH$_3$	25 °C		233
(SiH$_3$)$_3$P,H$_2$E(E = O,Se)	(SiH$_3$)$_2$E,PH$_3$	25 °C		233
(SiH$_3$)$_3$P,H$_2$S	no reaction	25 °C		233
(SiH$_3$)$_3$P,I$_2$	SiH$_3$I,P$_x$(red)	CH$_2$Cl$_2$, sealed tube		233
(SiH$_3$)$_3$P,RI (R = Me,SiH$_3$)	no reaction	sealed tube		233
(SiH$_3$)$_3$P,PX$_3$(X = Cl,Br)	SiH$_3$X,X$_2$,P$_4$(red)	25 °C		233
(SiH$_3$)$_3$P,S$_8$	(SiH$_3$)$_3$S	25 °C,CS$_2$		233
(SiH$_3$)$_3$P,FeCO)$_5$	no reaction	25 °C		233
(SiH$_3$)$_3$P,MeMn(CO)$_5$	no reaction	25 °C,CS$_2$		233
SiH$_3$Br,KAsH$_2$,H$_2$S	SiH$_3$AsH$_2$	H$_2$S added to solid from SiH$_3$Br/KAsH$_2$	bc	375
SiH$_3$X,'LiAl(AsH$_2$)$_4$'	SiH$_3$AsH$_2$	−78 °C, diglyme excess SiH$_3$X		381,731
SiH$_3$Br,'LiAl(AsH$_2$)$_4$'	(SiH$_3$)$_2$AsH	−112 °C,Et$_2$O excess LiAl(AsH$_2$)$_4$		70
Si$_2$H$_5$Br,LiAl(AsH$_2$)$_4$	Si$_2$H$_5$AsH$_2$	−45 °C,diglyme	bc	375
SiH$_3$AsH$_2$,MeGeH$_2$Cl	MeGeH$_2$AsH$_2$	0 °C	bc	731
SiH$_3$AsH$_2$,Me$_3$GeCl	Me$_3$GeAsH$_2$	0 °C	bc	731
Si$_2$Cl$_6$,Me$_2$MMMe$_2$ (M = P,As)	SiCl$_3$MMe$_2$	40 °C, M = P		731
		150 °C, M = As		369
SiCl$_3$(PEt$_2$)$_2$SiH$_3$PEt$_2$	SiCl$_2$(PEt$_2$)$_2$SiH$_3$Cl		59	372

Table 6.7 Preparations and reactions of silicon–oxygen compounds

Reactants	Products	Experimental conditions	Yield %	Physical data	References
Si,SiO$_2$	SiO	(a) 1250–1400 °C			732
		(b) 1350–1400 °C, 10^{-4} torr			733
SiO	(SiO)$_n$ $n = 2 \to 5$	Ne,Ar matrix		b	734
SiF$_4$,H$_2$O	SiO$_2$(films)				735
Si(OEt)$_4$	SiO$_2$	(a) (i) HCl		e	736
		(ii) NH$_4$OH			736
		(iii) 650–750 °C			736
		(b) oxygen plasma			737
(K$_2$SiO$_3$)$_n$,SO$_3$	K$_2$Si(S$_2$O$_7$)$_3$	95 °C		ei	738
(SiH$_3$)$_2$S,ROH (R = FCH$_2$CH$_2$, CF$_3$CH$_2$,Ph)	SiH$_3$OR,H$_2$S	25 °C, sealed tube	100	bc	422, 423
SiH$_3$OPh,HI	SiH$_3$I,PhOH	25 °C			422
SiH$_3$OPh,BX$_3$ (X = F,Cl)	SiH$_3$F, polymer	$-120 \to -64$ °C			422
SiH$_3$OMe	SiH$_{4-n}$(MeO)$_n$SiH$_{4-n}$ ($n = 2$–4)	(a) -46 °C, NH$_3$			443
	SiH$_3$F,MeOPF$_4$	(b) -96 °C, LiH,Me$_2$O			418
	SiH$_4$,SiH(OMe)$_3$	-78 °C			443
SiH$_3$OMe,PF$_5$	(SiH$_3$)$_2$O,Si$_2$H$_6$	Hg(^{3}P$_1$)		jk	739, 740
SiH$_2$(OMe)$_2$,NH$_3$	N$_2$O,N$_2$,H$_2$	Radiation			
SiH$_4$,NO	SiH$_{4-n}$(Si$_3$H$_4$O$_2$)$_n$	-196–25 °C	100	ij	443
(SiH$_3$)$_2$O	SiH$_{4-n}$(OSiH$_3$)$_n$ ($n = 2$–4)	NH$_3$			
(SiH$_3$)$_2$O,NH$_3$		-46 °C		f	443
(SiH$_3$)$_2$O,LiH	SiH$_4$,SiH$_{4-n}$(OSiH$_3$)$_n$ ($n = 2$–4)	-96 °C, Me$_2$O		f	443
(SiH$_3$)$_2$O,PF$_5$	SiH$_3$OSiH$_2$F,(SiH$_2$F)$_2$O PF$_5$$_{-n}H_n$ ($n = 1,2$)	-78 °C		abcf	418
(SiH$_3$SiH$_2$)$_2$O,PF$_5$	SiH$_3$SiH$_2$F,POF$_3$	-78 °C			418
SiF$_2$,H$_2$O	(SiHF$_2$)$_2$O, polymer	Warm from -196 °C		bcfj	418

262

Reactants	Products	Conditions	Yield (%)	Notes	Ref.
$(SiHF_2)_2O,Cl_2$	$(SiF_2Cl)_2O,HCl$	25°C		bf	468
$(SiHF_2)_2O,HCl$ or BF_3	No reaction	25°C			468
$(SiHF_2)_2O,MeOH$	$(MeOSiF_2)_2O$	25°C		f	468
$SiF_4,MeOH$	SiF_3OMe	25°C			466
$SiCl_3OMe,SbF_3$	SiF_3OMe				741
$SiCl_4,MeOH$	$SiCl_3OMe,HCl$	0°C, Et_2O		c	741
SiF_3OMe,HX	No reaction	25°C		c	741
SiF_3OMe,BX_3 ($X = F,Cl,Br$)	$F_3SiX,MeOBX_2$	25°C	~100	c	741
SiF_4,H_2O	$(SiF_3)_2O,SiF_3OSiF_2OSiF_3$	25°C		b	418
SiH_2X_2,H_2O ($X = Cl,Br$)	$(H_2SiO)_m,HCl$	25°C, Et_2O, hexane		b	742
$SiHCl_3,H_2O$	$(HSiO_{3/2})_n$, resin ($n = 8,10,12,14,16$)	H_2O from $NiCl_2 6H_2O$ 25°C, $PhH,SO_3/H_2SO_4$		bcfi	415
$Si/Cu,ROH$ ($R = Me,Et,n$-Pr,i-Bu)	$SiH(OR)_3,Si(OR)_4$	260–310°C silicone oil		j	284
$SiHCl_3,ROH$ ($R = Me,Et$)	$SiH(OR)_3$	-50°C	85–91	a	743, 743a
$SiCl_4,EtOH$	$SiCl_{4-n}(OEt)_n$ ($n = 1$–4)	0°C, hexane			744
$SiCl_4,ROH$ ($R = Me,Et$)	$SiCl(OR)_3$	Reflux hexane 1:3		ac	745
$SiCl_4,HCO_2Et,EtOAc$	$Si(OEt)_4$	0°C,PhH,NH_3		bc	746
$SiCl_4,(BuO)_3SiOH,py$	$[(BuO)_3SiO]_nSiCl_{4-n}$ py-HCl ($n = 1$–3)	-5–0°C, hexane	60–80	bc	747
$[(BuO)_3SiO]_nSiCl_{4-n}$ $H_2O(n = 1$–$3)$	$[(BuO)_3SiO]_nSi(OH)_{4-n}$	$-10 \to -20$°C hexane	20–85	bc	747
$(t$-$BuO)_nSi(NMe_2)_{4-n}$ ROH ($n = 2,3$; R = aryl)	$(t$-$BuO)_nSi(OR)_{4-n},Me_2NH$	Heat	75–98	ab	748
$(t$-$BuO)_3SNMe_2$; 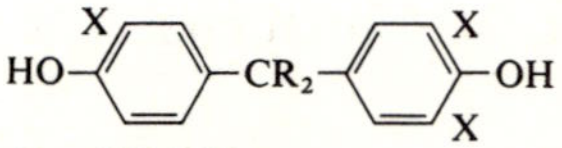 ![structure]HO—C6H4—CR2—C6H4—OH (R = H,Me,Ph; X = Cl,Br)	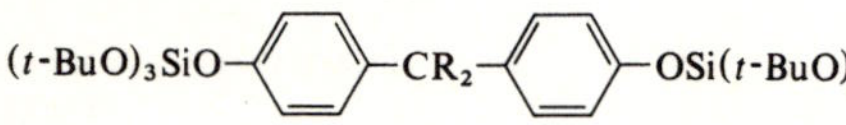 $(t$-$BuO)_3SiO$—C6H4—CR2—C6H4—$OSi(t$-$BuO)_3$	Heat	95	a	748
$R_nSi(OEt)_{4-n}$, $HO(CH_2)_2NRR'$ (R = R' = H,Me or R = H,R' = aryl)	$R_nSi[O(CH_2)_2NRR']_{4-n}$	Reflux PhH		ab	749

264

Table 6.7 Preparations and reactions of silicon–oxygen compounds *(continued)*

Reactants	Products	Experimental conditions	Yield %	Physical data	References
R_nSiX_{4-n},$ClCH_2CH(OH)CH_2X$ (R = H,Me,Et,Ph,CH_2 = CH; n = 1–3; X = Cl, OEt, OBu)	$R_nSi[OCH(CH_2Cl)CH_2X]_{4-n}$	Reflux Et_3N	53–93	a	750
R_nSiX_{4-n}, $H_2C\overset{O}{\frown}CHCH_2X$ (n = 1–3; X = Cl,OEt,OBu)	$R_nSi[OCH(CH_2Cl)CH_2X]_{4-n}$	Reflux PhH	56–89	aj	750
$Si(OEt)_4$,$(CH_2)_4(OH)_2$	$Si[O(CH_2)_4OH]_4$	190	93	a	751
$SiCl_4$,$CH_2\overset{O}{\frown}CH_2$	$Si[O(CH_2)_2Cl]_4$			a	752
$Si(OEt)_4$,$Cl(CH_2)_2OH$	$Si[O(CH_2)_2Cl]_4$	Reflux		a	752
$Si[O(CH_2)_2Cl]_2$, n-$C_7H_{15}OH$	$Si(n-C_7H_{15}O)_4$				752
$Si(OR)_4$,SO_3 (R = alkyl)	$Si(OR)_4$,SO_3	25 °C, CCl_4		ail	753
$(RO)_3SiCl$,$MeOSiMe_2Cl$ NH_3	$(RO)_3SiNHSiMe_2(OR)$ (R = Me,Et)	120 °C	35–40	ac	745
$[(MeO)_3Si]_2$,NH,$PhNCO$	$(MeO)_3SiNCO$ $(MeO)_3SiNHPh$	25 °C		a	754
$Si(OCMe_2CMe_2O)_2$, $MeOH$,Et_3N	$EtNH[MeOSi(OCMe_2CMe_2O)_2]$	25 °C		a	755
$Si(OCMe_2CMe_2O)_2$ $H_2N(CH_2)_3OH$	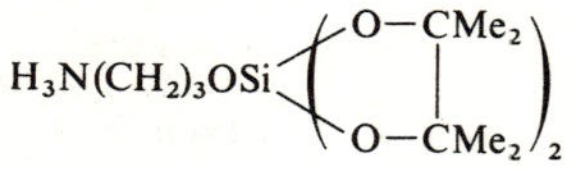 $H_3N(CH_2)_3OSi\!\left(\!\begin{array}{c}O-CMe_2\\O-CMe_2\end{array}\!\right)_{\!2}$	25 °C, PhH			755
$(EtO)_4Si$,$HO(CR_2)OH$ $HO(CH_2)_3NH_2$ (R = H,Me)	$H_3N(CH_2)_3OSi[O(CR_2)_2O]_2$	Reflux CH_3CN	64	a	755

$(EtO)_4Si, HO(CH_2)_2OH$	$H_3N(CH_2)_6NH_3$ [EtOSi(OCH_2CH_2O)_2]_2	Reflux CH_3CN	63		755
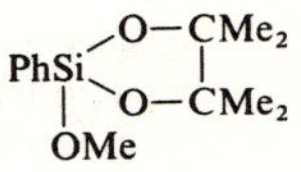 PhSi(OMe)$\begin{matrix}O-CMe_2\\O-CMe_2\end{matrix}$ (HOCH_2CH_2)_2NH	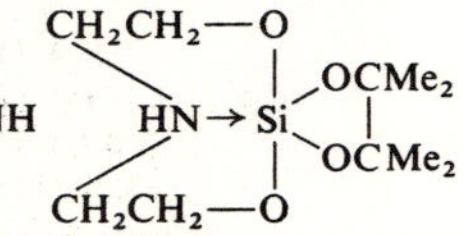 $\begin{matrix}CH_2CH_2-O\\HN\to Si\\CH_2CH_2-O\end{matrix}\begin{matrix}OCMe_2\\OCMe_2\end{matrix}$	Reflux			755
$(EtO)_4Si, PFP(H)_2, Et_3N$ (PFP = perfluoropinacol)	$Et_3NH[EtOSi(PFP)_2]$	140–150 °C	48	bc	755
$(EtO)_3SiOSi(OEt)_3$ $PFP(H_2)Et_3N$	$(Et_3NH)_2[(PFP)_2SiOSi(PFP)_2]$	Reflux	21	c	755
$(EtO)_4Si, PFP(H_2)$ $Me_2N(CH_2)_2OH$	$Me_2NH(CH_2)_2OSi(PFP)_2$	Reflux o-xylene	70	bc	755
$SiO_2, C_6H_4(OH)_2$	$(NH_4)_2SiOH(C_6H_5O_2)$ $(C_6H_4O_2)_2$ $(NH_4)_2[SiOH(C_6H_5O_2)]$ $C_6H_4O_2)_2]C_6H_6O_2$ $(NH_4)_2[SiOEt(C_6H_5O_2)]$ $(C_6H_4O_2)_2]$	25 °C, NH_4OH		bcei	756
$SiCl_4, C_6H_6O_2$	$(Et_4N)_2[Si(C_2O_4)_3]$			i	756
$SiCl_4, Ag(C_2O_4)$ Et_4NBr		25 °C, acetone	70	b	757
$(t-BuO)_3SiOH$ $(i-PrO)_4Ti$	$[(t-BuO)_3SiO]_nTi(i-PrO)_{4-n}$ $(n = 2,3)$	Reflux PhH	60	abc	758
$(t-BuO)_3SiOH$ $Ti(i-PrO)_2(acac)_2$ (acac = acetylacetonato)	$(acac)_2Ti[OSi(t-BuO)_3]_2$ i-PrOH	Reflux PhH	77	abc	758
$[(t-BuO)_3SiO]_2Si(OH)_2,$ $(i-PrO)_4Ti$	$(i-PrO)_3Ti-O-Si[OSi(i-BuO)_3]_2$ $\begin{matrix}&O&O\\&|&|\end{matrix}$ $[(t-BuO)_3SiO]_2Si-O-Ti(i-PrO)_2$	Reflux PhH	44	abc	758
$(RO)_3SiOH,(i-PrO)_3Sb$ ($R = sec-Bu, t-Bu, Me, Ph$)	$[(RO)_3SiO]_3Sb$	Reflux hexane	35–94	ab	759

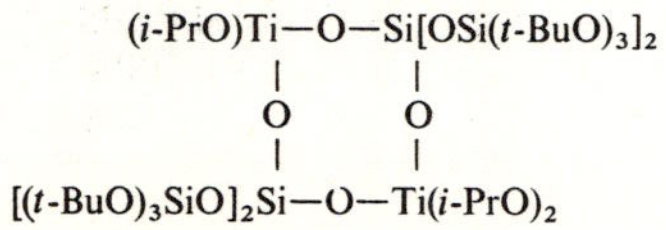

Table 6.8 Preparations and reactions of compounds of silicon with S, Se, Te

Reactants	Products	Experimental conditions	Yield %	Physical data	References
Si, S_8	SiS_2	700–800 °C, 1:1			760
$CaSi_2, S_2Cl_2$	$(SiS)_n, CaCl_2$	reflux CCl_4		do	761
SiH_4, H_2S	$SiH_3SH, (SiH_3)_2S$	electric discharge		cf	426
	$Si_nH_{2n+2}(n = 2-4)$				
$(SiH_3)_2S, H_2S$	SiH_3SH	25 °C			416
$Me_3NH(SSiH_3), BF_3$	SiH_3SH, Me_3NBF_3	25 °C			420
$SiH_4, MeSH$	$SiH_3SMe, (SiH_3)_2S$				
	$Si_nH_{2n+2}, Me_2S, Me_2S_2$	electric discharge		cf	426
	$n = 2-4$				
$NH_4(SSiH_3), MeI$	SiH_3SMe, NH_4I	25 °C	~100		420
$SiH_3Br, PhSK$	SiH_3SPh, KBr	−64 °C, Et_2O		bc	422
SiH_4, GeH_4, H_2S	SiH_3SGeH_3,	electric discharge		cf	426
	other products				
$(SiH_3)_3N, H_2S$	$(SiH_3)_2S, NH_4(SSiH_3)$	25 °C, sealed tube			420
SiH_3Br, Li_2S	$(SiH_3)_2S, LiBr$	−96 °C, Me_2O	89		416
$SiH_3Br, Me_3NH(SH)$	$(SiH_3)_2S$	−96 °C, $Me_2O, 1:1$	58		416
$SiH_3Br, Me_3NH(SH)$	$(SiH_3)_2S, Me_3NH(SSiH_3)$	−96 °C, Me_2O			420
		excess $Me_3NH(SH)$			
$Me_3N(SSiH_3)_2BCl_3$	$(SiH_3)_2S, B_2S_3$	25 °C		j	420
SiH_3SPh, HI	$SiH_3I, PhSH$	−120–0 °C			422
SiH_3SPh, BF_3	no reaction	−64 °C			422
SiH_3SPh, BCl_3	$SiH_3SPh \cdot BCl_3$	25 °C			422
$SiH_3SPh \cdot BCl_3$	$SiH_3SPh \cdot BCl_3$				422
$(SiH_3)_2S, AgNCS$	SiH_3NCS		86		416
$(SiH_3)_2S, BCl_3$	SiH_3Cl, B_2S_3	−96 °C			416
$(SiH_3)_2S, ClCN$	$SiH_3NCS, SiH_3CN,$	25 °C, sealed tube			416
	SiH_3Cl, S_8				

(SiH₃)₂S,ClOMe	SiH₃Cl,(MeCO)₂S	25 °C	416
(SiH₃)₂S,NOCl	SiH₃Cl,NO,S₈	−96 °C	416
(SiH₃)₂S,PhOH	SiH₃OPh,H₂S	25 °C	416
(SiH₃)₂S,PhSH	no reaction	25 °C	416
(SiH₃)₂S,H₂O₂	(SiH₃)₂O,H₂S unidentified solid	25 °C	416
(SiH₃)₂S,KPH₂	(SiH₃)₃P	−96 °C, Me₂O excess (SiH₃)₂S	416
(SiH₃)₂S,KPH₂,H₂S	(SiH₃)₂PH	H₂S added to solid from (SiH₃)₂S/KPH₂	416
SiH₄,H₂Se	SiH₃SeH,(SiH₃)₂Se Sᵢₙ H₂ₙ₊₂ (n=2-4)	electric discharge cf	426
NH₄(SeSiH₃),MeI	SiH₃SeMe,NH₄I	25 °C	420
SiH₄,GeH₄,H₂Se	SiH₃GeH₃ other products	electric discharge cf	426
SiH₃Br,Li₂Se	(SiH₃)₂Se,LiBr		417
(SiH₃)₃N,H₂Se	(SiH₃)₂Se,NH₄(SeSiH₃)	−96 °C, Me₂O	420
(SiH₃)₂Se,GeH₃Br	(GeH₃)₂Se,SiH₃Br		762
(SiH₃)₂Te,GeH₃Br	(GeH₃)₂Te,SiH₃Br		762

Table 6.9 Preparations and reactions of silicon halides and halogenosilanes

Reactants	Products	Experimental conditions	Yield %	Physical data	References
SiCl$_4$,CaF$_2$	SiF$_4$,CaCl$_2$	400–500 °C			763
SiF$_3$CCl$_3$,SbF$_3$	SiF$_4$,Sb(CCl$_3$)$_3$	500 °C			764
SiO$_2$,HF	SiF$_4$,H$_2$SiF$_6$,H$_2$SiO$_3$	360 °C			765
ferrosilicon, Cl$_2$	SiCl$_4$	130–290 °C	80–85		766
Si,Cl$_2$	SiCl$_4$	17–80 torr		k	767
		>2000 °C,Ar			
SiO$_2$,CaF$_2$,C	SiF$_3$,SiF$_2$SiF$_3$,SiF$_3$)$_2$O				768
	SiF$_3$OSiF$_2$OSiF$_3$,SiOF$_2$				
SiF$_3$Br,AgNCE	SiF$_3$NCE,AgBr	25 °C,carborundum		bcf	769
(E = O,S,Se)					
SiF$_3$Br,Hg(NCSe)$_2$	SiF$_3$NCSe	25 °C,carborundum		bcf	769
SiF$_2$,PhH	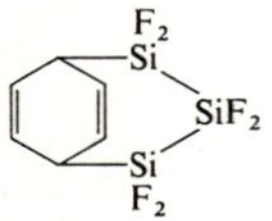 [bicyclic Si$_2$F$_4$ benzene adduct]	warm from −196 °C			770
SiF$_2$,C$_6$F$_6$	C$_6$F$_{6-n}$(SiF$_3$)$_n$ n = 1–3	warm from −196 °C			770
SiF$_2$,SiF$_4$,MeOH	SiF$_3$H,SiF$_3$OMe, SiF$_2$(OMe)$_2$	warm from −196 °C		bcj	466
SiF$_4$,MeOH	SiF$_3$OMe	25 °C			466
SiF$_2$,H$_2$O	(SiF$_2$H)$_2$O	warm from −196 °C		bcfj	468
SiF$_4$,H$_2$O	(SiF$_3$)$_2$O	25 °C		b	468
SiF$_2$,H$_2$S	SiF$_3$,OSiF$_2$OSiF$_3$ SiF$_2$H(SH),SiF$_2$HSiF$_2$(SH), Si$_2$F$_5$H,SiF$_2$HSSH, polymer	warm from −196 °C		cf	465
SiF$_2$H(SH),HCl	SiF$_3$H,SiF$_2$ClH, H$_2$S,SiF$_4$	25 °C, sealed tube			465
SiF$_2$,I$_2$	SiF$_2$I$_2$,SiF$_3$I, polymer	warm from −196 °C		bcf	467
SiCl$_4$,O$_2$	SiO$_2$	SiCl$_4$ injected into plasma	99		771

Reactants	Products	Conditions	Notes	Ref.
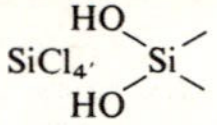 $SiCl_4$, $\begin{array}{c}HO\\ \diagdown\\ Si\\ \diagup\\ HO\end{array}$ (from silica surface)	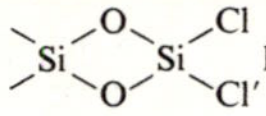 $\begin{array}{c}\diagdownO\diagup Cl\\ SiSi\\ \diagupO\diagdown Cl'\end{array}$ HCl	200–400 °C	jk	772
$SiCl_4,MoO_2Cl_2$	1:1→1:2 compound	–70–180 °C sealed tube		773
SiH_4,HCl	$SiH_{4-n}Cl_n,Si_2H_6$ ($n = 1$–3)	(a) fresh pyrex surface (b) Si coated pyrex surface	j	774
Si,HCl	$SiHCl_3,H_2$	(a) 250–300 °C; 35–115 torr (b) 350 °C,$NaAlCl_4$ melt	jk	775,776 777
$SiHCl_3$	Si_2Cl_6(major product)	0 °C, electric discharge 100 torr		778
SiH_2Cl_2	$SiHCl_2SiHCl_2$ (major product)	electric discharge 100 torr		778
SiH_3Cl	SiH_2ClSiH_2Cl (major product)	electric discharge 100 torr		778
$SiH_4,SiHCl_3$	SiH_3SiCl_3 (major product)	electric discharge 100 torr		778
SiH_4,SiH_3Cl	SiH_3SiH_2Cl (major product)	electric discharge 100 torr		778
SiH_3Cl,SiH_2Cl_2	$SiH_2ClSiHCl_2$ (major product)	electric discharge 100 torr		778
Si_2H_6,BX_3 (X = Cl,Br)	$Si_2H_{6-n}X_m,B_2H_6$ ($n = 1$–4)	0 °C	bc	492
SiH_3SiHCl_2,SbF_3	SiH_3SiHF_2,SiH_3SiF_3 SiH_4,SiF_4	25 °C	bc	492
SiH_3SiH_2Cl,BBr_3	$SiHBrClSiH_3$, SiH_3SiHCl_2 SiH_4,B_2H_6	0 °C	bc	492
SiH_3SiHCl_2,BBr_3	$SiH_3SiCl_2Br,SiH_3SiCl_3$	0 °C	bc	492
Si_2H_6,I_2	$Si_2H_{6-n}I_n$	25 °C, pentane	c	779
Si_3H_8,I_2	$SiH_3I_2,SiHI_3(n = 1,2)$ $(SiH_3)_2SiHI,$ $SiH_3SiH_2SiH_2I,SiH_3SiHISiH_2I$	25 °C, pentane	c	779
n-Si_4H_{10}	$SiH_3SiH_2SiHISiH_3,$ $SiH_3SiH_2SiH_2I$	25 °C, pentane	c	779
R_3SiX,NaH (R = H,OEt,alkyl)	$R_3SiH,NaCl$	20–45 °C		780
$SiHCl_3,RX$	$RH,SiCl_3^{\cdot}$	hν	jk	781
$SiHCl_3,R_3SiMe$ (R = alkyl)	$SiHCl_2Me,R_3SiCl$	150 °C, sealed tube Pt catalyst		485
$SiHCl_3,C_6H_{12}$	$Cl_3SiC_6H_{13}$	60 °C, sealed tube $(Ph_3P)_3RhCl$ catalyst		782

Table 6.9 Preparations and reactions of silicon halides and halogenosilanes *(continued)*

Reactants	Products	Experimental conditions	Yield %	Physical data	References
SiHCl$_3$,R$_2$S=O (R = aryl)	R$_2$S,HOSiCl$_3$	25 °C,Et$_2$O		jk	482,483
SiHCl$_3$,n-Pr$_3$N	n-Pr$_3$NH$^+$SiCl$_3^-$	−40–40 °C,MeCN		c	251
SiHCl$_3$,n-Pr$_3$N,R$_2$CO (R = aryl)	R$_2$CHSiCl$_3$,n-Pr$_3$NHCl (SiCl$_2$O)$_n$	50–80 °C	40–99		478
SiHCl$_3$,n-Pr$_3$N,RCOCl (R = alkyl, aryl)	RCH(SiCl$_3$)$_2$,n-Pr$_3$NHCl (SiCl$_2$O)$_n$	80 °C,MeCN	40–70		480
SiHCl$_3$,n-Pr$_3$N,RX (R = alkyl, aryl; X = Cl,Br)	RSiCl$_3$,n-Pr$_3$NHCl	35–150 °C	25–80		479
SiHCl$_3$,n-Pr$_3$N,RCO$_2$H (R = aryl)	RCH$_2$SiCl$_3$,n-Pr$_3$NHCl (SiCl$_2$O)$_n$	70–80 °C	50–70		481
SiF$_4$,(CH$_2$)$_4$NH	SiF$_4$·2(CH$_2$)$_4$NH	25 °C		b	783
SiF$_4$,(CH$_2$)$_5$NH	SiF$_4$·2(CH$_2$)$_5$NH	25 °C		b	783
SiF$_4$,C$_5$H$_5$N	SiF$_4$·2C$_5$H$_5$N	25 °C		b	783
SiF$_4$,α-nap (α-nap = α-naphthylamine)	SiF$_4$·α-nap	80 °C, PhH		e	784
SiF$_4$,α-nap	SiF$_4$·α-nap,2SiF$_4$·α-nap	45 °C, octane		e	784
SiF$_4$·2C$_5$H$_5$N,en (en = ethylenediamine)	SiF$_4$·en,C$_5$H$_5$N	25 °C		b	783
SiF$_4$,en	SiF$_4$·en	25 °C		b	783
SiF$_4$,bzd (bzd = benzidine)	SiF$_4$·bzd	25–110 °C		bei	785
SiF$_4$,PDA (PDA = o-, m-, or p-phenylenediamine)	SiF$_4$·PDA	PhH or toluene 25 °C,PhH			786
SiF$_4$,Me$_3$P SiF$_6^{2-}$,SiO$_2$	SiF$_4$·Me$_3$P,SiF$_4$·2Me$_3$P SiF$_4$·2H$_2$O	−78 °C high acid concentration		abi	502,503 787
silicic acid HF,Et$_4$NOH	Et$_4$NSiF$_5$			b	498
R$_4$MCl·SiO$_2$,HF (R = alkyl, aryl; M = N,As)	R$_4$MSiF$_5$	MeOH,HF	70	bil	497,498
n-Pr$_4$NSiF$_5$,NH$_3$ NH$_4$Cl,SiF$_6^{2-}$	n-Pr$_4$N[NH$_3$SiF$_5$] (NH$_4$)$_2$SiF$_6$	−78 °C		b	498 788

Compound	Product	Temp/Conditions	Yield (%)	Notes	Ref.
$SiF_4 \cdot 2C_5H_5N, HF$	$(C_5H_5NH)_2SiF_6$	25 °C		abj	783
SiO_2, NH_4HF_2	$(NH_4)_3(SiF_6)F$ $(NH_4)_2SiF_6$	100–140 °C		bem	789
$NH_4SiF_6 \cdot NH_4F$	$(NH_4)_3(SiF_6)F$	fuse		e	790
$SiF_6^{2-}, Sn(OX)_3^{2-}$ (OX = oxalate)	$SiF_2(OX)_2^{2-}, SiF_4(OX)^{2-}$ $SiO(X)_3^{3-}, SnF_6^{2-}$	25 °C, MeOH		cjkm	791
$SiF_6^{2-}, Sn(mal)_3^{2-}$ (mal = malonate)	$SiF_4(mal)^{2-}$	25 °C, MeOH		cj	791
$SiF_6^{2-}, Sn(Me-Mal)_3^{2-}$ (Me-Mal = methylmalonate)	$SiF_4(Me-Mal)^{2-}$	25 °C, MeOH		cj	791
$SiCl_4, HMPT$ (HMPT = $(Me_2N)_3P=O$)	$SiCl_4 \cdot 2HMPT$	25 °C			792
$SiCl_4, Me_3P$	$SiCl_4 \cdot 2Me_3P$	25 °C	100	abi	502, 503
$SiCl_3I, Me_3P, AgClO_4$	$[SiCl_3 \cdot 2Me_3P]ClO_4$	25 °C, PhH		b	502, 503
$[SiCl_3 \cdot 2Me_3N]ClO_4, Me_3P$	$[SiCl_3 \cdot 2Me_3P]ClO_4$ Me_3N	25 °C, MeCN		b	502, 503
$SiCl_4, PhCOCH_2CO_2Et$	$[SiPhCOCH_2CO_2Et)_3]HCl_2$	25 °C, CHCl_3	100	abdi	793
$SiCl_4, p\text{-}XC_6H_4COCH=PPh_3$ ($X = H, Cl, NO_2$)	$SiCl_4 \cdot 2p\text{-}XC_6H_4COCH=PPh_3$	25 °C, PhH or hexane		b	794
$SiCl_4, 5\text{-}Cl\text{-}2\text{-}OH\text{-}C_6H_3COPh$	$SiCl_4 \cdot 2(5\text{-}Cl\text{-}20H\text{-}C_6H_3COPh$	25 °C, PhH	98	abij	795
$SiCl_4, CDTA$ (CDTA = trans-cyclohexanediamine tetra-acetic acid)	$SiCl_n(CDTA)_mH$ (Si:CDTA:1:1.5→1:0.85)	25 °C, PhH		b	796
$SiCl_4, acac$ (acac = acetylacetone)	$Si(acac)_2Cl_2$ $Si(acac)_3HCl_2$	0 °C, CH_2Cl_2		abci	797
$SiCl_4, Et_4NCl$	$(Et_4N)SiCl_5$	25 °C, $PhNO_2$		b	479
$Si_2Cl_6, bipy$ (bipy = 2,2'-bipyridine)	$SiCl_2 \cdot 2bipy(SiCl_2)_n$ $SiCl_4 \cdot bipy$	25 °C, THF	90	ej	798
$R_2Si_2Cl_4 \cdot bipy$ (R = Cl, Me)	Cl R / Cl–Si–Si–Cl / Cl Cl (with N–N bipy)	25 °C, pentane	100	ce	799
$SiBr_4, Me_3N$	no reaction	−78–25 °C			502, 503
$SiBr_4, Me_3P$	$SiBr_4 \cdot 2Me_3P$	25 °C		ai	502, 503

271

Table 6.10 Preparations and reactions of compounds of silicon with metals and metalloids

Reactants	Products	Experimental conditions	Yield %	Physical data	References
$SiH_3K,K[2\text{-}1B_{10}H_{12}]$	$K[H_3SiB_{10}H_{12}],KI$	$-30\,°C$, monoglyme	96		800
$2\text{-}H_3SiB_5H_8$	$1\text{-}H_3SiB_5H_8$	$150\,°C$			801
$2\text{-}H_3SiB_5H_8,BX_3$	$2\text{-}(XH_2Si)B_5H_8$	$0\,°C$	58	abcf	512
$(X = Cl,Br)$	B_2H_6				
$2\text{-}H_3SiB_5H_8,HX$	$2\text{-}(XH_2Si)B_5H_8,H_2$	$25\,°C, AlX_3$	24	abcf	512
$(X = Cl,Br)$					
$\mu\text{-}H_3SiB_5H_8,BX_3$	$\mu\text{-}(XH_2Si)B_5H_8$	$25\,°C$	50	abcf	512
$(X = Cl,Br)$	B_2H_6				
$\mu\text{-}H_3SiB_5H_8,HX$	$\mu\text{-}(XH_2Si)B_5H_8,H_2$	$25\,°C, AlX_3$		acbf	512
$(X = Cl,Br)$					
$2\text{-}H_3SiB_5H_8,DCl$	$2\text{-}(XH_2Si)B_5H_{8-n}D_n$	$25\,°, AlCl_3$			512
Si,BF_3	SiF_3BF_2,SiF_4,Si_2F_6	warm reactants from $-196\,°C$		f	802
	B_2F_4, polymer				
Si,B_2F_4	$F_2Si(BF_2)_2,FSi(BF_2)_3$	warm reactants from $-196\,°C$		abcfj	802
	polymer				
SiF_2,BF,SiF_4,BF_3	$Si_2F_5BF_2,SiF_2(BF_2)_2$	warm reactants from $-196\,°C$		abcfj	802
SiF_2,B_2F_4	$Si_2F_5BF_2,F_2Si(BF_2)_2$	warm reactants from $-196\,°C$			802
	SiF_3BFBF_2, polymer				
SiF_4,B	BF_3, polymer	warm reactants from $-196\,°C$			802
SiF_4,BF	SiF_3BF_2,BF_3,B_2F_4	warm reactants from $-196\,°C$		f	802
	polymer				
$SiBr_3CH_2Br,LiAlH_4$	SiH_3CH_2Br	$0\,°C, n\text{-}Bu_2O$		abcf	803
SiH_3CH_2Cl,NaI	$SiH_3CH_2I,NaCl$	reflux acetone	80	abcf	803
SiH_3CH_2I,AgF	SiF_4,CH_2F_2	$25\,°C$		j	803
$SiH_3K,ClCH_2OMe$	SiH_3CH_2OMe	warm from $-78\,°C$		abcf	804
$SiH_3CH_2Cl,NaGeH_3$	$SiH_3CH_2GeH_3$	$25\,°C$	35	ac	805
$SiH_3CH_2GeH_3,HCl$	$SiH_2ClCH_2GeH_3,$	$25\,°C, AlCl_3$		ac	805, 806
	$SiHCl_2CH_2GeH_3$				
	$(SiH_3CH_2)_2GeH_2,GeH_4$				
$SiH_2ClCH_2GeH_3,H_2O$	$(GeH_3CH_2SiH_2)_2O$	$25\,°C$			806

Reactants	Products	Conditions	Yield	Notes	Ref
SiF_2,CF_3I	CF_3SiF_2I, other products	warm reactants from $-196\,°C$		bcfj	467
SiH_4,GeH_4	$SiH_3,SiH_2GeH_3,$	electric discharge		bcf	516
	$SiH_3,SiH_2GeH_2SiH_3,SiH_3GeH_2GeH_3$				
$MeSiH_2Cl,KGeH_3$	$MeSiH_2GeH_3,KCl$	$0\,°C$, HMPT	5		805
SiF_2,GeH_4	$SiF_2HGeH_3,SiF_3GeH_3,$	warm reactants from $-196\,°C$		bcf	523
	$SiF_2HSiF_2GeH_3,$				
	$SiF_2H(SiF_2)_2GeH_3$				
$SiH_3Br,K[M(CO)_3Cp]$	$H_3SiM(CO)_3Cp,KBr$	$25°, <1$ atm	25	ab	807, 808
(M = Cr,Mo)					
$SiHCl_3,(\pi\text{-}PhH)Cr(CO)_3$	$Cl_3SiCr(H)(CO)_2(\pi\text{-}PhH)$	u.v., hexane	>90	abcj	809
$SiH_3I,NaMn(CO)_5$	$H_3SiMn(CO)_5,NaI$	$-20\,°C, Et_2O$	80	abcdf	810
$SiF_3H,M_2(CO)_{10}$	$F_3SiM(CO)_5,H_2$	$160\,°C$, sealed tube		abf	811, 812
(M = Mn,Re)					
$SiF_3I,Mn_2(CO)_{10}$	$F_3SiMn(CO)_5,Mn(CO)_5I$	heat, sealed tube			811, 812
$SiCl_3H,\pi\text{-}CpMn(CO)_3$	$Cl_3SiMn(H)(CO)(Cp\text{-}\pi)$	u.v., hexane		abcj	809
$H_3SiMn(CO)_5$	$H_2Si[Mn(CO)_5]_2,SiH_4$	$25\,°C$, 14 days			810
$H_3SiMn(CO)_5,HCl$	$H_{3-n}Cl_nSiMn(CO)_5,H_2$	$75\,°C$, sealed tube	10	b	810
	(n = 1–3)				
$H_3SiMn(CO)_5,CO$	no reaction	$70\,°C$, 60 atm			810
$H_3SiMn(CO)_5,H_2O$	$(SiH_3)_2O,HMn(CO)_5$	$25\,°C$			810
$H_3SiMn(CO)_5,SO_2$	$(SiH_3)_2S,(SiH_3)_2O$	$25\,°C$			810
	$HMn(CO)_5$				
$H_3SiMn(CO)_5,NH_3$	$(SiH_3)_2NH,HMn(CO)_5$	$25\,°C$			260
$H_3SiMn(CO)_5,L$	$H_3SiMn(CO)_5\cdot2L$	$25\,°C$	72	bci	260
(L = Me_3N, pyridine)					
$H_3SiMn(CO)_5$·bipyr	$H_3SiMn(CO)_5$·bipyr	$25\,°C$, isopentane		b	260
$F_3SiM(CO)_5HBr$	no reaction	$165\,°C$			811, 812
(M = Mn,Re)					
$F_3SiM(CO)_5,PPh_3$	$F_3SiM(CO)_4PPh_3$	$155\,°C$, PhH			811, 812
(M = Mn,Re)					
$F_3SiMn(CO)_5,C_2F_4$	$F_3Si(CF_2CF_2)_nMn(CO)_5$	u.v., hexane			811, 812
	(n = 1–3)				
$SiH_3K,BrFe(CO)_2Cp\text{-}\pi$	$H_3SiFe(CO)_2Cp\text{-}\pi,KBr$	$-40\,°C$, monoglyme	70	bc	813
$SiH_3I,Na_2Fe(CO)_4$	$(H_3Si)_2Fe(CO)_4,NaI$	$25\,°C$, butane		abc	814
	$H_3SiFeH)(CO)_4$				
	$H_2FeCO)_4$				

273

Table 6.10 Preparations and reactions of compounds of silicon with metals and metalloids *(continued)*

Reactants	Products	Experimental conditions	Yield %	Physical data	References
$SiF_3H_2[\pi-CpFe(CO)_2]_2$	$F_3SiFe(CO)_2Cp$	160 °C, sealed tube		bcj	811, 812
$SiCl_3H_2Fe(CO)_5$	cis-$Cl_3SiFe(H)(CO)_4$	u.v., hexane			809
cis-$Cl_3SiFe(H)(CO)_4 \cdot C_2F_4$	$[Cl_3SiFe(CO)_4]_2$	u.v., hexane		abcj	809
$SiCl_3H, Cl_3SiFe(CO)_2Cp-\pi$	$'Cl_3Si)_2Fe(H)(CO)(Cp-\pi)$	>110 °C			809
$(H_3Si)_2Fe(CO)_4$	$H_3SiFe(H)(CO)_4SiH_4, H_2$	40 °C			814
$(H_3Si)_2Fe(CO)_4 \cdot HCl$	$SiH_3Cl, H_2Fe(CO)_4$				814
	$H_3SiFe(H)(CO)_4$				
$(H_3Si)_2Fe(CO)_4 \cdot H_2O$	$(SiH_3)_2O, H_2Fe(CO)_4$				814
$(H_3Si)_2Fe(CO)_4 \cdot Me_3N$	$(H_3Si)_2Fe(CO)_4 \cdot 2Me_3N$	25	75	b	814
$(H_3Si)_2Fe(CO)_4 \cdot 2Me_3N, HCl$	$SiH_3Cl, H_2Fe(CO)_4$ Me_3NHCl	25 °C			814
$(H_3SiFe(H)(CO)_4, Me_3N$	$[(H_3Si)Fe(H)(CO)_4]_n \cdot Me_3N$ $(n = ?)$			b	814
$SiH_3I, NaCo(CO)_4$	$H_3SiCo(CO)_4$	-23 °C, Et_2O	55	abcdf	815
$SiH_2I_2, NaCo(CO)_4$	$H_2Si[Co(CO)_4]_2$	-23 °C, Et_2O		ab	815
$SiH_4, HCo(CO)_4$	no reaction	25 °C, sealed tube			816
$SiF_3H, HCo(CO)_4$	$F_3SiCo(CO)_4, H_2$	25 °C, 2 atm		a b	534
$SiF_3H, Co_2(CO)_8$	$F_3SiCo(CO)_4, H_2$	25 °C, 40 atm		ab	534
$SiF_3I, Co_2(CO)_8$	$F_3SiCo(CO)_4, Co(CO)_4I$	heat sealed tube			811, 812
$SiCl_3H, HCo(CO)_4$	$Cl_3SiCo(CO)_4, H_2$	25 °C sealed tube	86		816
$SiCl_3H, CpCo(CO)$	$Cl_3CoI(H)(CO)Cp$	u.v., hexane		abcj	809
$H_3SiCo(CO)_4$	$[HSiCo(CO)_2]_n$	>90 °C		j	814
	$SiH_4, HCo(CO)_4, H_2CO$				
$H_3SiCo(CO)_4 \cdot NH_3$	$(SiH_3)_2NH, HCo(CO)_4$	25 °C			260
$H_3SiCo(CO)_4 \cdot L$ $(L = Me_3N, pyridine)$	$H_3SiCo(CO)_4 \cdot 2L$	25 °C	30	bi	260
$H_3SiCo(CO)_4 \cdot bipyr$	$H_3SiCo(CO)_4 \cdot bipyr$	-60 °C, Et_2O		b	260
$H_3SiCo(CO)_4 \cdot PH_3$	no reaction	25 °C			260
$H_3SiCo(CO)_4 \cdot PPh_3$	trans-$H_3SiCo(CO)_3PPh_3$, CO	60 °C, hexane			815
$H_3SiCo(CO)_4 \cdot PF_3$	$SiH_3F, HCo(CO)_{4-n}(PF_3)_n$	25 °C			260

Reactants	Products	Conditions	%	Notes	Ref.
H₃SiCo(CO)₄,HX (X = F,Cl)	SiH₃X,HCo(CO)₄	20 °C			815
H₃SiCo(CO)₄,H₂O	(SiH₃)₂O,HCo(CO)₄	20 °C			815
H₃SiCo(CO)₄,CO	no reaction	70 °C, 25 atm.			815
H₃SiCo(CO)₄,HgX₂ (X = Cl,I)	SiH₃X,HgX[Co(CO)₄] Hg[Co(CO)₄]₂	20 °C			815
H₂Si[Co(CO)₄]₂,HX (X = F,Cl)	SiH₂X₂,HCo(CO)₄	20 °C			815
H₂Si[Co(CO)₄]₂,H₂O	(SiH₂O)ₘ,HCo(CO)₄	20 °C			815
F₃SiCo(CO)₄	Co₄(CO)₁₂,Co₂(CO)₈ SiF₄,CO,Si	25 °C, 18 h			534
F₃SiCo(CO)₄	SiF₄,CO,Co,Si	138 °C, 2 h			534
F₃SiCo(CO)₄,HgCl₂	SiF₃Cl,Hg[Co(CO)₄]₂	60 °C	90		534
F₃SiCo(CO)₄,H₂O	SiO₂,HCo(CO)₄	25 °C			534
Cl₃SiCo(CO)₄	SiCl₄, solid	150 °C, 5 days	35		809
Cl₃SiCo(CO)₄·NMe₃	no reaction	25 °C			817
Cl₃SiCo(CO)₄·PF₃	trans-Cl₃SiCo(CO)₃PF₃	u.v., cyclohexane			817
Cl₃SiCo(CO)₄·PR₃ (R = Me,Et)	trans-Cl₃SiCo(CO)₃PR₃	25 °C			817
Cl₃SiCo(CO)₄·PF₅	no reaction	25 °C			816
Cl₃SiCo(CO)₄,I₂	Cl₃SiI, solid	48 °C	80		816
Cl₃SiCo(CO)₄,EtI	no reaction	25 °C			816
(Ph₃M)₃RhX,SiHR₃ (M = P,As,Sb; X = Cl,Br,I; R₃ = (OEt)₃,(RₙCl₃₋ₙ) (n = 0–3)	(Ph₃M)₂RhH(SiR₃),Ph₃M	excess silane		abc	782
Cl₃SiRh(H)(Cl)(PPh₃)₂,L (L = CO,PF₃)	SiCl₃H,trans-RhCl(L)(PPh₃)₂	25 °C, CH₂Cl₂		j	782
Cl₃SiRh(H)(Cl)(PPh₃)₂,CO	Cl₃SiRhCl(H)(CO)(PPh₃)₂	25 °C, 50 atm		b	782
Cl₃SiRh(H)(Cl)(PPh₃)₂,C₂H₄	SiCl₃H,(C₂H₄)RhCl(PPh₃)₂	25 °C, CH₂Cl₂ sealed tube			782
Cl₃SiRhCl(H)(CO)(PPh₃)₂	SiCl₃H,RhCl(CO)(PPh₃)₂	25 °C			782
Cl₃SiRh(H)(Cl)(PPh₃)₂,SiCl₃H	(Cl₃Si)₂RhCl(PPh₃)₂,H₂	60 °C, 30 days			782
(EtO)₃SiRh(H)(Cl)(PPh₃)₂,HCl	Si(OEt)₃H,RhCl₂H(PPh₃)₂	25 °C, CHCl₃		j	782
(EtO)₃SiRh(H)(Cl)(PPh₃)₂,LiBr	(EtO)₃SiRh(H)(Br)(PPh₃)₂,LiCl	25 °C, THF			782
SiR₃H,IrH(CO)(PPh₃)₃ (R = Cl, OEt, Ph)	R₃SiIrH₂(CO)(PPh₃)₂	25 °C, CH₂Cl₂ or PhH		abcjk	536,818

Table 6.10 Preparations and reactions of compounds of silicon with metals and metalloids *(continued)*

Reactants	Products	Experimental conditions	Yield %	Physical data	References
$SiR_3H,IrCl(CO)(PPh_3)_2$ (R = OEt,Et)	$R_3SiIrH(Cl)(CO)(PPh_3)_2$	25°C, PhH excess SiR_3H		c	819
$SiR_3H,R_3SiIrH(Cl)(CO)(PPh_3)_2$	$SiR_2Cl,(R_3Si)IrH(H_2)(CO)(PPh_3)_2$	25°C, CH_2Cl_2		c	819
$Me_n(EtO)_{3-n}SiH,$ [Ir(diphos)$_2$]$^+$ (n = 0–3)	$[Me_n(EtO)_{3-n}SiIrH(diphos)_2]^+$	25°C, THF		jk	820
$SiCl_3H,\pi\text{-}Cp_2Ni_2(Ph_2P)_2NiCl_2$	$Cl_3SiNiCp\text{-}\pi(PPh_3),HCl$	25°C, THF		bc	821
$SiH_3X,trans\text{-}PtH(X)(PEt_3)_2$ (X = Cl,Br,I)	$trans\text{-}XH_2SiPtX(PEt_3)_2$	25°C, PhH			822
$SiH_3X,trans\text{-}PtH(Y)(PEt_3)_2$ (X = Cl,Br,I; Y = Cl,Br,I; X ≠ Y) A = lightest halogen of X and Y; Z = heaviest halogen of X and Y	$trans\text{-}AH_2SiPt(Z)(PEt_3)_2$				822
$SiH_3Br,trans\text{-}PtCl_2(PEt_3)_2$	$SiH_3Cl,ClH_2SiPtBr(PEt_3)_2$ $trans\text{-}PtBr_2(PEt_3)_2,SiH_2Cl_2$				822
$ClH_2SiPtCl(PEt_3)_2,$ GeH_3Cl	$SiH_3Cl,ClH_2GePtCl(PEt_3)_2$	25°C, PhH			822
$trans\text{-}Cl_{3-n}H_nSiPtCl(PEt_3)_2,HCl$ (n = 1,2)	$Cl_{3-n}SiPtCl(PEt_3)_2,H_2$ (n = 2,3)	25°C, PhH		j	540
$trans\text{-}IH_2SiPtI(PEt_3)_2,HI$	$IH_2SiPtI_2(H)(PEt_3)_2$	25°C, PhH		bc	540
$IH_2SiPtI_2(H)(PEt_3)_2$	$trans\text{-}I_2HSiPtI(PEt_3)_2,H_2$	25°C, PhH		j	540
$SiCl_3H,trans\text{-}PtH(Cl)(PEt_3)_2$	$trans\text{-}Cl_3SiPtCl(PEt_3)_2$	25°C, PhH			540

Reactants	Products	Conditions	Yield (%)	Notes	Ref.
cis-(X₃Si)₂Pt(PPh₃)₂,Ph₃P (X = Cl,Br)	cis-(X₂Si)₂Pt(PPh₃)₂, Ph₃PCl₂	PhH	90	b	537
(Cl₂Si)₂Pt(PPh₃)₃, Ph—C≡C—Ph	$\begin{array}{c}Ph\\ \diagdownCl_2\\ C{-}Si\diagdown\\ Pt(PPh_3)_2\\ C{-}Si\diagup\\ \diagupCl_2\\ Ph\end{array}$	PhH	65		557
SiR₃H, trans-PtH(Cl)(PMe₂Ph)₂	R₃SiPtCl(PMe₂Ph)₂,H₂	25 °C		abc	823, 824
(R = Cl,OEt,alkyl,aryl)					
SiCl₃H,R₂Hg	Cl₃SiHgR,RH	25 °C, u.v.	72	abc	825
(R = Me,Et)					
(Cl₃Si)₂Hg, Et₂Hg	Cl₃SiHgEt	25 °C, CS₂	12		825
Cl₃SiHgEt,HCl	SiCl₄,EtH,Hg	25 °C			825
Cl₃SiHgEt,HX	Cl₃SiHgX,EtH	25 °C			825
(X = Br,I)					
Cl₃SiHgX	Cl₃SiX,Hg	160 °C			825
(X = Br,I)					
Cl₃SiHgEt,Cl₃SiH	(Cl₃Si)₂Hg,EtH	25 °C, u.v.			825
(Me₃SiCH₂)₂Hg,SiCl₃H	(Cl₃Si)₂Hg,Me₄Si	50 °C, u.v. sealed tube	~100	a	826
(Cl₃Si)₂Hg	SiCl₄,Si₂Cl₆,(SiCl₂)ₘ,Hg	250 °C			825
(Cl₃Si)₂Hg,HGeCl₃	Cl₃SiHgGeCl₃	25 °C, u.v. sealed tube		a	826
Cl₃SiHgGeCl₃	Cl₃SiGeCl₃,Hg	90 °C			826
(Cl₃Si)₂Hg,Mn₂(CO)₁₀	Cl₃SiMn(CO)₅	140 °C, 60 h	~100	b	826

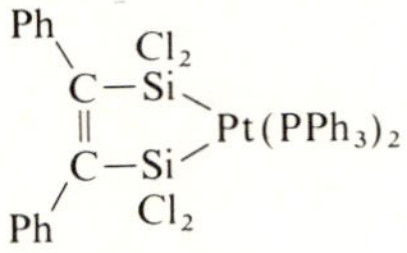

277

Table 6.11 Infrared and Raman studies on carbon and silicon compounds

Molecule	References	Molecule	References
$(CN)_2$	827	$Si(OMe)_4, PhH$	850
$HNCO$, Lewis bases	828	$Si(OEt)_4$, metal acetates	851, 852
$(CN)_2CO$	829	SiF_2	853
$CNX (X = F, Cl, Br, I)$	830–832	SiF_4	854
C_3O_2	159	$SiF_{4-n}Br_n (n = 1–3)$	855
COS	833	$SiCl_4$	856, 856a
$^{12}C^{32}S_2, ^{13}C^{32}S_2$	834	SiH_3CN	857
$SiNH$	835	SiH_2Cl_2	858
$(SiCl_3)_2NH$	836	$SiHCl_3, SiDCl_3$	859, 860
$(SiCl_3)_2NBCl_2$	837	$Si_2X_6 (X = H, Cl, Br, I)$	861–863
$SiCl_3NCS$	838	$[SiC_2O_4)_3]^{2-}$	757
$Si(NCS)_{4-n}(OR)_n$	839	$Si(OH)L_3$ (L = 2-5-dinitro salicylate)	864
$(R = aryl; n = O—3)$		$SiX_4 \cdot n(Me_3M)$	264
$(SiBr_3)_2NH$	836	$(X = F, Cl, Br; M = N, P; n = 1,2)$	
$SiH_3MH_2 (M = P, As)$	840	SiF_5^-	497
SiO	841	SiF_6^{2-}	865, 866
$(SiH_3)_2^{16}O, (SiH_3)_2^{18}O$	842	$SiCl_6^{2-}$	866
$(SiX_3)_2O (X = Cl, Br)$	836, 843	$(SiH_3)_2CH_2, (SiD_3)_2CH_2$	867
$HSiR_3$ (R = alkoxy, aryloxy amido)	844–847	SiH_3GeH_3, SiH_3GeD_3	868
$Cl_nSi(OR)_{4-n}$ (R = aryl, n = 1–3)	848	SiD_3GeH_3	
$Si(OR)_4$ (R = alkyl, aryl)	839, 849	$Cl_3SiCo(CO)_4$	869

Table 6.12 Nuclear magnetic resonance studies on silicon compounds

Molecule	References	Molecule	References
SiH_4	870	$(SiH_3)_2E, H_2E$ (E = S,Se,Te)	417
$PcSi(OR)_3$	871	SiF_4	874–876
(Pc = phthalocyanine; R = alkyl)		SiF_6^{2-}	877–879
$SiH_3PH_2BH_3, SiH_3PH_2BD_3$	730	$SiCl_4, Me_2NCOR$ (R = aryl)	880
$SiH(OR)_3$ (R = alkyl)	872	$Si(acac)_3^{2-}$	881
SiH_3EPh (E = O,S)	873	$X_3SiCo(CO)_4$	882
SiH_3EGeH_3 (E = S,Se)	437	(X = Cl,Br,I,Ph)	
$(SiH_3)_2E$ (E = S,Se)			

Table 6.13 Electronic spectra of carbon and silicon compounds

Molecule	References	Molecule	References
CS	883	SiX_2 (X = F,Cl,Br)	891, 893–895
SiH	884	$SiCl_4$, metal fluorides	896
SiN	885	$SiCl_4$, bipyr	897
SiO	886	$Si(OEt)_4, SnCl_4, PhH$	898
SiX (X = F,Cl,Br,I)	887–892	$PcSi(OR)_3$	871

Table 6.14 Crystal structures of carbon and silicon compounds

Molecule	References	Molecule	References
$(CN)_2$	899	$Li_6Si_{2.5}S_{14}$	904
$(CS_2)Pt(PPh_3)_2$	121	SiF_4, C_5H_5N	261
$\alpha\text{-}Si_3N_4$	900	$SiCl_4, 2Me_3P$	262
$Si_2P_2O_7$	901, 902	$SiO_5N_2C_{12}H_{16}$	905
SiO	903	SiH_3M (M = K, Rb, Cs)	507

Table 6.15 Mass and photoelectron spectra of carbon and silicon compounds

Molecule	References	Molecule	References
CNX (X = Cl, Br, I)	906*	$Si(NCO)_4$	914
CF_4	907*, 908*, 909	SiO	909a*, 915
Si	909a*	$(SiHO_{3/2})_n$	909a*
SiH_4	910*, 911, 912	SiO_2	909a*
Si_2H_6	911, 912	SiS_2	909a*
SiB_4	909a*	SiX_4 (X = F, Cl, Br, I)	907*, 908*, 909, 909a*
$(SiC)_n$	909a*	$X_3SiCo(CO)_4$	913, 917, 918
SiCN	913	(X = F, Cl, Me)	240, 245
$(SiN_4)_n$	909a*		

*References to photoelectron studies marked with an asterisk

Table 6.16 Thermodynamic data for carbon and silicon compounds

Molecule	References	Molecule	References
CH_4, CD_4	919	$SiHX_3$	923
Si^-, SiH^-, SiH_2^-	920	SiX_4	923, 924
SiH_4, SiD_4	919, 921	SiX_3Y	922
SiH_3X	922, 923	SiX_2YZ (X,Y,Z = halogen)	925
SiH_2X_2 (X = halogen)	923	$SiX_4, 2C_5H_5N$	926

Table 6.17 Molecular orbital calculations

Molecule	References	Molecule	References
H_2CN_2	927	SiH_4	931–933
CO_3	145	SiH_3NH_2	934
CO_3^-	928	$X_3SiCo(CO)_4$ (X = F, Cl)	240
SiH_n ($n = 1,2,3,5$)	929, 930		

Table 6.18 Miscellaneous studies

Molecule		References
$SiH_3\cdot$	e.s.r. spectra	935
$SiH_nO_mX_p$ (X = F,Cl,Br,I; $n+m+p = 4$)	microwave spectra	936
SiH_3CH_2X, CH_3SiH_2X (X = Cl,Br,I)	dipole moment	937
SiO	dipole moment	938
SiO	resonance fluorescence	939
$SiCl_4$	nuclear quadrupole resonance	940
$X_3SiCo(CO)_4$ (X = Cl, Ph)	nuclear quadrupole resonance	941

Table 6.19 Electron diffraction studies

Molecule	Bond lengths Å		Bond angles °		References
C_3O_2	C=O	1.163 ± 0.001			158
	C=C	1.289 ± 0.002			
SiH_3NMe_2	Si—N	1.75 ±	Si—N—C	120.0 ±	353
	C—N	1.462 ±	C—N—C	111.1 ±	
$(SiH_3)_2NH$	Si—N	1.725 ± 0.003	Si—N—Si	127.7 ± 0.1	350
	Si—H	1.484 ± 0.006	H—Si—H	108.0 ± 1.0	
	N—H	0.995 ± 0.036			
$(SiH_3)_2NMe$	Si—N	1.726 ± 0.003	Si—N—Si	125.4 ± 0.4	349
	C—N	1.465 ± 0.005			
$(SiH_3)_2NBF_2$	Si—N	1.737 ± 0.004	Si—N—Si	123.9 ± 0.3	354
	B—N	1.496 ± 0.017	F—B—F	123.2 ± 1.8	
	B—F	1.330 ± 0.006			
$(SiH_3)_3N$	Si—N	1.734 ± 0.002	Si—N—Si	119.7 ± 0.1	942
			H—Si—N	108.1 ± 0.6	
			H—Si—H	110.8 ± 0.6	
$(SiH_3)_4N_2$	Si—N	1.731 ± 0.004	Si—N—Si	129.5 ± 0.7	352
	N—N	1.457 ± 0.016	N—Si—H	109.0 ± 1.4	
	Si—H	1.487 ± 0.014			
$ClSi(NMe_2)_3$	Si—N	1.715 ± 0.002	Si—N—C	120.5 ± 1.0	943
	Si—Cl	2.082 ± 0.004	C—N—C	118.5 ± 1.5	
	C—N	1.462 ± 0.005	Cl—Si—N	113.5 ± 1.0	
$ClSi(NCO)_3$	Si—N	1.684 ± 0.005	Si—N—C	145 ± 2	360
	Si—Cl	2.020 ± 0.009			
	N=C	1.213 ± 0.005			
	C=O	1.144 ± 0.005			
$Cl_2Si(NCO)_2$	Si—N	1.687 ± 0.004	Si—N—C	136 ± 1	360
	Si—Cl	2.024 ± 0.005			
	N=C	1.217 ± 0.005			
	C=O	1.146 ± 0.005			
$Cl_3Si(NCO)$	Si—N	1.646 ± 0.008	Si—N—C	138.0 ± 0.4	360
	Si—Cl	2.014 ± 0.005			
	N=C	1.219 ± 0.007			
	C=O	1.139 ± 0.008			
$(SiH_3)_3Sb$	Si—Sb	2.337 ± 0.004	Si—Sb—Si	88.6	944
	Si—H	1.394 ± 0.007			
SiH_3OMe	SiO	1.640 ± 0.003	Si—O—C	120.6 ± 0.9	430
	C—O	1.418 ± 0.009			
SiH_3OPh	Si—O	1.648 ± 0.007	Si—O—C	121 ± 1	431
	C—O	1.357 ± 0.009			
$(SiF_3)_2O$	Si—O	1.580 ± 0.025	Si—O—Si	155.7 ± 2	433
	Si—F	1.554 ± 0.010	F—Si—F	108.8 ± 0.5	
$SiCl_4$	Si—Cl	2.019 ± 0.008			945
$SiF_3SiF_2BF_2$	Si—B	2.008 ± 0.017	Si—Si—B	125.0 ± 2.9	946
	Si—Si	2.361 ± 0.012	Si—B—F	120.6 ± 1.3	
	BF	1.309 ± 0.009	B—Si—F	109.1 ± 2.4	
	$SiF_{av.}$	1.575 ± 0.002	Si—Si—F	102.9 ± 1.7 (centre)	
			Si—Si—F	109.5 ± 1.0 (terminal)	
$SiH_3CH_2SiH_3$	Si—C	1.873 ± 0.002	Si—C—Si	114.4 ± 0.2	947
	C—H	1.11 ± 0.02	H—Si—H	108.1 ± 1.1	
	Si—H	1.572 ± 0.006			

Table 6.20   Supplementary references

Otani, S. Formation of Carbon (Carbonisation). (1970). *Tanso,* **61,** 60
Stoller, H. M. and Frye, E. R. Carbon–Carbon Materials for Aerospace Applications. (1969). *AIAA.ASME Struct. Dyn. Mater. Conf., Collect. Tech. Pap., 10th.,* 193
Bailey, J. E. and Clarke, A. J. Carbon Fibres. (1970). *Chem. Brit.,* **6,** 484
Mayumi, K. Recent Trends in Carbon Fibres. (1970). *Sen-i To Kogyo,* **3,** 309
Zhelikhovskaya, E. I. and Syskov., K. I. Optical Studies of the Structure of Pyrolytic Carbon. (1970). *Khim. Tverd. Topl.* 93
Maire, J. and Méring, J. Graphitization of Soft Carbons. (1970). *Chem. Phys. Carbon.,* **6,** 125
Puri, B. R. Surface Complexes on Carbons. (1970). *Chem. Phys. Carbon.,* **6,** 191
Gyarnati, E. and Nickel, H. Consolidation of Coated Fuel Particles with Pyrolytic Carbon or Silicon Carbide. (1970). *Carbon,* **8,** 400
Tesner, P. A., Rabinovich, E. Y. and Refal'kes, I. S. and Aref'eva, E. F. Formation of Carbon Fibres from Acetylene. (1970). *Carbon,* **8,** 433
Lukacs, J. and Hoffmann, K. Flexible Carbon Fibre Products. (1970). *Swiss Patent,* 494, 188
Hall, R. W. Carbon Fibre–Reinforced Plastics. (1970). *Brit. Patent,* 1, 205, 852
Carton, B. and Hérold, A. Recherches sur les Systemes Carbones Durs-Potassium. 1. Étude Thermogravimetrique. (1970). *Bull. Soc. Chim. France,* 4264
Kohler, V. H., Eichler, B. and Salewski, R. Untersuchungen zum Saurstoffanalogen Charakter der $C(CN)_2$ und NCN Gruppen. (1970). *Z. Anorg. Allg. Chem.,* **379,** 183
Kohler, V. H. and Seifert, B. Nitrosodicyanmethanid und Nitrodicyanmethanid Komplexe Zweiwertiger 3-d Metalle.(1970). *Z. Anorg. Allg. Chem.,* **379,** 1
Bartnitskii, I. N., Ayupov, B. M. and Kuryaeva, R. G. Possibility of Obtaining Silicon Nitride Films on Silicon During the Electrolysis of Liquid Ammonia. (1970). *Electrokhimiya,* **6,** 1227
Egorchkin, A. N., Vyazankin, N. S., Khorshev, S. Y. Chernysheva, T. I. and Kuz'min, O. V. d_π-p_π Interaction in Silicon–Halogen Bonds. (1970). *Izv. Akad. Nauk. SSSR Ser. Khim.,* 1651
Ali, S. I. Reactions of $SiBr_4$ and SiI_4 with Potassium Amide in Liquid Ammonia. (1970). *Z. Anorg. Allg. Chem.,* **379,** 68
Vandrish, G. and Onyszchuk, M. A Reinvestigation of the Enthalpies of Reaction of Silicon, Germanium and Tin Tetrahalides with Pyridine and Isoquinoline. (1970). *J. Chem. Soc. A.,* 3327

Acknowledgements

I am grateful to Mr. B. K. Nicholson and Dr. B. H. Robinson for valuable discussions and suggestions, and I am particularly indebted to Dr. Robinson for his critical reading of the manuscript. My thanks are also due to Mrs. I. C. Morrison and Miss L. Campbell for secretarial assistance, and to my wife for her patient encouragement.

References

1. *Gmelins Handbook of Inorganic Chemistry.* (1968). 8th Edition System No. 14 Carbon Part B. Sections 2 and 3
2. Roemer, W. (1968). *Goldschmiede (Stuttgart),* **66,** 906
3. 1969 *Book of ASTM Standards Pt. B.,* (1969 (ASTM Philadelphia, Pa.)
4. Newotny, C. (1969). *Magy. Kem. Lapja,* **24,** 423
5. Tolansky, S. (1969). *Lab. Pract.,* **18,** 544
6. Iguchi, S. (1970). *Kagaku No Ryoiki,* **24,** 170
7. Hall, H. T. (1961). *J. Chem. Educ.,* **38,** 2851
8. Tolansky, S. (1962). *History and Uses of Diamond,* (London: Methuen)
9. Giardini, A. A. and Kohn, J. A. (1962). *U.S. Dept. Com. Office Tech. Serv.,* A.D. 286, 642
10. Neuhaus, A. (1963). Umschau Wiss. Tech., **63,** 521
11. Milledge, H. J. (1963). *Sci. Progr.,* **51,** 540
12. Tomonari, T. (1964). *Kagaku No Ryoiki,* **18,** 383
13. Batuzov, V. P. (1964). *Issled. Prir. Tekhn. Mineraloobrazov, Mater. Sovetch,* 7th Lov., 10

14. Loens, H. H. (1968). *Metall. (Berlin)*, **22**, 1020
15. Nightingale, R. E. (Ed.). (1962). *Nuclear Graphite*, (New York: Academic Press)
16. Austerman, S. B. (1968). *Chem. Phys. Carbon*, **4**, 137
17. Noda, T. (1968). *Carbon*, **6**, 125
18. Ubbelohde, A. R. (1968). *Nature (London)*, **220**, 434
19. Toshiko, I. (1966). Nenryo Kyokaishi, **45**, 278
20. Thomas, J. M. and Roscoe, C. (1968). *Chem. Phys., Carbon.*, **3**. 1
21. Ergun, S. (1968). Ibid., **3**, 45.
22. Ergun, S. (1968). Ibid., **3**, 211
23. Ruland, W. (1968). Ibid., **4**, 1
24. Takahashi, Y., Inagaki, M. and Yamanchi, S. (1968). *Denki Kagaku*, **36**, 550
25. Richards, B. P. (1968). *J. Appl. Cryst.*, **1**, 35
26. Nightingale, R. E. (1968). *Fundam. Refract. Compounds*, 203
27. Kobayashi, K. (1970). *Tanso*, **60**, 21
28. Walker, P. L. (1962). *Amer. Scientist*, **50**, 259
29. Cahn, R. W. and Harris, B. (1969). *Nature (London)*, **221**, 132
30. Komatsu, Y. (1969). *Bussei*, **10**, 136
31. Vohler, O., Reiser, P. L., Martina, R. and Overhoff, D. (1970). *Angew. Chem. Int. Ed. Engl.*, **9**, 414
32. Palmer, H. B. and Cullis, C. F. (1965). *Chem. Phys. Carbon*, **2**, 265
33. Bokros, J. C. (1969). Ibid., **5**, 1
34. McLaughlin, L. M. (1969). *U.S. At. Energy Comm.*, Y-Da-2654
35. Brooks, J. D. and Taylor, G. H. (1968). *Chem. Phys. Carbon*, **4**, 243
36. Boeder, H. and Fitzer, E. (1970). *Naturwissenschaften*, **57**, 29
37. Graham, L. W. and Price, M. S. T. (1966). *Proc. Second Conf. on Industrial Carbon and Graphite*, 446 (London: Soc. Chem. Industry)
38. Goeddel, W. V. (1967). *Nucl. Appl.*, **3**, 599
39. Ubbelohde, A. R. (1965). *Endeavour*, **24**, 63
40. Whiffen, J. D., Dutton, R., Young, W. P. and Gott, V. L. (1964). *Surgery*, **56**, 404
41. Higgins, J. K. and Antill, J. E. (1966). *Proc. Second Conf. on Industrial Carbon and Graphite*, 269 (London: Soc. Chem. Industry).
42. Lewis, J. C. (1966). Ibid., 258
43. Crowland, F. C. and Lewis, J. C. (1967). *J. Mater. Sci.*, **2**, 507
44. Yamada, S. (1968). *U.S. Clearinghouse, Fed. Sci. Tech. Inform.*, 568, 465
45. Noda, T., Inagaki, M. and Yamada, S. (1969). *J. Non-Cryst. Solids*, **1**, 285
46. Lersmacher, B., Lydlin, H. and Knippenberg, W. F. (1970). *Chem. Ing. Tech.*, **42**, 659
47. Morgan, W. C. (1967). *J. Nucl. Materials*, **21**, 232
48. Thomas, J. M. (1970). *Chem. Brit.* **6**, 60.
49. Bacon, R. (1960). *J. Appl. Phys.*, **31**, 283
50. Tang, M. and Bacon, R. (1964). *Carbon*, **2**, 211.
51. Badami, D. V., Campbell, C., Davy, A. D. and Lindsay, M. J. (1966). *Proc. Second Conf. on Industrial Carbon and Graphite*, (London: Soc. Chem. Industry)
52. Shindo, A., Nakamishi, Y. and Soma, I. (1968). *Polym. Prepr., Amer. Chem. Soc. Div. Polym. Chem.*, **9**, 1333
53. Reinhart, F. (1968). *Glas-Email-Keramo-Tech.*, **19**, 425
54. Watt, W. and Johnson, W. (1969). *New Sci.*, **41**, 398
55. Otani, S. (1969). *Kagatau Kogyo*, **20**, 889
56. Morris, J. B. (1969). *Scu. Azione*, 39
57. Kotima, V. E., Konkin, A. A., Gorbacheva, V. O. and Erofeeva, N. F. (1969). *Khim. Volokna*, 1
58. Obuta, K. (1970). *Chem. Econ. Eng., Rev.*, **2**, 37
59. Otani, S. (1970). *Sekiyu Gakkai Shi.*, **13**, 438
60. Muscel, I. I. (1970). *Ind. Usoara*, **17**, 37
61. Coyle, R. A. and Gillin, L. M. (1969). *Aust. Aeronaut. Res. Lab. Met. Rep.* ARL/MET-63
62. Badami, D. V. (1970). *New Sci.*, **45**, 251
63. Peters, D. H. (1968). *Design Engineering (U.K.)*, 29
64. Phillips, L. N. (1968). *Fibre. Sci. Technol.*, **1**, 3
65. Taviere, J. A. (1969). *Plast. Mod. Elastomeres*, **21**, 77
66. Shaver, R. G. (1969). *Tech. Pap., Reg. Tech. Conf., Soc. Plast. Eng., Inc., West. N. Engl. Sect.*, 50

67. Crofts, R. C. (1960). *Quart. Rev. Chem. Soc.,* **14,** 1
68. Ubbelohde, A. R. and Lewis, F. A. (1960). *Graphite and its Crystal Compounds,* (Oxford: Clarendon Press)
69. French Carbon Study Group. (1965). *Les Carbonnes,* **2,** (Paris-Masson).
70. Rudorff, W., Stumpp, E., Spriester, W. and Siecke, F. W. (1963). *Angew. Chem. Int. Ed. Engl.,* **2,** 67
71. Ottmers, D. M. and Rase, H. F. (1966). *Ind. Eng. Chem. Fundam.,* **5,** 302
72. Latreille, H. (1969). *Electrochim. Metal.,* **4,** 111
73. *British Patents,* 912 823, 912 824, 912 825
74. New Scientist Award Winner, Hambling, J. K. (1968). *New Sci.,* **38,** 563
75. Palmer, H. B. and Shelef, M. (1968). *Chem. Phys. Carbon,* **4,** 85
76. Dobson, J., Tucker, P. M., Stone, F. G. A. and Schaeffer, R. (1969). *J. Chem. Soc. A,* 1882
77. McGlinchey, M. J., Odom, J. D., Reynoldson, T. and Stone, F. G. A. (1970). Ibid., 31
78. Binenboym, J. and Schaeffer, R. (1970). *Inorg. Chem.,* **9,** 1578
79. Tsuchida, E. and Kanetio, M. (1967). *Yuki Gosei Kagaku Kyokai Shi,* **25,** 465
80. Webb, R. L. (1967). *Encycl. Polym. Sci. Technol.,* **7,** 568
81. Nakamura, S. (1968). *Yuki Gosei Kagaku Kyokai Shi.,* **26,** 23
82. Rodewald, H. J. (1965). *Chem. Ztg.,* **89,** 522
83. Cullis, C. F. and Yates, J. G. (1964). *J. Chem. Soc.,* 2833
84. Argento, B. J., Fitton, P., McKeon, J. E. and Rick, E. A. (1969). *Chem. Commun.,* 1427
85. Sumner, P. D. and Phillips, J. G. (1963). *The Red System $(A^2 11\text{-}X^2\Sigma)$ of the CN Molecule,* (Berkeley: Univ. of California Press)
86. Cook, A. J., Johnson, W. B. and Parsons, M. L. (1964). NASA, N65–19752 *Rept. No.* NASA-CR-57334
87. Fawcett, F. S. and Lipscomb, R. D. (1964). *J. Amer. Chem. Soc.,* **86,** 2576
88. Thyagurajan, B. S. (1968). *The Chemistry of Cyanogen Halides,* (Santa Monica, Calif.: Intra Science Res. Found)
89. Durrel, W. S. and Suryanarayan, Y.S. (1969). *Fr. Pat.,* 1 550 717
90. Jean, M. A. (1969). *Ger. Offen.,* 1 809 194
91. Durrel, W. S. and Eckert, R. I. (1969). Ibid., 1 801 311
92. Ewers, W. J. (1969). *Fr. Patent,* **1** 555 980, 1 555 981
93. Trickey, E. B. (1969). Ibid., 1 555 979
94. Matasa, C. J. (1969). *Rom. Pat.,* 52 020
95. Mur, V. I. (1964). *Usp. Khim.,* **33,** 182
96. Wittig, G. (1951). *Ann. Chem.,* **573,** 209
97. Borch, R. F. and Durst, H. D. (1969). *J. Amer. Chem. Soc.,* **91,** 3996
98. Baird, W. C., unpublished observations quoted in Ref. 99
99. Berschied, J. R. and Purcell, K. F. (1970). *Inorg. Chem.,* **9,** 624
100. Lippard, S. J. and Welcker, P. S. (1970). *Chem. Commun.,* 515
101. *Gmelins Handbook of Inorganic Chemistry,* (1970). 8th Ed. System No. 14, Carbon Part. C, Sect. 1.
102. Ergun, S. and Menster, M. (1965). *Chem. Phys. Carbon,* **1,** 204
103. Thomas, I. M. (1965). Ibid., **1,** 122
104. Lang, F. M. and Magnier, P. (1968). Ibid., **3,** 121.
105. Walker, P. L., Shelef, M. and Anderson, R. A. (1968). Ibid., **4,** 287
106. Vacek, B. and Vostatch, M. (1963). *Zvaranie,* **12,** 11
107. Reed, R. M. and Comley, E. A. (1964). *Kirk-Othmer, Encycl. Chem. Technol.* 2nd Ed., **4,** 353
108. Sevin, P., Heurteaux, J. and Lecarme, J. (1970). *J. Chim. Ind., Genie Chim.,* **103,** 803
109. Folkins, H. O. (1964). *Kirk-Othmer, Encycl. Chem. Technol.* 2nd Ed., **4,** 370
110. Reddy, S. R., Gazen, A. K. and Naik, S. C. (1969). *Chem. Age. India,* **20,** 497
111. Thacker, C. M. (1970). *Hydrocarbon Process,* **49,** 124
112. Lappert, M. F. and Prokai, B. (1967). *Advan. Organometal. Chem.,* **5,** 225
113. Bradley, D. C. and Gitlitz, M. H. (1969). *J. Chem. Soc. A,* 1152
114. McCormick, B. J. and Kaplan, R. I. (1970). *Can. J. Chem.,* **48,** 1876
115. Baird, M. C. and Wilkinson, G. (1966). *Chem. Commun.,* 514
116. Idem. (1967). *J. Chem. Soc. A,* 865
117. Baird, M. C., Hartwell, G. and Wilkinson, G. (1967). Ibid., 2037
118. Yagupsky, M. P. and Wilkinson, G. (1968). Ibid., 2813

119. Deeming, A. J. and Shaw, B. L. (1969). Ibid., 1128
120. Commereuc, D., Douek, I. and Wilkinson, G. (1970). Ibid., 1771
120a. Ugo, R. (1968). *Coordin. Chem. Rev.*, **3**, 319
121. Mason, R. and Rae, A. I. M. (1970). *J. Chem. Soc. A*, 1767
122. Mason, R. (1968). *Nature (London)*, **217**, 543
123. McWeeny, R., Mason, R. and Towl, A. P. C. (1969). *Discuss. Faraday Soc.*, **47**, 20
124. Bennett, M. A. and Longstaff, P. A. (1965). *Chem. Ind.*, 846
125. Nyman, C. J., Wymore, C. E. and Wilkinson, G. (1967). *Chem. Commun.*, 407
126. Idem. (1968). *J. Chem. Soc. A*, 561
127. Hayward, P. J., Blake, D. M., Nyman, G. J. and Wilkinson, G. (1969). *Chem. Commun.*, 987
128. Idem., (1970). *J. Amer. Chem. Soc.*, **92**, 5873
129. Iwashita, Y. and Hayata, A. (1969). Ibid., **91**, 2525
130. Baird, M. C. and Wilkinson, G. (1966). *Chem. Commun.*, 267
131. Baddley, W. H. (1966). *J. Amer. Chem. Soc.*, **88**, 4545
132. de Boer, J. L., Rogers, D., Skapski, A. C. and Troughton, P. G. H. (1966). *Chem. Commun.*, 756
133. Klumpp, G. B. and Marko, L. (1968). *J. Organometal. Chem.*, **11**, 207
134. Busetto, L. and Angelici, R. J. (1968). *J. Amer. Chem. Soc.*, **90**, 3283
135. Busetto, L., Belluco, U. and Angelici, R. J. (1969). *J. Organometal. Chem.*, **18**, 213
136. Gilgert, J. D., Baird, M. C. and Wilkinson, G. (1968). *J. Chem. Soc. A*, 2198
137. Butler, I. S. and Fenster, A. E. (1970). *Chem. Commun.*, 933
138. Steudel, R. (1966). *Z. Naturforsch.*, **21b**, 1106
139. Idem., (1967). *Angew. Chem. Int. Ed. Engl.*, **6**, 635
140. Richards, W. G. (1967). *Trans. Faraday Soc.*, **63**, 257
141. Katakis, D. and Taube, H. (1962). *J. Chem. Phys.*, **36**, 412
142. Moll, N. G., Clutter, D. R. and Thomson, W. E. (1966). Ibid., **45**, 4469
143. Krishnamurty, K. V. (1967). *J. Chem. Educ.*, **44**, 594
144. Gimare, M. and Chou, T. S. (1968). *J. Chem. Phys.*, **49**, 4043
145. Cornille, M. and Horsley, J. (1970). *Chem. Phys. Letters*, **6**, 373
146. Batchelor, J. S. P. (1962). *The Preparation of Carbon Suboxide*, (London: H.M. Stationary Office).
147. Bukowski, A. and Porejko, S. (1969). *Wiad. Chem.*, **23**, 679
148. Koschel, D. (1970). *Chem. App.*, **94**, 249
149. Miller, F. A. and Fateley, W. G. (1964). *Spectrochim. Acta*, **20**, 253
150. Lafferty, W. J., Maki, A. G. and Plyler, E. K. (1964). *J. Chem. Phys.*, **40**, 224
151. McDougall, L. A. and Kilpatrick, J. E. (1965). Ibid., **42**, 2311
152. Miller, F. A., Lemmon, D. H. and Witowski, R. E. (1965). *Spectrochim. Acta*, **21**, 1709
153. Smith, W. H. and Leroi, G. E. (1966). *J. Chem. Phys.*, **45**, 1767
154. Pitzer, K. S. and Strickler, S. J. (1964). Ibid., **41**, 730
155. Bell, S., Varadarajan, T. S., Walsh, A. D. Warsop, P. A., Lee, J. and Sutcliffe, L. (1966). *J. Mol. Spectr.*, **21**, 42
156. Reddington, R. L. (1967). *Spectrochim. Acta*, **23A**, 1863
157. Almenningen, A., Arnesen, S. P., Bartiansen, O., Seip, H. M. and Seip, R. (1968). *Chem. Phys. Lett.*, **1**, 569
158. Tanimoto, M., Kuchitsu, K. and Morino, Y. (1970). *Bull. Chem. Soc. Japan*, **43**, 2776
159. Clark, A. and Seip, H. M. (1970). *Chem. Phys. Lett.*, **6**, 452
160. Smith, W. H. and Leroi, G. E. (1966). *J. Chem. Phys.*, **45**, 1778
161. Bates, J. B. and Smith, W. H., unpublished observations quoted in Ref. 164.
162. Blues, E. T. and Bryce-Smith, D. (1969). *Discuss. Faraday Soc.*, **47**, 190
163. Ginsberg, A. P. and Silverthorn, W. E. (1969). *Proc. XII Int. Conf. Coordination Chem.*, 16 (Marrickville, N.S.W.: Science Press)
164. Idem., (1969). *Chem. Commun.*, 823
165. Hardie, D. W. F. (1964). *Kirk-Othmer, Encycl. Chem. Technol. 2nd Ed.* **5**, 128
166. Miller, S. A. (1967). *Chem. Process Eng.*, **48**, No. 4, 79
167. Khazkinskaya, G. M. and Stetskaya, S. (1966). *Nauch. Soobshch, Inst. Gom. Dela., Akad. Nauk SSSR*, **36**, 62
168. Collinge, R. N., Nyholm, R. S. and Tobe, M. L. (1964). *Nature (London)*, **201**, 1322
169. Muller, U. and Dehnicke, R. (1966). *Angew. Chem. Int. Ed. Engl.*, **5**, 841
170. Anderson, P., Klewe, B. and Thom, E. (1967). *Acta Chem. Scand.*, **21**, 1530

171. Sass, R. L. and Bugg, C. (1967). *Acta Crystallogr.*, **23**, 282
172. Bekoe, D. A., Gantzel, P. K. and Trueblood, K. N. (1967). Ibid., **22**, 657
173. Dickens, B. (1967). *Chem. Commun.*, 246
174. Golovina, N. I. and Atovmyan, L. O. (1967). *Zh. Strukt. Khim.*, **8**, 307
175. McDaniel, D. H. and Deiters, R. M. (1966). *J. Amer. Chem. Soc.*, **88**, 2607
176. Stevenson, D. P. and Coppinger, G. M. (1962). Ibid., **84**, 149
177. Weiner, R. F. and Prausnitz, J. M. (1965). *J. Chem. Phys.*, **42**, 3643
178. Blandamer, M. J., Gough, T. E. and Symons, M. C. R. (1966). *Trans. Faraday Soc.*, **62**, 301
179. Morris, H. L., Kulevsky, N., Tamres, M. and Searles, S. (1966). *Inorg. Chem.*, **5**, 124
180. Cotton, F. A. and Wilkinson, G. (1966). *Advanced Inorganic Chemistry* 2nd Ed., Chapter 15, (London: Interscience).
181. Eaborn, C. (1960). *Organosilicon Compounds*, (London: Butterworths).
182. Gilman, H. and Schwebbe, G. L. (1964). *Advan. Organometal. Chem.*, **1**, 89.
183. Sommer, L. H. (1965). *Stereochemistry, Mechanism and Silicon*, (New York: McGraw Hill)
184. Gilman, H., Atwell, W. H. and Cartledge, F. K. (1966). *Advan. Organometal Chem.*, **4**, 1
185. Schott, G. (1967). *Fortsch. Chem. Forsch.*, **9**, 60
186. Schmeisser, M. and Voss, P. (1967). Ibid., **9**, 165
187. Sakurai, H. (1967). *Yukei Gosei Kagaku Kyokai Shi*, **25**, 555
188. Idem., (1967). Ibid., **25**, 642
189. Kumada, M. and Tamao, K. (1968). *Advan. Organometal Chem.*, **6**, 19
190. MacDiarmid, A. G. (1968). *New Pathways in Inorganic Chemistry, (Ebsworth, E.A.V. Ed.)* 149 (Cambridge University Press)
191. Eaborn, C. and Bott, R. W. (1968). *Organometallic Compounds of the Group IV Elements, (MacDiarmid, A.G. Ed.)*, **1**, 105 (New York: Dekker)
192. Mackay, K. M. (1969). *Organometal Chem. Rev.*, **4**, 137
193. Vyazankin, N. S. and Kruglaya, O. A. (1966). *Usp. Khim.*, **35**, 1388
194. Ebsworth, E. A. V. (1968). *Organometallic Compounds of the Group IV Elements, (MacDiarmid A.G. Ed.)*, **1**, 1 (New York: Dekker)
195. Connor, J. A., Finney, G., Leigh, G. J., Robinson, P. J., Sedgwick, R. D. and Simmons, R. F. (1966). *Chem. Commun.*, 178
196. Connor, J. A. Haszeldine, R. N. Leigh, G. J. and Sedgwick, R. D. (1967). *J. Chem. Soc. A*, 768
197. Hess, G. G., Lampe, F. W. and Sommer, L. H. (1965). *J. Amer. Chem. Soc.*, **87**, 5327
198. Eméleus, H. J. and Reid, C. (1935). *J. Chem. Soc.*, 1021
199. Steele, W. C., Nichols, L. D. and Stone, F. G. A. (1962). *J. Amer. Chem. Soc.*, **84**, 4441
200. Salfeld, F. E. and Svec, H. J. (1964). *Inorg. Chem.*, **3**, 1442
201. Band, S. J., Davidson, I. M. T., Lambert, C. A. and Stephenson, I. L. (1967). *Chem. Commun.*, 723
202. Davidson, I. M. T. and Stephenson, I. L. (1968). *J. Chem. Soc. A*, 282
203. Lappert, M. F., Simpson, J. and Spalding, T. R. (1969). *J. Organometal. Chem.*, **17**, P1
204. Lappert, M. F., Pedley, J. B., Simpson, J. and Spalding, T. R. (1971). *J. Organometal. Chem.*, **29**, 195
205. Band, S. J., Davidson, I. M. T. and Lambert, C. A. (1968). *J. Chem. Soc. A*, 2068
206. Barrow, R. F. and Rowlinson, H. C. (1954). *Proc. Roy. Soc. (London)*, **A224**, 374
207. Ogilvie, J. F. and Cradock, S. (1966). *Chem. Commun.*, 364
208. Fritz, G. and Grobe, J. (1962). *Z. Anorg. Allg. Chem.*, **315**, 157
209. Nefedov, O. M. and Manakov, M. N. (1966). *Angew. Chem. Int. Ed. Engl.*, **5**, 1021
210. Gusel'nikov, L. E. and Flowers, M. C. (1967). *Chem. Commun.*, 864
211. Kumada, M., Tamao, K., Ishikawa, M. and Matsumo, N. (1968). Ibid., 614
212. Douglas, B. E. and McDaniel, D. H. (1965). *Concepts and Models of Inorganic Chemistry*, 58. (Boston: Ginn).
213. Pitzer, K. S. (1948). *J. Amer. Chem. Soc.*, **70**, 2140
214. Mulliken, R. S., Riecke, C. A., Orloff, D. and Orloff, H. (1948). *J. Chem. Phys.*, **17**, 1248
215. Jaffé, H. H. (1953). Ibid., **21**, 258
216. Attridge, C. J. (1970). *Organometal. Chem. Rev.*, **5**, 323
217. Mulliken, R. S. (1950). *J. Amer. Chem. Soc.*, **72**, 4493
218. Craig, D. P., Maccoll, A., Nyholm, R. S., Orgel, L. E. and Sutton, L. E. (1954). *J. Chem. Soc.*, 332

219. Pauling, L. (1940). *Nature of the Chemical Bond*, 2nd Ed., 92 (New York: Cornell Press)
220. Szekely, T. (1961). *Magyar Kem. Lapja.*, **16**, 324
221. Idem., (1965). *Magyar. Tud. Akad. Kem. Oszt. Kozleneeny*, **24**, 271
222. Ebsworth, E. A. V. (1965). *Stereochim. Inorg. Acad. Naz. Lincei Corso. Estive Chim. 9th.*, (Rome: Accad Naz. Lincei.)
223. West, R. (1966). *Pure Appl. Chem.*, **13**, 1
224. Seyferth, D., Singh, G. and Suzuki, R. (1966). Ibid., **13**, 159
225. Mitchell, K. A. R. (1969). *Chem. Rev.*, **69**, 157
226. Hedburg, K. (1965). *J. Amer. Chem. Soc.*, **77**, 6491
227. Burg, A. B. and Kuljian, E. S. (1950). Ibid., **72**, 3103
228. Aronson, J. R., Lord, R. C. and Robinson, D. W. (1960). *J. Chem. Phys.*, **33**, 1004
229. McKean, D. C. (1959). *Proc. Chem. Soc.*, **321**
230. Beagley, B., Robiette, A. G. and Sheldrick, G. M. (1967). *Chem. Commun.*, 601
231. Idem., (1968). *J. Chem. Soc. A*, 3002
231a. Ebsworth, E. A. V. (1966). *Pure Appl. Chem.*, **13**, 189
232. Idem., (1966). *Chem. Commun.*, 530
233. Ebsworth, E. A. V., Glidewell, C. and Sheldrick, G. M. (1969). *J. Chem. Soc. A*, 352
234. Baird, M. C. (1968). *Progr. Inorg. Chem.*, **9**, 1
235. Young, J. F. (1968). *Advan. Inorg. Chem. Radiochem.*, **11**, 91
236. Hagen, A. P. and MacDiarmid, A. G. (1967). *Inorg. Chem.*, **6**, 686 and 1941
237. Jetz, W., Simons, P. B., Thompson, J. A. J. and Graham, W. A. G. (1966). Ibid., **5**, 2217
238. Patmore, D. J. and Graham, W. A. G. (1968). Ibid., **7**, 771
239. Dalton, J., Paul, I., Smith, J. G. and Stone, F. G. A. (1968). *J. Chem. Soc. A*, 1199
240. Berry, A. D., Corey, E. R., Hagen, A. P., MacDiarmid, A. G., Saalfeld, F. E. and Wayland, B. B. (1970). *J. Amer. Chem. Soc.*, **92**, 1940
241. Cotton, F. A. and McCleverty, J. A. (1967). *J. Organometal. Chem.*, **4**, 490
242. Cotton, F. A. and Wing, R. M. (1967). Ibid., **9**, 571
243. Robinson, W. T. and Ibers, J. A. (1967). *Inorg. Chem.*, **6**, 1208
244. Emerson, K., Ireland, P. R. and Robinson, W. T., unpublished observations, quoted in Ref. 240
245. Saalfeld, F. E., McDowell, M. V. and MacDiarmid, A. G. (1970). *J. Amer. Chem. Soc.*, **92**, 2324
246. Kuriyama, R. (1961). *Kagaku Kyoto*, **16**, 833
247. Aylett, B. J. (1969). *Progr. Stereochem.*, **4**, 213
248. Thompson, J. G. and Margrave, J. L. (1967). *Science*, **155**, 669
249. Atwell, W. H. and Weyenberg, D. R. (1969). *Angew. Chem. Int. Ed. Engl.*, **8**, 469
250. Morehouse, R. L., Cristiansen, J. J. and Gordy, W. (1966). *J. Chem. Phys.*, **45**, 1751
251. Benkeser, R. A., Foley, K. M., Grutzner, J. B. and Smith, W. E. (1970). *J. Amer. Chem. Soc.*, **92**, 697
252. Alpatova, N. M. and Kessler, Y. M. (1964). *Zh. Struct. Khim.*, **5**, 332
253. Muetterties, E. L. and Schunn, R. A. (1966). *Quart. Rev. Chem. Soc.*, **20**, 245
254. Zvyagintsev, O. E., Babaeva, A. V., Golovnya, V. A. and Sklyarenko, Y. S. (1967). *Razv. Obsch. Anal. Khim. SSSR 1917–1967, Akad. Nauk. SSSR Inst. Istor. Estestvozn. Tekh.*, 137
255. Campbell-Fergusson, H. J. and Ebsworth, E. A. V. (1967). *J. Chem. Soc. A*, 705
256. Fergusson, J. E., Grant, D. K., Hickford, R. H. and Wilkins, C. J. (1959). *J. Chem. Soc.*, 99
257. Beattie, I. R. (1963). *Quart. Rev. Chem. Soc.*, **17**, 382
258. Clark, H. C., Corfield, P. W. R., Dixon, K. R. and Ibers, J. A. (1967). *J. Amer. Chem. Soc.*, **89**, 3360
259. Beattie, I. R. and Gilson, T. (1965). *J. Chem. Soc.*, 6595
260. Aylett, B. J. and Campbell, J. M. (1969). *J. Chem. Soc. A*, 1920
261. Bain, V., Killean, R. C. G. and Webster, M. (1969). *Acta Crystallogr.*, **B25**, 156
262. Blayden, A. E. and Webster, M. (1970). *Inorg. Nucl. Chem. Lett.*, **6**, 703
263. Beattie, I. R., Gilson, T. R. and Ozin, G. A. (1968). *J. Chem. Soc. A*, 2772
264. Beattie, I. R. and Ozin, G. A. (1970). Ibid., 370
265. Gielen, M. and Spreckner, N. (1966). *Organometal. Chem. Rev.*, **1**, 455
266. Hengge, E. (1968). *Allg. Prakt. Chem.*, **19**, 310
267. Runyan, W. R. (1969). *Kirk-Othmer, Encycl. Chem. Technol* 2nd. Ed., **18**, 111
268. Simms, E. G. (1969). Ibid., **18**, 125

269. Spenke, E. (1969). *Semicond. Silicon, Int. Symp., Pap. 1st.,* (New York: Electrochem. Soc. Inc.)
270. Runyan, W. R. (1965). *Silicon Semiconductor Technology,* (New York: McGraw Hill)
271. Singh, K. and Pandey, L. P. (1968). *J. Sci. Ind. Res.,* **27,** 386
272. Runyan, W. R. (1966). *Vapour Deposition,* (New York: Wiley).
273. Noda, T. (1966). *Kagaku Kyoto,* **21,** 366
274. Healey, G. W. (1970). *Earth Miner. Sci.,* **39,** 46
275. Petrov, A. D., Mironov, V. F., Ponamerenko, V. A. and Chemyshev, E. A. (1964). *Synthesis of Organosilicon Monomers,* (New York: Consultants Bureau)
276. Zuckerman, J. J. (1964). *Adv. Inorg. Chem. Radiochem.,* **6,** 383
277. Bazant, V. (1964). *Periodica Polytech.,* **8,** 115
278. Idem., (1966). *Pure Appl. Chem.,* **13,** 313
279. Bonitz, E. (1966). *Angew. Chem. Int. Ed. Engl.,* **5,** 462
280. Voorhoeve, R. J. H. (1967). *Organohalosilanes, Precursors to Silicones,* (New York: Elsevier)
281. Bazant, V., Joklik, J. and Rathousky, J. (1968). *Angew. Chem. Int. Ed. Engl.,* **7,** 112
282. Meals, R. (1969). *Kirk-Othmer, Encycl. Chem. Technol., 2nd Ed.,* **18,** 221
283. Bonitz, E. (1961). *Chem. Ber.,* **94,** 220
284. Newton, W. E. and Rochow, E. G. (1970). *Inorg. Chem.,* **9,** 1071
285. MacDiarmid, A. G. (1961). *Advan. Inorg. Chem. Radiochem.,* **3,** 207
286. Stone, F. G. A. (1962). *Hydrogen Compounds of the Group IV Elements,* (Englewood Cliffs, N.J.: Prentice-Hall)
287. Ebsworth, E. A. V. (1963). *Volatile Silicon Compounds,* (London: Pergamon)
288. Mackay, K. M. (1966). *Hydrogen Compounds of the Metallic Elements,* (London: Spon.)
289. Hinckley, A. A. (1966). *Kirk-Othmer, Encycl. Chem. Technol. 2nd.,* **11,** 200
290. Shaw, B. L. (1967). *Inorganic Hydrides,* (London: Pergamon)
291. Armirotto, A. L. (1968). *Solid State Technol.,* **11,** 43
292. Aylett, B. J. (1968). *Advan. Inorg. Chem. Radiochem.,* **11,** 249
293. Van Dyke, C. H. (1969). *Kirk-Othmer. Encycl. Chem. Technol. 2nd. Ed.,* **18,** 172
294. Wiberg, E. and Amberger, E. (1970). *Hydrides of the Elements of the Main Groups I-IV.* (New York: Elsevier)
295. Stock, A. (1933). *Hydrides of Boron and Silicon,* (Ithaca, N.Y.: Cornell University Press)
296. Greenwood, N. N. (Senior Reporter) (1968, 1969). *Spectroscopic Properties of Inorganic and Organometallic Compounds, Volumes* 1 *and* 2. (London: Chem. Soc.)
297. Muetterties, E. L. and Phillips, W. D. (1962). *Advan. Inorg. Chem. Radiochem.,* **4,** 231
298. Jolly, W. L. and Norman, A. D. (1968). *Preparative Inorg. Reactions. (Jolly, W. L. Ed.),* **4,** 1
299. Finholt, A. E., Bond, A. C., Wilzbach, K. E. and Schlesinger, H. I. (1947). *J. Amer. Chem. Soc.,* **69,** 2692
300. Thompson, M. L.-R. (1964). *Dissertation Abst.,* **25,** 2836
301. Gaspar, P. P., Levy, C. A. and Adair, G. M. (1970). *Inorg. Chem.,* **9,** 1272
302. Ward, L. G. L. and MacDiarmid, A. G. (1960). *Abstr. 137th Meeting, Amer. Chem. Soc.,* 11M
303. Urry, G. (1963). *Inorg. Chem.,* **2,** 432
304. Feher, F., Kuhlborsch, G. and Luhleich, H. (1960). *Z. Anorg. Allg. Chem.,* **303,** 283
305. Idem., (1960). *Ibid.,* **303,** 294
306. Borer, K. and Phillips, C. S. G. (1959). *Proc. Chem. Soc.,* 189
307. Spanier, E. J. and MacDiarmid, A. G. (1962). *Inorg. Chem.,* **1,** 432
308. Gokhale, S. D. and Jolly, W. L. (1964). *Ibid.,* **3,** 946
309. Gokhale, S. D., Drake, J. E. and Jolly, W. L. (1965). *J. Inorg. Nucl. Chem.,* **27,** 1911
310. Andrews, T. D. and Phillips, C. S. G. (1966). *J. Chem. Soc. A,* 46
311. Pavlov, A. M., Bodyagin, G. N. and Agaforrov, I. L. (1967). *Tr. Khim. Khim. Technol.,* 175
312. Purnell, J. H. and Walsh, R. (1966). *Proc. Roy. Soc. Ser. A,* **293,** 543
313. Ring, M. A., Puentes, M. J. and O'Neal, H. E. (1970). *J. Amer. Chem. Soc.,* **92,** 4845
314. Ring, M. A., Beverley, G. D., Koester, F. H. and Hollandsworth, R. P. (1969). *Inorg. Chem.,* **8,** 2033
315. Tebben, E. M. and Ring, M. A. (1969). *Ibid.,* **8,** 1787
316. Estacio, P., Sefcik, M. D., Chan, E. K. and Ring, M. A. (1970). *Ibid.,* 1068
317. Bowrey, M. and Purnell, J. H. (1970). *J. Amer. Chem. Soc.,* **92,** 2594

318. Kennedy, R. C., Freeman, L. P., Fox, A. P. and Ring, M. A. (1966). *J. Inorg. Nucl. Chem.*, **28**, 1373
319. Morrison, J. A. and Ring, M. A. (1967). *Inorg. Chem.*, **6**, 100
320. Carey, N. A. D. and Ebsworth, E. A. V. (1969). *J. Inorg. Nucl. Chem.*, **31**, 2953
321. Ring, M. A., Baird, R. B. and Estacio, P. (1970). *Inorg. Chem.*, **9**, 1004
322. Benesovsky, F. (1967). *Kirk-Othmer, Encycl. Chem. Technol. 2nd. Ed.*, **13**, 814
323. Oliver, D. A. (1962). *High Temp. Aeron. Proc. Symp. Turin*, 377
324. Carruthers, J. R. (1968). *Kirk-Othmer, Encycl. Chem. Technol. 2nd Ed.*, **17**, 862
325. Langheinrich, W. and Eisbrenner, D. (1969). *MetalloBerflaeche*, **23**, 129
326. Kobayashi, K., Haneta, Y. and Nakanuma, S. (1969). *J. Electrochem. Soc. Jap. (Overseas Ed.)*, **37**, 87
327. Kuwemo, Y. (1969). *Jap. J. Appl. Phys.*, **8**, 876
328. Billy, M. and Goursat, P. (1969). *C.R. Acad. Sci. Ser. C*, **269**, 919
329. Idem., (1970). *Rev. Chim. Miner.*, **7**, 193
330. Fessenden, R. J. and Fessenden, J. S. (1961). *Chem. Rev.*, **61**, 361
331. Aylett, B. J. and Burnett, G. M. (1961). *Soc. Chem. Ind. (London)*, Monograph, **13**, 5
332. Haiduc, I. (1961). *J. Chem. Educ.*, **38**, 134
333. Wannagat, U. (1964). *Advan. Inorg. Chem. Radiochem.*, **6**, 225
334. Idem., (1966). *Pure Appl. Chem.*, **13**, 263
335. Fink, W. (1966). *Angew. Chem. Intern. Ed. Engl.*, **5**, 760
336. Wannagat, U. (1967). *Fortsch. Chem. Forsch.*, **9**, 102
337. Haiduc, I. (1967). *Stud. Cercet. Chim.*, **15**, 71
338. Scherer, O. J. (1968). *Organometal. Chem. Rev. Sect. A*, **3**, 281.
339. Drake, J. E. and Riddle, C. (1970). *Quart. Rev. Chem. Soc.*, **24**, 263
340. Haiduc, I. (1970). *The Chemistry of Inorganic Ring Systems*, (London: Wiley-Interscience)
341. Aylett, B. J. (1965). *Preparative Inorg. Reactions. (Jolly, W. L. Ed)*, **2**, 93
342. Aylett, B. J. and Hakim, M. J. (1966). *Inorg. Chem.*, **5**, 167
343. Idem (1969). *J. Chem. Soc. A*, 636
344. Wells, R. L. and Schaeffer, R. (1966). *J. Amer. Chem. Soc.*, **88**, 37
345. Aylett, B. J. (1956). *J. Inorg. Nucl. Chem.*, **2**, 325
346. Wannagat, U., Schmidt, P. and Schulze, M. (1967). *Angew. Chem. Int. Ed. Engl.*, **6**, 446
347. Idem. (1967). *Ibid.*, **6**, 447
348. Wannagat, U., Schulze, M. and Burger, H. (1970). *Z. Inorg. Allg. Chem.*, **375**, 157
349. Glidewell, C., Rankin, D. W. H., Robiette, A. G. and Sheldrick, G. M. (1969). *J. Mol. Struct.*, **4**, 215
350. Rankin, D. W. H., Robiette, A. B., Sheldrick, G. M., Sheldrick, W. S., Aylett, B. J., Ellis, I. A. and Monaghan, J. J. (1969). *J. Chem. Soc. A*, 1224
351. Vilkov, L. V. and Tarasenko, N. A. (1969). *Chem. Commun.*, 1176
352. Glidewell, C., Rankin, D. W. H., Robiette, A. G. and Sheldrick, G. M. (1970). *J. Chem. Soc. A*, 318
353. Idem. (1970). *J. Mol. Struct.*, **6**, 231
354. Robiette, A. G., Sheldrick, G. M. and Sheldrick, W. S. (1970). *Ibid.*, **5**, 423
355. Ebsworth, E. A. V., Jenkins, D. R., Mays, M. J. and Sugden, T. M. (1963). *Proc. Chem. Soc.*, 21
356. Ebsworth, E. A. V. and Mays, M. J. (1963). *J. Chem. Soc.*, 4844
357. Jenkins, D. R., Kewley, R. and Sugden, T. M. (1960). *Proc. Chem. Soc.*, 220
358. Idem. (1962). *Trans. Faraday Soc.*, **58**, 1284
359. Ebsworth, E. A. V., Mould, R., Taylor, R., Wilkinson, G. R. and Woodward, L. A. (1962). *Ibid.*, **58**, 1069
360. Hilderbrandt, R. L. and Bauer, S. H. (1969). *J. Mol. Struct.*, **3**, 325
361. Gillette, R. H. and Brockway, L. O. (1940). *J. Amer. Chem. Soc.*, **62**, 3236
362. Koster, D. F. (1968). *Spectrochim. Acta*, **24**, 395
363. Scherer, O. J. (1968). *Organometal. Chem. Rev. Sect. A*, **3**, 281
364. Sommer, L. H., Citron, J. D. and Parker, G. A. (1969). *J. Amer. Chem. Soc.*, **91**, 4729
365. Fritz, G. (1966). *Angew. Chem. Int. Ed. Engl.*, **5**, 53
366. Chemyshev, E. A. and Bugerenko, E. F. (1968). *Organometal. Chem. Rev. Sect. A*, **3**, 469
367. Fluck, E. and Novobilsky, V. (1969). *Fortsch: Chem. Forsch.*, **13**, 125
368. Abel, E. W. and Illingworth, S. M. (1970). *Organometal. Chem. Rev. Sect. A*, **5**, 143
369. MacDiarmid, A. G. (1969). *Abst. 158th. Meeting, Amer. Chem. Soc.*, INOR 122
370. Drake, J. E., Goddard, N. and Riddle, C. (1969). *J. Chem. Soc. A*, 2704

371. Fritz, G., Becker, G. and Kummer, D. (1970). *Z. Inorg. Allgem. Chem.*, **372**, 171
372. Fritz, G. and Becker, G. (1970). Ibid., **372**, 196
373. Fritz, G. (1953). *Z. Naturforsch.*, **8B**, 776
374. Drake, J. E. and Jolly, W. L. (1961). *Chem. Ind.*, 1470
375. Glidewell, C. and Sheldrick, G. M. (1969). *J. Chem. Soc. A*, 350
376. Amberger, E. and Boeters, H. D. (1964). *Chem. Ber.*, **97**, 1999
377. Norman, A. D. (1968). *Chem. Commun.*, 812
378. Idem. (1968). *J. Amer. Chem. Soc.*, **90**, 6556
379. Norman, A. D. and Wingeleth, D. C. (1970). *Inorg. Chem.*, **9**, 98
380. Norman, A. D. (1970). Ibid., **9**. 870
381. Anderson, J. W. and Drake, J. E. (1969). *Inorg. Nucl. Chem. Lett.*, **5**, 887
382. Crosbie, K. D., Glidewell, C. and Sheldrick, G. M. (1969). *J. Chem. Soc. A*, 1861
383. Moedritzer, K. (1966). *Organometal. Chem. Rev.*, **1**, 729
384. Idem. (1968). *Advan. Organometal. Chem.*, **6**, 171
385. Drake, J. E., Goddard, N. and Simpson, J. (1968). *Inorg. Nucl. Chem. Lett.*, **4**, 361
386. Cradock, S. Davidson, G., Ebsworth, E. A. V. and Woodward, L. A. (1965). *Chem. Commun.*, 515
387. Ebsworth, E. A. V., Rankin, D. W. H. and Sheldrick, G. M. (1968). *J. Chem. Soc. A*, 2828
388. Drake, J. E. and Riddle, C. (1968). Ibid., 1675
389. Mackay, K. M., Sutton, K. J., Stobart, S. R., Drake, J. E. and Riddle, C. (1969). *Spectrochim. Acta*, **25A**, 925
390. Idem. (1969). Ibid., **25A**, 941
391. Davidson, G., Ebsworth, E. A. V., Sheldrick, G. M. and Woodward, L. A. (1966). *Spectrochim. Acta*, **22**, 67
392. Cowley, A. H. and White, W. D. (1967). *Abstr. 153rd. Meeting, Amer. Chem. Soc.*, 145L
393. Beagley, B., Robiette, A. G. and Sheldrick, G. M. (1968). *J. Chem. Soc. A*, 3006
394. Rankin, D. W. H., Robiette, A. G., Sheldrick, G. M., Beagley, B. and Hewitt, T. G. (1969). *J. Inorg. Nucl. Chem.*, **31**, 2351
395. Drake, J. E. and Goddard, N. (1969). *J. Chem. Soc. A*, 662
396. Drake, J. E. and Simpson, J. (1967). *Chem. Commun.*, 249
397. Idem. (1967). *Inorg. Chem.*, **6**, 1984
398. Russ, C. R. and MacDiarmid, A. G. (1966). *Angew. Chem. Int. Ed. Engl.*, **5**, 418
399. Drake, J. E. and Simpson, J. (1968). *J. Chem. Soc. A*, 1039
400. Riddle, C. and Simpson, J., unpublished observations quoted in reference 339
401. Coyle, T. D. (1969). *Kirk-Othmer, Encycl. Chem. Technol.*, **18**, 46
402. Maher, P. K. (1969). Ibid., **18**, 61
403. Dumbaugh, W. H. and Schultz, P. C. (1969). Ibid., **18**, 73
404. Landise, R. A. and Ballman, A. A. (1969). Ibid., **18**, 105
405. Wills, J. H. (1969). Ibid., **18**, 134
406. Bass, R. L. (1961). *Rept. Progr. Appl. Chem.*, **46**, 286
407. Tamura, K. and Nakajima, I. (1963). *Kogyo. Kagaku. Zasshi*, **66**, 523
408. Hengge, E. (1967). *Fortschr. Chem. Forsch.*, **9**, 145
409. Noll, W. (1968). *Chemistry and Technology of Silicones*, (New York: Academic Press).
410. Andrianov, K. A., Khaiduk, I. and Khanaansgvili, L. M. (1963). *Usp. Khim.*, **32**, 539
411. Spialter, L. and Smith, J. S. (1967). *Decomp. of Organometal. Compds., Refract. Ceram. Metals and Metal Alloys, Proc. Int. Symp.*, 195
412. Haas, A. (1965). *Angew. Chem. Int. Ed. Engl.*, **4**, 1014
413. Millard, M. M., Pazdernick, L. J. and Doyle, J. R. (1969). *U.S. Clearinghouse, Fed. Sci. Tech. Inform.*, AD 1969. No. 700, 139
414. Benyon, J. (1970). *Vacuum*, **20**, 293
415. Frye, C. L. and Collins, W. T. (1970). *J. Amer. Chem. Soc.*, **92**, 5586
416. Glidewell, C. (1969). *J. Inorg. Nucl. Chem.*, **31**, 1303
417. Glidewell, C., Rankin, D. W. H. and Sheldrick, G. M. (1969). *Trans. Faraday Soc.*, **65**, 1409
418. Kifer, E. W. and Van Dyke, C. H. (1969). *Chem. Commun.*, 1330
419. Hengge, E. (1967). *Halogen Chem. (Gutmann, V. Ed.)*, **2**, 169
420. Angus, H. F., Cradock, S. and Ebsworth, E. A. V. (1969). *Inorg. Nucl. Chem. Lett.*, **5**, 717
421. Weiss, G. S. and Nixon, E. R. (1965). *Spectrochim. Acta*, **21**, 903
422. Glidewell, C. and Rankin, D. W. H. (1969). *J. Chem. Soc. A*, 753

423. Gibbon, G. A., Van Dyke, C. H., Sprecher, R., Wang, W. T. and Hembre, J. I. (1969). *Abstr.* 157*th Meeting, Amer. Chem. Soc.*, INOR 159
424. Sternbach, B. and MacDiarmid, A. G. (1961). *J. Amer. Chem. Soc.*, **83**, 3384
425. Idem. (1961). *J. Inorg. Nucl. Chem.*, **23**, 225
426. Drake, J. E. and Riddle, C. (1970). *J. Chem. Soc. A*, 3134
427. Ganckberg, A. and Vandervelde, J. (1964). *Ind. Chim. Belge.*, **29**, 591
428. Anderson, A. R. (1969). *Kirk-Othmer, Encycl. Chem. Technol.*, **18**, 216
429. Abel, E. W. and Armitage, D. A. (1967). *Advan. Organometal. Chem.*, **5**, 1
430. Glidewell, C., Rankin, D. W. H., Robiette, A. G., Sheldrick, G. M. Beagley, B. and Freeman, J. M. (1970). *J. Mol. Struct.*, **5**, 417
431. Idem. (1969). *Trans. Faraday Soc.*, **65**, 2621
432. Almenningen, A., Bastiansen, O., Ewing, V., Hedburg, K. and Traetteberg, M. (1963). *Acta Chem. Scand.*, **17**, 2455
433. Airey, W., Glidewell, C., Rankin, D. W. H., Robiette, A. G., Sheldrick, G. M. and Cruickshank, D. W. J. (1970). *Trans. Faraday Soc.*, **66**, 551
434. Almenningen, A., Hedberg, K. and Seip, R. (1963). *Acta Chem. Scand.*, **17**, 2264
435. Almenningen, A., Fernholt, L. and Seip, H. M. (1968). *Ibid.*, **22**, 51
436. Wang, J. T. and Van Dyke, C. H. (1967). *Chem. Commun.*, 612
437. Drake, J. E. and Riddle, C. (1970). *Inorg. Nucl. Chem. Lett.*, **6**, 713
438. Dreeskamp, H. and Pfisterer, G. (1968). *Mol. Phys.*, **14**, 295
439. West, R., Whatley, L. S. and Lake, K. J. (1961). *J. Amer. Chem. Soc.*, **83**, 761
440. Van Dyke, C. H. and MacDiarmid, A. G., unpublished observations quoted in Ref. 293
441. Onyszchuk, M. (1961). *Can. J. Chem.*, **39**, 808
442. Van Dyke, C. H. and MacDiarmid, A. G. (1964). *Inorg. Chem.*, **3**, 747
443. Yoshioka, T. and MacDiarmid, A. G. (1969). *Inorg. Nucl. Chem. Lett.*, **5**, 69
444. MacDiarmid, A. G. (1963). *J. Inorg. Nucl. Chem.*, **25**, 1534
445. *Gmelins Handbook of Inorganic Chemistry*. (1959). Silicon Part B.
446. Belyaev, A. I., Nisel'son, L. A. and Petrusevich, I. V. (1968). *Izv. Akad. Nauk. SSSR Metal.*, 65
447. Nagai, Y. (1970). *Kagaku Kyoto.*, **25**, 544
448. MacDiarmid, A. G. (1964). *Preparative Inorg. Reactions (Jolly, W. L. Ed.)*, **1**, 165
449. Kemmit, R. D. W. and Sharp, D. W. A. (1965). *Advan. Fluorine Chem.*, **4**, 142
450. Haas, A. (1969). *Chem. Unserer Zeit*, **3**, 17
451. Lappert, M. F. and Pyszora, H. (1966). *Advan. Inorg. Chem. Radiochem.*, **9**, 133
452. Prejzner, J. (1969). *Wiad. Chem.*, **23**, 601
453. Anderson, A. R. (1969). *Kirk-Othmer, Encycl. Chem. Technol.* 2*nd Ed.*, **18**, 166
454. Wayenberg, D. R., Mahone, L. G. and Atwell, W. H. (1969). *Ann. N.Y. Acad. Sci.*, **159**, 38
455. Margrave, J. L. (1963). *U.S. At. Energy Comm.*, AROD1428 : 9
456. Nefedov, O. M. and Manakov, M. N. (1966). *Angew. Chem. Int. Ed. Engl.*, **5**, 1021
457. Timms, P. L. (1968). *Endeavour*, **27**, 133
458. Idem. (1968). *Preparative Inorg. Reactions*, **4**, 59
459. Idem. (1969). *U.S. Clearinghouse, Fed. Sci. Tech. Inform.*, AD1969 No. 698476
460. Idem. (1968). *Inorg. Chem.*, **7**, 387
461. Timms, P. L., Ehlert, T. C., Margrave, J. L., Brinckmann, F. E., Farrar, T. C. and Coyle, T. D. (1965). *J. Amer. Chem. Soc.*, **87**, 3819
462. Timms, P. L., Stump, D. D., Kent, R. A. and Margrave, J. L. (1966). *Ibid.*, **88**, 940
463. Thompson, J. C. and Margrave, J. L. (1966). *Chem. Commun.*, 566
464. Solan, D. and Timms, P. L. (1968). *Inorg. Chem.*, **7**, 2157
465. Sharp, K. G. and Margrave, J. L. (1969). *Ibid.*, **8**, 2655
466. Margrave, J. L., Sharp, K. G. and Wilson, P. W. (1969). *Inorg. Nucl. Chem. Lett.*, **5**, 995
467. Idem. (1970). *J. Inorg. Nucl. Chem.*, **32**, 1813
468. Idem. (1970). *J. Amer. Chem. Soc.*, **92**, 1530
469. Massey, A. G. (1963). *J. Chem. Educ.*, **40**, 311
470. Jolly, W. L. (1963). *Tech. Inorg. Chem.*, **1**, 179
471. Kana'an, A. S. and Margrave, J. L. (1964). *Advan. Inorg. Chem. Radiochem.*, **6**, 143
472. Urry, G. (1970). *Accounts Chem. Res.*, **3**, 306
473. Kaczmarczyk, A., Millard, M., Nuss, J. W. and Urry, G. (1964). *J. Inorg. Nucl. Chem.*, **26**, 421
474. Joiner, J. and Urry, G., unpublished observations quoted in Ref. 472

475. Schmeisser, P. and Voss, P. (1967). *Fortschr. Chem. Forsch.*, **9**, 165
476. Calus, R. (1966). *Pure Appl. Chem.*, **13**, 61
477. Benkeser, R. A. (1966). Ibid., **13**, 133
478. Benkeser, R. A. and Smith, W. E. (1969). *J. Amer. Chem. Soc.*, **91**, 1566
479. Benkeser, R. A., Gaul, J. M. and Smith, W. E. (1969). Ibid., **91**, 3666
480. Benkeser, R. A., Foley, K. M. Gaul, J. M., Li, G. S. and Smith, W. E. (1969). Ibid., 4578
481. Benkeser, R. A. and Gaul, J. M. (1970). Ibid., 720
482. Chan, T. H. and Melnyk, A. (1970). Ibid., **92**, 3718
483. Chan, T. H., Montillier, J. P., Van Horn, W. F. and Harpp, D. N. (1970). Ibid., **92**, 7224
484. Beck, K. R. and Benkeser, R. A. (1970). *J. Organometal. Chem.*, **21**, P35
485. Aylett, B. J. and Ellis, I. A. (1960). *J. Chem. Soc.*, 3415
486. Abedini, M. and MacDiarmid, A. G. (1963). *Inorg. Chem.*, **2**, 608
487. Drake, J. E. and Simpson, J. (1966). *Inorg. Nucl. Chem. Lett.*, **2**, 219
488. Van Dyke, C. H. and MacDiarmid, A. G. (1963). *J. Inorg. Nucl. Chem.*, **25**, 1503
489. Hollandsworth, R. P., Ingle, W. M. and Ring, M. A. (1967). *Inorg. Chem.*, **6**, 844
490. Hollandsworth, R. P. and Ring, M. A. (1968). Ibid., **7**, 1635
491. Drake, J. E. and Goddard, N. (1968). *Inorg. Nucl. Chem. Lett.*, **4**, 385
492. Idem. (1970). *J. Chem. Soc. A*, 2587
493. Jensen, M. A. (1968). *J. Organometal. Chem.*, **11**, 423
494. Voronkov, M. G. (1966). *Pure Appl. Chem.*, **13**, 35
495. Mueller, R. (1965). *Z. Chem.*, **5**, 220
496. Clark, H. C. and Tsang, W. F. (1967). *J. Amer. Chem. Soc.*, **89**, 529
497. Clark, H. C., Dixon, K. R. and Nicolson, J. G. (1969). *Inorg. Chem.*, **8**, 450
498. Kleboth, K. (1970). *Monatsh. Chem.*, **101**, 357
499. Beattie, I. R. and Livingston, K. M. (1969). *J. Chem. Soc. A*, 859
500. Aylett, B. J. (1960). *J. Inorg. Nucl. Chem.*, **15**, 87
501. Ahrland, S., Chatt, J. and Davies, N. R. (1958). *Quart. Rev. Chem. Soc.*, **12**, 265
502. Ozin, G. A. (1969). *Chem. Commun.*, 104
503. Beattie, I. R. and Ozin, G. A. (1969). *J. Chem. Soc. A*, 2267
504. Stone, F. G. A. (1958). *Chem. Rev.*, **58**, 101
505. Ring, M. A. and Ritter, D. M. (1961). *J. Amer. Chem. Soc.*, **83**, 802
506. Cradock, S., Gibbon, G. A. and Van Dyke, C. H. (1967). *Inorg. Chem.*, **6**, 1751
507. Weiss, E., Hencken, G. and Kuhr, H. (1970). *Chem. Ber.*, **103**, 2868
508. Amberger, E. and Romer, R. (1966). *Z. Anorg. Allg. Chem.*, **345**, 1
509. Hagenmuller, P. and Pouchard, M. (1964). *Bull. Soc. Chim. France*, 1187
510. Gaines, D. F. and Iorms, T. V. (1968). *J. Amer. Chem. Soc.*, **90**, 6617
511. Dahl, L. F. and Calabrise, J., unpublished observations quoted in Ref. 510
512. Geisler, T. C. and Norman, A. D. (1970). *Inorg. Chem.*, **9**, 2167
513. Varma, R. and Cox, A. P. (1964). *Angew. Chem. Int. Ed. Engl.*, **3**, 586
514. Dutton, W. A. and Onyszchuk, M. (1968). *Inorg. Chem.*, **7**, 1735
515. Spanier, E. J. and MacDiarmid, A. G. (1963). Ibid., **2**, 215
516. Mackay, K. M., Horsfield, S. T. and Stobart, S. R. (1969). *J. Chem. Soc. A*, 2937
517. Timms, P. L., Simpson, C. C. and Phillips, C. S. G. (1964). *J. Chem. Soc.*, 1467
518. Royen, P. and Rocktaschel. (1964). *Angew. Chem. Int. Ed. Engl.*, **3**, 314
519. Timms, P. L. and Phillips, C. S. G. (1963). *Anal. Chem.*, **35**, 505
520. Phillips, C. S. G., Powell, P., Semlyen, J. A. and Timms, P. L. (1963). *Z. Anal. Chem.*, **197**, 202
521. Wiberg, E., Amberger, E. and Cambensi, H. (1967). *Z. Anorg. Allg. Chem.*, **351**, 164
522. Amberger, E. and Muhlhofer, E. (1968). *J. Organometal. Chem.*, **12**, 55
523. Solan, D. (1969). *U.S. Clearinghouse, Fed. Sci. Tech. Inform.*, PB-187819
524. Piper, T. S., Lemal, D. and Wilkinson, G. (1956). *Naturwiss.*, **43**, 129
525. Stone, F. G. A. (1968). *New Pathways in Inorganic Chemistry, (Ebsworth, E. A. V. Ed.)*, 283, (Cambridge: University Press)
526. Lewis, J. and Nyholm, R. S. (1964). *Sci. Progr.*, **52**, 557
527. Cross, R. J. (1967). *Organometal. Chem. Rev.*, **2**, 97
528. Sasaki, Y. (1968). *Kagaku No Ryoiki*, **22**, 906
529. Kolobora, N. E., Antonova, A. B. and Anisimov, K. N. (1968). *Usp. Khim.*, **38**, 1802
530. Vyazankin, N. S., Razuvaev, G. A. and Kruglaya, O. A. (1968). *Organometal. Chem. Rev. Sect. A*, **3**, 323

531. Czakis-Sulikowsa, H. and Kuznic, B. (1969). *Wiad. Chem.*, **23,** 17
532. Abel, E. W. and Stone, F. G. A. (1969). *Quart. Rev. Chem. Soc.*, **23,** 325
533. Aylett, B. J. and Campbell, J. M. (1965). *Chem. Commun.*, 217
534. Hagen, A. P. and MacDiarmid, A. G. (1970). *Inorg. Nucl. Chem. Lett.*, **6,** 345
535. Chalk, A. J. and Harrod, J. F. (1965). *J. Amer. Chem. Soc.*, **87,** 1133..
536. Harrod, J. F., Gilson, D. F. R. and Charles, R. (1969). *Can. J. Chem.*, **47,** 2205
537. Schmid, G. and Balk, H-J. (1970). *Chem. Ber.*, **103,** 2240
538. Aylett, B. J., Campbell, J. M. and Walton, A. (1968). *Inorg. Nucl. Chem. Lett.*, **4,** 79
539. Aylett, B. J. and Campbell, J. M. (1967). *Ibid.*, **3,** 137
540. Bentham, J. E. and Ebsworth, E. A. V. (1970). *Ibid.*, **6,** 145
541. Findiesen, B. (1969). Ger. (East) Patent, 68 685
542. Idem. (1970). *Carbon*, **8,** 396
543. Bokros, J. C. (1970). *Ger. Offen.*, 1 950 066
544. Inoue, K. (1970). *U.S. Patent*, 3 513 014
545. Leeds, D. H. and Heicklen, J. (1969). *Ind. Eng. Chem. Prod. Res. Develop.*, **8,** 233
546. Inoue, K. (1970). *Japan. Patent*, 7 011 003
547. Diefendorf, R. J. (1969). *Fr. Patent*, 1 552 357
548. Turkal, H. and Robba, W. A. (1969). *Brit. Patent*, 1 173 573
549. Mayr, K. A. M. and Flam, A. J. (1969). *Ibid.*, 1 158 637
550. Clark, T. J. and Ettinger, B. L. (1969). *U.S. Patent*, 3 462 522
551. Emyashev, A. V. (1969). *Khim. Tverd. Topl.*, 127
552. General Electric Co. (1970). *Brit. Patent*, 1 206 118
553. Kovalevskii, N. N., Rogailin, M. I. and Farbarov, I. L. (1970). *Khim. Tverd. Topl.*, 141
554. Ettinger, B. L. (1969). *Brit. Patent*, 3 457 042
555. Volkov, G. M. and Perevezentsev, V. P. (1969). *Zavod. Lab.*, **35,** 48
556. Patel, A. R. and Deshapande, S. V. (1970). *Carbon*, **8,** 242
557. Duerrwacchter, E. (1969). *Ger. Patent*, 1 533 230
558. Baker, C. (1969). *Carbon*, **7,** 293
559. Shindo, A. (1970). *U.S. Patent*, 3 529 934
560. Skyoryhina, I. R., Gusev, S. S., Vorobeva, N. K. and Ermolenko, I. N. (1970). *Vesti. Akad. Navuk. Belarus. SSR Khim. Navuk.*, 29
561. Miyarmich, K. (1970). *Ger. Offen.*, 1 955 474
562. Shindo, A., Nakamishi, Y. and Sormea, I. (1969). *Appl. Polym. Symp.*, **9,** 271
563. Mikawa, S. (1970). *Chem. Econ. Eng. Rev.*, **2,** 43
564. Fukuoko, Y. (1969). *Jap. Chem. Quart.*, **5,** 63
565. Ishikawa, T. and Morishita, M. (1969). *Japan. Patent*, 6 902 510
566. Kavamura, K. and Jenkins, G. M. (1970). *J. Mater. Sci.*, **5,** 262
567. Idem. (1970). *Ger. Offen.*, 1 944 908
568. Accountius, O. E. (1969). *Fr. Patent*, 1 535 800
569. Idem. (1969). *U.S. Patent*, 3 443 899
570. Logsdail, D. H. (1969). *Appl. Polym. Symp.*, **9,** 245
571. Watt, W. and Johnson, W. (1969). *Ibid.*, **9,** 215
572. Turner, W. N. and Johnson, F. C. (1969). *J. Appl. Polym. Sci.*, **13,** 2073
573. Moreton, R. (1969). *U.S. Clearinghouse, Fed. Sci. Tech. Inform.*, AD1969 No. 699492
574. Lundquist, B. C. (1969). *Ny. Tek.*, 12
575. Johnson, D. J. (1970). *Nature*, **226,** 750
576. Astin, M. and Barker, A. J. (1970). *Birmingham Univ., Chem. Eng.*, **21,** 45
577. Thorne, D. J., Gough, V. J. and Hippkiss, G. (1970). *Fibre Sci. Technol.*, **3,** 90
578. Johnson, W., Phillips, L. N. and Wa, H. W. (1969). *Brit. Patent*, 1 148 874
579. Cuckson, K. and Chapman, G. W. (1969). *Ibid.*, 1, 165 307
580. Gresham, H. E., Hall, D. W., Hannah, C. G., Phillips, D. J. and Hewitt, J. D. (1969). *Ibid.*, 1 174 868
581. Rulison, R. N. (1969). *Fr. Patent*, 1 581 203
582. Standage, A. E. (1969). *Ibid.*, 1 580 443
583. Courtaulds Ltd. (1968). *Ibid.*, 1 541 287
584. Whitney, I. and Johnson, J. W. (1969). *Fr. Demande.*, 2 003 352
585. Courtaulds Ltd. (1969). *Ibid.*, 2 002 722
586. Dixon, K. G. O., Gill, R. M. and Lovell, D. R. (1969). *Ger. Offen.*, 1 904 944
587. Tatchell, J. R. (1969). *Ibid.*, 1 904 943
588. Woemer, H. J., Lukacs, J. and Hoffman, K. (1969). *Ibid.*, 1 900 243

589. Johnson, W., Lloyd, T. and Watt, W. (1970). *Brit. Patent*, 1 193 263
590. Higgins, F. J. (1970). *Ger. Offen.*, 1 806 399
591. Whitney, I. (1970). *Ibid.*, 1 919 393
592. Fujiwara, S., Nagae, K. and Okuhashi, T. (1970). *Ibid.*, 1 928 330
593. Bunning, R. C., Rodenburgh, M. L., Parts, L. P. and Peresic, R. J. (1970). *Ibid.*, 1 945 145
594. Cooper, G. A. and Mayer, R. M. (1970). *Ibid.*, 1 949 830
595. Atkins, R. A. (1970). *Ibid.*, 1 959 600
596. Scragg, E. (1970). *Ibid.*, 1 960 344
597. Moutand, G., Loisseau, J. P. and Desmicht, D. (1970). *Ibid.*, 1 963 718
598. Young, M. A. (1970). *S. African Patent*, 6 905 450
599. Ezekiel, H. M. (1969). *Appl. Polym., Symp.*, **9**, 315
600. Johnson, W., Watt, W., Phillips, L. N. and Moreton, R. (1969). *Brit. Patent*, 1 166 251
601. Stuetz, D. E. (1969). *U.S. Patent*, 3 449 077
602. Noss, W. J. (1970). *Ibid.*, 3 488 151
603. Shindo, A., Nakanishi, Y. and Somua, I. (1969). *Appl. Polym. Symp.*, **9**, 305
604. Araki, T. and Gomi, S. (1969). *Ibid.*, **9**, 331
605. Ezekiel, H. M. and Spain, R. G. (1970). *U.S. Patent*, 3 528 774
606. Yomeshige, K. and Teranishi, H. (1970). *Japan. Patent*, 7 002 774
607. Ermolenko, I. N. and Sviridova, R. N. (1969). *Vesti. Akad. Navauk. Belarus. SSR Ser. Khim. Navuk.*, 40
608. Krugler, A. H. and Massie, J. E. (1969). *Fr. Patent*, 1 574 297
609. Jackson, D. W. (1969). *Metals Eng. Quart.*, **9**, 22
610. Miura, I. and Hirano, S. (1969). *Iyo. Kizai Kenkyusho Hokuku, Tokyo Ika Shika Daigaku*, **3**, 17
611. Segal, C. L. and Boyle, G. (1969). *Tech. Pap. Reg. Tech. Conf., Soc. Plast. Eng., S. Calif. Sect.*, II–1.
612. Ellein, R. A., Fust, G. and Hanley, D. P. (1969). *Amer. Soc. Test. Mater. Spec. Tech. Publ.*, ASTMSTP **460**, 321
613. Owen, M. J. (1970). *Mod. Plast.*, **47**, 158
614. McLaughlin, J. R. (1970). *Nature (London)*, **227**, 701
615. Scott, D., Blackwell, J., McCullagh, P. J. and Mills, G. H. (1970). *Wear*, **15**, 257
616. Browning, C. E. and Marshall, J. A. (1970). *Proc. Anniv. Conf. S.P.I. Reinf. Plast./Compos. Div. 25th.*, 19–C
617. Owen, M. J. and Morris, S. (1970). *Ibid.*, 8–E
618. Lubowitz, H. R., Kendrick, W. P., Jones, J. F., Thorpe, R. S. and Burns, E. A. (1969). *Fr. Patent*, 1 580 456
619. Sara, R. V. (1969). *Ibid.*, 1 561 254
620. Courtaulds Ltd. (1969). *Ibid.*, 1 544 188
621. Idem. (1969). *Ibid.*, 1 554 188
622. Bowen, D. H., Sambell, R. A. J., Lambe, K. A. D. and Mattingley, N. J. (1969). *Ger. Offen.*, 1 925 009
623. Rohl, C. W. and Robinson, J. H. (1969). *S. African Patent*, 6 901 542
624. Wavlett, C. E. (1969). *U.S. Patent*, 3 470 003
625. Desai, R. R. (1970). *Ger. Offen.*, 2 001 018
626. Gresham, H. E., Hannah, C. G. and Hall, M. B. (1970). *Ibid.*, 1 923 622
627. Young, M. A. (1970). *S. African Patent*, 6 905 451
628. Buschow, A. G., Esola, C. H., Hess, I. J., Kreitz, D. B. and Schmidt, F. J. (1969). NASA Contract Rep., NASA-CR-66763
629. Baker, A. A., Harris, S. J. and Youden, G. H. (1970). *Ger. Offen.*, 1 939 339
630. Donovan, P. D. and Watson-Adams, B. R. (1969). *Metals Mater.*, **3**, 443
631. Wadsworth, N. J. and Watt, W. (1969). *Fr. Patent*, 1 564 708
632. Fourdeaux, A., Herinckx, C., Perret, R. and Ruland, W. (1969). *C.R. Acad. Sci. Ser. C*, **269**, 1597
633. Brydges, W. T., Badami, D. V., Jourer, J. C. and Jones, G. A. (1969). *Appl. Polym. Symp.*, **9**, 255
634. Yamamoto, M. and Yamada, S. (1969). *Ibid.*, **9**, 263
635. Hugo, J. A., Phillips, V. A. and Roberts, B. W. (1970). *Nature (London)*, **226**, 144
636. Coyle, R. A., Gillin, I. M. and Wicks, B. J. (1970). *Ibid.*, **226**, 257
637. Johnson, D. J. and Tyson, C. N. (1970). *J. Phys. D.*, **3**, 526
638. Gyanmati, E. and Nickel, H. (1969). *Chem. Ber. Kernforschungsunlage Juelich*, JUL 615 RW

639. Taylor, R. E. and Kline, D. E. (1970). *Chem. Phys., Carbon.*, **6**, 283
640. United Kingdom Atomic Energy Authority (1969). *Brit. Patent*, 1 144 320
641. Horsley, G. W. (1969). *Ibid.*, 1 146 015
642. Societe Anon. Belgonucleaire (1969). *Fr. Patent*, 1 577 044
643. Beutler, H., Hammer, R. L. and Robbins, I. M. (1969). *Ger. Offen.*, 1 807 667
644. Schwartz, A. S. (1969). *Ibid.*, 1 902 344
645. Beatty, R. L. and Kiplinger, D. V. (1969). *U.S. Patent*, 3 471 314
646. Beutler, H. and Payne, M. C. (1969). *Ibid.*, 3 472 677
647. Fitzer, E. and Vohler, O. (1970). *Ger. Patent*, 1 439 115
648. Reagen, P. E., Long, E. L., Morgan, J. G. and Coob, J. H. (1970). *Nucl. Appl. Technol.*, **8**, 417
649. Fitzer, E., Schaeffer, V. and Yamada, S. (1969). *Carbon*, **7**, 643
650. Otterbein, H., Brousse, E., Bonnetain, L. and Lespinasse, B. (1970). *C.R. Acad. Sci. Ser. C*, **270**, 662
651. Halpin, M. K. and Jenkins, G. M. (1969). *Proc. Roy. Soc. Ser. A*, **313**, 421
652. Furdin, G., Buck, B. and Herold, A. (1970). *C.R. Acad. Sci. Ser. C*, **271**, 683
653. Aronson, S. and Salzano, F. J. (1969). *Nucl. Sci. Eng.*, **38**, 187
654. Billand, D. and Herold, A. (1969). *C.R. Acad. Sci. Ser. C*, **269**, 490
655. Co-Minh-Duc., Rose, H. and Pascault, J. P. (1970). *Ibid.*, **270**, 569
656. Charbonages de France (1969). *Fr. Patent*, 1 566 796
657. Sudo, M., Ichikawa, M., Soma, M., Onishi, T. and Tamaru, K. (1969). *J. Phys. Chem.*, **73**, 1174
658. Daumas, N. and Herold, A. (1969). *C.R. Acad. Sci. Ser. C*, **268**, 373
659. Co-Minh-Duc, Prost, M., Rose, M. and Pascault, J. P. (1970). *Ibid.*, **270**, 961
660. Co-Minh-Duc, Rose, M. and Pascault, J. P. (1970). *Ibid.*, **270**, 657
661. Hishiyama, Y. and Inamura, T. (1970). *Tanso*, **61**, 42
662. Ginderow, D. and Setton, R. (1970). *C.R. Acad. Sci. Ser. C*, **270**, 135
663. Kotosonov, A. S., Demin, A. V., Polozhikhin, A. I., Nikol'skii, I. F. and Rakcheeva, V. I. (1970). *Khim. Tverd. Topl.*, 115
664. Hooley, J. G. (1970). *Carbon*, **8**, 333
665. Takahashi, Y., Miyauchi, K. and Sasa, T. (1970). *Tanso*, **60**, 8
666. Hooley, J. G. and Sormassy, R. N. (1970). *Carbon*, **8**, 191
667. Hohlwein, D., Grigutsch, F. D. and Knappwost, A. (1969). *Angew. Chem. Int. Ed. Engl.*, **8**, 382
668. Knappwost, A. and Grigutsch, F. D. (1969). *Z. Naturforsch. A*, **24**, 601
669. Grigutsch, F. D., Hohlweim, D. and Knappwost, A. (1969). *Z. Phys. Chem. (Frankfurt)*, **65**, 322
670. Ganz, S. N., Krasnokutskii, Y. I. and Ashkimazi, L. A. (1969). *Z. Prikl. Chim.*, **42**, 761
671. Landt, U. (1970). *Angew. Chem. Int. Ed. Engl.*, **9**, 780
672. Mayer, E. (1969). *Monatsh.*, **100**, 462
673. Koehler, H. (1969). *Wiss. Z. Martin Luther Univ., Halle-Wittenberg. Math. Naturwiss. Reihe.*, **18**, 33
674. Mayer, E. (1969). *Angew. Chem. Int. Ed. Engl.*, **8**, 601
675. Golov, V. G., Kuznetsova, L. V., Vodop'yanov, V. G. and Ivanov, M. G. (1970). *Khim. Prom. (Moscow)*, **46**, 198
676. Kreevoy, M. M. and Hutchins, J. E. C. (1969). *J. Amer. Chem. Soc.*, **91**, 4329
677. Uppal, S. S. and Kelly, H. C. (1970). *Chem. Commun.*, 1619
678. Mayer, E. and Kleboth, K. (1969). *Angew. Chem. Int. Ed. Engl.*, **8**, 444
679. Mayer, E. (1970). *Monatsh. Chem.*, **101**, 834
680. Lur'e, Y. Y. and Belevtsev, A. N. (1969). *Ochistka. Proizvod. Stochnykh Vod.*, 55
681. Vast, P. (1970). *C.R. Acad. Sci. Ser. C*, **270**, 811
682. Riethmann, J. A. and Scheck, L. (1969). *Ger. Offen.*, 1 900 972
683. Varma, R. and Signorelli, A. J. (1969). *Inorg. Nucl. Chem. Lett.*, **5**, 1017
684. Lavrov, N. V. (1970). *Khim. Tverd. Topl.*, 54
685. Dalton, R. F. and Jones, K. (1970). *J. Chem. Soc. A*, 590
686. Markovskii, L. Y., Vekshina, N. V. and Voevodskaya, T. K. (1970). *Zh. Prikl. Khim. (Leningrad)*, **43**, 1149
687. Dautzenberg, H. and Phillip, B. (1969). *Faserforsch. Textiltech.*, **20**, 213
688. Olszyna, K. J. and Heicklen, J. (1970). *J. Phys. Chem.*, **74**, 4188
689. Gelius, R. and Kirbach, E. (1970). *Z. Chem.*, **10**, 117

690. Lindner, E., Grimmer, R. and Weber, H. (1970). *Angew. Chem. Int. Ed. Engl.*, **9**, 639
691. Idem. (1970). *J. Organometal. Chem.*, **23**, 209
692. Carriel, J. T. and Hewitt, E. N. (1969). *U.S. Patent*. 3 472 887
693. International Nickel Ltd. (1970). *Brit. Patent*, 1 196 520
694. Kaliya, O. L., Temkin, O. N., Kirchenkova, G. S., Smirnova, E. M., Kimel'fel'd, Y. M. and Flid, R. M. (1969). *Izv. Akad. Nauk. SSSR Ser. Khim.*, 2854
695. Vit, J., Casensky, B., Kuhl, J., Prochazka, V. and Donnorova, Z. (1969). *Czech. Patent*, 126 672
696. Hauslik, T. (1969). *Ibid.*, 133 017
697. Antipin, L. H., Sobelov, E. S. and Miranov, V. F. (1969). *Zh. Prikl. Khim. (Leningrad)*, **42**, 451
698. Benzing, W. C. (1969). *U.S. Patent*, 3 484 311
699. Strater, K. and Mayer, A. (1969). *Semicond. Silicon Int. Symp. Pap. 1st.*, 469 (New York: Electrochem. Soc., Inc.)
699a. Swan, R. C. G. and Pyne, R. E. (1969). *J. Electrochem. Soc.*, **116**, 1014
700. Cetini, G., Castiglioni, M., Volpe, P. and Gambino, D. (1969). *Ric. Sci.*, **39**, 392
701. Gasper, P. P. and Markusch, P. (1970). *Chem. Commun.*, 1331
702. Petrik, A. G., Falkevich, E. S. and Ustinova, N. K. (1969). *Kremnii Germanii*, **1**, 7
703. Petrik, A. G. and Ustinova, N. K. (1969). *Ibid.*, **1**, 14
704. Schmidt, J. F. and Lampe, F. W. (1969). *J. Phys. Chem.*, **73**, 2706
705. Milligan, D. E. and Jacox, M. E. (1970). *J. Chem. Phys.*, **52**, 2594
706. Morris, E. R. and Thynne, J. C. (1969). *J. Phys. Chem.*, **73**, 3294
707. Strausz, O. P., Jakubowski, E., Sandhu, H. S. and Gunning, H. E. (1969). *J. Chem. Phys.*, **51**, 552
708. Ring, M. A., Beverley, G. D. and Koester, F. H. (1969). *Abstr. 157th Meeting, Amer. Chem. Soc.*, INOR 65
709. Hengge, E. and Olbrich, G. (1970). *Monatsh. Chem.*, **101**, 1068
710. Sakai, M. and Soka, S. (1969). *Japan Patent*, 6 901 700
711. Carborundum Co. (1969). *Brit. Patent*, 1 164 796
712. Gaude, J. and Lang, J. (1969). *C.R. Acad. Sci. Ser. C*, **268**, 1785
713. Marchand, R. (1970). *Rev. Chim. Miner.*, **7**, 87
714. Coles, N. G. and Glasson, D. R. (1969). *J. Appl. Chem.*, **19**, 178
715. Horton, R. M. (1969). *J. Amer. Ceram. Soc.*, **52**, 121
716. Hagen, A. P. and Russo, P. J. (1970). *Inorg. Nucl. Chem. Lett.*, **6**, 507
717. Glidewell, C. and Rankin, D. W. H. (1970). *J. Chem. Soc. A*, 279
718. Aylett, B. J. and Hakim, M. J. (1969). *Ibid.*, 636
719. Idem. (1969). *Ibid.*, 800
720. Washburne, S. S. and Peterson, W. R. (1969). *Inorg. Nucl. Chem. Lett.*, **5**, 17
721. Wannagat, U. and Schulze, M. (1969). *Ibid.*, **5**, 789
722. Schmutz-Du Mont, O. and Jansen, W. (1969). *Z. Anorg. Allg. Chem.*, **371**, 113
723. Summerford, C. and Wade, K. (1969). *J. Chem. Soc. A*, 1487
724. Manoussakis, G. E. and Tossidis, J. A. (1969). *Inorg. Nucl. Chem., Lett.*, **5**, 733
725. Takiguchi, T. and Suzuki, M. (1969). *Bull. Chem. Soc. Jap.*, **42**, 2708
726. Eméleus, H. J., Spaziente, P. M. and Williamson, S. M. (1969). *Chem. Commun.*, 768
727. Idem. (1970). *J. Inorg. Nucl. Chem.*, **32**, 3219
728. Paul, R. C., Aggarwal, V. K., Ahluwalia, S. C. and Narula, S. P. (1970). *Inorg. Nucl. Chem. Lett.*, **6**, 487
729. Fritz, G. and Becker, G. (1970). *Z. Anorg. Allg. Chem.*, **372**, 180
730. Davis, J., Drake, J. E. and Goddard, N. (1970). *J. Chem. Soc. A*, 2962
731. Anderson, J. W. and Drake, J. E. (1970). *Ibid.*, 3131
732. Krikorov, V. S. (1970). *Ger. Patent*, 1 792 339
733. Idem. (1969). *Brit. Patent*, 1 159 415
734. Hastie, J. W., Hauge, R. H. and Margrave, J. L. (1969). *Inorg. Chim. Acta*, **3**, 601
735. Arslambekov, V. A., Gorbunova, K. M. and Karateeva, V. I. (1970). *Izv. Akad. Nauk. SSSR Neorg. Mater.*, **6**, 1625
736. Lastovskii, R. P., Stepin, B. B., Blyum, G. Z., Shvarto, M. M. and Lavhinker, S. M. (1969). *Khim. Prom. (Moscow)*, **45**, 477
737. Osipov, K. A. and Folmanis, G. E. (1970). *Izv. Akad. Nauk. SSSR Neorg. Mater.*, **6**, 1167
738. Thilo, E. and Winkler, A. (1969). *Z. Anorg. Allg. Chem.*, **365**, 180

739. Varma, R., Ray, A. K. and Sahay, B. K. (1969). *Inorg. Nucl. Chem. Lett.*, **5**, 497
740. Kamaratos, E. and Lampe, F. W. (1970). *J. Phys. Chem.*, **74**, 2267
741. Airey, W. and Sheldrick, G. M. (1970). *J. Inorg. Nucl. Chem.*, **32**, 1827
742. Fischer, C. and Kriezsman, H. (1969). *Z. Inorg. Allg. Chem.*, **367**, 233
743. Rotte, W. (1969). *Ger. Patent*, 1 298 972
743a. Wojnowska, M. and Wojnowski, W. (1970). *Rocz. Chem.*, **44**, 1019
744. Dewitt, N. P. M. (1969). *Fr. Patent*, 1 527 418
745. Goldsbury, R. E. and Weibrecht, W. E. (1970). *Org. Prep. Proced.*, **2**, 1
746. Anand, S. K., Singh, J. J., Multani, R. K. and Jain, B. D. (1969). *J. Inst. Chem.*, **41**, 79
747. Yoshimoto, A. and Ichiro, K. (1969). *Bull. Chem. Soc. Jap.*, **42**, 1118
748. Ismail, R. M. (1969). *Z. Anorg. Allg. Chem.*, **371**, 23
749. Mehrotra, R. C. and Bajaj, P. (1970). *J. Organometal. Chem.*, **24**, 612
750. Idem. (1970). *Ibid.*, **22**, 41
751. Kuznetsova, V. P. and Belogovina, G. N. (1969). *Zh. Obsch. Khim.*, **39**, 547
752. Emblem, H. G. (1969). *J. Prakt. Chem.*, **311**, 970
753. Paul, R. C., Narula, S. P. and Makhri, H. S. (1970). *J. Inorg. Nucl. Chem.*, **32**, 3122
754. Zhimkin, D. Y. and Morgunova, M. M. (1969). *Zh. Obsch. Khim.*, **39**, 552
755. Frye, C. L. (1970). *J. Amer. Chem. Soc.*, **92**, 1205
756. Barnum, D. W. (1970). *Inorg. Chem.*, **9**, 1942
757. Dean, P. A. W. and Phillips, R. F. (1969). *J. Chem. Soc. A*, 363
758. Abe, Y. and Kijima, I. (1970). *Bull. Chem. Soc. Jap.*, **43**, 466
759. Yoshimoto, A. and Ichiro, K. (1969). *Ibid.*, **42**, 1148
760. Emons, H. H., Mochlhenrich, S. and Theisen, L. (1969). *Ger. (East) Patent*, 65 914
761. Hengge, E. and Olbrich, G. (1969). *Z. Anorg. Allg. Chem.*, **365**, 321
762. Cradock, S., Ebsworth, E. A. V. and Rankin, D. W. H., unpublished observations quoted in Ref. 417
763. Boehm, H. P. (1969). *Z. Anorg. Allg. Chem.*, **365**, 176
764. Mueller, R., Reichel, S. and Dathe, C. (1969). *J. Prakt. Chem.*, **311**, 930
765. Becher, W. and Massonne, J. (1970). *Ger. Offen.*, 1 906 843
766. Mukherjee, A. K., Chunhan, R. B. and Sharma, S. D. (1970). *Labdev. Part A*, **8**, 17
767. Fel'dshtein, N. S., Gorbunov, A. I., Belyi, A. P., Golubtsov, S. A. and Sharafamov, V. I. (1969). *Z. Fiz. Khim.*, **43**, 747
768. Langer, H. G. (1969). *U.S. Patent*, 3 453 079
769. Airey, W. and Sheldrick, G. M. (1969). *J. Chem. Soc. A*, 2865
770. Margrave, J. L. and Timms, P. L. (1969). *U.S. Patent*, 3 485 862
771. Audsley, A. and Bayliss, R. K. (1969). *J. Appl. Chem.*, **19**, 33
772. Hair, M. L. and Hirtl, W. (1969). *J. Phys. Chem.*, **73**, 2372
773. Korshunov, B. G. and Nirsha, B. H. (1969). *Zh. Neorg. Khim.*, **14**, 1971
774. Gorse, R. A., DuVigneaud, J. and Ring, M. A. (1969). *Inorg. Chem.*, **8**, 1530
775. Fel'dshtein, N. S., Belyi, A. P., Gorbunov, A. I. and Golubstov, S. A. (1969). *Z. Fiz. Khim.*, **43**, 1112
776. Belyi, A. P., Gorbunov, A. I., Flid, R. M. and Golubstov, S. A. (1969). *Ibid.*, **43**, 1144
777. Groslev, G-L., Danov, S. M., Yurlova, Z. I. and Shilova, A. V. (1969). *Ibid.*, **43**, 786
778. Drake, J. E. and Westwood, N. P. C. (1969). *Chem. Ind. (London)*, 24
779. Fehér, F., Plichta, P. and Guillery, R. (1970). *Chem. Ber.*, **103**, 3028
780. Chalk, A. J. (1970). *Ger. Offen.*, 1 929 902
781. Cadman, P., Tilsley, G. M. and Trotman-Dickenson, A. F. (1969). *J. Chem. Soc. A*, 1370
782. Haszeldine, R. N., Parish, P. V. and Parry, D. J. (1969). *J. Chem. Soc. A*, 683
783. Guertin, J. P. and Onyszchuk, M. (1969). *Can. J. Chem.*, **47**, 1275
784. Ennan, A. A., Kats, B. M. and Novikova, A. A. (1969). *Izv. Vyssh. Ucheb. Zaved. Khim., Khim. Technol.*, **12**, 1630
785. Ennan, A. A., Kats, B. M., Anisomov, Y. N. and Yur'eva, E. I. (1969). *Zh. Neorg. Khim.*, **14**, 3172
786. Ennan, A. A. and Kats, B. M. (1970). *Ibid.*, **15**, 2161
787. Kleboth, K. (1969). *Monatsh. Chem.*, **100**, 1057
788. Barker, J. E. and Robinson, J. H. (1969). *U.S. Patent*, 3 462 242
789. Bratishko, V. D., Rabov, E. G., Sudarikov, B. N., Cherkasov, V. A. and Kulyado, Y. M. (1969). *Tr. Mosk. Khim. Tekhnol. Inst.*, **60**, 111
790. Hajek, B. and Benda, F. (1970). *Coll. Czech. Chem. Commun.*, **35**, 2494
791. Dean, P. A. W. and Evans, D. F. (1970). *J. Chem. Soc. A*, 2569

792. Brini-Fritz, M., Geistel, M. M. and Pousse, A. (1969). *C.R. Acad. Sci. Ser. C*, **268**, 2040
793. Schott, G., Kibbel, H. V. and Hildebrandt, W. (1969). *Z. Anorg. Allg. Chem.*, **371**, 81
794. Shelepina, V. L., Shelepin, O. E. and Osipov, O. A. (1969). *Zh. Neorg. Khim.*, **14**, 1427
795. Pinkus, A. G. and Ku, A. T. Y. (1969). *J. Org. Chem.*, **34**, 1094
796. Dhar, S. K. and Tomau, J. A. (1969). *J. Inorg. Nucl. Chem.*, **31**, 2787
797. Thompson, D. W. (1969). *Inorg. Chem.*, **8**, 2015
798. Kummer, D. and Koester, H. (1969). *Angew. Chem. Int. Ed. Engl.*, **8**, 878
799. Kummer, D., Koester, H. and Speck, M. (1969). Ibid., 599
800. Amberger, E. and Leidl, P. (1969). *Chem. Ber.*, **102**, 2764
801. Gaines, D. F. and Iorns, T. V. (1970). *U.S. Clearinghouse, Fed. Sci. Tech. Inform.*, AD1970 No. 707, 436
802. Kirk, R. W. and Timms, P. L. (1969). *J. Amer. Chem. Soc.*, **91**, 6215
803. Bellama, J. M. and MacDiarmid, A. G. (1969). *J. Organometal. Chem.*, **18**, 275
804. Varma, R. (1970). *Inorg. Nucl. Chem. Lett.*, **6**, 9
805. Gibbon, G. A., Kifer, E. W. and Van Dyke, C. H. (1970). Ibid., **6**, 617
806. Kifer, E. W. and Van Dyke, C. H. (1969). *Abstr. 158th Meeting, Amer. Chem. Soc.*, INOR 125
807. Hagen, A. P. and Russo, P. J. (1969). Ibid., INOR 75
808. Idem. (1969). *Inorg. Nucl. Chem. Lett.*, **5**, 885
809. Jetz, W. and Graham, W. A. G. (1969). *J. Amer. Chem. Soc.*, **91**, 3375
810. Aylett, B. J. and Campbell, J. M. (1969). *J. Chem. Soc. A*, 1916
811. Schrieke, R. R. and West, B. O. (1969). *Proc. XII Int. Conf. Coordination Chem.*, 59, (Marrickville, N.S.W.: Science Press)
812. Idem. (1969). *Inorg. Nucl. Chem. Lett.*, **5**, 141
813. Amberger, E., Muehlhofer, E. and Stern, H. (1969). *J. Organometal. Chem.*, **17**, P5
814. Aylett, B. J., Campbell, J. M. and Walton, A. (1969). *J. Chem. Soc. A*, 2110
815. Aylett, B. J. and Campbell, J. M. (1969). Ibid., 1910
816. Baay, Y. L. and MacDiarmid, A. G. (1969). *Inorg. Chem.*, **8**, 986
817. Bald, J. F. and MacDiarmid, A. G. (1970). *J. Organometal. Chem.*, **22**, C22
818. Harrod, J. F. and Smith, C. A. (1970). *Can. J. Chem.*, **48**, 870
819. Chalk, A. J. (1969). *Chem. Commun.*, 1207
820. Harrod, J. F. and Smith, C. A. (1970). *J. Amer. Chem. Soc.*, **92**, 2699
821. Glockling, F., McGregor, A., Schneider, M. E. and Shearer, H. M. H. (1970). *J. Inorg. Nucl. Chem.*, **32**, 3101
822. Bentham, J. E., Cradock, S. and Ebsworth, E. A. V. (1969). *Chem. Commun.*, 528
823. Chatt, J., Eaborn, C., Ibekwe, S. D. and Kapoor, P. N. (1970). *J. Chem. Soc. A*, 1343
824. Idem. (1970). *Brit. Patent*, 1 177 702
825. Marano, G. A. and MacDiarmid, A. G. (1969). *Inorg. Nucl. Chem. Lett.*, **5**, 621
826. Bettler, C. R., Sendra, J. C. and Urry, G. (1970). *Inorg. Chem.*, **9**, 1060
827. Sawodny, W. and Ruoff, A. (1970). *J. Mol. Spectr.*, **33**, 173
828. Nelson, J. (1970). *Spectrochim. Acta*, **26A**, 109
829. Bates, J. B. and Smith, W. H. (1970). Ibid., **26A**, 455.
830. Ruoff, A. (1970). Ibid., **26A**, 545
831. Murchison, C. B. and Overend, J. (1970). Ibid., **26A**, 599
832. Bundy, A. R., Friedrich, H. B. and Person, W. P. (1970). *J. Chem. Phys.*, **53**, 674
833. Darmon, I., Gerschel, A. and Brot, C. (1970). *Chem. Phys. Letters*, **7**, 53
834. Foss-Smith, D. and Overend, J. (1970). *Spectrochim. Acta*, **26A**, 2269
835. Ogilvie, J. F. and Newlands, M. J. (1969). *Trans. Faraday Soc.*, **65**, 2602
836. Burger, H., Burezyk, K., Hofler, F. and Sawodny, W. (1969). *Spectrochim. Acta*, **25A**, 1891
837. Burger, H. and Hofler, F. (1970). Ibid., **26A**, 31
838. Lele, A. B. and Sathianandan, K. (1969). *Indian J. Pure Appl. Phys.*, **7**, 647
839. Rodziewicz, W. and Michalowski, Z. (1969). *Rocz. Chem.*, **43**, 465
840. Drake, J. E. and Riddle, C. (1970). *Spectrochim. Acta*, **26A**, 1697
841. Hastie, J. W., Hauge, R. H. and Margrave, J. L. (1969). *Inorg. Chim. Acta*, **3**, 601
842. McKean, D. C. (1970). *Spectrochim. Acta*, **26A**, 1833
843. Durig, J. R. and Hellams, K. L. (1969). *Inorg. Chem.*, **8**, 944
844. Reich, P. (1969). *Exp. Tech. Phys.*, **7**, 329
845. Razuvaev, G. A., Egorchkin, A. N., Khorshev, S. Y., Vyazankin, N. S. and Mironov, V. F. (1969). *Dokl. Akad. Nauk. SSSR*, **185**, 100

846. Newton, W. E. and Rochow, E. G. (1969). *U.S. Clearinghouse, Fed. Sci. Tech. Inform.*, AD 1969 No. 701762
847. Idem. (1970). *J. Chem. Soc. A*, 2664
848. Rodziewicz, W. and Michalowski, Z. (1969). *Rocz. Chem.*, **43**, 267
849. Ozolins, L., Kovalev, I. F., Arbuzova, V. A., Shevchenko, I. V., Voronkov, M. G. and Lukevics, E. (1970). *Latv. PSR. Zinat. Akad. Vestis. Kim. Ser.*, **47**, 849
850. Raynes, W. T. and Raza, M. A. (1969). *Mol. Phys.*, **17**, 157
851. Kreshkov, A. P., Kirichenko, E. A. and Davydov, V. D. (1969). *Tr. Mosk. Khim.-Tekhnol. Inst.*, **61**, 63
852. Kirichenko, E. A. and Davydov, V. D. (1969). *Izv. Vyssh. Ucheb. Zaved., Khim. Khim. Tekhnol.*, **12**, 1381
853. Hastie, J. W., Hauge, R. H. and Margrave, J. L. (1969). *J. Amer. Chem. Soc.*, **91**, 2536
854. Borsette, F., Cabana, A., Fournier, R. and Savoie, R. (1970). *Can. J. Chem.*, **48**, 410
855. Dubois, M., Delhayne, M. B. and Wallart, F. (1969). *C.R. Acad. Sci. Ser. B*, **269**, 260
856. Levin, I. N. (1969). *Spectrochim Acta*, **25A**, 1157
856a. Thomas, T. E., Orville-Thomas, W. J., Chamberlain, J. and Gebbie, M. A. (1970). *Trans. Faraday Soc.*, **66**, 2710
857. Rao, D. V. R. A. and Rai, D. K. (1969). *Indian J. Appl. Phys.*, **7**, 276
858. Christensen, D. H. and Nielsen, O. F. (1970). *J. Mol. Spectr.*, **33**, 425
859. Bürger, H. and Ruoff, A. (1970). *Spectrochim. Acta*, **26A**, 1449
860. Aleshenkova, Y. A. and Plotnikova, A. D. (1970). *Zh. Prikl. Spektrosk.*, **12**, 1038
861. Ozin, G. A. (1969). *J. Chem. Soc. A*, 2952
862. Griffiths, J. E. (1969). *Spectrochim. Acta*, **25A**, 965
863. Hofler, F., Sawodny, W. and Hengge, E. (1970). Ibid., **26A**, 819
864. Dhar, S. K. and Wesolowski, D. J. (1970). *Abstr. 160th Meeting, Amer. Chem. Soc.*, INOR 99
865. Trefler, M. and Wilkinson, G. R. (1969). *Discuss. Faraday Soc.*, **48**, 108
866. Rao, D. V. R. A., Thakur, S. N. and Rai, D. K. (1970). *Proc. Indian Acad. Sci. Sect. A*, **71**, 42
867. McKean, D. C., Davidson, G. and Woodward, L. A. (1970). *Spectrochim. Acta*, **26A**,

868. Lannon, J. A., Weiss, G. S. and Nixon, E. R. (1970). Ibid., **26A**, 221
869. Walters, K. L., Brittain, J. R. and Risen, W. M. (1969). *Inorg. Chem.*, **8**, 1347
870. Tigelaar, H. L. and Flygare, W. H. (1970). *Chem. Phys. Lett.*, **7**, 254
871. Kane, A. R., Sullivan, J. F., Kenny, D. H. and Kenny, M. E. (1970). *Inorg. Chem.*, **9**, 1445
872. Newton, W. E. and Rochow, E. G. (1970). *Inorg. Chim. Acta*, **4**, 133
873. Glidewell, C., Rankin, D. W. H. and Sheldrick, G. M. (1969). *Trans. Faraday Soc.*, **65**, 2801
874. Hindemann, D. K. and Williams, L. L. (1969). *J. Chem. Phys.*, **50**, 2839
875. Mohanty, S. and Bernstein, H. J. (1970). Ibid., **53**, 461
876. idem. (1970). *Chem. Phys. Lett.*, **4**, 575
877. Van Duyneveldt, A. J., Tromp, H. R. C. and Gorten, C. J. (1969). *Physica (Utrecht)*, **45**, 272
878. Afanas'ev, M. L., Gabuda, S. P., Davidovich, R. L. and Matsutsyn, A. A. (1969). *Spectrosc. Lett.*, **2**, 1924
879. Haque, R. and Cyr, N. (1970). *Trans. Faraday Soc.*, **66**, 1848
880. Matsubayishi, G. and Tanaka, T. (1969). *J. Inorg. Nucl. Chem.*, **31**, 1963
881. Thompson, D. W. (1969). *J. Magn. Resonance*, **1**, 606
882. Spiess, H. W. and Sheline, R. K. (1970). *J. Chem. Phys.*, **53**, 3036
883. Donovan, R. J., Husain, D. and Stevenson, C. D. (1970). *Trans. Farday Soc.*, **66**, 1
884. Herzberg, G., Lagerquist, A. and McKenzie, B. J. (1969). *Can. J. Phys.*, **47**, 1889
885. Dunn, T. M., Rao, K. M., Nagaraj, S. and Verma, R. D. (1969). Ibid., 47, 2128
886. Nagaraj, S. and Verma, R. D. (1970). Ibid., **48**, 1436.
887. Kuzmenko, N. E., Smirnov, A. D. and Kuzyakov, Y. Y. (1970). *Vestn Mosk. Univ. Khim.*, **11**, 357
888. Mishra, R. K. and Khanna, B. N. (1969). *Curr. Sci.*, **38**, 361
889. Rae, K. B. and Haramath, P. B. V. (1969). *Proc. Phys. Soc. (London)*, **2**, 1381
890. Kuznetsova, L. A. and Kuzyakov, Y. Y. (1969). *Vestn. Mosk. Univ. Khim.*, **24**, 103
891. idem. (1969). *Zh. Prikl. Spektrosk.*, **10**, 413

CARBON AND SILICON

892. Lakshminarayana, A. and Haranath, P. B. V. (1969). *Curr. Sci.*, **38**, 136
893. Dixon, R. N. and Halle, M. (1970). *J. Mol. Spectr.*, **36**, 192
894. Ramachandra, R. D. (1970). Ibid., **34**, 284
895. Hastie, J. W., Hauge, R. H. and Margrave, J. L. (1969). Ibid., **29**, 152
896. McLean, R. R., Sharp, D. W. A. and Winfield, J. M. (1970). *Chem. Commun.*, 52
897. Tanaka, T., Matsubayashi, G., Shimizu, A. and Matsuo, S. (1969). *Inorg. Chim. Acta*, **3**, 187
898. Dice, A. Y. and Lyubimova, G. A. (1969). *Latv. PSR. Zinat. Akad. Vestris. Kim. Ser.*, 502
899. Bel'skii, V. K. and Zorkii, P. M. (1970). *Zh. Strukt. Khim.*, **11**, 564
900. Marchand, R., Laurent, Y., Lang, J. and Le Bihan, M. T. (1969). *Acta Crystallogr. Sect. B*, **B25**, 2157
901. Liebeau, F. and Hesse, K. F. (1969). *Naturwissenschaften*, **56**, 634
902. Bissert, G. and Liebeau, F. (1969). Ibid., **56**, 212
903. Lin, S. C. H. and Joshi, M. (1969). *J. Electrochem. Soc.*, **116**, 1740
904. Michelet, A. and Flahant, J. (1969). *C.R. Acad. Sci. Ser. C*, **268**, 326
905. Turley, J. W. and Boer, F. P. (1969). *J. Amer. Chem. Soc.*, **91**, 4129
906. Heilbronner, E., Hornung, V. and Muszkat, K. A. (1970). *Helv. Chim. Acta*, **53**, 347
907. Bassett, P. J. and Lloyd, D. R. (1969). *Chem. Phys. Lett.*, **3**, 22
908. Bull, W. E., Pullen, B. P., Grimm, F. A., Moddeman, W. E., Schweitzer, G. K. and Carlson, T. A. (1970). *Inorg. Chem.*, **9**, 2474
909. MacNeil, K. A. G. and Thynne, J. C. J. (1970). *Int. J. Mass. Spectrom. Ion. Phys.*, **3**, 455
909a. Norberg, R., Brecht, H., Albridge, R. C., Fahlman, A. and Van Wazer, J. R. (1970). *Inorg. Chem.*, **9**, 2469
910. Pullen, B. P., Carlson, T. A., Moddeman, W. A., Schweitzer, G. K., Bull, W. E. and Grimm, F. A. (1970). *J. Chem. Phys.*, **53**, 768
911. Potzinger, P. and Lampe, F. W. (1970). *J. Phys. Chem.*, **74**, 719
912. Idem. (1969). Ibid., **73**, 3912
913. Muenow, D. W. and Margrave, J. L. (1970). Ibid., **74**, 2577
914. Wilkerson, B. E. and Dillard, J. G. (1969). *Chem. Commun.*, 212
915. Hildebrand, D. L. and Muran, E. (1969). *J. Chem. Phys.*, **51**, 807
916. Zubkov, V. I., Tikhmirov, M. V., Andrianov, K. A. and Golubstov, S. A. (1969). *Dokl. Akad. Nauk. SSSR*, **188**, 594
917. Thynne, J. C. J. and MacNeil, K. A. G. (1970). *Inorg. Chem.*, **9**, 1946
918. Svee, H. J. and Sparrow, G. R. (1970). *J. Chem. Soc. A*, 1162
919. Ramaswamy, K. and Ranganathan, V. (1969). *Indian. J. Phys.*, **43**, 177
920. Potzinger, P. and Lampe, F. W. (1970). *J. Phys. Chem.*, **74**, 587
921. Klein, M. L. and Morrison, J. A. (1969). *Disc. Faraday Soc.*, **48**, 93
922. Reference not allocated
923. Mueller, A., Kebabeioglu, R., Krebs, B. and Glemser, O. (1969). *Z. Phys., Chem. (Leipzig)*, **240**, 92
924. Lapidus, I. I., Nisel'son, L. A. and Seifer, A. L. (1969). *Teplofiz. Kharakter Veshchestu.*, **1**, 703
925. Gross, P., Hyman, C. and Mwroka, S. (1969). *Trans. Faraday Soc.*, **65**, 2856
926. Maslov, P. G., Usryattseva, T. R., Boiko, V. G., Karetnikova, N. I. and Engalychev, Y. S. (1970). *Zh. Fiz. Khim.*, **44**, 825
927. Moffat, J. B. and Vogt, C. (1970). *J. Mol. Spectr.*, **33**, 494
928. Olsen, J. F. and Burnelle, L. (1970). *J. Amer. Chem. Soc.*, **92**, 3659
929. Cade, P. E., Bader, R. F. W., Henneker, W. H. and Keaveny, I. (1969). *J. Chem. Phys.*, **50**, 5313
930. Hartmann, H., Papula, L. and Strecht, W. (1970). *Theor. Chim. Acta*, **17**, 131
931. Cook, D. B. and Palmieri, P. (1969). *Chem. Phys. Lett.*, **3**, 219
932. Boer, F. P. and Lipscomb, W. N. (1969). *J. Chem. Phys.*, **50**, 989
933. Rothenburg, A., Young, R. H. and Schaefer, H. F. (1970). *J. Amer. Chem. Soc.*, **92**, 3243
934. Lehn, J. M. and Munsch, B. (1970). *Chem. Commun.*, 994
935. Bennet, S. W., Eaborn, C., Hudson, A., Jackson, R. A. and Root, K. D. J. (1970). *J. Chem. Soc. A*, 348
936. Kewley, R., McKinney, P. M. and Robiette, A. G. (1970). *J. Mol. Spectr.*, **34**, 390
937. Bellama, J. M. and MacDiarmid, A. G. (1970). *J. Organometal. Chem.*, **24**, 91
938. Raymonda, J. W., Muenter, J. S. and Klemperer, W. A. (1970). *J. Chem. Phys.*, **52**, 3458

939. Shirh, J. S. and Bass, A. M. (1969). *Anal. Chem.*, **41,** 103A
940. Kaplansky, M. and Whitehead, M. A. (1969). *Mol. Phys.*, **16,** 481
941. Brown, T. L., Edwards, P. A., Harris, C. B. and Kirsch, J. L. (1969). *Inorg. Chem.*, **8,** 763
942. Beagley, B. and Conrad, A. R. (1970). *Trans. Faraday Soc.*, **66,** 2740
943. Vilkov, L. V. and Tarasenko, N. A. (1969). *Chem. Commun.*, 1176
944. Rankin, D. W. H., Robiette, A. G., Sheldrick, G. M., Beagley, B. and Hewitt, T. G. (1969). *J. Inorg. Nucl. Chem.*, **31,** 2351
945. Ryan, R. R. and Hedberg, K. (1969). *J. Chem. Phys.*, **50,** 4986
946. Chang, C. H., Porter, R. F. and Bauer, S. H. (1970). *J. Phys. Chem.*, **74,** 1363
947. Almenningen, A., Seip, H. M. and Seip, R. (1970). *Acta Chem. Scand.*, **24,** 1697

7
Germanium, Tin and Lead

J. E. DRAKE and J. W. ANDERSON
University of Windsor, Ontario

7.1	REVIEWS	304
7.2	HYDRIDES	304
	7.2.1 *Binary hydrides*	304
	7.2.2 *Hydride derivatives*	305
	7.2.3 *Group V derivatives*	306
	7.2.4 *Group VI derivatives*	306
	7.2.5 *Group VII derivatives*	307
	7.2.6 *Transition metal derivatives*	308
7.3	ORGANO-DERIVATIVES	308
	7.3.1 *Halogen and pseudohalogen derivatives*	310
	7.3.2 *Chalcogen derivatives*	313
	7.3.3 *Group V derivatives*	315
	7.3.4 *Group III derivatives*	318
7.4	HALOGEN COMPOUNDS	319
	7.4.1 *Complex ions of the halides*	320
	7.4.2 *Pseudohalogen complex ions*	323
7.5	REACTIONS OF THE HALOGENS, PSEUDOHALOGENS AND ORGANOHALIDE DERIVATIVES	324
	7.5.1 *Oxidation state IV*	324
	7.5.1.1 Complex formation	324
	7.5.1.2 Organo-halide and pseudohalide complexes	327
	7.5.1.3 Other reactions	329
	7.5.2 *Oxidation state II*	330
	7.5.2.1 Complex formation	330
7.6	CHALCOGEN COMPOUNDS	331
	7.6.1 *Chalcogenides of germanium, tin and lead in both II and IV oxidation states*	331
	7.6.2 *Miscellaneous chemistry of oxidation state IV*	334
	7.6.3 *Alkoxides, carboxylates and their derivatives*	335
7.7	COMPOUNDS CONTAINING A GROUP IV—TRANSITION METAL BOND	337

7.1 REVIEWS

The general chemistry of germanium[1] and tin[2] prior to 1960 has been covered in reviews. Recently, the first book to cover all aspects of the inorganic and organic chemistry of germanium was published[3]. The chain compounds of Group IV [4] and the chemistry of tin(II) [5] have also received general coverage.

More specialised reviews have covered the germanium[6,7], tin[6-8], and lead[6,7] amines, phosphines, arsines, stibines, and bismuthines[9,10]; the sulphides[11,12], selenides and tellurides[12], and pseudohalides[13,14], of Group IV; oxygen, hydrogen and nitrogen compounds of lead[15]; and stannoxanes[16]. The volatile compounds of the hydrides of germanium with elements of Groups V and VI have been discussed[17], and there have been several reviews of compounds containing a Group IV element–transition metal bond[18-20].

The structural aspects of tin compounds with coordination numbers greater than four[21] and the acceptor properties of germanium, tin and lead[22] have been discussed. A recent account has dealt with the vibrational spectra of tin and lead compounds[23], while a useful collection of ^{119}Sn Mössbauer data has been compiled[24].

7.2 HYDRIDES

7.2.1 Binary hydrides

The feature of germanium chemistry that provides a marked contrast with that of tin and lead is the chemistry of the hydrides. Binary germanes were first prepared by hydrolysis of the germanide[25], but of the various preparative methods that are well known, hydroborate reduction of the oxide or halide has been particularly popular[26-28]. The latter method was re-examined recently to obtain the conditions of optimum yields[29]. Deuteriated germane has been obtained by hydrogen isotope exchange between germane and solutions of KOD in D_2O [30], although good yields may be obtained by the hydrolysis of magnesium germamide[25] with perdeuteriophosphoric acid[31]. Germane is produced by direct heating of germanium solid in the presence of hydrogen by a ruby laser[32], while germanium-atom reactions with germane can be used to give specifically $^{75}GeH_4$ and $^{75}GeH_3 \cdot GeH_3$ [33]. The germylene, GeH_2, radical is proposed as an intermediate in the co-pyrolysis of SiH_4 and GeH_4 [34], and of a $Ge_2H_6/EtGeD_3$ mixture[35]. An electron diffraction study of Ge_2H_6 indicates that the Ge–Ge bond is shorter than that in solid germanium[36]. The photoelectron spectrum of GeH_4 gives its adiabatic ionisation energy as 11.31 eV which agrees reasonably with the appearance potential for GeH_4^+ in its mass spectrum[37].

LCAO–MO–SCF calculation of the wave-functions of GeH_4 indicates that chemical bonding is in accord with a very simple mode of bond formation involving essentially sp^3 hybridisation and no d-orbital participation[38]. A double-resonance study of geminal spin coupling constants in the ^1H n.m.r. spectrum of a series of Group IV hydrides, gives the sign of the direct J_{GeH}, J_{SnH} (and J_{SiH}) coupling constants as negative, relative to J_{CH} as positive, with that of J_{PbH} positive[39].

By contrast, the study of the chemistry of stannane, SnH_4, is limited. It does not react with various amines or amides but reduces nitrobenzene to aniline, benzaldehyde to benzyl alcohol and acetone to isopropyl alcohol. On treatment with $BF_3 \cdot Et_2O$, stannane is converted to SnF_4 [40]. The mechanism of the sodium hydroborate reduction of Na_2SnO_2 is examined[41].

7.2.2 Hydride derivatives

In 1962, only 24 volatile compounds containing the germyl group (GeH_3) had been reported including the five binary hydrides, Ge_nH_{2n+2} ($n = 1$ to 5) and the four monohalides, GeH_3X (X = F, Cl, Br, I)[42]. Two recent reviews quite extensively cover the work, since that time, on germanium hydride derivatives of the main Group elements of Groups IV [4], V and VI [17]. The force field in the analysis of the i.r. spectra of silylgermane analogues, $GeH_3 \cdot SiH_3$, $GeH_3 \cdot SiD_3$ and $GeD_3 \cdot SiH_3$ [43], are compared with those in earlier work and those of $MeSiH_3$ and $MeGeH_3$ [44]. Recently a reassignment of the Si—Ge stretch is suggested from a Raman study[45]. Silylgermane was first prepared by subjecting a SiH_4/GeH_4 mixture to an ozoniser-type silent electric discharge[46] and a multitude of silylgermanes, with up to six heavy-atom skeletons, were prepared soon after by the hydrolysis of an MgSiGe alloy[47]. A further examination of the discharge of SiH_4–GeH_4 and also Si_2H_6–GeH_4 leads to the characterisation by ^1H n.m.r., infrared, and Raman spectroscopy, of $GeH_3Si_2H_5$, $(GeH_3)_2SiH_2$, $GeH_3 \cdot GeH_2 \cdot SiH_3$ and $GeH_2(SiH_3)_2$ [48]. In 1967, the extremely unstable stannanes, $SiH_3 \cdot SnH_3$ and $GeH_3 \cdot SnH_3$ were prepared[49], thus completing the series of ten mixed-hexahydrides, $MH_3M'H_3$ (M,M' = C, Si, Ge, Sn). The decrease in stability with increasing molecular weight is very marked, and for lead, even CH_3PbH_3 is, as yet, unknown.

Methylsilylgermanes, $(GeH_3)_nSiMe_{4-n}$ ($n = 1$–4), can be formed by the general reaction of germyl-sodium or -potassium with chloro(methyl)silanes, $Me_{4-n}SiCl_n$ [50]. The vibrational spectra of $GeH_3 \cdot SiMe_3$, $GeD_3 \cdot SiMe_3$, $GeH_3 \cdot GeMe_3$ and $GeD_3 \cdot GeMe_3$ are assigned and the mass spectra of the silyl-derivatives demonstrate the exchange of methyl groups and hydrogen between silicon and germanium in the course of fragmentation[45]. The vibrational spectrum of $Ge(SiMe_3)_4$ is studied along with those of other $M(SiMe_3)_4$ (M = C, Si, Sn), species and the results discussed in terms of possible (p–d)π interactions[51]. The reactions of the germyl anion, GeH_3^-, with $ClCH_2SiH_3$ and $ClSiH_2 \cdot CH_3$ give $GeH_3 \cdot CH_2 \cdot SiH_3$ and $GeH_3 \cdot SiH_2 \cdot CH_3$ respectively, both of which are characterised by ^1H n.m.r. and i.r. spectroscopy[52]. Both silanes[53] and germanes[54] have been chlorinated by the action of hydrogen chloride in the presence of aluminium trichloride. A similar treatment of $GeH_3 \cdot CH_2 \cdot SiH_3$ gives no Ge–Cl containing species, but $GeH_3 \cdot CH_2 \cdot SiH_2 \cdot Cl$, $GeH_3 \cdot CH_2 \cdot SiHCl_2$, and $GeH_2(CH_2 \cdot SiH_3)_2$ are all identified by ^1H n.m.r., i.r., and mass spectroscopy[52]. $GeH_3^-K^+$ and $GePh_3^-K^+$ react with Me_2GaCl to give $K[Ga(GeR_3)ClMe_2]$, R = H or Ph. In non-polar solvents these decompose to $Me_2Ga \cdot GeR_3$. The Ga—Ge stretching mode is assigned at 306 cm^{-1} for $Me_2Ga \cdot GeH_3$ and 314 cm^{-1} for $Me_2GaGePh_3$ [55]. The B-chloro-B,N-methylborazines, $Cl_nMe_{3-n}B_3N_3Me_3$, $n = $ 1,2,3, also react with $GeH_3^-K^+$ and $GePh_3^-K^+$ to give $R_3Ge_nMe_{3-n}B_3N_3$

Me$_3$. Their i.r. spectra are interpreted as indicating a diminishing π-bonding in the B—N ring relative to the corresponding Me and H derivatives[56]. Further, GeH$_3^-$K$^+$ and GePh$_3^-$K$^+$ react with MCl$_2$ (M = Zn, Cd, Hg) to give the donor-solvent stabilised complexes K[M(GeR$_3$)Cl$_2$] and K$_2$[M(GeR$_3$)$_2$Cl$_2$]. These, by heating or by displacement of solvent with toluene, give R$_3$GeMCl or (R$_3$Ge)$_2$M. GePh$_3^-$K$^+$ reacts with MeHgI to give Ph$_3$Ge·HgMe which undergoes dismutation to (Ph$_3$Ge)$_2$Hg and Me$_2$Hg [57]. The Raman spectra of GeH$_3^-$ and GeD$_3^-$ in liquid ammonia or dimethoxyethane solution are consistent with the expected pyramidal, C_{3v}, structure[58], and ^{1}H n.m.r. parameters and dipole studies of PhGeH$_2^-$ in liquid ammonia suggest there is no delocalisation of electronic charge into the benzene ring[59]. It is predicted on the basis of the importance of differences in electronegativity that the radicals GeH$_3$ and SnH$_3$ should be pyramidal, unlike the effectively planar CH$_3$ radical[60].

7.2.3 Group V derivatives

The preparation[61] and characterisation[62] of trigermylamine, (GeH$_3$)$_3$N, is reported. Despite preliminary suggestions from vibrational spectra to the contrary, an electron diffraction examination indicates it has the planar Ge$_3$N skeleton[63], with the Ge—N—Ge angle of 120 degrees, most favourable to (p–d)π-bond formation[64]. By contrast, the electron diffraction study of trigermylphosphine, (GeH$_3$)$_3$P, shows it to be pyramidal with a Ge—P—Ge angle of 95.4 degrees[65]. Thus, as with the corresponding silyl-derivatives, (p–d)π-bonding is apparently structurally important in the amine but not the phosphine. The Raman and solid-state i.r. spectra of germyl-isocyanate, GeH$_3$NCO, seem to confirm a non-linear heavy atom skeleton[66]. The inference is taken that (p–d)π-bonding is not as important here as in the silyl analogue which has a linear skeleton. Calculations on the previously recorded vibrational spectra of HNGe and HNSi[67] suggest that the N—Ge bond order is c. 2.3, compared with 2.7 for N—Si [68].

Monogermylphosphine, GeH$_3$PH$_2$, and -arsine, GeH$_3$AsH$_2$, which were originally prepared by the action of a discharge on GeH$_4$/PH$_3$ and GeH$_4$/AsH$_3$ mixtures, are prepared by the reaction of monohalogenogermanes on lithium tetraphosphino- and tetraarsino-aluminates[69, 70], and by exchange reactions between silyl-phosphine and -arsine with monohalogenogermanes. The latter method is particularly useful for the formation of specifically deuteriated species[71] and has led to a thorough analysis of their vibrational spectra[72, 73]. The vibrational spectra of digermyl-phosphine and -arsine are also reported[74]. Exchange reactions of silyl-phosphine with dichlorogermane and chloro(methyl)germane give GeH$_2$ClPH$_2$ and MeGeH$_2$PH$_2$ respectively, characterised mainly by their ^{1}H n.m.r. spectra[75]. The only stibine to be characterised is trigermylstibine, (GeH$_3$)$_3$Sb which was prepared by exchange of GeH$_3$Br with (SiH$_3$)$_3$Sb [76].

7.2.4 Group VI derivatives

These are now known for the sequence (GeH$_3$)$_2$E (E = O, S, Se, Te). Two preparative routes are generally applicable; one involves reactions of germyl compounds with H_2S, S, H_2Se [77], and the second involves exchange reactions[78]. The former reagents react with either germyl-phosphine or -arsine to give $(GeH_3)_2S$ or $(GeH_3)_2Se$ via GeH_3SH and GeH_3SeH, as is proved by ^1H n.m.r. spectroscopy[77]. The equilibrium $(GeH_3)_2E + H_2E \rightleftharpoons 2GeH_3EH$ (E = S, Se, Te) has also been studied by ^1H n.m.r., and satellites attributable to ^{77}Se and ^{125}Te at natural abundance are noted[79]. The exchange reactions of $(SiH_3)_2E$ (E = Se, Te), with GeH_3Br give $(GeH_3)_2Se$ and $(GeH_3)_2Te$ which are characterised by ^1H n.m.r., i.r., and Raman spectroscopy. The speculation as to the Ge—E—Ge bond angles for E = O, S, from calculations based on the i.r. spectra[80, 81], have been clarified by an electron-diffraction examination which places the Ge—O—Ge angle at 126 degrees and that of Ge—S—Ge at 99 degrees[82, 83]. Thus the earlier suggestion that there is evidence for extensive π-bond character in the oxide but not the sulphide, in common with silyl analogues, has been justified. A ^1H n.m.r. examination of GeH_3—O—Ph and GeH_3—S—Ph suggests that any involvement of germanium d-orbitals in the bond to oxygen, or indeed sulphur, does not strongly influence the bonding in the aromatic ring[84], while the mass spectrum of GeH_3OPh suggests that loss of O is a significant fragmentation pathway[85]. Long-range H—H coupling, noted for several sulphur derivatives of Group IV hydrides $MH_3SM'H_3$ (M and M' = C, Si, Ge), is also noted for SiH_3—S—GeH_3 and SiH_3—Se—GeH_3 [86]. The latter compounds are among the products of electrical discharge reactions of silane and germane with some volatile Group VI species[87].

7.2.5 Group VII derivatives

The halogeno-derivatives of germanium hydrides are important starting points for many syntheses. With monogermane, all the possible halides, GeH_3X, GeH_2X_2, and $GeHX_3$ (X = F, Cl, Br, I), have been known for some time and their spectral properties extensively studied. More recent studies include further calculations on the vibrational spectra of the GeH_3X species[88, 89]. Different approaches give best force fields for X = F, Cl and for X = Br, I [89]. The thermodynamic functions are calculated for GeH_3X and $GeHX_3$ [90]. Much less work has been carried out on the vibrational analysis of the C_{2v} symmetry GeH_2X_2 species than the C_{3v} symmetry GeH_3X or $GeHX_3$. However, such analyses are reported for GeH_2Cl_2 [91] and GeH_2X_2 (X = F, Br, I)[92] and their deuteriated analogues. The force fields are reported for GeH_3CN and GeD_3CN [93]. The reaction of germanium metal with HCl to give $GeHCl_3$ has been found to be first-order in HCl and is strongly catalysed by the addition of copper[94]. Of the possible halogenodigermanes, only three are definitely known, Ge_2H_5Cl, Ge_2H_5Br, and Ge_2H_5I [95, 96], and a tentative identification of Ge_2H_5F has been reported[96]. Trends in the chemical shifts of their ^1H n.m.r. spectra are discussed along with that of $GeH_3 \cdot SiH_3$ [97]. Only Ge_3H_7I has been definitely prepared of the possible halogenotrigermanes and from its reaction it appears that the 2-iodotrigermane isomer is preferred[98]. The reaction of trigermane with iodine at −63 °C, followed by addition of MeMgI, gives 1- and 2-methyltrigermane in a ratio of 1:6. The products are characterised by ^1H n.m.r. and vibrational spectroscopy[99]. Synthetic routes to the series of halogeno(methyl)germanes, $MeGeH_nX_{3-n}$ (X = F, Cl, Br, I; n = 0, 1, 2), have been compared and the compounds characterised by their ^1H n.m.r., i.r., and mass spectra[100].

7.2.6 Transition-metal derivatives

Germanium hydride derivatives of the transition metals are relatively rare considering the wide range of germanium–transition metal compounds. Prior to 1969 the only derivatives characterised were $H_2Ge[Mn(CO)_5]_2$[101] and $H_2Ge[Fe(CO)_2(\pi-C_5H_5)]_2$[102]. The first germyl derivative, $H_3Ge\cdot Mn(CO)_5$, which is prepared by the action of GeH_3Br or $NaMn(CO)_5$, is stable to 100 °C in the dark[103]. By similar methods H_3Ge—$Re(CO)_5$ and —$Co(CO)_4$ are prepared and characterised by vibrational and mass spectroscopy[104, 105]. The reaction of monohalogenogermanes with trans-$XPt(PEt_3)_2H$ results in the formation of trans-$XPt(PEt_3)_2GeH_2Y$ (X,Y = Cl, Br, I) with the release of H_2[106]. Alternatively, the corresponding —SiH_2Y derivatives can be converted to —GeH_2Cl derivatives by exchange with GeH_3Cl. A ^1H n.m.r. spectroscopic study of the reaction of trans-$ClPt(PEt_3)_2H$ with excess GeH_3Cl in benzene leads to the identification of at least four species in which Pt is octahedrally surrounded by trans-PEt_3 groups, H, and various —GeH_2Cl and —$GeHCl_2$ groups[107]. The first digermyl transition metal derivative was also reported in 1970; $Ge_2H_5Mn(CO)_5$ is formed from the reaction of Ge_2H_5I with $NaMn(CO)_5$[108].

7.3 ORGANO-DERIVATIVES

In a discussion of organo-germanes, -stannanes and -plumbanes, a balance had to be reached between topics appropriate to a 'germanium–tin–lead' section and those appropriate to an 'organo-metallic' section. We have chosen to include organo-derivatives where the structure and chemical properties are apparently of interest as much for the presence of a M—X bond (X ≠ C) as for the organo-groups. Clearly this is a very personal choice. Where in doubt we have decided to 'overlap' with the organometallic sections.

The 'parent' organogermane, methylgermane, $MeGeH_3$, has come under some examination after a period of being ignored. It can be halogenated, to yield the various halides $MeGeH_2X$, $MeGeHX_2$, and $MeGeX_3$ (X = F, Cl, Br, I)[100]. The nature of the products resulting from the pyrolysis of $MeGeH_3$ are compared with those from $MeSiH_3$ and it is apparent that the Ge—C bond is more readily broken than the Ge—H, whereas the Si—C bond is less readily cleaved[109]. Methylgermane and digermane when reacted with iodine at −63 °C, produce compounds which on treatment with MeMgI yield a variety of methyldigermanes[110]. Mass, ^1H n.m.r., and vibrational spectra confirm the formation of $MeGeH_2\cdot GeH_2Me$, $Me_2GeH\cdot GeH_3$ and $Me_2GeH\cdot GeH_2Me$, while ^1H n.m.r. tentatively suggests the presence of $Me_2GeH\cdot GeHMe_2$, $Me_3Ge\cdot GeH_2Me$ and $Me_3Ge\cdot GeMe_2H$. The signs of the direct coupling constants of J_{GeH} and J_{SnH} are negative and of J_{PbH} is positive relative to J_{CH} as positive from a double resonance study of the

^{1}H n.m.r. spectra of GeH_4, Me_2GeH_2, SnH_4, $PhSnH_3$, Me_3SnH, R_2SnH_2 (R = Me, Et, ^{i}Pr, ^{i}Bu, Ph), Me_3PbH and Me_2PbH_2 [39].

The mass spectra of a large number of methylpolygermanes, $Me_{2n+2}Ge_n$ (n = 4–10), prepared from the action of GeI_2 with Me_3Al, indicate the importance of Ge—Ge bond-cleavage[111]. The ^{1}H n.m.r. and Raman spectra of several methylpolygermanes are reported[112]. The far i.r. spectrum[113] of solid Me_4Ge and Me_4Sn suggests that the CH_3-torsional barriers are greater than previously estimated from ^{1}H n.m.r. spectra[114]. The vibrational modes of Me_4M (M = Si, Ge, Sn, Pb) are re-examined, particularly the C—H stretches[115], and force-constant calculations reported on germylacetylene[116]. The i.r. ^{1}H, ^{13}C and ^{207}Pb n.m.r. spectra of Me_6Pb_2 are reported. The vibrational spectrum is consistent with D_{3d} symmetry (i.e., a staggered ethane-type structure) as has Ph_6Pb_2. The value of J_{CPb} is surprisingly small implying a high degree of s-character in the Pb—Pb bond[117]. The electron-impact ionisation-potentials of the catenates of the Group IV elements, $R(R_2M)_nR$ (where R = alkyl, M = Si, Ge, Sn), show a systematic decrease as the chain length (n) increases and as Group IV is descended[118a]. The i.r. spectra of mixed metal organometallic compounds are reported for species such as $Et(Et_3Ge)_2SiH$, $Et_2(Et_3Ge)SiH$, $(Et_3Ge)_3SiBr$, $Et(Et_3Ge)_2Si\cdot SnEt_3$, $Et(Et_3Ge)_2Si\cdot Si(GeEt_3)_2Et$, $Et(Et_3Ge)_2Si\cdot Hg\cdot Si(GeEt_3)_2Et$, $Et_2(Et_3Ge)Si\cdot Si(GeEt_3)Et_2$ and $Et_2(Et_3Ge)Si\cdot Hg\cdot Si(GeEt_3)Et_2$. The inductive effect in these and related compounds is greater for $GeEt_3$ than $SiEt_3$ [118b]. The enthalpy and entropy of sublimation of Ph_4Sn and Ph_6Sn_2 lead to an average bond dissociation energy of c. 55.8 kcal mol^{-1} for Sn—C and c. 36.3 kcal mol^{-1} for Sn—Sn [119]. High yields of hexa-aryldilead, Ar_6Pb_2, compounds result from the reaction of $PbCl_2$ with ArMgX in the presence of 1,2-dichloro- or 1,2-dibromo-methane, where Ar = Ph, o-, m- and p-tolyl; p-methoxy-phenyl, and 1- and 2-naphthyl [120]. Well-crystallised cyclogermanes $(p-tol_2Ge)_n$ (n = 4,5,6) are formed from $(p-tol)_2GeCl_2$ and lithium, sodium, or sodium–naphthalene. The mass spectra of these relatively unusual catenated Ge species are discussed[121].

The main and ^{13}C, ^{117}Sn and ^{119}Sn satellite ^{1}H n.m.r. spectra of tetra-vinyltin are completely analysed and J_{Sn-H} cis, trans, and geminal reported along with their relative signs[122]. Another study of both tetravinyl-tin and -lead also reports the coupling constants associated with ^{207}Pb [123].

The microwave spectrum of Me_2GeH_2 is consistent with essentially all tetrahedral angles[124], while a microwave study of germacyclopentane, $C_4H_8GeH_2$, gives a C—Ge—C angle in the five-membered ring system of 98 degrees[125]. Two papers on the i.r. spectra of several R_3GeH species (R = alkyl, aryl or halogeno) suggest that the Ge—H stretching vibration is (a) raised by a $-I$ substituent bound to Ge and lowered by a $+I$ substituent with a steric effect superimposed for several bulky ligands on Ge [126], and (b) more or less independent of solvent presumably because the polarisation of the bond (Ge^+—H^-) prevents the formation of a true hydrogen bond with proton acceptors[127]. In another study of a series of Group IV hydrides, Ph_3GeH and Ph_3SnH are used as radical-generating and -trapping agents in the cyclisation or reduction of di-iodoalkalenes[128].

Ph_6Pb_2 and finely divided sodium in THF gives Ph_3PbNa, while $PbCl_2$ reacts with PhMgX to give Ph_3PbMgX (X = Cl, Br). The compounds usually couple readily with reactive halides but exchange can also occur, the tendency increasing in the order $Ph_3PbNa < Ph_3PbLi < Ph_3PbMgX$ [129]. The action of Ph_3MLi (M = Si, Ge, Sn, Pb) and of Me_3SnLi with NOCl leads to the formation of the nitroso-derivatives Ph_3MNO and Me_3SnNO. The assignment of the absorption spectra is discussed [130]. The i.r. and Raman spectra of $CF_3COSnMe_3$, prepared by the action of Me_3SnLi with CF_3COCl, is reported [131].

The reactions of Bu_2^tHg with R_3MH (M = Si, Ge, Sn), proceed smoothly in the absence of solvent to give $(R_3M)_2Hg$ (R = Me, Et, Bu^t) [132]. $(Ph_3Ge)_2Hg$, which is prepared by the action of Ph_3GeH on Ph_2Hg, is converted under u.v. light to $Ph_3Ge \cdot GePh_3$ [133]. An equilibrium is rapidly established in solution between $(Me_3Si)_2Hg$ and $(Me_3Ge)_2Hg$ to give $Me_3Si \cdot Hg \cdot GeMe_3$ [134]. $(Me_3Si)_2Hg$ reacts with R_3SnX compounds (where X = OMe, OEt, $OSnEt_3$, $OSnBu_3$, $OSiMe_3$ and NEt_2) to give Me_3SiX and R_3Sn—SnR_3 [135], and $(Et_3Ge)_2Hg$ is readily converted to $Et_3Ge \cdot HgR$ on treatment with RHgX (R = Me, Et, Pr^n, Pr^i, Ph; X = Cl, I [136]. Preliminary studies of the reaction of Ph_3PbCCl_2Li include its conversion, in 16% yield, to $Ph_3Pb \cdot CCl_2 \cdot PbPh_3$ by hydrolysis [137]. The ionisation potentials of the Me_3M (M = Si, Ge, Sn) moieties are measured to obtain the M—X bond dissociation energies of compounds of the type Me_3M—X [138]. The e.s.r. spectrum of Me_3Ge is reported along with other related free radicals in solution [139].

7.3.1 Halogen and pseudohalogen derivatives

Organo-germanium, -tin and -lead halides have been studied extensively. The i.r. and Raman spectra of several R_3GeX species have been reported recently. These include the re-assignment of several modes in the spectrum of Me_3GeCl [140], the far i.r. and Raman spectra of solid Me_3GeCl and Me_3GeBr [141], the assignment of the vibrational spectra of Ph_3GeX (X = F, Cl, Br) [142], the assignment of the low wave-number vibrations in Ph_3GeCl, Ph_3GeBr, Ph_3GeI, Ph_3GeH and Ph_3GeD [143], which are basically in agreement with those of related silane [144] and stannane [145] molecules, the i.r. and Raman spectra of Et_3GeF [146] and the effect of the electronegativity of X on the i.r. of the Et_3Ge moiety in Et_3GeX (X = F, Cl, Br, I, and also O, S, Se, Te, Sb, Bi, H, In, Tl, Zn, Cd, Hg, Li) [147]. The spectra are calculated for methyl-chlorogermanes [148, 149] and the i.r. and Raman spectra are assigned for the methylbromogermanes, Me_3GeBr, Me_2GeBr_2, $MeGeBr_3$ [150] and the ethyl-chlorostannanes, Et_3SnCl, Et_2SnCl_2 and $EtSnCl_3$ [151]. The vibrational spectra and thermodynamic functions of $MeSnCl_3$ and $MeSnH_3$ are reported [152], and the vibrations in the far i.r. spectrum of Ph_3SnCl and $NH_4SnCl_2Ph_3$ [153] are re-assigned [154]. The i.r. and Raman spectra of organodilead and organolead halide species indicate that Ph_6Pb_2 has a staggered ethane-type structure; that in benzene solution, Ph_3PbX (X = Cl, Br, I) and Ph_2PbI_2 are all monomeric and that Ph_3PbF and Ph_2PbX_2 (X = Cl, Br), which are not soluble in benzene, as well as the other organolead halides, probably have halogen bridges in the solid with the lead atom 5-coordinated in the triphenyl derivatives and 6-coordinated in the diphenyls [155].

The kinetics and synthesis mechanism of dimethyldichlorogermane[156] and the thermodynamic data for the halogenation of Ph_4Pb to Ph_2PbX_2 (X = Br, I) in chloroform[157] are reported. A preliminary kinetic study is reported of the cleavage by IBr of C—Sn bonds in R_4Sn, which takes place to give exclusively R_3SnBr [158], and insertion by $PhHg \cdot CClBr_2$ and $PhHg \cdot CBr_3$ into the Ge—H bond results in the formation of dibromomethyl-, and bromochloromethyl-derivatives[159]. The reaction of $PhHg \cdot CCl_2Br$ with Me_3SnCl gives both $Me_3Sn \cdot CCl_3$ and $Me_3Sn \cdot CCl_2Br$, suggesting that there is CCl_2 insertion into Sn—Cl and substituent exchange between mercury and tin[160], while the compounds $Me_3Sn \cdot CX_2 \cdot SnMe_3$ ($X_2 = Cl_2$, ClBr, Br_2), are formed by insertion of CX_2 into the Sn—Sn bond[161]. The reaction of $HGeCl_3$ with $Hg[CH_2SiMe_3]_2$ gives $Cl_3Ge \cdot Hg \cdot GeCl_3$, when a 2:1 mixture is irradiated at $-45\,°C$ and $Me_3Si \cdot CH_2 \cdot Hg \cdot GeCl_3$, when a 1:1 mixture is left at room temperature. On pyrolysis, the latter compound yields $Me_3Si \cdot CH_2 \cdot GeCl_3$ and Hg [162]. The preparation of a variety of alkyl- or aralkyl-trihalogermanes by reaction of $HGeCl_3$ with alcohol is reported[163a]; $[(C_6F_5)_2TlBr]_2$ reacts with $SnCl_2$ in benzene to give $(C_6F_5)_2SnCl_2$ [163b].

Organolead chlorides are conveniently prepared by the action of $SOCl_2$ on R_4Pb (R = Me, Et) to give R_3PbCl and R_2PbCl_2, both of which disproportionate to give $PbCl_2$ [164]. Ph_3PbF and $MePbF_3$ are prepared by the action of $PhSiF_3$, HF and NH_4F, on lead tetra-acetate[165], and Bu_3SnF by the action of tetrafluorohydroquinone on Bu_3SnCl or $(Bu_3Sn)_2O$ [166]. Organotin halides are converted to R_3SnF and R_2SnF_2 by KHF_2, and on addition of BF_3OEt_2, the fluoroborates are formed and i.r. and n.m.r. spectra reported[167].

The crystal structure of $Ph_2SnI(CH_2)_4SnIPh_2$ shows slightly distorted tetrahedral coordination about tin[168] so it is a reasonably good standard for comparing the Mössbauer quadrupole splittings of compounds supposed to be tetrahedral. On this basis, many R_3SnX (R = neophyl, 2-methyl-2-phenyl-1-propyl; X = F, Cl, Br, I, CH_3CO_2; and R = Ph; X = Cl, Br, I, OCOCEtBu) derivatives are unassociated, truly tetrahedral tin compounds, whereas Me_3SnX (X = F, Cl, Br, I, CH_3CO_2) derivatives are associated with a 5-coordinate tin[169]. This suggests caution is necessary in interpreting the data on several compounds that are probably 5-coordinate but were assumed to be tetrahedral, namely R_3SnX (R = Et, Pr, Bu, Bu^i; X = F, Cl, Br)[170], and $(PrF_3)_3SnCl$ and Et_3SnBr [171]. The dihalides, R_2SnX_2 (X = Cl, Br, I), although also associated, do not have the same structures as Me_2SnF_2 [172]. The Mössbauer isomer shifts and quadrupole splittings for 23 compounds of the type A_mSnB_{4-m} are examined, where A and B represent groups such as Me, Et, Ph, OMe, Cl, Br, I and 3,3,3-trifluoropropyl. The authors conclude that electronegative ligands reduce the population of the tin hybrid orbital directed towards it and increase the population of those directed away. This implies that in 'losing' electrons to the electronegative ligand, the tin atom compensates by withdrawing electrons from the remaining ligands. The influence of π-back bonding may modify this conclusion[171]. The validity of the generalisation, that if a compound shows a Mössbauer effect at room temperature then its structure is polymeric, has been tested for a number of organotin derivatives known, from other evidence, to be polymeric[173].

A kinetic study of rapid chloride exchange in R_3GeCl (R = Bu^n, n-hexyl,

Ph, cyclohexyl) with $Li^{36}Cl$ in acetone–dioxan at 25 °C suggests steric factors influence the rate particularly with cyclohexyl substitution[174]. In addition, bond-energy and bond-polarity considerations rationalise why germanium compounds exchange more rapidly than the silicon analogues. A steric effect, or possibly an increase in (d–d)π-bonding between Sn and I, could account for the large differences in the strength of the Sn—I bond in Ph_3SnI relative to Me_3SnI based on appearance potentials[175]. The equilibrium constants for the exchange of chlorine and iodine in the system $MeGeI_3$ v. $MeP(S)Cl_2$, are measured by use of ^1H n.m.r. data[176]. At equilibrium, as was previously predicted, Cl is preferably associated with P [177].

Triorganotin chlorides are readily converted to diorganotin perfluorocarboxylates[178] by the reaction:

$$R_3SnCl + R'CO_2H \rightarrow R'CO_2SnClR_2 + RH$$
(R = Me, Ph, Bu; R' = Me, CF_3, C_2F_5, C_3F_5, C_3F_7, CF_2Cl)

The reaction of Bu^t_3SnX (X = Cl, Br, I), with Bu^tLi gives a coupling leading to $Bu^t_6Sn_2$, rather than the expected substitution reaction to give Bu^t_4Sn. To establish the structural requirements for normal substitution to occur, two series of organotin halides are examined. For $R_2R'SnX$, the series R = Bu^t; R' = Ph, p-MeC_6H_4, Bu^n gives coupling, while the series R = Ph, $PhCH_2$, Bu^n; R' = Bu^t gives normal substitution[179]. A ^1H n.m.r. spectroscopic study of the redistribution equilibria of three Me_2GeX_2 systems: (a) with $X_2 = Cl_2$, Br_2, I_2; (b) with $X_2 = Cl_2$, I_2 $(OPh)_2$; and (c) with $X_2 = Cl_2$, Br_2, I_2, $(OPh)_2$, indicates that all possible mixed species are also formed; e.g., in (a), $Me_2GeClBr$, Me_2GeClI, and Me_2GeBrI are also present at equilibrium[180]. The exchange reactions of the diorganotin chlorides, R_2SnCl_2 (R = Me, Ph), with triorganotin carboxylates, $R'_3Sn(OCR'')$ (R' = Me, Ph, Bu^n and R'' = Me, CH_2Cl, Ph), are followed by ^1H n.m.r. spectroscopy[181].

Earlier conclusions[182] regarding the nature of the charge distribution about the tin atom in Me_2SnCl_2 and Me_2SnMoO_4 are shown to be in error[183, 184].

Calculations of the expected Mössbauer spectra suggest that there is an excess of negative charge in the Me—Sn—Me bond direction rather than the SnX_4 plane[185]. The crystal structure of dimethyltin dichloride shows that the environment of the tin atoms is considerably distorted from tetrahedral towards octahedral because of association of neighbouring molecules. The Mössbauer spectra of the dialkyltin dichlorides are discussed in terms of this information, and of the dialkyltin difluorides in terms of a *trans*-octahedral configuration[186]. The i.r. and Raman spectra of Me_2SnF_2 are consistent with this *trans*-octahedral arrangement[187].

The dilithio-salts of *o*-, *m*- and *p*-carborane react with various alkyl-substituted chloro-germanium and -tin derivatives. In the germanium series, the disubstituted *o*-carborane derivatives formed are similar to those in the analogous silicon reactions, but the germanium *m*- and *p*-carborane products are mixtures of disubstituted monomers and related polymers. In the tin series, the tendency to polymerise is more marked[188]; $R_3SnCB_{10}H_{10}CPh$ results when R_3SnCl (R = Me, Ph) reacts with phenylneocarboranyl-lithium[189a], and 1-Et_3Ge-*o*-carborane is formed from the action of $(Et_3Ge)_2NH$ with *o*-carborane[189b].

Dimethyltin di-isothiocyanate forms orthorhombic crystals in which infinite chains of weakly linked $Me_2Sn(NCS)_2$ molecules have a strongly distorted tetrahedral arrangement about tin with linear N–C–S groups[190, 191]. The crystal structure of $Me_3Sn(NCS)$ consists of zig-zag —S····Sn—N≡C=S····Sn— chains, bent only at sulphur with nearly planar trimethyltin groups to give essentially a 5-coordinated, trigonal-bipyramidal arrangement about tin[192, 193]. The same planar trimethyltin groups are found in the crystal structure of $(Me_3Sn)_2N_2C$, in which the infinite helical network is linked by linear N—C—N units[194]. The trigonal-bipyramidal arrangement is also suggested from the i.r. and Mössbauer spectra of Me_3SnN_3, in which there are presumably azide bridges. The species R_3SnN_3 and $R_2Sn(N_3)_2$ (R = Et, Pr, Bu^n), are monomeric[195]. The aryllead(IV) thiocyanate system is re-examined and $Ph_2Pb(NCS)_2$, $Ph_3Pb(NCS)$, $Me_4M_2[Ph_2Pb(NCS)_4]$ and $Me_4M[Ph_3Pb(NCS)_2]$ (M = N, As) are analysed[196]. Two preparative routes, one involving M—N cleavage with diazomethane and the other involving the action of diazomethyl-lithium on a chloro(tri-organo)-derivative, are reported for the formation of the Group IV organometallic diazomethanes, $(Me_3M)_2CN_2$ (M = Sn or Pb); $(Ph_3Ge_2)_2CN_2$ and $(R_3Sn)_2CN_2$ (R = Et, Bu^n, Ph)[197].

7.3.2 Chalcogen derivatives

The organo-germanium, -tin and -lead sulphides, Ph_3M—S—$M'Ph_3$ (M, M' = Ge, Sn, Pb), can be prepared by the action of Ph_3MSLi on $Ph_3M'Cl$. The mixed M—S—M' species tend to re-arrange to $(Ph_3M)_2S$ and $(Ph_3M')_2S$[198]. Organogermanium sulphides are also produced by the action of sulphur on $(R_3Ge)_2Hg$[199] and the action of H_2S on the germylated amide of carboxylic acid[200]. The assignments of M—S vibrations in the spectra of Ph_3Ge—S—MPh_3 (M = Ge, Sn, Pb) are discussed with reference to other organometallic sulphides[201]. The alkylation of organotin sulphides, $(R_3Sn)_2S$ (R = Bu, Ph), is compared with that of the corresponding oxides[202]. $(Ph_3Sn)_2O$ reacts with thiourea in refluxing acetonitrile to give $(Ph_3Sn)_2$-sulphide, -carbodi-imide and -dicyandiamide. With 1,3-disubstituted thioureas the sulphide and, with the exception of di-t-butylthiourea, the 1,3-disubstituted urea are formed[203]. The electron diffraction of $(Me_3Ge)_2O$ and $(Me_3Sn)_2O$ give both the Ge—O—Ge and Sn—O—Sn angles as c. 141 degrees indicating a similar degree of multiple bonding[204]. Pyrolysis of $Et_3Ge·Hg·O·GeEt_3$ gives $(Et_3Ge)_2O$[205]. The di-n-butyltin oxide dimer, $[Bu_2SnO]_2$, reacts with phenyl benzoate to give 1:1 as well as the reported 2:1 adducts[206]. The frequency shifts of OH in methanol and NH in pyrrole of solutions of R_3MOMR_3, R_3MOR' and $R_2M(OR')_2$ (M = Si, Ge, Sn; R = Me, Bu^n; R' = Me), lead to an ordering of their relative basicity which increases with mass, e.g., from SiOSi—GeOGe—SnOSn[207]. R_3SnSR compounds react readily with sulphenyl halides to give R_3SnX and diphenyl disulphides in high yield[208]. The molecular spectra of Ph_3MSPh (M = Ge, Sn, Pb and $Ph_2Sn(SPh)_2$ are reported and the M—S vibrational frequencies assigned. Comparisons are made with the Sn—O and Sn—Se frequencies in Me_3SnEPh (E = O, Se)[209]. R_3GeER species are prepared by the action of the appropriate magnesium, R_3MgER (E = S, Se) salt on R_3GeBr to give R_3GeER [210]. Vibrational assignments are proposed, based on group frequencies and Raman polarisation data, for Me_3MSMe (M = Ge or Sn) [211].

The preparation, n.m.r. and i.r. spectra of the series of pentafluorophenyl derivatives $C_6F_5EMR_3$ (R = Me, Ph; M = Si, Ge, Sn, Pb; E = O, S, NH) are described[212]. A relationship between π-interactions and coupling constants is established[213] and the ^1H n.m.r. parameters for C_6F_5—E—MR_3 (E = O, S; M = Si, Ge, Sn, Pb; R = Me, Ph) interpreted as indicating that the tendency to form π-bonds to O and S decreases in the order Si > Ge > Sn > Pb [214]. The Mössbauer spectra of $(Bu_3^tSn)_2SO_4$, $(Bu_3^tSn)_2O$, $(Bu_3^tSn)_2S$, $(Me_3Sn)_2SO_4$ and $(Me_3Sn)_2SeO_4$ are discussed and it is concluded that in R_3SnX species, the Mössbauer data, even when considered with other spectroscopic data, must be examined very carefully before structural inferences are drawn[215]. Bis(tri-n-butyltin) sulphide is readily oxidised to the sulphate by hydrogen peroxide[216]. In the presence of light $(Me_3Sn)_2Se$ and $(Me_3Sn)_2Te$ react with metal hexacarbonyls $M(CO)_6$ (M = Cr, Mo, W) to give $(Me_3Sn)_2E$—$M(CO)_5$ derivatives (E = Se or Te) [217].

A large range of tetraorganotin compounds undergo insertion of SO_2 into the C—Sn bond. Vibrational, i.r. and mass spectra suggest the products are O-sulphinate in type, but aggregate in the solid phase and solution so that tin generally achieves 5-coordination with an essentially planar R_3Sn moiety[218]. Similar insertion occurs with C—Pb bonds and the behaviour of analogous tin and lead systems are compared[219]. With alkylic, allenylic and 2-propynylic (R') derivatives, R_3SnR' insertion is in the Sn—R' bond followed by rearrangement[220]. One mole of SO_2 is absorbed by Me_4Sn and Et_4Sn at $-20\,°C$ to give the 5-coordinated trialkyltin alkenesulphinates, R_3SnO_2SR [221]. The i.r. and Raman spectra of Me_3SnO_2SMe are assigned[222]. At $-60\,°C$ Me_4Sn reacts with liquid SO_2 to give $(Me_3Sn)_2SO_4$ but Et_4Sn gives a mixture of 6-coordinated diethyltin bis(ethane sulphinate) and $(Et_3Sn)_2SO_4$. ^1H n.m.r., Mössbauer and mass spectra indicate they are double O-sulphinato- and sulphato-complexes[221]. With SO_2, Ph_4Sn gives $Ph_2Sn(O_2SPh)_2$ and Ph_3MCl (M = Sn, Pb) gives $Ph_2M(O_2SPh)_2$ and Ph_2MCl_2 [223]. $SnCl_2 \cdot 2H_2O$ reacts with NaO_2SR (R = p-$CH_3C_6H_4$, Ph) in ethanol to give $Sn(O_2SR)_2$. 5- and 6-coordinated double O-sulphinato complexes, Ph_3SnO_2SR, $Ph_2Sn(O_2SR)_2$ and $Me_2Sn(O_2SR)_2$ are formed when Ph_3SnCl, Ph_2SnCl_2 and Me_2SnCl_2 react with NaO_2SR (R also = Me) in THF. Structural conclusions are based on i.r., Raman, Mössbauer and mass spectra[224]. Attempts to prepare tetrasulphonates of tin or lead failed but compounds of the type $R'_nM(OSO_2R)_{4-n}$ (R = CF_3, C_2F_5; R' = Me or Ph; n = 1,2,3) are produced[225].

The i.r. spectra of several new thiocarbonate derivatives, Ph_3M—S—C(O)—OEt, Ph_3M—S—C(S)—OEt and $Ph_2M(S$—C(S)—$OEt)_2$ (M = Ge, Sn, Pb), are discussed[226]. Studies of displacement reactions revealed a preferred tendency of formation of Sn—S bonds[227]. The preparation and spectroscopic studies of two N,N-dimethylthioselenocarbamate(dmtsc) complexes of dimethyltin(IV), $Me_2ClSn(dmtsc)$ and $Me_2Sn(dmtsc)$, are reported in which dmtsc is acting as a bidentate ligand[228]. Except that the N—Me proton signals are splitting in these complexes, the ^1H n.m.r. and i.r. spectra are similar to those[229] of the corresponding dithiocarbamate complexes,

Me$_2$Sn(dmdtc)$_2$ and Me$_2$ClSn(dmdtc), which were assumed to have distorted *trans*-octahedral and trigonal-bipyramidal configurations respectively. The x-ray determination on the latter confirms the appropriate trigonal-bipyramidal arrangement[230]. The crystal structures of the monoclinic[231] and orthorhombic[232] modifications of *N,N*-dimethyldithiocarbamatotrimethylstannane are reported and the Sn·S·C(S)·N skeleton is planar.

As part of a study of a series of silicon phthalocyanines, the electronic spectra of two germyl analogues are briefly reported[233]. The oxime derivatives, Me$_3$M—ON=C$_6$H$_{10}$ (M = Ge, Sn, Pb) are all monomeric in solution. Cyclohexanone reacts with Me$_3$GeCl, and Ph$_3$PbOH or (Ph$_3$Pb)$_2$O to give the germanium and lead derivatives, but the lithium salt is used in reaction with Me$_3$SnCl [234]. Further studies of R$_3$SnON=C$_6$H$_{10}$ suggest that for R = Et, Pr^n, Bu^n or Ph, the liquid oxime derivatives are monomeric in dilute solution of benzene. The solid derivative obtained for R = Me apparently has cyclic Sn—O—Sn—O structures[235].

Three different synthetic routes to the formation of phosphinous acid esters of the type (R$_2$P—O—)$_n$MR$'_{4-n}$ (M = Si, Ge, Sn) are described, as well as their reactions with H$_2$O$_2$, S, H$_2$O, CH$_3$I, HgBr$_2$, SOCl, CS$_2$ and isothiocyanates[236]. Four types of Sn—O—P compounds are characterised. All are oligomeric with phosphinato- or phosphonato-bridges. Structures, which are proposed on the basis of i.r., n.m.r. and Mössbauer spectral data, indicate 5-coordinate tin for R$_3$SnOP(O)R$_2$, 4- and 5-coordinate tin for (R$_3$SnO)$_2$P(O)R, 5-coordinate tin for poly(diorganotin organophosphates) and 6-coordinate tin for R$_2$Sn(OP(O)R$_2$)$_2$ [237]. R$_3$GeCl (R = lower alkyl), on treatment with NaO(RO)P(O)H, gives R$_3$GeO(RO)P(O)H [238]. The oxidation of Ph$_3$MAsPh$_2$ (M = Ge, Pb) results in the formation of Ph$_3$M—O—As(O)Ph$_2$ [239], while with Me$_3$SnAsR$_2$ (R = Me or Ph), Me$_3$SnOH and R$_2$As(O)OH are formed[240]. However, the action of O$_2$ gives insertion to give R$_3$Sn—O—As(O)R$_2$ [240]. Me$_3$Sn—SBu^n can be prepared by the cleavage of the Sn—P bond in Me$_3$Sn—PPh$_2$ with Ph$_2$BS Bu^n [241]. 1,3-Bis(diethylgermyl)propane or 1,4-bis(diethylgermyl)-butane react with diethylmercury. Pyrolysis of the resulting mercury derivative gives a five-membered ring containing a Ge—Ge bond. Reaction with oxygen, sulphur and selenium takes place readily to give a six-membered ring containing a Ge—E—Ge link (E = O, S, Se) [242]. The x-ray structure of 1-ethylgermatrane is reported[243].

7.3.3. Group V derivatives

The preparation and ^{1}H n.m.r. spectra of a number of (dialkylamino)stannanes, R$_{4-n}$Sn(NR$'_2$)$_n$ (R = Me, Et, Bu, Ph; R' = Me, Et), are reported, as well as their alkylation which proceeds via Sn—N bond cleavage[244]. Further reactions of these amines, as well as analogous tin phosphines with acetylenes, are also reported[245]. The electric dipole moments in solution of several dialkylaminostannanes, R$_3$SnNR$'_2$, R$_2$Sn(NR$'_2$)$_2$, RSn(NR$'_2$)$_3$ and Sn(NMe$_2$)$_4$ (R = Me, Et, Bu^n, Ph; R' = Me, Et), are recorded. The data are interpreted as supporting the concept of π-bonding in these compounds[246]. In support of this, the electron diffraction of Sn(NMe$_2$)$_4$, while showing the expected tetrahedral configuration about Sn, shows nearly co-planar bonds about nitrogen[247]. The Mössbauer spectra are reported[248] of several aminostannanes, $Sn(NR_2)_4$ (R = Me, Et, Me_3Sn—NHPh, —NMePh, —NR'_2, —NHC_6H_4Cl-p, —$NPhCONMe_2$, —NMe_2, $(Me_3Sn)_2NMe$, $(Me_3Sn)_3N$, $Me_2Sn(NEt_2)_2$, $Ph_{4-n}Sn(NMe_2)_n$ (n = 1,2,3). The i.r. spectra of Me_3MNMe Bu^t[M = (Si), Ge, Sn, Pb] and $(Me_3M)_2NMe$ (M = Ge, Sn, Pb), are reported, as well as those of mixed amines such as $Me_3GeN(MMe_3)Me$ (M = Sn, Pb)[249]. A mixed amine is also produced by reaction of $BrN(SiMe_3)_2$ with Me_3SnLi to give $Me_3SnN(SiMe_3)_2$[250]. The azides Ph_3MN_3 (M = Si, Ge, Sn) react with the corresponding hydrides Ph_3MH, to allow isolation of $Ph_3Si\cdot NH\cdot GePh_3$ and $(Ph_3Ge)_2NH$, but not compounds containing Sn—N bonds, presumably because cleavage of that bond by the hydride is favoured under these conditions[251]. However, Me_3SnLi reacts with PhN_3, to evolve N_2 and produce lithio(trimethylstannyl)phenylamide, $Me_3SnNPhLi$. This reacts with Me_3MCl (M = Si, Ge, Sn, Pb) to give the phenylamines $(Me_3M)_2NPh$. Cleavage of the Sn—H bond in R_3SnH (R = Me, Bu, Ph) by PhN_3 gives $R_3SnNHPh$; the i.r. spectra are discussed[252]. Me_3MNR_2 (M = Si, Ge, Sn) form 1 : 1 adducts with thiobenzoyl isocyanate. With silyl- and germyl-amines, the Me_3M group can migrate between an oxygen and nitrogen atom. The stannylamines prefer the S-metallated structure consistent with their 'softer' acid character[253]. The [(trialkylgermyl)immino]trialkylphosphoranes, R_3Ge—$N{=}PR_3$, like their silicon analogues, form 1:1 addition compounds with Me_3M (M = Al, Ga, In). In the corresponding tin species, R_3Sn—$N{=}PR_3$, there is cleavage of the Sn—N bonds, which are more labile towards electrophilic reagents[254]. $(Et_3Sn)_2O$, on treatment with cyanamide, gives the carbodiimide $Et_3SnN{:}C{:}NSnEt_3$[255] and Me_3GeSPh reacts with PhSNSO to give Me_3GeNSO[256].

The reactions of Bu_3SnNRR' and Et_3GePEt_2 with a variety of ketones are reported[257, 258], as well as the reaction of R_3SnNMe_2 (R = Me or Bu) and Et_3GePPh_2 with a variety of C=O and S=O containing species[259, 260]. An extensive study shows that 1,2-dipoles, A—B, represented by PhNCO, PhNCS, MeNCS, p-$MeC_6H_4\cdot N{:}C{:}N\cdot C_6H_4Me$-p, $C_6H_{11}\cdot N{:}C{:}N\cdot C_6H_{11}$, $MeO_2C\cdot C{:}C\cdot CO_2Me$, CO_2, CS_2 $CH_2{:}CH\cdot CN$, PhCN, p-$MeC_6H_4\cdot CN$, $CH_2{:}CCl(CN)$, $C_6F_5\cdot CN$, $CCl_3\cdot CN$, $\overline{RCH\cdot CH_2\cdot S}$, $\overline{CH_2\cdot CH_2\cdot CO\cdot O}$, and $MeCH{:}CH\cdot CHO$ insert into the M—N bond of Me_3M—NMe_2 [M = (Si), Ge, or Sn]. The stoichiometry of reactions, structure of adducts and relative reactivities are discussed. Extensive references to earlier work are given[261]. Aminostannanes, stannazanes and stannylamines react with CO_2 to give organotin carbamate addition products but with CS_2, organotin dithiocarbamate adducts are not always isolated[262]. The general reaction

$$R_3M\text{—}NMe_2 + LM'H \rightarrow HNMe_2 + R_3M\text{—}M'L$$

[where M = (Si), Ge, Sn; M' = metal, and L are ligands attached to M' other than hydrogen] is applicable for the formation of germanium- or tin–metal bonds as in:

$$(\pi\text{-}C_5H_5)(CO)_3M'\text{—}H + Me_3M\text{—}NMe_2 \rightarrow$$
$$(\pi\text{-}C_5H_5)(CO)_3M'\text{—}MMe_3 + HNMe_2$$
$$(M' = Cr, Mo, \text{ or } W)$$

Mechanisms are discussed as well as some properties of the products[263]. A large number of cleavage reactions of organotin amines are reported[241], typical of which are the following:

$$3Me_3SnNMe_2 + MF_3 \rightarrow 3Me_3SnF + M(NMe_2)_3 \ (M = P, As, Sb),$$
$$Et_3SnNMe_2 + Cl_2 \rightarrow Et_3SnCl + ClNMe_2$$
$$2Me_3SnNMe_2 + (AlEt_3)_2 \rightarrow 2Me_3SnEt + (Et_2Al \cdot NMe_2)_2$$

Me_3SnNMe_2 reacts with C_6F_5H in refluxing benzene to give $Me_3SnC_6F_5$, providing the first example of such an amine elimination reaction involving an aromatic hydrocarbon[264]. Cleavage of the Ge—N bond, but not the Sn—N bond, results from reactions with chloramine to give the organo-(halogeno)germane, amine (or ammonia) and nitrogen[265]. The methyl (chloro)germanes, Me_3GeCl, Me_2GeCl_2 and $MeGeCl_3$ react with $[Me_2S(O)N]^-$ to yield $Me_3GeNS(O)Me_2$, $Me_2Ge[NS(O)Me_2]_2$, and $MeGe[NS(O)Me_2]_3$, respectively, which are identified by their i.r. spectra[266].

Ammonolysis of various organotin compounds with KNH_2 in liquid ammonia is examined. With $Sn(CH=CH_2)_4$ and $Sn(CH_2C_6H_5)_4$ ammonolysis is complete; with $Sn(C_nH_{2n+1})_4$ ($n = 1,2,3,4,6,8,12$) the ease of ammonolysis decreases with increasing n, and with $(C_6H_{11})_{4-n}SnPh_{3n}$ ($0 \leqslant n \leqslant 3$); the products are $[(C_6H_{11})_3Sn]_2NH$, $\{K_2[NH(C_6H_{11})_2Sn]_2NH\}_x$, and $K_2[C_6H_{11}Sn(NH_2)_5]$, respectively, for $n = 1$, 2 or 3 [267]. With $(R_3Sn)_2$ ($R = C_6H_{11}$, Ph, Me, Et), cleavage of the Sn—Sn bond leads to the formation of two salts, e.g., $K_2[Sn(NH_2)_6]$ and $K[Sn(NH_2)_3]$; while $(C_6H_{11})_3Sn$—$SnPh_3$ also gives $[(C_6H_{11})_3Sn]_2NH$ and $[(C_6H_{11})_3Sn]_2$ [268]. $(C_6H_{11})_3PbI$ with KNH_2–liq. NH_3 yields $[(C_6H_{11})_3Pb]_2NH$ [269]. Ph_2PbI_2 reacts with KNH_2–liq. NH_3 to give $K_2[Pb(NH_2)_6]$. The intermediates, $Ph_2PbI_2 \cdot 3NH_3$, Ph_2PbNH, and $K_2[Ph_2Pb(NH_2)_4]$ were isolated[270]. The reaction of MCl_2 or Bu_2^hM ($M = Ge$, Sn) with 1,3-bis(methylamino)pentamethyldisilazane, after metallation or in the presence of triethylamine, yields novel six-membered rings containing $\overline{N—Si—N—M—N—Si}$ [271]. The insertion of phenyl isocyanate into the Sn—O bonds of 2,2′-biphenylenedioxytin(II) may constitute the formation of the first compound containing N—Sn^{II} bonds[272]. However, evidence from vibrational spectra of $Sn(NCS)_2$ and some complex ions suggests these also contain N—Sn^{II} bonds[273].

Organogermanium-, organotin- and organolead-phosphines can be prepared by the reaction of the appropriate halide with a phosphine in the presence of base to eliminate HX (X = Cl, Br). Thus Ph_3MX (M = Ge, Sn, Pb) reacts with Ph_2PH to give Ph_3M—PPh_2, and with $PhPH_2$ to give $(Ph_3M)_2PPh$, while R_3MCl (R = Me or Ph) reacts with PH_3 to give $(R_3M)_3P$ [274]. The latter reaction can also be used for the formation of the corresponding $(R_3Pb)As$ and $(R_3Pb)Sb$ species[275], and all the reactions are applicable to the formation of organotinarsines[276]. Phenyltin arsines have been prepared by the action of phenyltin chlorides with the $AsPh_2^-$ ion[277] and similarly $Ph_3MM'Ph_2$ (M = Ge, Pb; M′ = As, Sb) are obtained from the action of Ph_3MCl with $NaM'Ph_2$ [239]. The action of $(Ph_3Sn)_2PPh$ on Bu^nLi gives the $(Ph_3Sn)PPhLi$ salt which on treatment with Ph_3MX (M = Ge, Pb), gives the 'mixed' P—M bond compounds, $(Ph_3Sn)PPh(MPh_3)$ [274].

Me_3SnNMe_2 reacts with Bu_2^tPH to produce $Me_3SnPBu_2^t$ [278], while Me_3

SnCl reacts with MePH$_2$ in the presence of Et$_3$N to produce (Me$_3$Sn)$_2$ PMe [279]. These in turn displace a molecule of CO from Ni(CO)$_4$, Fe(CO)$_5$ or Mo(CO)$_6$ to give Me$_3$SnP Bu$_2^t$ and (Me$_3$Sn)$_2$PMe—Ni(CO)$_3$,—Fe (CO)$_4$ or —Mo(CO)$_5$. The i.r. and n.m.r. spectra are assigned[278, 279]. The phosphines (Me$_3$M)$_3$P [M = (Si), Ge, Sn] react with the iron carbonyls, Fe(CO)$_5$ and Fe$_2$(CO)$_9$, to give (CO)$_4$FeP(MMe$_3$)$_3$ [280]. Me$_3$SnPPh$_2$, Me$_2$Sn(PPh$_2$)$_2$ and their arsine analogues, react with nickel, chromium, and molybdenum carbonyls to give species such as Me$_3$SnP(Ph$_2$)Ni(CO)$_3$ or

$$\text{Me}_2\text{Sn}\begin{array}{c}\text{PPh}_2\\\text{PPh}_2\end{array}\text{Cr(CO)}_4$$

which are characterised by i.r. and ^{1}H n.m.r. spectra[281]. Apparently, the lower electronegativity of tin, relative to carbon, allows stronger donation by P or As to the transition metals.

The phenyltin arsines, Ph$_3$SnAsPh$_2$ and (Ph$_3$Sn)$_2$AsPh, react with PhN$_3$ to evolve N$_2$ and give Ph$_3$Sn—N(Ph)—As(=NPh)Ph$_2$ and [Ph$_3$Sn—N (Ph)]$_2$As(=NPh)Ph [282]. The removal of Me$_3$SnCl from mixtures of Me$_3$ SnAsPh$_2$, (Me$_3$Sn)$_2$MPh and (Me$_3$Sn)$_3$M (M = P or As) by the action of Ph$_2$ClM gives several new phosphines or arsines such as (Ph$_2$P)$_3$P or Ph$_2$PAsPh$_2$ [283].

To obtain primary organogermylphosphines, Me$_3$GeCl and Me$_2$GeCl$_2$ are reacted with LiAl(PH$_2$)$_4$ in triglyme to give Me$_3$GePH$_2$ and Me$_2$Ge (PH$_2$)$_2$, respectively[284]. The products are characterised by their n.m.r., i.r. and mass spectra. An interesting cage molecule, (Me$_2$Ge)$_6$P$_4$, in which the P and Me$_2$Ge groups are apparently equivalent, is obtained from the mercury-catalysed thermal decomposition of Me$_2$Ge(PH$_2$)$_2$ [285]. The formation of (PhGeP)$_7$ is also reported[286]. The primary organogermylarsines, MeGeH$_2$ AsH$_2$ and Me$_3$GeAsH$_2$, are prepared either by treatment of LiAl(AsH$_2$)$_4$ with MeGeH$_2$Br or Me$_3$GeBr or by exchange between the halides and SiH$_3$AsH$_2$. The germylarsines are characterised by cleavage with HCl or BCl$_3$ to give the chlorogermane, and water catalyses the condensation of Me$_3$GeAsH$_2$ to (Me$_3$Ge)$_3$As and AsH$_3$ [287].

7.3.4 Group III derivatives

A germanium atom can be incorporated as part of a carborane cage. MeGeCl$_3$ is refluxed with Na$_3$B$_{10}$H$_{10}$CH in THF to give a sublimable solid which on refluxing with piperidine leads to the isolation and identification by ^{1}H n.m.r. and mass spectra of 1,2-B$_{10}$H$_{10}$CHGeMe and Me$_4$N[B$_{10}$H$_{10}$ CHGe] [288]. The reaction of [(3)-1,2-B$_9$C$_2$H$_{11}$]$^{2-}$ with GeI$_2$, SnCl$_2$, or Pb(CH$_3$CO$_2$)$_2$ leads to the formation of MB$_9$C$_2$H$_{11}$ (M = Ge, Sn, Pb) in which M is probably incorporated in an icosahedral tricarbaborane[289, 290].

The Mössbauer, n.m.r. and mass spectra are reported for the first Sn—Group III complex, (C$_5$H$_5$)$_2$Sn—BF$_3$ [291]. There is no Sn—Al bond in (C$_6$H$_6$)Sn(AlCl$_4$)$_2$·C$_6$H$_6$ but essentially 7-coordinated pentagonal-bi-pyramidal Sn^{II}; of the five chlorine atoms in the plane, two pairs form bridges to Al, the fifth Cl being one of the pair of chlorine atoms to a third

Al with the other Cl occupying an axial position. The Sn^{II} polyhedron is completed by a symmetrical axial coordination to one of the benzene rings[292].

7.4 HALOGEN COMPOUNDS

Recent data are available on the binary halides of germanium, tin and lead of both oxidation states II and IV.

The i.r. spectra of GeF_2 as a gas and as a solid in neon or argon matrices give the two Ge—F stretching modes at 692 and 663 cm^{-1}, respectively, and the bending mode at 263 cm^{-1}. From isotopic shift measurements, the F—Ge—F angle is calculated to be c. 94 degrees[293, 294]. The Raman spectrum of gaseous $SnCl_2$ is the first reported of a discrete bent MX_2 species (X = halogen)[295]. The gas-phase Raman spectra of $GeCl_2$, $SnCl_2$, $PbCl_2$ and $PbBr_2$ in the presence of excess halogen[296], the i.r. spectrum of solid $GeCl_2$, $SnCl_2$ and $PbCl_2$ in an argon matrix[297], and the Raman spectrum of SnClBr [296] are all consistent with a bent molecular structure. Force-constant calculations are interpreted as indicating a weaker M—Cl bond in MCl_2 than MCl_4 (M = Ge, Sn, Pb) possibly due to a decrease in the s-character of the bonds[297]. The bending modes in $GeCl_2$, $SnCl_2$ and $PbCl_2$ are assigned to bands at 159, 120 and 99 cm^{-1}, respectively[296], which compare reasonably well with 162, 120 and 110 cm^{-1} suggested for the ground states from a study of their electronic spectra[298]. In the latter study band assignments are made from a semi-empirical extended Hückel calculation of the molecular orbitals. A new weak and diffuse emission band-system of $GeCl_2$ is reported in the yellow–red region[299] and a tentative model for the electronic states of SnI_2 is proposed[300].

A mass spectral examination of the vapour species of the systems Ge(s) + $GeX_4(g) \rightleftharpoons 2GeX_2(g)$ (X = Cl, Br) indicates that these are the only equilibria involved. Values for ΔH_f^0 (298 K) are obtained for $GeCl_2(g)$, 42 and $GeBr_2(g)$, 13 kcal mol^{-1} [301]. The crystal structure of SnI_2 is re-examined and can be best described as $Sn(SnI_3)_2$ with infinite $(SnI_3)_n^{n-}$ trigonal prisms with essentially 7-coordinated tin atoms, since each has a close iodine neighbour from an adjacent prism in addition to the six in its own prism. The prism pairs are connected to adjacent pairs by other tin atoms which are 6-coordinated in a distorted octahedron giving twice as many 7- as 6-coordinated tin atoms. There are no discrete molecular units[302]. In a single-crystal Raman study of orthorhombic $PbCl_2$, the frequencies are interpreted in terms of symmetry coordinates and by comparison with the modes of the isomorphous $PbBr_2$ [303]. The mixed tin(II) halides SnXF (X = Cl, Br, I), SnIX (X = Cl, Br), Sn_2XF_3 (X = Cl, I) and Sn_2BrCl_3 and Sn_3BrF_5 are examined by x-ray and ^{119}Sn Mössbauer spectroscopy[304]. With the SnXF species, it is deduced that the pyramidal environment for each tin atom consists of two bonds to bridging fluorine atoms and one terminal bond to the other halogen. With the SnXI compounds, the iodines probably act as the bridges. For Sn_2XF_3 (X = Cl, I), one terminal bond to fluorine, one bridging Sn—F bond and a bridging Sn—X bond are consistent with the Mössbauer data, but for Sn_3BrF_5 there is no ready interpretation in terms of a simple structure.

The vacuum u.v. photolysis of mixtures of (a) $GeCl_4/H_2GeCl_2$ and (b) $GeCl_4/HGeCl_3$ in argon, nitrogen or carbon monoxide matrices at 4 K, is followed by an i.r. spectra analysis. No new Ge—H features are observed, but the new features in the i.r. spectrum of (a) suggest the formation of $GeCl_2$, while those of (b) suggest the formation of $GeCl_3$ as a pyramid molecule with a Cl—Ge—Cl angle of c. 111 degrees[305]. The pyramidal structure is also indicated by e.s.r. examination of the $GeCl_3$ radical trapped in a $GeCl_4$ matrix following the γ-irradiation of the latter[306]. The u.v. absorption spectrum of GeCl, which is observed during flash photolysis of $GeCl_4$ [307], is compared with that of SnCl [308]. The vibrational spectrum of GeI is reported[309]. Electron impact studies of $SnCl_4$ and $SnCl_2$ give an ionisation potential for the SnCl radical which suggests the Sn—Cl bond is essentially covalent[310] and also that PbCl is covalent rather than ionic[311]. The value of $D_{(Sn-Cl)}$ 3.6 eV is indicated although another study suggested the value 3.2 eV [312]. Electron impact studies of $SnBr_2$ and $SnBr_4$ lead to ionisation potentials of the parent molecules and some of the radical fragments. It is suggested that in MX_4 compounds, the first electron comes from a lone-pair orbital of the halogen, but for MX_2 from a non-bonding metal orbital[313].

Relatively large values for the asymmetry parameters of the electric-field gradients at the chlorine in a polycrystalline Zeeman analysis of the n.q.r. transitions in $GeCl_4$ and $SnCl_4$ (and $SiCl_4$) are interpreted as being indicative of (d–p)π-bonding in these molecules[314]. As just mentioned, analysis of the i.r. spectrum of $GeCl_2$ indicates the Ge—Cl bond there is weaker than in $GeCl_4$ [297]. The pure quadrupole resonance spectrum of ^{127}I in solid SnI_4 indicates that several of the low-frequency vibrational modes of SnI_4 contribute to the temperature dependence of the quadrupole resonance frequency in the solid[315]. The optical absorption of both SnI_4 and SnI_2 is examined in the visible and near u.v. [300] and the absorption spectrum of SnI_4 in dioxane is discussed in connection with earlier work where discrepancies may result from peroxide impurities in the solvent[316]. The electron diffraction of $SnCl_4$ gives a more accurate value for $r_g(SnCl)$ of 2.280 ± 0.003 Å [317]. Dielectric relaxation studies of $SnCl_4$ in some non-polar solvents confirm the zero dipole of $SnCl_4$ in carbon tetrachloride; they are consistent with the previous values in benzene and support the value of 4.1 D, rather than an earlier one of 3.4D, in dioxan[318].

Liquid-state dispersion curves of some XY_4 molecules, including X = Ge, Sn; Y = Cl, Br, are obtained between 5000 and 20 cm^{-1}. Vibrational band intensities are reported and trends discussed[319]. The utility of a low-temperature Raman cell is demonstrated by an examination of the v_1 vibrations in $GeCl_4$ and $SnCl_4$ [320].

7.4.1 Complex ions of the halides

The formation of complex ions of tin(IV) halides has continued to be studied. The $[SnF_6]^{2-}$ ion is isolated in the form of the compounds M_2SnF_6 and $M_2SnF_6 \cdot MF$ (M = Cs, Rb) as solid phases in the system M_2SnF_6—HF—H_2O [321]. The Sn Mössbauer and vibrational spectra of $[SnF_6]^{2-}$ are studied. With M_2SnF_6 (M = K, Cs, NO), a trigonal K_2GeF_6 structure (D_{3d} site symmetry) is indicated, and with Na_2SnF_6, an orthorhombic structure

(D_{2h} symmetry). For $[ClO_2]_2SnF_6$, strong anion–cation interaction is suggested from the vibrational analysis and its isomer shift, which is the highest in the series, suggests an increase in s-electron density about tin[322]. No unusually low isomer shifts are noted in contrast to a previous report[323]. In the series $(Me_4N^+)_2$-$(SnCl_6)^{2-}$, -$(SnBr_6)^{2-}$, -$(SnCl_4Br_2)^{2-}$, -$(SnCl_2Br_4)^{2-}$ -$(SnCl_2I_4)^{2-}$ and -$(SnBr_2I_4)^{2-}$ [324] and also $(Et_4N)_2(SnX_4Y_2)$ (X = Cl, Br, I and Y = F, Cl, Br, I)[325], isomer shifts are related to increasing electronegativity (Pauling's[324] and Mulliken's[325]) of the ligand. Departure from linearity is noted for $(SnBr_4I_2)^{2-}$ and $(SnCl_4I_2)^{2-}$. The Mössbauer data are rationalised in terms of asymmetry in the stereochemistry of the tin coordination[325], e.g., distortion from octahedral symmetry with two bulky ligands in the cis-positions[324].

As part of general studies of the ^{35}Cl and ^{81}Br n.q.r. spectra of $[MX_6]^{2-}$ species, those of Cs_2SnX_6 (X = Cl, Br) at 270 °C [326] and their temperature dependence[327] are reported. The resonance frequency trends in the ^{35}Cl and ^{37}Cl n.q.r. spectra of ten $(SnCl_6)^{2-}$ salts are interpreted in terms of lattice effects. Differences in the crystal lattice induced up to 10% variation in the ^{35}Cl resonance frequency, and for M_2SnCl_6 (M = K, Rb, Cs) it is clear that the cations cannot be treated as point charges but their sizes must be considered[328]. It is also noted that the n.q.r. resonance is at a lower frequency for the complex ion $(SnCl_6)^{2-}$ than in $SnCl_4$, which could result from an increase in either (or both) ionic character or π-bond character[329].

The vibrational frequencies[330] and force constants[331] of $[MF_6]^{2-}$ (M = Ge, Sn) are calculated as are the force constants for $[SnX_6]^{2-}$ (X = Cl, Br)[332]. The far i.r. spectra of M_2SnI_6 (M = Cs, Rb) powder leads to the assignment of the SnI stretch at c. 166 cm^{-1} [333]. In more extensive studies of many $[MX_6]^{2-}$ containing species, including $SnCl_6^{2-}$, and $SnBr_6^{2-}$, a relationship is suggested between the degree of covalency of the M—X bond and the stretching force-constant[334]. A comparison of the far i.r. and Raman spectra of solids containing $[SnX_6]^{2-}$ (X = Cl, Br, I; $[SnX_3]^{-}$, X = F, Cl, Br, I) and SnX_4 (X = Cl, Br, I) leads to the suggestion that for a related series of compounds the primary stretching force constants are approximately proportional to the metal oxidation state divided by the coordination number[335]. The preparation of $(C_6H_5S_2)_2SnCl_6$ is reported[336]. The low-temperature ^1H and ^{119}Sn n.m.r. spectra of aqueous acetone solutions of $SnCl_4$ and $SnBr_4$ are interpreted as indicating the presence of $Sn(H_2O)_6^{4+}$ and SnX_6^{2-} as the dominant species between which there is rapid intermolecular exchange[337]. ^1H n.m.r. spectral studies, involving ^{19}F and ^{119}Sn double-irradiation, on $(PhMe_2CCH_2)_2SnF_2$ and $[SnF_6]^{2-}$ confirm that the sign of J_{SnF} is negative[338].

By contrast, recently, much fewer studies of complex ions of germanium(IV) and lead(IV) halides have been reported. The formation of GeF_5^- is reported as is that of SnX_5^- [339], by use of various large cations, Pr_4N^+ Bu_4N^+, and Ph_4As^+; i.r. spectra are reported. The ion forms a 1:1 adduct $[GeF_5NH_3]^-$ with ammonia and its heat of formation suggests that the acceptor power of GeF_5^- is greater than that of SiF_5^- [340]. GeF_5^- is also produced when GeF_4 is bombarded by low-energy electrons in the ion source of a time-of-flight mass spectrometer; GeF_2^-, GeF_3^-, $Ge_2F_4^-$ and $Ge_2F_8^-$ are also observed[341]. Anions of the type $[GeF_{6-2x}B_x]^{2-}$ (B = bidentate ligand, e.g. oxalate, malonate, methylmalonate, dimethylmalonate) are characterised in solution by ^{19}F n.m.r. spectroscopy as are anions of the type $[GeF_5OR]^{2-}$ and $[GeF_5O_2CR]^{2-}$. Redistribution reactions of $M(ox)_3^{2-}$ and MF_6^{2-} are studied quantitatively for M = Ge and Sn [342a]. The crystal lattice of $(NH_4)_3(GeF_6)F$ contains isolated (GeF_6) octahedra[342b] as do two forms of Li_2GeF_6 [342c].

The preparation, i.r. and Raman spectra are reported for the tetra-alkylammonium salts of $[Ph_2SnCl_3]^-$, $[Ph_2SnCl_4]^{2-}$, and $[Ph_3SnCl_2]^-$, but no interpretation of structures was attempted[343]. The i.r. and Raman spectra of $[Me_2SnX_4]^{2-}$ species (X = F, Cl, Br, NCS) are consistent with *trans*-octahedral structures[187]. Structure assignments of these types of octahedral tin(IV) complex ions have also been attempted on the basis of the degree of quadrupole splitting in their Mössbauer spectra, e.g., $[Me_2SnCl_4]^{2-}$ $[pyH]_2^+$ has ΔE of c. 4.0 and is deemed *trans*[344].

The preparation is reported of new $MPbF_6$ compounds (M = Mg, Ni, Cd, Hg). With M = Mg and Ni, they have the $LiSbF_6$ type structure, while Cd and Hg are isotypic with VF_3 [345]. The vibrational spectrum of the $[PbCl_6]^{2-}$ ion has been discussed, along with thirty other related species, in terms of the relationship between the degree of covalency of the M—X bond and the stretching force-constant[334].

A series of hexachloroplumbate(II,IV) salts is prepared, characterised by analysis and x-ray powder diffraction, and the two-probe D.C. conductivity measurements are reported[346]. The specific conductances of $PbCl_2/MCl$ (M = Na, Rb, Cs) mixtures as a function of composition from 800 °C to a few degrees of the liquidus suggest the presence of the complex $[PbCl_4]^{2-}$ ion in the RbCl and CsCl systems[347]. The conductance of molten $PbCl_2/KCl$ in the temperature range 500–800 °C and 20–100 mol % in $PbCl_2$ is reported[348].

Complex halide ions have also been more extensively studied for tin(II) than for germanium(II) and lead(II). In addition to the study mentioned above, the n.q.r. of the phase transition in $CsPbCl_3$ is reported[349]. The far i.r. spectrum of the species $M^+GeCl_3^-$ ($M^+ = Me_4N^+$, Me_3NH^+, $Me_2NH_2^+$, and $MeNH_3^+$), is consistent with the pyramidal, C_{3v} symmetry of the $GeCl_3^-$ anion[350]. Reactions of trichlorogermanate(II) salts, $MGeCl_3$, with aromatic substrates are reported[351] and their hydrolysis to yield $Ge(OH)_2$ is possibly the only chemical reaction of $MGeCl_3$ that produces a Ge^{II} species[350].

On refluxing $Me_3NHGeCl_3$ with $[\pi\text{-}C_5H_5Fe(CO)_2]_2$ in THF, the previously reported $[\pi\text{-}C_5H_5Fe(CO)_2]_2GeCl_2$ is formed in good yield. The i.r. spectrum of the latter suggest the existence of two conformers which may account for disagreements with conclusions from x-ray and i.r. studies of this compound. $Me_3NHGeCl_3$ also reacts with $\pi\text{-}C_5H_5Fe(CO)_2Cl$ to give good yields of $\pi\text{-}C_5H_5(CO)_2FeGeCl_3$ and with $Fe(CO)_5$ to give low yields of $Ge[Fe(CO)_4]_4$ [352]. A re-examination of the interaction of BF_3 with MCl_3^-, where M = Ge or Sn, suggests that, contrary to earlier reports, Cl_3MBF_3 complexes are not formed but rather the reaction is $SnCl_3^- + BF_3 \rightarrow SnCl_2 + BF_3Cl^-$ [353]. The $SnCl_3^-$ ion is proposed as a possible intermediate in several processes involving the oxidation of Sn^{II} to Sn^{IV}. These include the reduction of copper(II) by $SnCl_2$ [354], reduction of Mo^V to Mo^{III} by Sn^{II} in 9 M- and 12 M-HCl solution, where the data are consistent with attack of $SnCl_3^-$ on $MoOCl_5^{2-}$ and of $SnCl_4^{2-}$ on the open position *trans* to the Mo=O bond in $MoOCl_4^{-}$ [355], and the reduction of NO by aqueous $SnCl_2$ which is strongly catalysed by Cu^I possibly by formation of an intermediate of the type $N-CuCl_2-SnCl_3$ [356]. While Sn^{II} reduces vanadium(V) to predominately vanadium(III) in hydrochloric acid, in perchloric acid the predominant reduced species is vanadium(IV) [357, 358]. Some plausible mechanisms are considered[358]. Potentiometric and polarographic methods show that in its reaction with $SnCl_2$, platinum(II) is reduced to the zero-valent state and complex formation then occurs[359].

The crystal structure of $CsSnCl_3$ at room temperature is monoclinic with pyramidal $SnCl_3^-$ groups packed with Cs^+ cations in a layer structure. There is an irreversible phase transition at 117 °C to two cubic forms[360]. The pyramidal group is also seen for SnI_3^- in the crystal structure of SnI_2 where $(SnI_3)_n^{n-}$ units are present[302]. In the corresponding SnI_2—MI systems, there is evidence for the formation of $KSnI_3$, $RbSnI_3$, $Rb_2Sn_2I_5$ and $CsSnI_3$[361]. In the systems, SnF_2—MF_2 (M = Fe, Co, Ni), the ^{119}Sn Mössbauer parameters are obtained for materials isolated from aqueous and molten mixtures. The SnF_3^- unit is apparently present in $M(SnF_3)_2·6H_2O$, $M(Sn_2F_5)·2H_2O$, $8SnF_2$—MF_2, $4SnF_2$—MF_2 and $2SnF_2$—MF_2, while in $2SnF_2·2MF_2$ and $SnF_2·MF_2$ there is high coordination of fluorine about tin[362a]. $Na_4Sn_3F_{10}$ contains $[Sn_3F_{10}]^{4-}$ units in which there are distorted tetragonal SnF_4 pyramids with tin at the top[362b]. The far i.r. and Raman spectra of SnX_3^- (X = F, Cl, Br, I) are studied along with those of $(SnX_6)^{2-}$ and SnX_4 (X = Cl, Br, I). It is suggested that the primary stretching force-constants are approximately proportional to the metal oxidation state divided by the coordination number[335]. In a series of tin(II) salts of the type $Et_4N^+SnX_3^-$ and $Et_4N^+SnX_2Y^-$ (X,Y = Cl, Br, I), the stability to oxygen decreases from $SnCl_3^-$ through SnX_2Y^- to SnI_3^-. The i.r. and Raman spectra are recorded in the solid state and, where possible, in nitromethane solution. The solution spectra clearly indicate discrete pyramidal anions. The isomer shifts in the Mössbauer spectra of the SnX_3^- ions increase as the sums of the electronegativities of the attached halogen atoms decrease[363]. The chemical shifts and quadrupole splittings are also correlated for several triligand stannates including SnX_3^- (X = F, Cl, Br) [364a]. The absorption spectra of mixed KCl–KI solutions containing lead show the formation of mixed complexes of the type $[PbCl_{6-n}I_n]^{4-}$ ($n = 1-5$) [364b].

7.4.2 Pseudohalogen complex ions

The electronic spectrum of $Pb(SCN)_2$ is more complex than those of comparable thiocyanates. Possibly this results from a polymeric, bridged thiocyanate structure in the solid state which is broken down in alcohol solution[365a]. The solubility of $Pb(SCN)_2$ in $LiClO_4/SCN$ solution is discussed along with the fundamental thermodynamic functions for the formation of $Pb(SCN)_n^{2-n}$ ($n = 1 \to 5$) [365b]. The i.r. spectra of $Sn(NCS)_2$ and several of its complexes, $Me_4NSn(NCS)_3$, $Et_4NSn(NCS)_3$, $KSn(NCS)SO_4$ and $CsSn(NCS)_3$, indicate Sn—N bonding with extensive bridging[365c].

7.5 REACTIONS OF THE HALOGENS, PSEUDOHALOGENS AND ORGANOHALIDE DERIVATIVES

7.5.1 Oxidation state IV

7.5.1.1 Complex formation

Germanium(IV) and tin(IV) halides readily act as Lewis acids to give molecular complexes. GeX_4, SnX_4 (and SiX_4) (where X = halogen) react with pyridine and isoquinoline to give the 1:2 complexes, $MX_4 \cdot 2L$. Recently, the enthalpies of reaction were re-investigated and it is reported that, contrary to previous results, values do not vary greatly in each series of related adducts except that ΔH_f for $GeF_4 \cdot 2L > SiF_4 \cdot 2L$ [366]. Many vibrational spectroscopy studies have been carried out. The conclusions vary from determining the position of coordination to attempts to distinguish isomers, particularly *cis–trans* octahedral isomers. The x-ray determination[367] of $SnCl_4 \cdot 2py$, $SnBr_4 \cdot 2py$ and $GeCl_4 \cdot 2py$ confirms the *trans*-structure predicted by vibrational spectroscopy[368]. The expected contrast is noted between the Raman spectra of these *trans*-systems and the *cis*-octahedral structures resulting from the bidentate ligands in $SnCl_4 \cdot L$ (L = 1,10-phenanthroline, 2,2'-bipyridyl and 2,5-dithiahexane)[369]. The structure of $SnCl_4 \cdot 3PhNH_2$ is confirmed as $PhNH_3^+ [SnCl_4(PhNH_2)(PhNH)]^-$ [370]. $SnCl_4$ forms a 1:1 adduct with $(BuO)_3SiNH_2$ [371] as does $GeCl_4$ with trimethylamine[372]. The vibrational spectrum of the latter is best assigned on the basis of C_{3v} symmetry, i.e., a trigonal bipyramid with the NMe_3 group in axial position. With trimethylphosphine both $GeCl_4$ and $SnCl_4$ form 1:2 adducts which are deemed *trans*-octahedral from their vibrational spectra[372]. I.R. and Raman studies of the bis-phosphine adducts $SnCl_4 \cdot 2PR_3$ [PR_3 = PPh_3, PEt_3, PMe_2Ph, $PMePh_2$, PEt_2Ph, $PEtPh_2$ and $PPh_2(C{\equiv}CMe)$] indicate *trans*-stereochemistry, while the Raman of the chelate $Ph_2P \cdot CH_2 \cdot CH_2 \cdot PPh_2 \cdot SnCl_4$ suggests the *cis*-conformation as expected[373]. Isomer shifts in the Mössbauer spectra are more positive than in analogous O,N or S complexes indicating higher s-electron density at the tin nucleus and a relatively weak Sn—P interaction. This is interpreted as indicating that sp_z-orbitals of tin are used for bonding to P and three-centre two-electron bonds, involving its p_x- and p_y-orbitals are used for bonding to Cl [373]. The far i.r. and Raman spectra of $SnI_4 \cdot L$ (L = tetramethylethylenediamine, 2,2'-bipyridyl, 1,10-phenanthroline) and of $SnI_4 \cdot 2L$ (L = pyridine, trimethylamine, triphenylphosphine oxide) were studied. Those with bidentate ligands are presumably *cis*-octahedral but no firm conclusions could be reached from the spectra[374]. Tetrachlorotin(IV)-bis-acetonitrile has been examined by single-crystal x-ray diffraction which shows that the $SnCl_4 \cdot 2MeCN$ molecules are *cis*-octahedral. The i.r. and Raman spectra, of which there have been several conflicting reports, are assigned on the basis of C_{2v} symmetry[375]. The i.r. spectra are used to estimate the equilibrium constants for the 1:1 and 1:2 complexes formed in various solvents in the $SnBr_4$—PhCN, $SnBr_4$—MeCN, and $SnCl_4$—PhCN systems[376]. In other work involving CN-containing ligands, the studies have more or less been limited to determining the position of coordination. SnX_4 (X = Cl, Br) form 1:3 adducts with cyanamide (cm) but 1:2 adducts with diethylcyanamide (decm). With $SnX_4 \cdot 2decm$, the i.r. spectra seems to indicate that addition is through the CN-group rather than the amino and it is cis-octahedral. With $SnX_4 \cdot 3$ cm, the structure could be polymeric or of the form $[Sn(cm)_3X_3]^+X^-$ [377]. The i.r. spectra of the adducts $SnCl_4 \cdot fan$ and $SnCl_4 \cdot dmaf$ (fan = ferrocene acetonitrile; dmaf = N,N-dimethylaminomethyl-ferrocene) suggest that in the former coordination is through the CN-group but in the latter through the amino-group with little or no effect on the ferrocene part of the ligand[378]. The i.r. spectra of $SnX_4 \cdot L$ and $SnX_4 \cdot 2L$ [X = Cl, Br; L = $Hg(SCN)_2$ and $Hg(CN)_2$] have been interpreted as indicating that in the 1:1 complexes, coordination is via both the CN-groups to give bridges leading to 6-coordination about tin and with 1:2 complexes only one group is functional again to give octahedral complexes[379]. $SnCl_4$ and $SnBr_4$, but not $GeCl_4$, react with azomethane and azoethane to give 1:2 adducts[380].

Complexes via O—N, O—P and O—S linkages have been extensively studied. In the far i.r. spectra of a series of $SnF_4 \cdot 2L$ (L = 4-substituted pyridine- and quinoline 1-oxides), the N—O stretching frequencies are relatively easy to assign but not those of Sn—O. In the series $SnF_4 \cdot 2(4Z \cdot C_5H_5NO)$ (Z = NO_2, CN, Cl, H, CH_3, OCH_3), the back-donation from tin to ligand apparently increases with electron-withdrawing ability of the substituent[381]. The N—O stretching frequencies are also discussed for another series of amino-oxides[382]. The ^{119}Sn Mössbauer parameters of 30 tin(IV) chloride complexes of the type $SnCl_4 \cdot 2L$ and $SnCl_4 \cdot L'$ (where L and L' are monodentate and bidentate ligands, respectively, and in general contain >SO, >SO_2, >PO and >CO groups) are in contrast to similar complexes formed by nitrogen-donor molecules, in that nearly all show quadrupole splittings of the order 0.5–1.6 mm s^{-1}. The splitting is rationalised in terms of the weaker Lewis acid-base interactions expected for these adducts and in terms of steric hindrance caused by a bulky ligand group. The limitations of the use of Mössbauer spectroscopy in definitively distinguishing between cis- and trans-isomers is discussed[383]. The dipole moments, i.r. spectra and structures are discussed for adducts of SnX_4 (X = Cl, Br) with Me_2CO, EtCOMe and PhCOMe [384], while an n.q.r. study of complexes of $SnCl_4$ with $ClCH_2OMe$ and benzyl chloride suggest the 1:2 complexes are cis[385]. The isomer shifts in the Mössbauer spectra of complexes of $SnCl_4$ with various O-containing aromatic compounds are similar to those of aliphatics[386]. The dipole moment of the complex $SnCl_4 \cdot 2H_2O \cdot 2$-dioxane is reported[387]. Dipole-moment studies of the 6-coordinated halogeno- and organo-tin chelates R_2SnL [R = Ph, Me, Et, Bu; L = acetylacetonate (acac), dibenzoylmethanate, and hexafluoroacetylacetonate (hfacac)] and $X_2Sn(acac)_2$ (X = F, Cl, Br, I) support cis-octahedral structures[388, 389]. With $Cl_2Sn(hfacac)$, the Sn—O bonds are apparently less polar than in the corresponding organo-derivatives, possibly because of (p–d) π-bond character[390]. $GeCl_4$ forms 1:4 (and 1:6) complexes with p-aminobenzoic acid, ethyl p-aminobenzoate[391], and o- and m-$H_2N \cdot C_6H_4 \cdot CO_2H$ [392]. I.R. spectroscopy suggests that in the 1:4 complexes the two Cl atoms in the inner coordination sphere are trans[393]. The complexes of $SnCl_4$ with gluconic and saccharic acids or their salts are formulated as species such as $[SnCl_2(C_6H_9O_7)]^-$ and $[SnCl_2(C_6H_6O_8)]^{2-}$ [394].

The i.r. spectra provide evidence for Sn—O bonding in the complexes formed between SnX_4 (X = Cl, Br) and cyanoacetamide, while their molar conductance confirms their non-ionic nature[395]. The i.r. spectra of the adducts $SnCl_4·2L_A$ and $SnCl_4·L_B$, $SnCl_4·L_C$, and $SnCl_4·L_D$ (where L_A = phthalimide, L_B = phthalamide, L_C = 1,2-cyclohexaphthalimide, and L_D = succinimide) indicate that the bidentate ligand L_B gives a cis-octahedral structure, while the adducts $SnCl_4·L_C$ and $SnCl_4·L_D$ are possibly cyclic low-polymeric compounds[396]. On the other hand, other workers suggest that the succinimide adducts, as well as SnX_4·is (is = isatin) and SnX_4·ph (ph = phthalimide; X = Cl, Br), are all cis-octahedral[397]. A ^1H n.m.r. study of p-substituted N,N-dimethylbenzamides suggests that $SnCl_4$ interacts only via the carbonyl O-atom, while $GeCl_4$ (and $SiCl_4$) interact also via the N-atom[398]. However, a study of the amide adducts, $SnX_4·2L$ (L = acrylamide, acrylamide-d_2, and N,N-dimethylacylamide), concludes that while the N,N-dimethylacylamide ligand exhibits 'normal' O-coordination, the acrylamide complexes from i.r., n.m.r., mass and photoelectron spectroscopy, apparently provide the first examples of N-coordinated amides[399]. In the tetraethyldithio-oxamide complex, $Sn(tedto)Cl_4$, the ligand is bidentate and probably coordinates through the thiocarbonyl groups[400]. $SnCl_4$ forms 1:1 adducts with benzo- and 2-aminobenzo-thiazole, and 2-benzimidazole- and 2-benzothiazole-thiol[401].

Tin(IV) chloride and bromide, but not iodide, form addition compounds with several aldehydes. I.R. spectra were not conclusive in determining cis- or trans-isomers. Dipole-moment measurements indicate that of $SnX_4·2L$ species (where L = naphthaldehyde, cinnamaldehyde, furanaldehyde, benzaldehyde, salicylaldehyde). the naphthaldehyde complex may be trans-octahedral and those of furanaldehyde and benzaldehyde possibly cis. A 1:1 complex with phthalaldehyde is also reported[402].

$SnCl_4$ reacts with the chelating Schiff's base salicylidene–aniline, L, to give $SnCl_2L_2$ [403].

The i.r., n.m.r. and Mössbauer spectra of several tin(IV)–Schiff's base complexes are reported, including $SnCl_4·2(sa1H-N-p-MeC_6H_4)$, $SnCl_4·(salenH_2)$, $SnCl_2·(salen)$ and $SnCl_3[salH-N-(2-OC_6H_4)]Et_3N$, where salenH$_2$ = N,N$^-$ethylenebis(salicylaldimine). The ^1H n.m.r. spectra of R_2Sn(salen) (R = Me, Ph), suggest cis-R groups with the salen group 'twisted'[404]. $PbCl_4$ apparently forms non-electrolyte, probably octahedral, complexes with monodentate Schiff's bases of benzaldehyde and anisaldehyde with various amines[405]. The Mössbauer spectra are reported of tin(IV) complexes with the chelating agents 8-hydroxyquinoline (oxH) and salicylaldehyde (salH), namely $R_2Sn(ox)_2$ and $R_2Sn(sal)_2$ (R = alkyl, aryl, or halogen), and $SnCl_{4-n}(ox)_n$(n = 0–4), as well as $SnX_4·2oxH$ and $SnX_4·2salH$ (X = halogen). Quadrupole splitting suggests the C-atoms are cis in $R_2Sn(ox)_2$ (R = alkyl or aryl). The O-atoms in $SnX_4·2oxH$ are also cis but the trans-configuration is preferred in $SnX_4·2salH$. The other species are apparently polymeric[406]. The Mössbauer and i.r. spectra of various 6-coordinated oxinate complexes of tin(IV) containing Sn—C bonds, $R_2Sn(oxin)_2$, R_2SnX(oxin), $RSnX(oxin)_2$, $RSn(oxin)_3$, $R_3Sn(oxin)$ (R = alkyl or aryl; X = halogen or isothiocyanate) are reported[407]. The $R_2Sn(oxin)_2$ complexes (R = Me, Pr, Ph, Bu^n) are probably cis with $\Delta E \sim 2$. A linear correlation is reported between Mössbauer chemical shifts and 4d-binding energies of tin, as measured by high-energy photoelectron spectroscopy for the series of tin(IV) octahedral complexes $Y_2Sn(ox)_2$ (ox = 8-quinolinolato, and Y = Et, Ph, Cl, Br, I)[408].

The i.r. spectra of complexes formed by reaction of $GeCl_4$ with o-hydroxy-anils of aromatic aldehydes suggest a phenolate structure[409]. Some complexes are isolated in the reaction of $SnCl_4$ with acetylacetone anils[410], and an unstable 1:1 complex is formed with benzil[411]. Weak complex formation of the types Sn····ClCOR and Sn····OCRCl are suggested from the Raman spectra of mixtures of $SnCl_4$ with acyl chlorides[412].

Both $GeCl_4$ and $SnCl_4$ form 1:2 complexes with acylphosphinomethyl-enes, p-XC_6H_4COC=PPh_3 (X = H, Cl, NO_2, OMe), and i.r. spectra are discussed[413a, 413b]. $SnCl_4$ forms bis(methylphosphonato)-complexes with di-isopropyl methylphosphonate in which the tin atom has a distorted octahedral symmetry[414]. A number of new complexes in the pyrazolone series are reported among which is $[Ge(pyramidone)_3]Cl_4$ [415].

The rather unusual complex $SnCl_4 \cdot 2S_4N_4$ is re-examined and its Mössbauer and i.r. spectra[416] indicate the *trans*-octahedral structure originally suggested[417]. The Raman spectrum of the solid adduct $SnCl_4 \cdot 2SCl_4$ is reported along with several of ECl_4 (E = S, Se, Te)[418]. The u.v. spectrum suggests that $SnCl_4 \cdot 2Si(OEt)_4$ is formed when the two components are mixed in benzene[419]. The adduct $GeF_4 \cdot 2BrF_3$ is characterised by Raman and low-temperature i.r. spectroscopy[420].

7.5.1.2 Organohalide and pseudohalide complexes

The stereochemistry of 6-coordinated complexes of diorganotin dihalides has continued to be studied. Building on earlier Mössbauer studies[421], the data, particularly the value of the quadrupole splitting, are used to predict *cis*- and *trans*-isomerism relative to the C—Sn—C bond[344, 422]. For *trans*-species, i.e., C—Sn—C angle of 180 degrees, ΔE is c. 4 mm s^{-1} and in this class are Me_2SnCl_2(bipy), Me_2SnCl_2(phen), Bu_2SnX_2(phen) (X = Cl, Br, I), Bu_2SnX_2(bipy) (X = Cl, Br), Ph_2SnCl_2(bipy), Ph_2SnCl_2(phen), and $R_2SnX_2 L_2$ (R = Ph; X = Cl, Br; L = py, ½bipy, ½dipyam, ½tripyam; and R = Et; X = Cl, Br; L = ½dipyam, ½tripyam). With ΔE of c. 3.6 mm s^{-1}, it is suggested that there is considerable distortion resulting in the C—Sn—C angle becoming less than 180 degrees, and the low value for Ph_2SnX_2 (X = Cl, Br) is interpreted as indicating a *cis*-arrangement of the halogens[422]. The *cis*-Cl, *trans*-R geometry is also suggested for the complexes $R_2SnCl_2 \cdot L$ (R = Me, Et, Pr, Bu, C_5H_{11}, Ph, $C_{12}H_{25}$, Cl; L = 2,7-dimethyl-1,8-naphthyridine) for which assignments are also made of their i.r. spectra and of $J(^{117}Sn—H)$ and $J(^{119}Sn—H)$ for R = Me and Et [423]. A conductiometric and u.v. spectroscopic study is reported of the non-ionic 1:1 adducts of Me_2SnCl_2, Me_3GeCl and Me_3SnCl; and the equilibrium mixture of 1:1 and 1:2 adducts of Me_2GeCl_2 and Me_2SnI_2 formed with 2,2′-bipyridyl[424]. $R_2SnCl_2 \cdot 2py$ (R = Me, Et, Pr) complexes are prepared by treating R_2SnO–R_2SnCl_2 with pyridine[425]. The i.r. spectra of the adducts of $PhBuSnCl_2$ with a series of amines such as pyridine, quinoline and aniline, are consistent with distorted octahedral structures[426]. These amines react with $PrSnCl_3$ to give complexes whose molar conductances suggest the formulations (2L·PrSnCl$_2$)$^+$Cl$^-$ and (4L·PrSnCl$_2$)$^+$Cl$^-$ [427]. Methyltin halides form complexes with dibenzyl, diethyl, and methylbenzyl-sulphoxides. The larger organo-substituents lead to the isolation of complexes of different stoichiometries and structures[428]. X-ray analysis shows the structure cis-dichloro-cis-bis(dimethyl sulphoxide)-trans-dimethyltin(IV) [429]. The crystal structure of Me$_2$SnCl$_2$(C$_5$H$_5$NO)$_2$ shows trans-C—Sn—C, Cl—Sn—Cl, and O—Sn—O groups and the sign of its quadrupole interaction is positive so that V_{zz} is negative[430] in contrast to earlier conclusions. The bromine atoms are cis in dibromo-bis(1,2-bisethoxycarbonyl)-ethyltin[431]. Stability constants of the complexes of Me$_2$SnCl$_2$ and SnCl$_4$ with XC$_6$H$_4$·COR (R = Me, H, or NHMe$_2$; X = NMe$_2$, MeO, Me, H, Cl, or NO$_2$) become larger for Me < H < NMe$_2$ [432]. 1:1 Adducts are reported for the interaction of R$_2$MCl$_2$ (R = Me, Ph; M = Sn, Pb) with the tetradentate Schiff's bases, bis(acetylacetone) ethylenedi-imine and bis(acetylacetone)-1,2-di-iminopropane[433].

The Mössbauer parameters of trimethyltin chloride and adducts, Me$_3$SnCl·L (L = hexamethylphosphoramide, N,N-dimethylacetamide, p-methylpyridine-N-oxide, and triphenylphosphine oxide) are obtained[434]. The model proposed to explain the quadrupole splitting is consistent with previous explanations of changes in $J(^{119}$Sn—C—H) and suggests that π-bonding is not important in these systems, contrary to earlier suggestions. Similar conclusions as to the unimportance of π-bonding in influencing quadrupole splittings in tin(IV) compounds are drawn from a general study of such compounds including Me$_3$SnCl·py [435]. The Mössbauer parameters of the complexes Bu$_3$SnCl·2L (2L = 2Ph$_3$PO, 2dma, 2py, 2pyO, 2dmso, bipy, phen), Me$_2$SnCl$_2$·2L (L = dmso, pyO), oct$_2$SnCl$_2$·L (L = bipy, phen), [BuSnCl$_5$]$^{2-}$ [Et$_4$N]$^+_2$, and [BuSnCl$_3$Br$_2$]$^{2-}$ [Et$_4$N]$^+_2$, and of a series of species [SnX$_2$Y$_4$]$^{2-}$ (X = Cl, Br, I), are reported. In the latter species, there is a linear relationship between the isomer shift and the sum of the (Mulliken) electronegativities of the ligands[436].

The Mössbauer and i.r. spectra, as well as electric dipole moments, of several complexes of the type RSnX$_3$L (R = Bu, Ph; X = Cl (or NCS); L = α,α'-bipyridyl, o-phenantholine, 8-aminoquinoline) and RSnXL$'_2$ (L = oxinate, 2-pyridinethiol-1-oxide) are reported. Quadrupole splittings suggest a similar octahedral structure for all complexes with two X-groups trans in RSnX$_3$L and an R and X group cis in RSnXL$'_2$. Isomer shifts indicate the order of bond polarity is SnO ~ Sn—NCS > SnCl > SnS [437]. The 7-coordinate T$_3$SnCl and T$_3$SnOH (T$^-$ = C$_7$H$_5$O$_2^-$, the bidentate tropolonate ion) are isostructural in the dissimilar crystal structures observed for T$_3$SnCl·CHCl$_3$ and T$_3$SnOH·(3−2x)H$_2$O·xCH$_3$OH ($x \sim \frac{1}{2}$). The arrangement about Sn is approximately a pentagonal-bipyramid with Sn—Cl (or Sn—O) axial[438]. 1:2 Complexes of PhSnCl$_3$ with pyridine, β-picoline, isoquinoline, piperidine, aniline and benzylamine and 1:4 complexes with picoline and morpholine are reported[439]. The trends in the i.r., Raman, and ^{1}H n.m.r. spectral parameters of Et$_3$SnCl, Et$_3$SnCl·2py, Et$_2$SnCl$_2$, Et$_2$SnCl$_2$·2py, EtSnCl$_3$ and EtSnCl$_3$·2py are compared[440].

The i.r. spectra of the 1:1 adducts formed by diaryltin di-isothiocyanates, Ph$_2$Sn(NCS)$_2$ and (o-C$_6$H$_4$·CH$_3$)$_2$Sn(NCS)$_2$ with 2,2'-bipyridyl and 1,10-phenanthroline are discussed[441]. MeSn(NCS)$_3$ and R$_2$Sn(NCS)$_2$ (R = Me,

Et, Bu, octyl), form 6-coordinated adducts with some or all of 1,10-phenanthroline, 2,2′-bipyridyl, pyridine-1-oxide, quinoline-1-oxide, triphenylphosphine oxide and triphenylarsine oxide. The stability diminishes with increase in size of R. For $Me_2Sn(NCS)_2 \cdot 2L$, the methyl groups are *trans*. $Me_3Sn(NCS)$ forms 1:1 complexes with the monodentate ligands to give a trigonal bipyramidal arrangement about Sn with the three Me groups planar. Bipyridyl apparently acts as a monodentate ligand in $Me_3Sn(NCS) \cdot bipy$ and $2Me_3Sn(NCS) \cdot bipy$, but $Me_3Sn(NCS) \cdot phen$ is 6-coordinated as expected[442]. The Mössbauer parameters of a series of diorgonotin bisdithiocarbamates are interpreted as *cis*-octahedral for $Ph_2Sn[S_2CNR_2]_2$ and *trans*-octahedral for $R'_2Sn[S_2CNR_2]_2$ (R′ = alkyl)[443].

7.5.1.3 Other reactions

A novel reaction of germanium(IV) chloride involves the reaction with carbon vapour, generated from a carbon arc. When co-condensed with $GeCl_4$, mainly $(Cl_3Ge)_2CCl_2$, and smaller amounts of $GeCl_3 \cdot CCl_3$ and GeC_2Cl_6 (possibly of $Cl_3Ge \cdot CCl=CCl_2$), are produced[444].

The kinetics of the hydrolysis of GeI_4 is reported[445] and the rather complex hydrolysis of $SnCl_4$ has been studied by Mössbauer spectroscopy. The nature of the first product, $SnCl_4 \cdot 5H_2O$, is discussed[446]. The bond-character of Sn—Cl in various hydrates is discussed also in terms of n.q.r. data[447]. Conductivity measurements of the solvolysis of $GeCl_4$ in acetic acid and acetone indicate that solvation is greater in the former[448]. Similar measurements are made for the reactions of $GeCl_4$ with nicotinic acid and with *p*-aminobenzoic acid and its ethyl ester[449,450].

An interesting gradation in products is noted for the reaction of MCl_4 (M = Si, Ge, Sn) with enamines to form organometallic immonium type salts(I) and organometallic enamines(II):

$$MCl_4 + CH_2{=}C(NMe_2)R \rightleftharpoons (Cl_3M \cdot CH_2 \cdot C(NMe_2)R)^+ Cl^- \xrightarrow{base}$$
$$(I) \qquad Cl_3MCH{=}C(NMe_2)R$$
$$(II)$$

With $SiCl_4$, only the enamine(II) can be isolated; with $GeCl_4$ the intermediate immonium salt(I) can also be isolated; and with $SnCl_4$, the 'intermediate' cannot be converted with base. Thus silicon shows a preference for the unsaturated substituent while tin will not form it. The gradation is consistent with a decreasing ability from Si—Ge—Sn to form (p–d) π-bonds[451].

$SnCl_4$ reacts with $S_2O_6F_2$ to give $Sn(SO_3F)_4$ and chlorine. In excess chloride the following equilibrium is established:

$$SnCl_4 + Sn(SO_3F)_4 \rightleftharpoons 2SnCl_2(SO_3F)_2$$

Mössbauer and vibrational spectra suggest a polymeric-chain or a sheet-type structure with bridging fluorosulphate groups[452].

The reaction of tin fluoride with liquid NO_2 leads to the formation of an unstable oily solid which evolves NO_2 in vacuum. On heating, it becomes colourless and more stable. I.R. spectral analysis suggests[453] that the overall reaction can best be represented as:

$$3SnF_4 + 4NO_2 \rightarrow 2SnF_3(NO_3) + (NO)_2^+[SnF_6]^{2-}.$$

The reaction of $RPFS_2H$ (R = Me, Et) with SnX_4 (X = Cl, Br) yields $(RPFS_2)_2SnX_2$. Analysis and i.r., ^1H and ^{19}F n.m.r. spectra are reported and structures proposed[454]. Tetrakis(pyrrolidino)-germanium and -tin, $M(NC_4H_8)_4$, are prepared by the action of the tetrachlorides on pyrrolidine in the vapour phase in sealed ampoules ($SnCl_2$ gives $SnL_2Cl_2·2HCl$, L = NC_4H_8) [455]. Alkali tin(IV) hexaoxoiodates(VII), $MSnIO_6$ (M = Na, K, Rb, Cs, NH_4), isomorphous with the corresponding lead(IV) compounds[456], are prepared by treating $SnCl_4$ with the appropriate nitrate and hexaoxoiodic acid[457].

7.5.2 Oxidation state II

7.5.2.1 Complex formation

Adducts of tin(II) halides have been examined mainly by Mössbauer spectroscopy. For a number of complexes, $SnX_2·$piperazine, $·2$morpholine, $·2$piperidine, $·PPh_3$, $·PBu_3^t$ (X = Cl, Br) and $SnI_2·PBu_3^t$, the isomer shifts are lower than for the parent halides as expected for a bridging atom being replaced by a stronger Sn–ligand bond (Sn—N being much stronger than Sn—P). Conclusions based on quadrupole splittings are less precise[458]. Mössbauer data are more useful when combined with other studies such as vibrational spectra. Thus for the compounds $SnX_2·OPPh_3$ (X = Cl, Br, NCS, ClO_4), $SnX_2·2OPPh_3$ (X = Cl, Br), $SnX_2·ONC_5H_5$, $SnX_2·2OSMe_2$ (X = Br, NCS), $SnCl_2·OC(NH_2)_2$, $SnCl_2·OAsPh_3$, $SnX_2·2L$ (L = tetrahydrofuran), $2SnX_2L$ (L = diglyme), SnX_2L (L = 1,4-dioxan, 1,4-thioxan, or 1,4-dithian) (X = Cl, Br), the Mössbauer shifts suggest an order of donor strengths of E=O as As=O > P=O > N → O > S=O > C=O, while the shifts in the E=O stretching modes to lower frequencies suggest acceptor strengths in the order $SnCl_2 > SnBr_2 > Sn(NCS)_2$ [459]. The crystal structure of $GeCl_2·C_4H_8O_2$ indicates an infinite chain structure with the Cl—Ge—Cl angles c. 94 degrees[460].

Sparingly soluble molecular compounds are formed by lead halides with thiourea, $PbX_2·2(NH_2)_2CS$ (X = Cl, Br), $PbBr_2·(NH_2)_2CS$ and $3PbI_2·2(NH_2)_2CS$ [461]. The i.r. and Mössbauer spectral data of a series of thiourea (tu) complexes with tin indicate that the Sn—S bonds are strong in halide or sulphate complexes (e.g., $Sn(tu)X_2$, X = Cl, Br, I, SO_4) but weak in carboxylate, sulphamate, borofluoride, or perchlorate complexes such as $Sn(tu)_6(BF_4)_2$ or $Sn(tu)(HCO_2)_2$ [462].

Several coordination compounds of the types $2SnCl_2·L$, $SnCl_2·L$ and $SnCl_2·2L$ are obtained with organic ligands L containing C=O, CO_2^-, C=S, CN=C, NH, or NH_2 groups[463]; 1:1 adducts are formed between $SnCl_2$ and benzo- and 2-aminobenzo-thiazole as well as 2-benzimidazole- and 2-benzothiazole-thiol[407]; anhydrous $SnCl_2$, when dissolved in $(Me_2N)_3P$:O, hmpt, to which ligroin is added, gives the air-sensitive adduct $SnCl_2·(hmpt)$[464]; and $SnCl_2$ reacts with pyrrolidine (L) to give $SnCl_2L_2·2HCl$ [455].

Conductiometric determination of complexes of $SnCl_2$ with pyridine, quinoline, β-picoline and piperidine suggest ionic structures, possibly $[SnClL]^+[SnCl_3]^-$ [465]. The i.r. and Mössbauer spectra of several tin(II) complexes, e.g., SnX_2L (X = Cl, Br; L = bipy, p-toluidine), SnXL (L = oxine), and SnL_2 (L = oxine, anthranilate) are reported. From assignments of the NH_2 stretching and deformation modes, the order of nitrogen to metal dative bond strengths is assumed to be N—Cd > N—Sn^{II} > N—Pb^{II} [466]. The absorption spectral characteristics of several $SnCl_2$ complexes are studied in solution[467].

The formation and i.r. spectra of compounds with a six-membered ring containing a germanium or tin atom are reported. $[N(H)(Me)Si(Me)_2]O$ and MCl_2 (M = Ge or Sn) in the presence of butyl-lithium and triethylamine yield $Cl_2M[N(H)(Me)Si(Me)_2]_2O$. Analogous compounds with O replacing NMe are also reported[468].

7.6 CHALCOGEN COMPOUNDS

7.6.1 Chalcogenides of germanium, tin and lead in both II and IV oxidation states

The esterification reactions of SnO are reviewed[469]. The dipole moments of all the ME species (M = ^{28}Si, ^{74}Ge, ^{120}Sn, ^{208}Pb; E = ^{16}O, ^{32}S, ^{80}Se, ^{130}Te) are tabulated[470]; the molecular beam electric resonance spectrum of ^{74}GeO is measured for the lower vibrational states[471]; and the absorption spectrum of SnSe at c. 1500 °C is re-examined and compared with the system SnE (E = O, S, Te) in the visible region[472]. Electric dipole moments and microwave rotational spectra of GeO and GeS give the values GeO, 3.28 ± 0.10 D and GeS, 2.00 ± 0.6 D and the bond length of GeS is 2.012 Å [473]. The i.r. spectra of matrix isolated vapour-phase germanium[474] and tin[475] oxides leads to the identification of eight distinct species: MO, M_2O_2, M_3O_3 and M_4O_4 (M = Ge, Sn). Normal coordinate analysis and ^{18}O-substitution allow estimates of their molecular dimensions in which M_2O_2 have V_h symmetry (i.e., a planar four-membered ring), M_3O_3 have D_{3h} symmetry, and Sn_4O_4 has D_{4h} or T_d symmetry (possibly a distorted cubic structure[475]), higher symmetry than in Ge_4O_4 [474]. The ^{119}Sn Mössbauer spectra of SnE (E = O, S, Se, Te) and of SnE_2 ($E_2 = O_2, S_2, Se_2$, SSe) are reported[476, 477]. The isomer shifts reflect a linearly dependent decrease in electron-density about tin with increasing electronegativity of E. The occupancy of the s- and p-orbitals is discussed[476]. The Raman spectra of single crystals of both the trigonal and tetragonal forms of GeO_2 are reported[478], as well as that of SnO_2 [479]. The Stark components of some microwave rotational transitions of SnS_2 and PbS_2 in the 20–36 GHz frequency range at 600–800 °C give the dipole moments 3.38 ± 0.07 D for SnS, 4.02 ± 0.07 D for PbS, with an estimate of 2.25 D for GeS [480]. The thermal decompositions of α- and β-PbO_2 and Pb_3O_4 show that molecular oxygen is involved with β-PbO_2 but some liberation of atomic oxygen occurs with Pb_3O_4 [481, 482]. SnO reacts with excess Cl_2 on heating, to give $SnCl_4$ and SnO_2; with a limited supply of Cl_2, to give $SnCl_2$ and SnO_2; with $SnCl_4$, at over 100 °C, to give $SnCl_2$ and SnO_2; and with $COCl_2$, at over c. 150 °C, to give $SnCl_2$ and CO_2 [483]. SO_2 reacts with powdered germanium in the range 620–720 °C to give almost equimolar amounts of GeO_2 and GeS_2. The ratio GeS_2/GeO_2 decreases with increasing temperature[484]. The kinetics of the reactions $MO_2 + 4HCl \rightarrow MCl_4 + 2H_2O$ (M = Ge, Sn) are examined. Water evolved accelerates the reactions and lowers their activation energies[485]. Various studies of the solubility of GeO_2 include its solubility in aqueous tartaric acid solution[486]; the relationship between the limiting solubility in aqueous solution of weak bases and their dissociation constants[487] and the addition of Na_2S to solutions in sulphuric acid to precipitate GeS_2 [488] have been studied.

A new series of orthogermanates, containing the $(GeO_4)^{4-}$ anion, i.e., Li_2MGeO_4 (M^{II} = Mg, Zn, Mn, Co, Fe), is synthesised[489]. In $MnGeO_3$, the GeO_4 tetrahedra form chains bound together by Mn octahedra[490], while the crystal structure of $Li_2(Ge_7O_{15})$ contains puckered layers of GeO_4 tetrahedra linked by GeO_6 octahedra[491]. The i.r. absorption spectrum of $PbGeO_3$ suggests that symmetry of the crystals and configuration of the chains is similar to alamosite[492]. The crystal structure of $Na_2(TiO)GeO_4$ shows tetrahedrally coordinated germanium atoms and 5-coordinated titanium[493]. The formation of $BaSnGe_3O_9$ and $BaTi_{0.5}Sn_{0.5}Si_3O_9$ is reported: both are isostructural with benitoite[494]. The new germanates, $MM'Ge_2O_6$ (M ≠ M' = Ca, Sr, Ba), may be considered as individual compounds as well as intermediate links in the solid solution series[495]. In $Pb_3GeAl_{10}O_{20}$ there is a network of AlO_4 tetrahedra and AlO_6 octahedra with the lead atoms located within channels but with those of germanium randomly distributed[496]. Vibrational spectral examinations of many systems, including Zn_2GeO_4, $Ca_3Cr_3Ge_3O_{12}$, Ni_2GeO_4 and GeO_2, illustrate that many cations participate in a given vibrational frequency, so that, e.g., in Ni_2GeO_4 only the asymmetric stretch, of those frequencies assignable to the GeO_4 tetrahedra, is free of interactions[497]. The force constants for the Ge—O bond in a series of anions $(Si_2GeO_9)^{6-}$, $(SiGe_2O_9)^{6-}$, and $(Ge_3O_9)^{6-}$ are estimated[498]. The i.r. spectra of the rare-earth orthogermanates, $Ln_2O_3 \cdot 9GeO_2$, $Ln_2O_3 \cdot GeO_2$ and $2Ln_2O_3 \cdot GeO_2$, indicate a similarity to the corresponding silicates. The stability decreases with decreasing GeO_2 concentration[499]. The crystal structure of $Er_2Ge_2O_7$ is reported[500]. The structures of the stannates Ca_2SnO_4, Cd_2SnO_4 [501] and Li_8SnO_6 [502] all show the octahedral SnO_6 unit. Results on $Mg_3Li_2SnO_6$ suggest that differences in the octahedral arrangements of SnO_6 and ReO_6 in Li_8SnO_6 and Li_5ReO_6 [503] depend on the number of lithium atoms[504]. Rb_2PbO_3, Rb_2SnO_3 and orthorhombic Rb_2ZrO_3 are isotypic. The M^{4+} ions have an abnormal coordination number of five caused by the type of formula[505, 506]. The monoclinic lattice constants of Li_2PbO_3 are reported[507]. The free reaction energies, ΔG_T^0, have been measured for the reactions $CaO + SnO_2 \rightarrow CaSnO_3$ and $2CaO + SnO_2 \rightarrow Ca_2SnO_4$ [508]. The orthorhombic lattices Ca_2SnO_4, Cd_2SnO_4 and also Sr_2PbO_4 and Ca_2PbO_4 contain chains of MO_6 (M = Sn or Pb) octahedra similar to those of Pb_3O_4 [501]; the rhombohedral Li_8SnO_6 has Li ions occupying octahedral and tetrahedral holes of the slightly distorted hexagonal close-packing of the oxygens[502], and tetragonal body-centred Na_6PbO_5 has chains of highly distorted PbO_6 octahedra sharing opposite corners[509]. The usual preference of 6-coordination for Sn^{IV} and 4-coordination for Ge^{IV} in oxide systems is nicely illustrated by the crystal structure of $Na_4Sn_4Ge_4O_{12}(OH)_4$ in which the GeO_4 tetrahedra form strings by sharing corners, and these chains are bound together by SnO_6 octahedra[510]. The formation of $Pb_2Sn_2O_6 \cdot xH_2O$ ($x = 0 \rightarrow 1$)[511], Li_2PbO_3 (yellow α-form isotypic with Li_2ZrO_3, yellow β-form isotypic with Li_2SnO_3), pale-yellow Li_4PbO_4 and Li_8PbO_6 (the most thermally stable in the Li_2O/PbO_2 system) and Li_2SnO_3 and Li_8SnO_6 (isotypic with Li_8PbO_6) are also reported[512]. The crystal structure of potassium metastannate, K_2SnO_3, shows chains of base-edged SnO_5 square-pyramids[513].

Sodium thiogermanates, containing $(GeS_3)^{2-}$ and, at least in solution, $(Ge_2S_5)^{2-}$, which may be $GeS_2 + GeS_3^{2-}$, are obtained with the following compositions: $Na_2GeS_3 \cdot H_2O$, $Na_4GeS_4 \cdot 7H_2O$, $Na_2Ge_2S_5 \cdot 11H_2O$, and $Na_8GeS_6 \cdot 23H_2O$ [514]. The thiogermanates, Sr_2GeS_4, Ba_2GeS_4, $BaGeS_3$ and $BaGe_2S_5$, are examined crystallographically[515, 516] and M_2GeS_4 (M = Ca, Sr, Ba) are formed in the GeS_2/MS system[517]. The thiogermanates M_4GeS_6 (M = Co, Ni, Mn) are formed in the Na_8GeS_6—MCl_2—H_2O systems[518]. The x-ray analysis and vibrational spectra of hexathio-digermanates and -distannates of alkali metals suggest that the $[M_2S_6]^{4-}$ group in, e.g., $Na_4M_2S_6 \cdot 14H_2O$ (M = Ge, Sn) has two bridging S-groups (i.e., a diborane-type structure). In $Na_4SnS_4 \cdot 14H_2O$ there are discrete tetrahedrally coordinated $[SnS_4]^{4-}$ units[519]. SnS_2 reacts with solutions of alkali metals in liquid NH_3 to give $MSnS_2$, all of which, except $CsSnS_2$, have the structure of $NaHF_2$ [520].

Much less is known of systems with oxidation state (II). The solidification diagram of liquid PbO—$PbCl_2$ mixtures, for molar fractions of PbO smaller than $\frac{2}{3}$, identifies three compounds: $3PbCl_2 \cdot 2PbO$, $PbCl_2 \cdot PbO$ and $PbCl_2 \cdot 2PbO$ [521]. The PbS—$PbCl_2$ system is eutectic, but a metastable Pb_4SCl_6 can be prepared as well as a homologous Pb_4SeBr_6. The PbS—PbX_2 (X = Br, I) systems are pseudobinary peritectic with the ternary compounds $Pb_5S_2I_6$ and $Pb_7S_2Br_{10}$ [522]. The crystal structures of Pb_4SeBr_6 and $Pb_5S_2I_6$ are discussed in relation to other lead compounds[523]. The action of PbS on sulphides of the titanium group yield $PbMS_3$ (M = Ti, Zr, Hf). $PbZrS_3$ and $PbHfS_3$ are isotypic, but $PbTiS_3$ shows markedly more metallic properties[524].

PbO reacts with strongly basic metal oxides to form ternary oxides, e.g., $M_2Pb_2O_3$ (M = Na, Li), M_2PbO_2 (M = Na, Li), $M'PbO_2$ (M' = Ba, Sr), Na_4PbO_3 and Ba_3PbO_4, which are extremely sensitive to moisture and solvent. Thus $M'PbO_2$ with gaseous methanol give $Ba[Pb(OH)_2(OCH_3)_2]$ and $Sr[Pb(OH_{1.5}(OCH_3)_{2.5}]$, which compares with the action of absolute methanol on PbO to give $Pb(OH)(OCH_3)$ [525]. X-ray analysis shows that the crystal structure of the $[Pb_6O(OH)_6]^{4+}$ cluster has the six lead atoms in a three face-sharing tetrahedral arrangement[526]. This has necessitated a revision of the previously assigned[527] Raman cluster frequencies[528]. Alternative force-fields are proposed in the normal coordinate analyses of bridged polynuclear metal atom complexes, such as $[Pb_4(OH)_4]^{4+}$ [529]. The i.r. spectrum of PbOHX (X = Cl, Br, I) is interpreted as indicating there are folded (PbOH) bands linked by halide ions. Thermal analyses of PbOHX show that dehydroxylation is the initial mode of decomposition[530].

A re-examination of the reaction of germanium(IV) oxide with phosphorous acid suggests that germanium(IV) species $Ge(HPO_3)_2$ and $Ge(HPO_3)_2 \cdot$

$2H_2O$ [531] are formed, rather than germanium(II) [532]. Solutions of tin(II) phosphite in the pH range 4.5–8 contain $SnHPO_3$, $Sn(HPO_3)_2^{2-}$ and $Sn(HPO_3)_3^{4-}$. Salts of the latter with the M^I cations (M = Na, K, Rb, Cs, NH_4) have Mössbauer and i.r. spectra consistent with a pyramidal triphosphitostannate(II) ion[533]. The Raman and i.r. spectra of tin(II) orthophosphate indicate that this compound, which has been cited as $SnHPO_4 \cdot \frac{1}{2}H_2O$, is in fact, $Sn_2O(H_2PO_4)_2$ with an Sn—O—Sn bridge[534]. The crystal structure of $Pb(HPO_4)_2 \cdot H_2O$ shows PbO_6 octahedra and $HOPO_3$ tetrahedra[535]. Phosphato-lead(IV) acids are obtained by the action of phosphoric acid on lead(IV) acetate[536]. Anhydrous acid gives $H_2[Pb(H_2PO_4)_6]$ and acid solutions give $H_2[Pb(H_2PO_4)_2(HPO_4)_2]$. Intermediates, such as $H_2[Pb(H_2PO_4)_2(OCOCH_3)_2]$ are isolated, suggesting coordination precedes ligand exchange. Reduction to the lead(II) phosphate $Pb(H_2PO_4)_2$ is achieved by treating $H_2[Pb(H_2PO_4)_6]$ or $H_2[Pb(H_2PO_4)_2(HPO_4)_2]$ with a glycol[537]. In the Pb^{2+}—$Na_5P_3O_{10}$—H_2O system, potentiometric studies suggest that a complex ion such as $[Pb_3(P_3O_{10})_4]^{4-}$ predominates at low concentration of the polyphosphate, and $[Pb(P_3O_{10})]^{3-}$ at higher concentrations[538]. The synthesis of the lead(II) phosphate, arsinate and vanadate, $Pb_5(XO_4)_3OH$ (X = P, As, V) are reported[539], as is a series of niobates, $Pb(NbO_3)_2$, $Pb_3(NbO_4)_2$ and $K_4[Pb(NbO_3)_6]$ [540].

The i.r. and Raman spectra of $(Pb(NO_3)_2$ are reported along with those of other anhydrous divalent metal nitrates[541] and it is suggested that basic lead nitrate is probably better represented by the conventional formula $Pb(NO_3)_2 \cdot 5PbO \cdot xH_2O$ than by a form containing $Pb(OH)_2$ as has been suggested[542]. 1:1 Complexes are formed in dilute solutions between lead nitrate and sodium tartrate or trihydroxyglutarate[543]. The reaction of tetraethyllead with N_2O_4 gives $Et_2Pb(NO_3)_2$, whose i.r. spectrum is consistent with monodentate NO_3 ligands[544]. In $MeSn(NO_3)_3$, which is prepared by the reaction of $MeSnCl_3$ with N_2O_5, the vibrational spectrum[545] indicates that the NO_3 groups are bidentate, as in $Sn(NO_3)_4$ [546]. However, in the pyridine complex $MeSn(NO_3)_3 \cdot 2py$, the NO_3 groups are apparently monodentate[545], and covalent complexes are formed between $Ph_2Sn(NO_3)_2$ and 2,2'-bipyridyl, 1,10-phenanthroline and Me_2SO [547].

7.6.2 Miscellaneous chemistry of oxidation state IV

Complexes of composition $7SnO_2 \cdot L \cdot \sim 10H_2O$ and $8SnO_2 \cdot L' \cdot \sim 11H_2O$, prepared by reaction of α-stannic acid in aqueous solution with piperidine (L) and 2-methylpiperidine (L'), are molecular-insertion type compounds[548]. Even over a wide range of pH in dilute solutions, germanium(IV) forms only a 1:1 complex with ethylenediaminetetraacetic[549] and nitrilotriacetic[550] acids. Complexes of germanium(IV) with mandelic acid of compositions $M[Ge(PhHCO \cdot CO_2)_2(PhHCOH \cdot CO_2)H_2O]$ (M = K, Cs, NH_4), $[CoCl_2(en)_2]^+$, $[Pb(NH_3)_4]^{2+}$, or $[Co(NH_3)_6]^{3+}$, and also $[Co(en)_3]_2[Ge(PhHCO \cdot CO_2)_3]_3$ are isolated and suggestion made as to their structures[551]. Correlations are established between stability constants of GeL_3^{2-} complexes and the acidity constants of some o-diphenols, H_2L [552]. The reaction of germanium(IV) with 3,5-dinitropyrocatechol is studied spectrophotometrically and potentiometrically[553]. The i.r. and Raman spectra of the oxalate complex ions $[Ge(C_2O_4)_3]^{2-}$ are assigned on the basis of D_3 symmetry[554]. A study of hydroxy lead oxalates[555] and the preparation, thermogravimetric and i.r. analysis of trihydroxy- and monohydroxy-lead ferrocyanide are reported[556]. The x-ray diffraction powder pattern of the tetramethylammonium hexaazidostannate(IV) is indexed in the hexagonal system[557], and an x-ray powder determination shows the nitride $MgGeN_2$ is of the würtzite-type structure belonging to the same family as $BeSiN_2$ and $MgSiN_2$ [558].

In the Mössbauer spectra of sixteen compounds of the general type SnX_4Y_2 (where X = link through S, and Y = link through O or N), the quadrupole splittings for compounds with cis-Y groups are c. 1 and with trans-Y groups c. 2 mm s^{-1} [559]. The i.r. and Mössbauer spectra of the series of tin(IV) complexes of the bidentate ligand 2-pyridinethiol-1-oxide, $R_2Sn(2\text{-SpyO})_2$ (R = Bu, Ph, F, Cl, Br, I), $Sn(2\text{-SpyO})_4$ and $Ph_2Sn(NCS)(2\text{-SpyO})$, are interpreted as indicating, (a) a trans-Bu but cis-Ph arrangement, and a cis-F arrangement, (b) a greater electron release by the sulphur atom to tin than by the oxygen, and (c) that in $Sn(2\text{-SpyO})_4$ there could be high-coordination with all four S and all four O coordinating[560]. Complexes $R_2Sn\cdot L$ (R = Me, Ph; L = tridentate ONO and SNO ligands) are synthesised and their vibrational, ^1H n.m.r. and electronic spectra examined. The $R_2Sn(IV)$ moieties probably assume a bent configuration in R_2SnL. The ligands H_2L[2-pentanone-4-(2-benzothiazolinyl), 3-(o-hydroxyphenylamino)crotonphenone, 2-(o-hydroxyphenyl)benzothiazoline and 2,2'-methylidynenitrilo-diphenol] seem to coordinate to Sn^{IV} as planar tridentates in the dianionic Schiff's base form[561]. New examples of complexes of types $Me_2SnL_2(BPh_4)_2$ and $Me_3SnL_2BPh_4$ are reported, where L = Ph_3PO, Ph_3AsO and $(p\text{-}CH_3\cdot C_6H_4)_3AsO$. In solution, the ^1H n.m.r. spectrum indicates Ph-transfer from boron to tin can occur[562]. Organotin complexes $[R_2SnL\cdot H_2O]_n$ and $[Bu(OH)SnL\cdot 3H_2O]_n$ are prepared, where H_2L = bis(8-hydroxy-5-quinolyl)[563]. Dielectric relaxation techniques indicate that the octahedral complex bis(2,4-pentanedionato)diphenyltin(IV) prefers the cis form in benzene solutions[564]. Solvolysis with HSO_3X (X = F, CF_3, Cl, Me, Et), of Me_2SnCl_2 and Me_3SnCl gives $Me_2Sn(SO_3X)$. Vibrational spectra suggest there is only one type of SO_3X group and that $[OS(O)XO]_2SnMe_2$ is probably a repeating unit[565]. Organogermanes Et_3GeR react with SO_3 to give germyl sulphonates Et_3GeO_3SR. These exchange with organosilicon halides[566].

7.6.3 Alkoxides, carboxylates and their derivatives

A general review of alkoxides and alkylalkoxides of metals and metalloids includes a discussion of those of germanium and tin[567]. The organomethoxy- and organomethoxy-halogermanes, $R_2(MeO)GeH$, $R(MeO)GeH_2$, $R(MeO)_2GeH$ and $R(MeO)(X)GeH$ (R = alkyl or phenyl, X = halogen) are synthesised by direct reaction at low temperature of sodium or lithium methoxides with the appropriate organohalogermanes and by exchange reactions[568]. The compounds $R_2Sn(OR)_2$ are prepared by action of alcohols on the appropriate dialkylstannoxanes[569]. The i.r. and n.m.r. spectra and several methods for the preparation of the halogenogermanium methoxides and ethoxides,

$GeX_n(OR)_{4-n}$ (X = Cl, Br; R = Me, Et; n = 1,2,3) are discussed. All products are monomeric colourless liquids at room temperature[570]. The i.r. spectra of several alkoxyalkyl(aryl)germanes are reported, including $BuGe(OMe)_3$, $Bu_2Ge(OR)_2$ (R = Me, Et, Pr^i, Bu^s) and $Bu_3Ge(OR')$ (R = Me, Et)[571]. Mass spectra of $Me_nGe(OMe)_{4-n}$ indicate that the primary reactions involve splitting off of Me, MeO, or H; the Me groups come from Ge—C, and fragmentation of the H atoms from OMe follows[572]. The dipole moments of various tetra-alkoxygermanes are reported[573]. The reactions of dialkyltin diisopropoxides with 2-(dimethylamino)- and 2-(diethylamino)-ethanol[574] and with 1-nitroso-2-naphthol[575] are reported. $Me_3Pb(OMe)$ adds CO_2, CS_2, MeNCS and $H_2C{:}CO$ to the Pb—O bond and reacts with PhC≡CH, Ac_2CH_2 and ROH to give $Me_3PbC{\equiv}CPh$, dimethyllead bis(acetylacetonate) and $Me_3Pb(OR)$ (R = Et, Pr, Bu)[576]. $R_3Sn(OR)$ or $R_3SnY(Y = NEt_2$, $OSnBu_3$ and $OSiMe_3$) react with $(Me_3Si)_2Hg$ to give $Me_3Si(OR)$ or Me_3SiY, and R_3SnSnR_3[577]. $Et_2Sn(OEt)_2$, when heated with Et_2SnF_2, produces $Et_2Sn(OEt)F$[578]. $SnCl_3(OMe)$ and $SnCl_3$(2-chloroethoxide) form complexes with amides, ureas, dimethyl sulphoxide and amines. Possible structures suggested by their i.r. spectra have Sn^{IV} octahedrally coordinated through Sn—O(Me)—Sn bridging[579, 580].

The reaction of $Sn(OMe)_2$ with PhNCO is highly exothermic to give bis(MeN-phenylcarbamate)tin(II); with PhNCS it gives an oil which probably has N—Sn rather than SSn bonding present; with chloral it gives $Sn[OCH(CCl_3)OMe]_2$ quantitatively; with PhSH it gives $Sn(SPh)_2$ which reacts further on refluxing to give $Sn(SPh)_4$; and with BzCl gives $SnCl_2$ and 82% BzOMe[581].

Trivinyltin carboxylates $(CH_2{=}CH)_3SnOOCR$ (R = Me, Et, CH_2Cl, CF_3) and trimethyltin carboxylates are prepared by reacting mercury(I) ions, generated electrolytically, with tetravinyltin in methanol containing the sodium carboxylate as supporting electrolyte[582, 583]. Me_4Sn reacts with $Hg(O_2CCH_3)_2$ in methanol at room temperature to give a soluble form of trimethyltin acetate in high yield[584]. The Raman and i.r. spectra of the solid trimethyltin carboxylates, Me_3SnO_2CR, (R = H, Me, CH_2Cl, Et) are analysed. The i.r. spectra in chloroform solution indicate solution depolymerisation of carboxylate-bridged solid-state structures[585]. Trimethyltin haloacetates are shown to be linear polymers with pentacoordinated tin atoms and bridging OCO groups[586]. The preparation and properties of several triphenyllead carboxylates $Ph_3Pb\ O_2CR$ [R = C_6F_5, 4-MeO·C_6F_4, 4-EtO·C_6F_4, 2,4-$(NO_2)_2C_6F_3$, and C_6Cl_5] are reported. The first four are monomeric in chloroform but all presumably contain bridging carboxylate groups in the solid[587]. A Mössbauer and i.r. spectral study of triphenyltin carboxylates and the novel compounds $RSn(O)OCOR'$ is reported[588], as is a Mössbauer study of 4- and 5-coordination in branched-chain triphenyltin carboxylates[589]. The skeletal vibrations are assigned for the far i.r. spectra of twenty-one tricarboxylatostannate(II) complexes. A distorted pyramidal SnO_3 skeleton is indicated[590], and the crystal structure of potassium triformatostannate(II) shows that the tin atom in $[Sn(HCO_2)_3]^-$ is pyramidal with the Sn—O bond distances, 2.13, 2.17 and 2.18 Å[591]. The stability constants for a series of 1:1 complexes between divalent metals and the carboxylate anions, acetic, benzoic, formic, chloroacetic, acetoxyacetic, and N-.

acetylglycine, suggest the Pb^{2+} complexes are the most stable in the following series: $Pb > Cu > Cd > Zn > Ni \sim Co > Ca = Mg$ [592]. The electronic spectra of Me_3GeCO_2H and $Me_3GeCO_2^-$ are reported and features are discussed[593]. Among the topics covered in the solution chemistry of tin(II) and lead(II) compounds are the following: potentiometric studies of complex formation by Sn^{2+} in solutions of citric, tartaric, malic, succinic and acetic acids[594, 595]; a potentiometric study of Pb^{2+} with 3,5-dinitrosalicylic acid showing 1:1 adduct formation[596]; and a polarographic study in aqueous ethanol solution showing the formation of $[Pb(L)]^{2+}$ and $[Pb(L)_2]^{2+}$ (L = ethylenediamine or diethylenetriamine) and $[Pb(L')_3]^{2+}$ (L' = triethylenetetramine) [597]. The bis-chelate $Pb(RH_2)(OH)(H_2O)$ is formed by the reaction of N-phenylsulphonyl-L-glutamic acid (RH_3) with Pb^{2+} in aqueous solution[598]. Stability constants of Pb^{II} complexes with N-(8-quinolyl)glycine are recorded[599] and the standard enthalpies of formation of lead(II) dithiocarbamate, $Pb(CS_2NH_2)_2$, was found to be -24 kcal mol^{-1} [600]. Insertion of phenyl isocyanate into the Sn^{II}–oxygen bond in 2,2'-biphenylenedioxytin(II) gives 2,2'-biphenylenedicarbamatotin(II)[601]. O,O-diethyl diselenophosphate complexes of Sn^{II} and Pb^{II} are prepared along with several others and the nature of the bonding between the —P(Se)Se group and the central metal ion is discussed[602]. X-ray studies of triclinic tin(II) phthalocyanine show the tin atom is appreciably out of the plane of the 4-coordinating N atoms, suggesting stereochemical activity of the lone-pairs pointing away from the plane; Mössbauer data are consistent[603]. A neutron-diffraction study of lead(II) azide indicates that there are four different types of azide structure per unit cell[604]. The thermal dissociation of $PbCO_3$ under reduced pressure leads to the formation of the intermediates $5PbCO_3 \cdot PbO$, $2PbCO_3 \cdot PbO$, $7PbCO_3 \cdot 5PbO$, $PbCO_3 \cdot PbO$ and $PbCO_3 \cdot 2PbO$ [605]. The formation of $Pb(CO_3)_2^{2-}$ is noted in a study of the solubility of $PbCO_3$ in Na_2CO_3 [606] and the existence of the complex ions $Pb(CO_3)_2^{2-}$, $Pb(CO_3)_3^{4-}$, $Pb(HCO_3)_4^{2-}$ and $Pb(NCO_3)_2$ is shown by polarography and ion exchange on anionic resins in the presence of HCO_3^- and CO_3^{2-} ions[607]. $PbCO_3$ or $2PbCO_3(PbOH)_2$, respectively, are precipitated from lead nitrate solutions containing ammonium chloride by treatment with ammonium carbonate at $pH > 7.1$ or 7.1, respectively[608]. The $(CS_3)^{2-}$ ion in $PbCS_3$ occupies similar sites as the $(CO_3)^{2-}$ in $PbCO_3$ but there is considerable distortion of the anion[609].

7.7 COMPOUNDS CONTAINING A GROUP IV—TRANSITION METAL BOND

During 1969–1970 the study of transition metal–Group IV complexes proceeded at such a pace that nearly all transition metals are represented. Compounds containing X_3Ge and X_3Sn (X = usually a halogeno- or organo-group) are far more common than those containing X_3Pb. However, all three types are covered, in part, in general reviews[610–612].

E.S.R. evidence is presented for the formation of $[\pi\text{-}C_5H_5)_2Ti(MPh_3)_2]^-$ (M = Ge, Sn) and of $(\pi\text{-}C_5H_5)_2Ti\cdot PbPh_3$ by the addition of titanocene dichloride to excess $Ph_3M^-K^+$ (M = Ge, Sn, Pb) in 1,3-dimethoxyethane[613]; the 7-coordinated $Ph_3Sn\cdot M(CO)_6$ and $Ph_3Sn\cdot M(CO)_5PPh_3$ (M = V, Nb, Ta) have been isolated[614].

The reaction of $(MeCN)_3M(CO)_3$ (M = Cr, Mo, W) with $C_5H_5M'Me_3$ (M' = Ge, Sn) results in the formation of $(\pi\text{-}C_5H_5)(CO)_3M\cdot M'Me_3$ by insertion into the metal–carbon bonds[615]. The crystal structure of 2,2'-bipyridyl$(CO)_3$BrW·GeBr$_3$ shows no bridging of W—Br—Ge, but is best considered as having a definite GeBr$_3$ entity[616]. Heating $(\pi\text{-}C_5H_5)Mo(CO)_2$(PPh$_3$)SnMe$_3$ results in evolution of SnMe$_4$ and the formation of $[(\pi\text{-}C_5H_5)Mo(CO)_2(PPh_3)]_2SnMe_2$ [617]. The formation and far i.r. spectra of the series of compounds $(\pi\text{-}C_5H_5)Mo(CO)_2(PR_3)SnX_3$ (R = OMe, OPh, and Ph; X = Cl, Br, I) are reported[618a]. In a study by Mössbauer and vibrational spectroscopy of metal carbonyl complexes of tin containing arene ligands, all are found to have a tetrahedral environment about tin except $(Me_2SnPh_2)[Cr(CO)_3]_2$ in which the unique *trans*-square planar Me$_2$SnPh$_2$ moiety is suggested[618b].

The Mössbauer, n.m.r. and i.r. spectra of the series of compounds Me$_{4-n}$Sn[Mn(CO)$_5$]$_n$(n = 0–3) are discussed in terms of several linear relationships including those relating Mössbauer isomer shifts to (a) the number of Mn(CO)$_5$ ligands and (b) the methyl ^{1}H n.m.r. chemical shifts[619]. A relationship is established between π-interactions and coupling constants in several pentafluorophenyltin derivatives including $(C_6F_5)_3SnMn(CO)_5$, $(C_6F_5)_2SnPh\cdot Mn(CO)_5$, $(C_6F_5)SnPh_2\cdot Mn(CO)_5$, and $(C_6F_5)_2Sn[Mn(CO)_5]_2$ [213]. A re-evaluation of the CO stretching force-constants in ^{13}CO-enriched Ph$_3$M— and Br$_3$M—Mn(CO)$_5$(M = Ge, Sn) suggests that the groups Ph$_3$Ge and Ph$_3$Sn are even better π-acceptors than was previously suggested, possibly better than CO [620]. The molecular structure of Br$_3$GeMn(CO)$_5$, which is determined by electron diffraction[621], shows that in fact the Ge—Mn bond is considerably shorter than in Ph$_3$GeMn(CO)$_5$ [622]. The cleavage reactions of Me$_3$SnMn(CO)$_5$ with HX (X = Cl, Br, I), ICl, and ClF$_3$ are described as well as some for Et$_3$PbMn(CO)$_5$. [Me$_2$SnMn(CO)$_5$]BF$_4^-$ is formed on reaction with BF$_3$ [623]. Na[Mn(CO)$_4$(tdp)] [(tdp) = tris(dimethylamino)-phosphine] reacts with Me$_3$SnCl to give Me$_3$Sn·Mn(CO)$_4$(tdp) [624].

Passage of butadiene into $(\pi\text{-}C_5H_5)Fe(CO)_2MR_3$ (MR$_3$ = GeMeCl$_2$, Ph$_3$Ge, Ph$_3$Sn, Ph$_3$Pb) in u.v. light leads to the displacement of both CO groups to give $(\pi\text{-}C_5H_5)Fe(\pi\text{-}C_4H_6)MR_3$ [625]. A short Ge—Fe bond is noted from the x-ray crystallographic study of two modifications of π-butadiene-π-cyclopentadienyldichloromethylgermyliron[626]. The x-ray analysis shows that $[Me_2Ge]_3Fe_2(CO)_6$ has the Fe$_2$(CO)$_9$ structure with the three bridging CO groups replaced by Me$_2$Ge [627], whereas $(Ph_2Ge)_2Fe_2(CO)_7$ has two Fe(CO)$_3$ groups linked by an Fe—Fe bond and two (Ph$_2$Ge)$_2$ bridges[628]. The addition of a 10% excess of $(\pi\text{-}C_5H_5)Fe(CO)_2R$ in dioxane to GeCl$_2$·C$_4$H$_8$O$_2$ gives reasonable yields of $(\pi\text{-}C_5H_5)Fe(CO)_2GeRCl_2$ (R = Me, Et, Pr, Pr^i, Ph or Cl)[629]. The kinetics of the thermal insertion of SnX$_2$ (X = Cl, Br) into $[(\pi\text{-}C_5H_5)Fe(CO)_2]_2$ indicate that the activation energy to break the Fe—Fe bond is c. 32 kcal mol^{-1} [630]. SnCl$_2$ reacts with FeCl$_2$(p-MeO·C$_4$H$_3$·NC)$_4$ to give 1:1 and 1:2 octahedral adducts containing the SnCl$_3$ groups as ligands[631]. In the Mössbauer centre shifts and quadrupole splittings for *trans*- and *cis*-FeX$_2$L$_4$ (X = Cl, SnCl$_3$; L = p-methylphenyl isocyanide) and *cis*-FeCl(SnCl$_3$)L$_4$, the *trans*:*cis* 2:1 quadrupole splitting holds for SnCl$_3^-$ and ligands such as Cl$^-$ [632]. Random exchange of halogen is found when

(π-C$_5$H$_5$)Fe(CO)$_2$SnCl$_3$ is treated with a large excess of SnBr$_2$ or SnI$_2$ in methanol at room temperature[618a]. The Sn—Fe bond is slightly shorter in crystals of (π-C$_5$H$_5$)Fe(CO)$_2$SnX$_3$ (X$_3$ = Cl$_3$, Br$_3$, PhCl$_2$, Ph$_2$Cl) than is observed in other species containing that bond but evidence is presented to suggest it is essentially a σ-bond[633-636]. BrMn(CO)$_5$ and (π-C$_5$H$_5$)FeCl(CO)$_2$ react directly with excess magnesium in THF to give air-sensitive solutions. Addition of Ph$_3$SnBr, followed by hydrolysis, gives Ph$_3$Sn·Mn(CO)$_5$ and (π-C$_5$H$_5$)Fe(CO)$_2$SnPh$_3$ in good yield[637]. SnCl$_4$, SnBr$_4$, and PhSnCl$_3$ react rapidly in dry benzene with Hg[Fe(CO)$_3$(NO)]$_2$ and Hg[Fe(CO)$_3$NOL]$_2$ to give species such as X$_3$SnFe(CO)$_2$(NO)L [X$_3$ = Cl$_3$, Br$_3$, PhCl$_2$; L = R$_3$P, (RO)$_3$P, R$_3$As, R$_3$Sb]; this apparently general reaction for tin does not work with PhSiCl$_3$ or GeCl$_4$ [638]. By use of the Cotton and Kraihanzel force field, a more exact direct method is applied for calculating the CO stretching force constants in [R$_2$MFe(CO)$_4$]$_2$ (M = Ge, Sn, Pb), (R$_3$Sn)$_2$Fe(CO)$_4$, R$_4$M$_3$[Fe(CO)$_4$]$_4$ (M = Sn, Pb), and M[Fe(CO)$_4$]$_4$ (M = Ge, Sn, Pb)[639]. Reactions of Me$_3$M·Fe(CO)$_2$(π-C$_5$H$_5$) (M = Ge, Sn) with perfluorocyclobutene, perfluoro-2-butyne and 3,3,3-trifluoropropyne lead to the characterisation of the insertion products such as Me$_3$M(CF$_3$)C=C(CF$_3$)Fe(CO)$_2$(π-C$_5$H$_5$) and Me$_3$M·C$_2$H(CF$_3$)Fe(CO)$_2$(π-C$_5$H$_5$), although elimination of Me$_3$SnF also occurs[640]. Reaction with SO$_2$ readily occurs, although not with the silicon analogue, to give products formulated as Me$_3$Ge—S(O$_2$)—Fe(CO)$_2$(π-C$_5$H$_5$) and a polymer containing Sn—O—S(O)—Fe units[641]. Reactions of Cl$_3$Sn—R$_M$, where R$_M$ = —Fe(CO)$_2$(π-C$_5$H$_5$), —Mo(CO)$_3$(π-C$_5$H$_5$) and —Co(CO)$_3$L with bipyridyl and 8-hydroxyquinoline give compounds of the general formulae R$_M$SnCl(ox)$_2$ or (π-C$_5$H$_5$)(CO)$_2$FeSnCl$_3$·bipy [642]. (Ph$_3$Pb)$_2$Fe(CO)$_4$ and o-phenanthroline react with homolytic fission of the Pb—Fe—Pb bonds to form Fe(phen)$_3$Cl$_2$ and Ph$_3$PbCl. Subsequently, adduct formation occurs between the products[643].

The anion [Me$_3$Si·Ru(CO)$_4$]$^-$ is used to prepare Me$_3$Si·Ru(CO)$_4$—GeBu$_3$ and —SnR$_3$ (R = Me, Ph). With Me$_2$SnCl$_2$, the anion gives [Me$_3$Sn·Ru(CO)$_4$]$_2$ [644]. X-ray crystallographic study of [Me$_3$Sn(CO)$_3$·Ru·SnMe$_2$]$_2$ indicates that, despite the presence of the two Ru—SnMe$_2$—Ru bridges, strong Ru—Ru interaction distorts the structure. The terminal Me$_3$Sn groups are in a '*trans*'-configuration[645]. Both Ru atoms have distorted octahedral configurations in (SnCl$_3$)Ru$_2$Cl$_3$(CO)$_5$ [646]. [Me$_3$Ge·Ru(CO)$_4$]$^-$ reacts with various metal halides to give, e.g., [Me$_3$Ge·Ru(CO)$_4$]$_2$Hg, while the pyrolysis of (Me$_3$Ge)$_2$M(CO)$_4$ (M = Ru, Os) leads to a polynuclear complex, [Me$_2$Ge·M(CO)$_3$]$_3$, containing a heterocyclic Ge—M—Ge—M ring of six metal atoms[647]. Me$_3$SnH reacts with Os$_3$(CO)$_{12}$ to give Me$_3$Sn·Os(CO)$_4$H and (Me$_3$Sn)$_2$Os(CO)$_4$ in which the Me$_3$Sn and H or the Me$_3$Sn groups are *cis* and it reacts with Me$_3$Si·Os(CO)$_4^-$ to give a 1:1 mixture of the *cis*- and *trans*-isomers of Me$_3$Si·Os(CO)$_4$·SnMe$_3$ [648]. SnCl$_4$ with appropriate transition metal carbonyls gives Cl$_3$Sn·Co(CO)$_4$ [649], *cis*-(Cl$_3$Sn)$_2$Fe(CO)$_4$ [650], and Cl$_3$Sn·Ru$_2$(CO)$_5$Cl$_3$ [651]. With H$_2$Os(CO)$_4$, the new complex *cis*-(OC)$_4$Os(H)·SnCl$_3$ readily forms[652]. Its methyl analogue is prepared by reaction of Me$_3$SnH with Os$_3$(CO)$_{12}$ [653]. With H$_2$Os$_2$(CO)$_8$, tin tetrachloride gives (OC)$_4$(H)Os—Os(SnCl$_3$)(CO)$_4$, which reacts with CX$_4$ to give Os$_2$X(SnCl$_3$)(CO)$_8$ (X = Cl, Br) [652].

Conductivity and i.r. spectral data in N,N-dimethylformamide suggest that

Ph_3Sn—$Co(CO)_4$ is dissociated into ions more readily than Ph_3Sn—$Mn(CO)_5$ [654]. $Ph_3SnCo(CO)_4$ is not dissociated in acetonitrile, but nucleophilic displacement of the transition metal carbonyl anion occurs on addition of Et_4NBr. Such displacements occur in DMF for $Ph_3SnMn(CO)_5$ and $Ph_3SnMn(CO)_3(\pi-C_5H_5)$ but not for $Ph_3SnFe(CO)_2(\pi-C_5H_5)$ [655]. ^{59}Co and ^{35}Cl n.q.r. spectra of $X_2Sn[Co(CO)_4]_2$ ($X_2 = Ph_2$, PhCl, Cl_2) indicate that the inductive effect of a chlorine atom bound to tin increases the electron-affinity of the vacant 4d-orbitals of tin, which drains electron density from the filled 3d-orbitals of cobalt which, in turn, removes electron density from the π^*-orbital of CO [656]. The increase in the ^{59}Co n.q.r. frequencies with increased halogen and decreased methyl substitution in the series $X_nSn[Co(CO)_4]_{4-n}$ (X = Cl, Br, Me; n = 0–3) supports the hypothesis that the populations of the cobalt d_{z^2}, d_{xz} and d_{yz}-orbitals are lower than for $d_{x^2-y^2}$ and d_{xy} [657]. The ^{59}Co n.q.r. spectral data are tabulated for the following, where Y = $Co(CO)_4$: $GeCl_3Y$, Cl_2GeY_2, Ph_3GeY, Ph_3SnY, $PhCl_2SnY$, Ph_2ClSnY, Me_3SnY, Cl_2SnY_2, $ClSnY_3$ and also $R_3MCo(CO)_3PPh_3$ (R_3M = Ph_3Ge, Ph_3Sn, Cl_3Sn), and $Ph_2BrGeCo(CO)_4$; some multiplicity in the Sn—Co bond is apparently indicated[658]. For the series of cobalt complexes, $X_3MCo(CO)_4$ (M = Si, Ge, Sn, Pb; X = Cl, Br, I, Ph) variations in paramagnetic shifts are attributed to differences in π-bonding in the metal-metal bond which is strongest for Sn—Co [659]. The x-ray structure of PhGe$Co_3(CO)_{11}$, which is prepared from Ph_3GeH and $Co_2(CO)_8$, can be described as a $PhGe \cdot Co(CO)_4$ entity replacing a bridging CO group in $Co_2(CO)_8$ [660]. In $Co(GePh_3)(CO)_3(PPh_3)$, there is a trigonal bipyramidal arrangement about Co with equatorial CO groups. The Co—Ge bond length is consistent with a degree of multiple bonding[661]. The reaction of Ph_2GeH_2 with $Co_2(CO)_8$ gives a compound in high yield, whose i.r. spectrum is interpreted as indicating a Ph_2Ge bridge between two cobalt atoms and so can be described as μ-diphenylgermyl-μ-carbonyl-bis(tricarbonylcobalt)[662]. The Cl_2Ge moiety acts as a bridge between a cobalt and iron atom in $(\pi-C_5H_5)Co(CO)_2(GeCl_2)_2Fe(CO)_4$, which is prepared by the reaction between $(\pi-C_5H_5)Co(CO)(GeCl_3)_2$ and $Fe(CO)_5$ [663]. The reaction of R_3GeH with $(Ph_3P)_2Ir(CO)Cl$ yields $(Ph_3P)_2Ir(CO)(H)_2GeR_3$ (R = Me, Et, Cl), whereas Ph_3GeH gives $(Ph_3P)Ir(CO)(H)ClGePh_3$ [664, 665]. The i.r. and ^1H n.m.r. spectra of the solution and the x-ray analysis of the solid are consistent with a cis-$(Ph_3P)_2$, cis-$(H)_2$-octahedral structure for $(Ph_3P)_2Ir(CO)(H)_2GeR_3$. The 5-coordinate $(Ph_3P)Ir(CO)(H)ClGePh_3$ molecule probably has the H and CO groups in the axial positions of a trigonal bipyramid[665]. The complexes $Ir(SnCl_3)CO(PPh_3)_2(C_2H_2)$ and $Ir(SnCl_3)CO(PPh_3)_2(C_2H_4)$ are obtained by adding $SnCl_2$ and C_2H_2 or C_2H_4 to $IrCl(CO)(PPh_3)_2$ [666]. Hydrostannation by R_3SnH (R = Me, Et, Ph) of the iridium(I) complexes, trans-$IrX(CO)L_2$ (X = Cl, Br, I; L = PPh_3, PPh_2Me, PPh_2Et) yields the iridium(III) species $R_3SnIr(H)X(CO)L_2$. On the basis of i.r. and n.m.r. spectral data with Ph_3Sn, the complex is formulated as the isomer in which the phosphine ligands, L, are trans to each other with Ph_3Sn trans to H; with Me_3Sn there are two isomers in which Me_3Sn is trans to H or trans to X, and with Et_3Sn it is tentatively assumed that Et_3Sn is trans to X [667]. The following rhodium-(I) and -(III) complexes containing a Rh—Ge bond have been reported: $(Ph_3P)_2RhH(GeR_3)Cl$ (R = Me, Et, or Cl), $(Ph_3As)_2RhH$

(GeR$_3$)Cl (R = Me, Et), CODRhH(GeCl$_3$)Cl, (Ph$_3$P)$_2$Rh(CO)$_2$GeEt$_3$, (Ph$_3$PH)$_3$[Rh(GeCl$_3$)$_6$], and (Me$_3$NH)$_3$[Rh(GeCl$_3$)$_3$Cl$_3$] [668].

A series of compounds containing a Ge—Ni bond, (π-C$_5$H$_5$)Ni(L)GeX$_3$ (L = PR$_3$, AsR$_3$; X = Cl, Br), are isolated from the reaction of Cs$^+$GeX$_3^-$ with (π-C$_5$H$_5$)Ni(L)X. Insertion of SnX$_2$ into the Ni—X bond in the latter compound, produces the analogous tin compounds. The crystal structure of (π-C$_5$H$_5$)Ni(PPh$_3$)GeCl$_3$ shows a Ge—Ni bond-length considerably shorter than the sum of the covalent radii, raising the possibility of (d–d)π-bonding[669]. The reactions of LiSnPh$_3$ or KSnCl$_3$ with metal carbonyls in THF lead to the formation of [Ni(CO)$_3$SnPh$_3$]$^-$, [Fe(CO)$_4$SnPh$_3$]$^-$, Co(CO)$_4$SnCl$_3$ and [Co(CO)$_3$(SnCl$_3$)$_2$]$^-$. Extensive exposure of Ni(CO)$_4$ to KSnCl$_3$ gives the anion [Ni(SnCl$_3$)$_4$]$^{4-}$ [670]. Some π-allyl(phosphine)palladium–trichlorotin complexes are reported[671,672]. The Sn—Pd bond is shorter than the sum of the covalent radii in π-allyl-trichlorotin(triphenylphosphine)palladium suggesting appreciable π-bond character. The arrangement of the ligands also suggests that the *trans*-influence of SnCl$_3$ is approximately the same as that of Ph$_3$P [672]. Carbon monoxide insertion into tin-containing palladium and platinum complexes is reported[673]. Spectroscopic and chemical evidence suggest that the 1:1 addition of HCl to [(Ph$_2$P·CH$_2$·CH$_2$·PPh$_2$)(Et$_3$P)Pt·GeMe$_3$]$^+$Cl$^-$ gives the HCl$_2^-$ ion and not a 6-coordinated Pt^{IV} cation as previously suggested[674]. Exchange reactions between methyl-germanes or -stannanes and Pt—MMe$_3$ (M = Si or Ge) gives two octahedral platinum–germanium and –tin compounds, (Ph$_2$P·CH$_2$·CH$_2$·PPh$_2$)ClPtM'Me$_3$ and (Ph$_2$P·CH$_2$·CH$_2$·PPh$_2$)Pt(M'Me$_3$)$_2$ (M' = Ge or Sn) [675]. A novel route to *trans*-Pt(PEt$_3$)$_2$(PbPh$_3$)Cl involves reaction of *trans*-Pt(PEt$_3$)$_2$Cl$_2$ with Hg(PbPh$_3$)$_2$ in benzene[676]. The reaction of Pt(C$_2$H$_4$)(PPh$_3$)$_2$ with Me$_6$Sn$_2$ gives Pt(SnMe$_3$)$_2$(PPh$_3$)$_2$ [677]. The Mössbauer spectrum of the methanol solution of the Pt—Sn chloride complex (H$_2$PtCl$_6$·SnCl$_2$) is reported[678]. SnX$_2$ adds to *trans*-[PtX(SiPh$_3$)(PMe$_2$Ph)$_2$] to give *trans*-[Pt(SnX$_3$)(SiPh$_3$)(PMe$_2$Ph)$_2$] (X = Cl, Br) [679]. SnCl$_2$ reacts with [Pt(CNR)$_4$][PtCl$_4$] and *cis*-Pt(CNR)$_2$Cl$_2$ (R = cyclohexyl, *p*-tolyl, *p*-anisyl) to yield *trans*-[(RNC)$_2$Pt(SnCl$_3$)$_2$] [680]. The complexes [Pt$_2$(SnCl$_3$)$_2$(SO$_2$tol)$_2$(PR$_3$)$_2$] and [Pt$_2$(SnCl$_3$)Cl(SO$_2$tol)$_2$(PR$_3$)$_2$] are characterised by ^{1}H n.m.r. and i.r. spectral data. The *p*-toluenesulphinate ion takes up bridging positions only when Cl$^-$ is not present, as it is more weakly bridging than chlorine[681].

In conclusion, two general methods available for forming Sn-transition metal bonds must be mentioned. In the first, the formation of Sn-M (M = Mn, Re, Mo, W, Fe, Co, and Ni) results from reaction with Me$_6$Sn$_2$ on the appropriate transition metal complex[682]. In the second, Me$_3$SnC$_5$H$_5$ oxidatively cleaves metal-metal bonds in bi- and poly-nuclear carbonyl complexes to give SnMe$_3$M', where M' = Mn(CO)$_5$, Re(CO)$_5$, Co(CO)$_4$, Mo(CO)$_3$(π-C$_5$H$_5$), Fe(CO)$_2$(π-C$_5$H$_5$), NiCO(π-C$_5$H$_5$) and Fe(CO)$_4$ [683].

The relatively few complexes that can be considered as derivatives of Group IV hydrides in that they contain the entity M—GeR$_3$, where M = a transition metal and R = H as at least *one* of the Groups R, are discussed under transition metal derivatives of the hydrides.

Finally, it is claimed that those tin–transition metal complexes which have been previously designated as containing tin(II) ligands[19,20], are actually derivatives of tin(IV)[684]. The isolation of a true tin(II)–transition metal complex by reaction of $(C_5H_5)_2Sn$ [685] with $FeCl_3$ in THF is reported in a note on the formation of an Sn—B bond[291].

References

1. Quane, D. and Bottei, A. S. (1963). *Chem. Rev.*, **63**, 403
2. Ingham, R. K., Rosenberg, S. D. and Gilman, H. (1960). *Chem. Rev.*, **60**, 459
3. Glockling, F. (1969). *The Chemistry of Germanium*, (London: Academic Press)
4. Mackay, K. M. and Watt, R. (1969). *Organometal. Chem. Rev. A*, **4**, 137
5. Donaldson, J. D. (1967). *Prog. Inorg. Chem.*, **8**, 287
6. Luijten, J. G. A., Rijkens, F. and Van der Kerk, G. J. M. (1965). *Advan. Organometal. Chem.*, **3**, 397
7. Sherer, O. J. (1969). *Angew. Chem.*, **81**, 871
8. Jones, K. and Lappert, M. F. (1966). *Organometal. Chem. Rev.*, **1**, 67
9. Schumann, H. (1969). *Angew. Chem.*, **81**, 970
10. Abel, E. W. and Illingworth, S. M. (1970). *Organometal. Chem. Rev. A*, **5**, 143
11. Abel, E. W. and Armitage, D. A. (1967). *Advan. Organometal. Chem.*, **5**, 1
12. Schumann, H. and Schmidt, M. (1965). *Angew. Chem.*, **77**, 1049
13. Lappert, M. F. and Pyszora, H. (1966). *Advan. Inorg. Chem. Radiochem.*, **9**, 133
14. Thayer, J. S. and West, R. (1967). *Advan. Organometal. Chem.*, **5**, 169
15. Neumann, W. P. and Kuhlein, K. (1968). *Advan. Organometal. Chem.*, **7**, 241
16. Harrison, P. G. (1969). *Organometal. Chem. Rev. A*, **4**, 379
17. Drake, J. E. and Riddle, C. (1970). *Quart. Rev. Chem. Soc.*, **24**, 263
18. Stone, F. G. A. (1968) in *New Pathways in Inorganic Chemistry*, (Cambridge: Cambridge University Press)
19. Donaldson, J. D. (1967). *Progr. Inorg. Chem.*, **8**, 287
20. Young, J. F. (1969). *Advan. Inorg. Chem. Radiochem.*, **11**, 91
21. Okawara, R. and Wada, M. (1967). *Advan. Organometal. Chem.*, **5**, 137
22. Beattie, I. R. (1963). *Quart. Rev. Chem. Soc.*, **17**, 382
23. Tanaka, T. (1970). *Organometal. Chem. Rev. A*, **5**, 1
24. Smith, P. J. (1970). *Organometal. Chem. Rev. A*, **5**, 373
25. Dennis, L. M., Corey, R. B. and Moore, R. W. (1924). *J. Amer. Chem. Soc.*, **46**, 657
26. Macklen, E. D. (1959). *J. Chem. Soc.*, 1989
27. Piper, T. S. and Wilson, M. K. (1957). *J. Inorg. Nucl. Chem.*, **4**, 22
28. Drake, J. E. and Jolly, W. L. (1962). *J. Chem. Soc.*, 2807
29. Zorin, A. D., Frolov, I. A., Galkin, P. N. and Skachkova, I. N. (1970). *Russian J. Inorg. Chem.*, **15**, 1048
30. Allred, A. L. and Deming, R. L. (1970). *Inorg. Nucl. Chem. Lett.*, **6**, 39
31. Drake, J. E. and Riddle, C. (1969). *J. Chem. Soc. A*, 2114
32. Verdick, J. F. and Mau, A. W. (1969). *Chem. Commun.*, 226
33. Gaspar, P. P., Levy, C. A., Frost, J. J. and Bock, S. A. (1969). *J. Amer. Chem. Soc.*, **91**, 1574
34. Timms, P. L., Simpson, C. C. and Phillips, C. S. G. (1964). *J. Chem. Soc.*, 1467
35. Estacio, P., Sefcik, M. D., Chan, E. K. and Ring, M. A. (1970). *Inorg. Chem.*, **9**, 1068
36. Beagley, B. and Monaghan, J. J. (1970). *Trans. Faraday Soc.*, **66**, 2745
37. Pullen, B. P., Carlson, T. A., Moddeman, W. E., Schweitzer, G. K., Bull, W. E. and Grimm, F. A. (1970). *J. Chem. Phys.*, **53**, 768
38. Stevenson, P. E. and Lipscomb, W. N. (1970). *J. Chem. Phys.*, **52**, 5343
39. Schumann, C. and Dreeskamp, H. (1970). *J. Mag. Res.*, **3**, 204
40. Reifenberg, G. H. and Considine, W. J. (1969). *J. Amer. Chem. Soc.*, **91**, 2401
41. Khain, V. S. (1969). Ref. *Zh. Khim.*, **17B**, 815
42. Srivastava, T. N., Griffiths, J. E. and Onyszchuk, M. (1962). *Can. J. Chem.*, **40**, 739
43. Lannon, J. A., Weiss, G. S. and Nixon, E. R. (1970). *Spectrochim. Acta*, **26A**, 221
44. Griffiths, J. E. (1963). *J. Chem. Phys.*, 2879
45. George, R. D., Mackay, K. M. and Stobart, S. R. (1970). *J. Chem. Soc. A*, 3250
46. Spanier, E. J. and MacDiarmid, A. G. (1963). *Inorg. Chem.*, **2**, 215

47. Timms, P. L., Simpson, C. C. and Phillips, C. S. G. (1964). *J. Chem. Soc.*, 1467
48. Mackay, K. M., Hosfield, S. T. and Stobart, S. R. (1969). *J. Chem. Soc. A*, 2937
49. Wiberg, E., Amberger, E. and Cambensi, H. (1967). *Z. Anorg. Allgem. Chem*, **351**, 164
50a. Dutton, W. A. and Onyszchuk, M. (1968). *Inorg. Chem.*, **7**, 1735; (b) Amberger, E. and Muhlkofer, E. (1968). *J. Organometal. Chem.*, **12**, 55
51. Bürger, H., Goetze, U. and Sawodny, W. (1970). *Spectrochim. Acta*, **26A**, 685
52. Gibbon, G. A., Kifer, E. W., Van Dyke, C. H. (1970). *Inorg. Nucl. Chem. Lett.*, **6**, 617
53. Eméleus, H. J., Maddock, A. G. and Reid, C. (1941). *J. Chem. Soc.*, 353
54. Dennis, L. M. and Judy, P. R. (1929). *J. Amer. Chem. Soc.*, **51**, 2321
55. Amberger, E., Stoeger, W. and Honigschmid, J. (1969). *J. Organometal. Chem.*, **18**, 77
56. Amberger, E. and Stoeger, W. (1969). *J. Organometal. Chem.*, **17**, 287
57. Amberger, E. and Stoeger, W. (1969). *J. Organometal. Chem.*, **18**, 83
58. Birchall, T. and Drummond, I. (1970). *J. Chem. Soc. A*, 1859
59. Birchall, T. and Drummond, I. (1970). *J. Chem. Soc. A*, 1401
60. Pauling, L. (1969). *J. Chem. Phys.*, **51**, 2767
61. Rankin, D. W. H. (1969). *Chem. Commun.*, 194
62. Rankin, D. W. H. (1969). *J. Chem. Soc. A*, 1926
63. Glidewell, C., Rankin, D. W. H. and Robiette, A. G. (1970). *J. Chem. Soc. A*, 2935
64. Ebsworth, E. A. V. (1966). *Chem. Commun.*, 530
65. Rankin, D. W. H., Robiette, A. G., Sheldrick, G. M., Beagley, B. and Hewitt, T. G. (1969). *J. Inorg. Nucl. Chem.*, **31**, 2351
66. Mackay, K. M. and Stobart, S. R. (1970). *Spectrochim. Acta*, **26A**, 373
67. Cradock, S. and Ogilvie, J. F. (1966). *Chem. Commun.*, 364
68. Ogilvie, J. F. and Newlands, M. J. (1969). *Trans. Faraday Soc.*, **65**, 2602
69. Norman, A. D. and Wingleth, D. C. (1970). *Inorg. Chem.*, **9**, 98
70. Anderson, J. W. and Drake, J. E. (1969). *Inorg. Nucl. Chem. Lett.*, **5**, 881
71. Drake, J. E. and Riddle, C. (1968). *J. Chem. Soc. A*, 1675 and 2452
72. Mackay, K. M., Sutton, K. J., Stobart, S. R., Drake, J. E. and Riddle, C. (1969). *Spectrochim. Acta*, **25A**, 925
73. Drake, J. E., Riddle, C., Mackay, K. M., Stobart, S. R. and Sutton, K. J. (1969). *Spectrochim. Acta*, **25A**, 941
74. Drake, J. E. and Riddle, C. (1969). *Inorg. Chim. Acta*, **3**, 136
75. Drake, J. E. and Riddle, C. (1969). *J. Chem. Soc. A*, 2704
76. Ebsworth, E. A. V., Rankin, D. W. H. and Sheldrick, G. M. (1968). *J. Chem. Soc. A*, 2828
77. Drake, J. E. and Riddle, C. (1969). *J. Chem. Soc. A*, 1573
78. Cradock, S., Ebsworth, E. A. V. and Rankin, D. W. H. (1969). *J. Chem. Soc. A*, 1628
79. Glidewell, C., Rankin, D. W. H. and Sheldrick, G. M. (1969). *Trans. Faraday Soc.*, **65**, 1409
80. Goldfarb, T. D. and Sujishi, S. (1964). *J. Amer. Chem. Soc.*, **86**, 1679
81. Cradock, S. (1968). *J. Chem. Soc. A*, 1426
82. Glidewell, C., Rankin, D. W. H., Robiette, A. G., Sheldrick, G. M., Beagley, B. and Cradock, S. (1970). *J. Chem. Soc. A*, 315
83. Glidewell, C., Rankin, D. W. H., Robiette, A. G., Sheldrick, G. M., Cradock, S. and Ebsworth, E. A. V. (1969). *Inorg. Nucl. Chem. Lett.*, **5**, 417
84. Glidewell, C., Rankin, D. W. H. and Sheldrick, G. M. (1969). *Trans. Faraday Soc.*, **65**, 2801
85. Glidewell, C. and Rankin, D. W. H. (1969). *J. Chem. Soc. A*, 753
86. Drake, J. E. and Riddle, C. (1970). *Inorg. Nucl. Chem. Lett.*, **6**, 713
87. Drake, J. E. and Riddle, C. (1970). *J. Chem. Soc. A*, 3134
88. Yarandina, V. N. and Svedlov, L. M. (1969). *Isv. Vysch. Ucheb. Zaved. Fiz.*, **12**, 138
89. Ramasuamy, K. and Balasubramanian, V. (1969). *Indian J. Phys.*, **43**, 454
90. Mueller, A., Kebabuoglu, R., Krebbs, B. and Glemser, O. (1969). *Z. Phys. Chem.*, **240**, 92
91. Drake, J. E. and Riddle, C. (1969). *J. Chem. Soc. A*, 910
92. Drake, J. E. and Riddle, C. (1969). *J. Chem. Soc. A*, 2114
93. Venkateswarlu, K. and Bhamambal, P. (1970). *Acta. Phys. Pol. A*, **37**, 661
94. Belyi, A. P., Gorbunov, A. I. and Golubtsov., S. A. (1969). *Russ. J. Phys. Chem.*, **43**, 1145
95. Mackay, K. M. and Roebuck, P. J. (1964). *J. Chem. Soc.*, 1195
96. Mackay, K. M., Robinson, P., Spanier, E. J. and MacDiarmid, A. G. (1966). *J. Inorg. Nucl. Chem.*, **28**, 1377

97. Spanier, E. J. and MacDiarmid, A. G. (1969). *J. Inorg. Nucl. Chem.*, **31**, 2976
98. Mackay, K. M. and Robinson, P. (1965). *J. Chem. Soc.*, 5121
99. Hosfield, S. T. and Mackay, K. M. (1970). *J. Organometal. Chem.*, **24**, 107
100. Drake, J. E., Hemmings, R. T. and Riddle, C. (1970). *J. Chem. Soc. A*, 3359
101. Massey, A. G., Park, A. J. and Stone, F. G. A. (1963). *J. Amer. Chem. Soc.*, **85**, 2021
102. Flitcroft, N., Harbourne, D. A., Paul, I., Tucker, P. M. and Stone, F. G. A. (1966). *J. Chem. Soc. A*, 1130
103. Mackay, K. M. and George, R. D. (1969). *Inorg. Nucl. Chem. Lett.*, **5**, 797
104. Mackay, K. M. and Stobart, S. R. (1970). *Inorg. Nucl. Chem. Lett.*, **6**, 687
105. Mackay, K. M. and George, R. D. (1970). *Inorg. Nucl. Chem. Lett.*, **6**, 289
106. Bentham, J. E., Cradock, S. and Ebsworth, E. A. V. (1969). *Chem. Commun.*, 528
107. Bentham, J. E. and Ebsworth, E. A. V. (1970). *Inorg. Nucl. Chem. Lett.*, **6**, 671
108. Stobart, S. R. (1970). *Chem. Commun.*, 999
109. Kohanek, J. J., Estacio, P. and Ring, M. A. (1969). *Inorg. Chem.*, **8**, 2516
110. George, R. D. and Mackay, K. M. (1969). *J. Chem. Soc. A*, 2122
111. Glockling, F., Light, J. R. C. and Stafford, R. G. (1970). *J. Chem. Soc. A*, 426
112. Glockling, F., Light, J. R. C., Walker, J. and Mackay, K. M. (1970). *J. Chem. Soc. A*, 432
113. Durig, J. R., Craven, S. M. and Bragin, J. (1970). *J. Chem. Phys.*, **52**, 2046
114. Smith, G. W. (1965). *J. Chem. Phys.*, **42**, 4229
115. Graham, S. G. (1970). *Spectrochim. Acta*, **26A**, 345
116. Parker, J. and Ladd, J. A. (1970). *Trans. Faraday Soc.*, **66**, 1907; and Ventkateswarlu, K., Devi, V. and Natorajan, A. (1969). *Proc. Indian Acad. Sci. A*, **70**, 126
117. Clark, R. J. H., Davies, A. G., Puddephatt, R. J. and McFarlane, W. (1969). *J. Amer. Chem. Soc.*, **91**, 1334
118a. Pitt, C. G., Bursey, M. M. and Rogerson, P. F. (1970). *J. Amer. Chem. Soc.*, **92**, 519
118b. Egorochin, A. N., Korshev, S., Vyazankin, N. S. and Gladyshev, E. N. (1969). *Isv. Acad. Nauk SSSR, Ser. Khim.*, 969 and 1863
119. Keiser, D. and Kana'an, A. S. (1969). *J. Phys. Chem.*, **73**, 4264
120. Willemsens, L. C. and Van der Kerk, G. J. M. (1970). *J. Organometal. Chem.*, **21**, 123
121. Richter, M. and Neumann, W. P. (1969) *J. Organometal. Chem.*, **20**. 81
122. Lunazzi, L. and Taddei, F. (1969). *Spectrochim. Acta*, **25A**, 611
123. Krebs, P. and Dreeskamp, H. (1969). *Spectrochim. Acta*, **25A**, 1399
124. Thomas, E. C. and Laurie, V. W. (1969). *J. Chem. Phys.*, **50**, 3512
125. Thomas, E. C. and Laurie, V. W. (1969). *J. Chem. Phys.*, **51**, 4327
126. Mathis, R., Barthelat, M. and Mathis, F. (1970). *Spectrochim. Acta*, **26A**, 1993
127. Mathis, R., Barthelat, M. and Mathis, F. (1970). *Spectrochim. Acta*, **26A**, 2001
128. Kaplan, L. (1969). *Chem. Commun.*, 106
129. Willemsens, L. C. and Van der Kerk, G. J. M. (1969). *J. Organometal. Chem.*, **19**, 81
130. Jappy, J. and Preston, P. N. (1969). *J. Organometal. Chem.*, **19**, 196
131. Lindner, E. and Kunze, V. (1970). *J. Organometal. Chem.*, **21**, P19
132. Neumann, W. P. and Blanket, U. (1969). *Angew. Chem. Int. Ed. Engl.*, **8**, 611
133. Vyazankin, N. S., Bychkov, V. T., Linzina, O. V. and Razuvaev, G. A. (1969). *Zh. Obshch. Khim.*, **39**, 979
134. Bennett, S. W., Clare, H. J., Eaborn, C. A. and Jackson, R. A. (1970). *J. Organometal. Chem.*, **23**, 403
135. Mitchell, T. N. and Neumann, W. P. (1970). *J. Organometal. Chem.*, **22**, C25
136. Vyazankin, N. S., Gladyshev, E. N., Arkhangel'skaya, E. A. and Razuvaev, G. A. (1969). *J. Organometal. Chem.*, **17**, 340.
137. Warner, C. M. and Noltes, J. G. (1970). *J. Organometal. Chem.*, **24**, C4
138. Lappert, M. F., Simpson, J. and Spalding, T. R. (1969). *J. Organometal. Chem.*, **17**, P1
139. Bennett, S. W., Eaborn, C., Hudson, A., Hussain, H. A. and Jackson, R. A. (1969). *J. Organometal. Chem.*, **16**, P39
140. Durig, J. R., Lam, K. K., Turner, J. B. and Bragin, J. (1969). *J. Mol. Spectrosc.*, **31**, 419
141. Durig, J. R., Craven, S. M. and Bragin, J. (1969). *J. Chem. Phys.*, **51**, 5663
142. Mackay, K. M., Sowerby, D. B. and Young, W. C. (1968). *Spectrochim. Acta*, **24A**, 611
143. Durig, J. R., Sink, C. W. and Turner, J. B. (1969). *Spectrochim. Acta*, **25A**, 629
144. Smith, A. L. (1969). *Spectrochim. Acta*, **25A**, 1075
145. Poller, R. C. (1966). *Spectrochim. Acta*, **22**, 935
146. Peuker, C., Licht, K. and Kriegsmann, H. (1970). *Z. Phys. Chem.*, **244**, 61
147. Egorochkin, A. N., Korshev, S. and Vyazankin, N. S. (1969). *Dokl. Acad. Nauk. SSSR*, **185**, 353

148. Yarandina, V. N. and Sverdlov, L. M. (1969). *Isv. Vyssh. Ucheb. Zaved. Fiz.*, **12**, 80
149. Yarandina, V. N. and Sverdlov, L. M. (1969). *Isv. Vyssh. Ucheb. Zaved Fiz.*, **12**, 157
150. Van de Vondel, D. F., Van der Kelen, G. P. and Van Hooydonk (1970). *J. Organometal. Chem.*, **23**, 431
151. Kriegsmann, H., Peuker, C., Hees, R. and Geissler, H. (1969). *Z. Naturforsch A*, **24**, 778
152. DeAlti, G., Galasso, V. and Bigotto, A. (1968). *Corsi. Semin. Chim.*, **14**, 46
153. Srivastava, T. S. (1967). *J. Organometal. Chem.*, **10**, 373
154. Srivastava, T. S. (1969). *J. Organometal. Chem.*, **16**, P53
155. Clark, R. J. H., Davies, A. G. and Puddephatt, R. J. (1969). *Inorg. Chem.*, **8**, 457
156. Belij, A. P., Gorbunov, A. T., Golubtsov, S. A. and Feldshtein, N. S. (1969). *J. Organometal. Chem.*, **17**, 485
157. Butler, R. A., Carson, A. S., Laye, P. G. and Steele, W. V. (1970). *J. Organometal. Chem.*, **24**, C11
158. Redl. G., Altner, B., Anker, D. and Minot, M. (1969). *Inorg. Nucl. Chem. Lett.*, **5**, 861
159. Seyferth, D. and Hopper, S. P. (1970). *J. Organometal. Chem.*, **23**, 99
160. Seyferth, D. and Armbrecht, F. M. (1969). *J. Organometal. Chem.*, **16**, 249
161. Seyferth, D., Armbrecht, F. M. and Schneider, B. (1969). *J. Amer. Chem. Soc.*, **91**, 1954
162. Bettler, C. K., Sendra, J. C. and Urry, G. (1970). *Inorg. Chem.*, **9**, 1060
163a. Nefedov, O. M., Kolesnikov, S. R. and Perl'mutter, B. L. (1969). *Isv. Acad. Nauk SSSR, Ser. Khim.*, **11**, 2574
163b. Nyholm, R. S. and Royo, P. (1969). *Chem. Commun.*, 421
164. Gelius, R. (1970). *Z. Anorg. Allgem. Chem.*, **374**, 297
165. Muller, R., Reichel, S. and Dathe, C., *Brit.*, **1**, 165, 195
166. Dunn, P. and Oldfield, D. (1970). *J. Organometal. Chem.*, **23**, 459
167. Lorberth, J. (1969). *J. Organometal. Chem.*, **17**, 151
168. Cody, V. and Corey, E. R. (1969). *J. Organometal. Chem.*, **19**, 359
169. Platt, R. H. (1970). *J. Organometal. Chem.*, **24**, C23
170. Devooght, J., Gielen, M. and Lejeune, S. (1970). *J. Organometal. Chem.*, **21**, 333
171. Williams, D. E. and Kocher, C. W. (1970). *J. Chem. Phys.*, **52**, 1480
172. Parish, R. V. and Platt, R. H. (1970). *Inorg. Chim. Acta*, **4**, 65
173. Poller, R. C., Ruddick, J. N. R., Taylor, B. and Toley, D. L. B. (1970). *J. Organometal. Chem.*, **24**, 341
174. Grant, M. W. and Prince, R. H. (1969). *J. Chem. Soc. A*, 1138
175. Chambers, D. B. and Glockling, F. (1970). *Inorg. Chim. Acta*, **4**, 150
176. Moedritzer, K. (1970). *J. Inorg. Nucl. Chem.*, **32**, 2529
177. Moedritzer, K. and Van Wazer, J. R. (1969). *Rev. Chim. Minerale*, **6**, 293
178. Wang, C. S. and Shreeve, J. M. (1970). *Chem. Commun.*, 151
179. Kandil, S. A. and Allred, A. L. (1970). *J. Chem. Soc. A*, 2987
180. Moedritzer, K. and Van Wazer, J. R. (1969). *J. Chem. Soc. A*, 1124
181. Cohen, A. D. and Dillard, C. R. (1970). *J. Organometal. Chem.*, **25**, 421
182. Goodman, B. A. and Greenwood, N. N. (1969). *Chem. Commun.*, 1105
183. Erickson, N. E. (1970). *Chem. Commun.*, 1349
184. Gibbs, T. C., Goodman, B. A. and Greenwood, N. N. (1970). *Chem. Commun.*, 774
185. Kazimir, E. O. (1969). Ph.D. Thesis, Fordham University
186. Davies, A. G., Milledge, H. J., Puxley, D. C. and Smith, P. J. (1970). *J. Chem. Soc. A*, 2862
187. Hobbs, C. W. and Tobias, R. S. (1970). *Inorg. Chem.*, **9**, 1037
188. Schroeder, H., Papetti, S., Alexander, R. P., Sieckhaus, J. F. and Heying, T. L. (1969). *Inorg. Chem.*, **8**, 2444
189a. Bresadola, S., Plazzogna, G., Cecchin, G. and Tagliavini, G. (1970). *Gazz. Chim. Ital.*, **100**, 175
189b. Svitsyn, R. A., Zhigach, A. F., Sobolev, E. S., Antipin, L. M. and Mironov, V. F. (1970). *Khim. Geterotsikl. Soedin.*, **1**, 127
190. Chow, Y. M. (1970). *Inorg. Chem.*, **9**, 794
191. Forder, R. A. and Sheldrick, G. M. (1970). *J. Organometal. Chem.*, **22**, 611
192. Forder, R. A. and Sheldrick, G. M. (1969). *Chem. Commun.*, 1125
193. Forder, R. A. and Sheldrick, G. M. (1970). *J. Organometal. Chem.*, **21**, 115
194. Forder, R. A. and Sheldrick, G. M. (1970). *Chem. Commun.*, 1023
195. Cheng, H. S. and Herber, R. A. (1970). *Inorg. Chem.*, **9**, 1686
196. Bertazzi, N. and Barbieri, R. (1969). *Inorg. Nucl. Chem. Lett.*, **5**, 591
197. Lappert, M. F., Lorberth, J. and Poland, J. S. (1970). *J. Chem. Soc. A*, 2954

198. Schumann, H. and Schumann-Ruidisch, I. (1969). *J. Organometal. Chem.*, **18**, 355
199. Gladyshev, E. N., Andreevichev, V. S., Vyazankin, N. S. and Razuvaev, G. A. (1970). *Zh. Obshch. Khim.*, **40**, 939
200. Dinh-Huu-Nguyea, Baukov, I. and Lutsenko, I. F. (1969). *Zh. Obshch. Khim.*, **39**, 922
201. Schumann, H. and Reich, P. (1970). *Z. Anorg. Allgem. Chem.*, **377**, 63
202. Damle, S. B. and Considine, W. J. (1969). *J. Organometal. Chem.*, **19**, 207
203. Cardona, R. A., Kupchik, E. J. and Hanke, H. E. (1970). *J. Organometal. Chem.*, **24**, 371
204. Vilkov, L. V. and Tarasenko, N. A. (1969). *Zh. Strukt. Khim.*, **10**, 1102
205. Razuvaev, G. A., Aleksandrov, Y. A., Figurova, G. N. and Glushakova, V. N. (1969). *Zh. Obshch. Khim.*, **39**, 2499
206. Stapfer, C. H., Dworkin, R. D. and Weisfeld, L. B. (1970). *J. Organometal. Chem.*, **24**, 355
207. Marchand, A., Mendelsohn, J., Lebedeff, M. and Valade, J. (1969). *J. Organometal. Chem.*, **17**, 379
208. Wardell, J. L. and Grant, D. W. (1969). *J. Organometal. Chem.*, **20**, 91
209. Schumann, H. and Reich, P. (1970). *Z. Anorg. Allgem. Chem.*, **375**, 72
210. Lapkin, I., Dumler, V. A. and Ponosova, E. S. (1970). *Zh. Obshch. Khim.*, **40**, 1063
211. Van de Vondel, D. F., Van den Berghe, E. V. and Van der Kelen, G. P. (1970). *J. Organometal. Chem.*, **23**, 105
212. Oliver, A. J. and Graham, W. A. G. (1969). *J. Organometal. Chem.*, **19**, 17
213. Hogben, M. G. and Graham, W. A. G. (1969). *J. Amer. Chem. Soc.*, **91**, 283
214. Hogben, M. G., Gay, R. G., Oliver, A. J., Thompson, J. A. J. and Graham, W. A. G. (1969). *J. Amer. Chem. Soc.*, **91**, 291
215. Stapfer, C. H., Leung, K. L. and Herber, R. H. (1970). *Inorg. Chem.*, **9**, 970
216. Stapfer, C. H. and Dworkin, R. D. (1970). *Inorg. Chem.*, **9**, 421
217. Schumann, H. and Weis, R. (1970). *Angew. Chem. Int. Ed. Engl.*, **9**, 246
218. Fong, C. W. and Kitching, W. (1970). *J. Organometal. Chem.*, **22**, 95
219. Fong, C. W. and Kitching, W. (1970). *J. Organometal. Chem.*, **21**, 365
220. Fong, C. W. and Kitching, W. (1970). *J. Organometal. Chem.*, **22**, 107
221. Lindner, E., Kunze, U., Ritter, G. and Haag, A. (1970). *J. Organometal. Chem.*, **24**, 119
222. Vitzthum, G., Kunze, U. and Lindner, E. (1970). *J. Organometal. Chem.*, **21**, P38
223. Lindner, E. and Kunze, U. (1970). *J. Organometal. Chem.*, **23**, C53
224. Lindner, E., Kunze, U., Vitzthum, G., Ritter, G. and Haag, A. (1970). *J. Organometal. Chem.*, **24**, 131
225. Schmeisser, M., Sartori, P. and Lippsmeier, B. (1970). *Chem. Ber.*, **103**, 868
226. Schmidt, M., Schumann, H., Gliniecki, F. and Jaggard, J. F. (1969). *J. Organometal. Chem.*, **17**, 277
227. Schmidt, M. and Jaggard, J. F. (1969). *J. Organometal. Chem.*, **17**, 283
228. Kamitani, T. and Tanaka, T. (1970). *Inorg. Nucl. Chem. Lett.*, **6**, 91
229. Honda, M., Komura, M., Kawasaki, Y., Tanaka, T. and Okawara, R. (1968). *J. Inorg. Nucl. Chem.*, **30**, 3231
230. Furue, K., Kimura, T., Yusuoka, N., Kasai, N. and Kakudo, M. (1970). *Bull. Chem. Soc. Jap.*, **43**, 1661
231. Sheldrick, G. M., Sheldrick, W. S., Dalton, R. F. and Jones, K. (1970). *J. Chem. Soc. A*, 493
232. Sheldrick, G. M. and Sheldrick, W. S. (1970). *J. Chem. Soc. A*, 490
233. Kane, A. R., Sullivan, J. F., Kenny, D. H. and Kenney, M. E. (1970). *Inorg. Chem.*, **9**, 1445
234. Harrison, P. G. and Zuckerman, J. J. (1970). *Inorg. Nucl. Chem. Lett.*, **6**, 5
235. Harrison, P. G. and Zuckerman, J. J. (1970). *Inorg. Nucl. Chem.*, **9**, 175
236. Issleib, K. and Walther, B. (1970). *J. Organometal. Chem.*, **22**, 375
237. Ridenour, R. E. and Flagg, E. E. (1969). *J. Organometal. Chem.*, **16**, 393
238. Brazier, J. F., Houalla, D. and Wolf, R. (1970). *Bull. Soc. Chim. Fr.*, 1079
239. Schumann, H. and Schmidt, M. (1965). *Inorg. Nucl. Chem. Lett.*, **1**, 1
240. Schumann, H. and Roth, A. (1969). *Chem. Ber.*, **102**, 3725
241. George, T. A. and Lappert, M. F. (1969). *J. Chem. Soc. A*, 992
242. Mazerolles, P., Lesbre, M. and Joanny, M. (1969). *J. Organometal. Chem.*, **16**, 227
243. Atovmyan, L. O., Bleidelis, J., Kemme, A. A. and Shibaeva, R. P. (1970). *Zh. Strukt. Khim.*, **11**, 318
244. Lorberth, J. (1969). *J. Organometal. Chem.*, **16**, 235
245. Lorberth, J. (1969). *J. Organometal. Chem.*, **16**, 327

246. Lorberth, J. and Nöth, H. (1969). *J. Organometal. Chem.*, **19**, 203
247. Vilkov, L. V., Tarasenko, N. A. and Prokofev, A. K. (1970). *Zh. Strukt. Khim.*, **11**, 129
248. Dalton, R. F. and Jones, K. (1969). *Inorg. Nucl. Chem. Lett.*, **5**, 785
249. Schumann, H., Schumann-Ruidisch, I. and Ronecker, S. Z. (1970). *Naturforsch, B*, **25**, 565
250. Lorberth, J. (1969). *J. Organometal. Chem.*, **19**, 435
251. Tsai, T. T., Lehn, W. L. and Marshall, C. J. (1970). *J. Organometal. Chem.*, **22**, 387
252. Schumann, H. and Ronecker, S. (1970). *J. Organometal. Chem.*, **23**, 451
253. Itoh, K., Matsuda, I., Katsuura, T. and Ishii, Y. (1969). *J. Organometal. Chem.*, **19**, 347
254. Wolfsberger, W. and Schmidbaur, H. (1969). *J. Organometal. Chem.*, **17**, 41
255. Gerega, V. F., Derganov, Y. I., Pavlycheva, A. V., Mushkan, Y. I. and Aleksandrov, Y. A. (1970). *Zh. Obshch. Khim.*, **40**, 1099
256. Armitage, D. A. and Clark, M. J. (1970). *J. Organometal. Chem.*, **24**, 629
257. Pommier, J. C. and Robineau, A. (1969). *J. Organometal. Chem.*, **17**, P25
258. Satgé, J., Couret, C. and Escudié, J. (1970). *Compt. Rend., C*, **270**, 351
259. Chandra, G., George, T. A. and Lappert, M. F. (1969). *J. Chem. Soc. C*, 2565
260. Satgé, J., Couret, C. and Escudié, J. (1970). *J. Organometal. Chem.*, **24**, 633
261. Chandra, G., Jenkins, A. D., Lappert, M. F. and Srivastava, R. C. (1970). *J. Chem. Soc. A*, 2550
262. Dalton, R. F. and Jones, K. (1970). *J. Chem. Soc. A*, 590
263. Cardin, D. J., Keppie, S. A. and Lappert, M. F. (1970). *J. Chem. Soc. A*, 2594
264. Jenkins, A. D., Lappert, M. F. and Srivastava, R. C. (1970). *J. Organometal. Chem.*, **23**, 165
265. Highsmith, R. E. and Sisler, H. H. (1969). *Inorg. Chem.*, **8**, 1029
266. Schmidbaur, H. and Kimmel, G. (1969). *Chem. Ber.*, **102**, 4118
267. Schmitz-DuMont, O., Götze, H. J. and Götze, H. (1969). *Z. Anorg. Allgem. Chem.*, **366**, 180
268. Schmitz-DuMont, O. and Götze, H. J. (1969). *Z. Anorg. Allgem. Chem.*, **371**, 38
269. Schmitz-DuMont, O. and Jansen, W. (1969). *Z. Anorg. Allgem. Chem.*, **370**, 140
270. Schmitz-DuMont, O. and Jansen, W. (1970). *Z. Anorg. Allgem. Chem.*, **375**, 98
271. Wannagat, U., Bogusch, E. and Braun, R. (1969). *J. Organometal. Chem.*, **19**, 367
272. Harrison, P. G. and Zuckerman, J. J. (1969). *Inorg. Nucl. Chem. Lett.*, **5**, 545
273. Chamberlain, B. R. and Moser, W. (1969). *J. Chem. Soc. A*, 354
274. Schumann, H., Schwabe, P. and Stelzer, O. (1969). *Chem. Ber.*, **102**, 2900
275. Schumann, H., Roth, A., Stelzer, O. and Schmidt, M. (1966). *Inorg. Nucl. Chem. Lett.*, **2**, 311
276. Schumann, H. and Roth, A. (1969). *Chem. Ber.*, **102**, 3713
277. Schumann, H., Ostermann, T. and Schmidt, M. (1966). *Chem. Ber.*, **99**, 2057
278. Schumann, H., Rösch, L. and Stelzer, O. (1970). *J. Organometal. Chem.*, **21**, 351
279. Schumann, H. and Arbenz, U. (1970). *J. Organometal. Chem.*, **22**, 411
280. Schumann, H., Stelzer, O., Niederreuther, U. and Rösch, L. (1970). *Chem. Ber.*, **103**, 2350
281. Abel, E. W., Crow, J. P. and Illingworth, S. M. (1969). *J. Chem. Soc. A*, 1631
282. Schumann, H. and Roth, A. (1969). *Chem. Ber.*, **102**, 3731
283. Schumann, H., Roth, A. and Stelzer, O. (1970). *J. Organometal. Chem.*, **24**, 183
284. Norman, A. D. (1970). *Inorg. Chem.*, **9**, 870
285. Dahl, A. R. and Norman, A. D. (1970). *J. Amer. Chem. Soc.*, **92**, 5525
286. Schumann, H. and Benda, H. (1970). *J. Organometal. Chem.*, **21**, P12
287. Anderson, J. W. and Drake, J. E. (1970). *J. Chem. Soc. A*, 3131
288. Todd, L. J., Burke, A. R. Silverstein, H. T., Little, J. L. and Wikholm, G. S. (1969). *J. Amer. Chem. Soc.*, **91**, 3376
289. Voorhees, R. L. and Rudolph, R. W. (1969). *J. Amer. Chem. Soc.*, **91**, 2173
290. Rudolph, R. W., Voorhees, R. L. and Cochoy, R. E. (1970). *J. Amer. Chem. Soc.*, **92**, 3351
291. Harrison, P. G. and Zuckerman, J. J. (1970). *J. Amer. Chem. Soc.*, **92**, 2577
292. Luth, H. and Amma, E. L. (1969). *J. Amer. Chem. Soc.*, **91**, 7515
293. Hastie, J. W., Hauge, R. H. and Margrave, J. L. (1969). *Chem. Commun.*, 1452
294. Hastie, J. W., Hauge, R. H. and Margrave, J. L. (1968). *J. Phys. Chem.*, **72**, 4492
295. Perry, R. O. (1969). *Chem. Commun.*, 889
296. Beattie, I. R. and Perry, R. O. (1970). *J. Chem. Soc. A*, 2429
297. Andrews, L. and Frederick, D. L. (1970). *J. Amer. Chem. Soc.*, **92**, 775

298. Hastie, J. W., Hauge, R. H. and Margrave, J. L. (1969). *J. Mol. Spectrosc.*, **29**, 152
299. Pathak, C. M. and Palmer, H. B. (1969). *J. Mol. Spectrosc.*, **31**, 170
300. Zollweg, R. J. and Frost, L. S. (1969). *J. Chem. Phys.*, **50**, 3280
301. Uy, M., Muenow, D. W. and Margrave, J. L. (1969). *Trans. Faraday Soc.*, **65**, 1296
302. Moser, N. and Trevena, I. C. (1969). *Chem. Commun.*, 25
303. Ozin, G. A. (1970). *Can. J. Chem.*, **48**, 2931
304. Donaldson, J. D. and Senior, B. J. (1969). *J. Chem. Soc. A*, 2358
305. Guillory, W. A. and Smith, C. E. (1970). *J. Chem. Phys.*, **53**, 1661
306. Roncin, J. and Debuyst, R. (1969). *J. Chem. Phys.*, **51**, 577
307. Oldershaw, G. A. and Robinson, K. (1970). *Trans. Faraday Soc.*, **66**, 532
308. Oldershaw, G. A. and Robinson, K. (1969). *J. Mol. Spectrosc.*, **32**, 469
309. Murphy, A. and Haronath, P. (1970). *Curr. Sci.*, **39**, 132
310. Buchanan, A. S., Knowles, D. J. and Swinger, D. L. (1969). *J. Phys. Chem.*, **73**, 4394
311. Hastie, G. J. W., Bloom, H. and Morrison, J. D. (1967). *J. Chem. Phys.*, **47**, 1580
312. Kuryakov, Y. Y. (1968). *Vestu. Mosk. Univ. Khim.*, **23**, 21
313. Knowles, D. J., Nicholson, A. J. C. and Swinger, D. L. (1970). *J. Phys. Chem.*, **74**, 3642
314. Graybeal, J. D. and Green, P. J. (1969). *J. Phys. Chem.*, **73**, 2948
315. Ward, R. W., Williams, C. D. and Tipsword, R. F. (1969). *J. Chem. Phys.*, **51**, 823
316. Mishra, B. and Ramakrishna, V. (1969). *Spectrochim. Acta*, **25A**, 288
317. Fujii, H. and Kimura, M. (1970). *Bull. Chem. Soc., Jap.*, **43**, 1933
318. Crump, R. A. and Price, A. H. (1969). *Chem. Commun.*, 254
319. Thomas, T. E., Orville-Thomas, W. J., Chamberlain, J. and Gebbie, H. A. (1970). *Trans. Faraday Soc.*, **66**, 2710
320. Levin, I. W. (1969). *Spectrochim. Acta*, **25A**, 1160
321. Tychinskaya, I. I., Yudanov, N. F. and Opalovskii, A. A. (1969). *Russ. J. Inorg. Chem.*, **14**, 1636
322. Carter, H. A., Quershi, A. M., Sams, J. R. and Aubke, F. (1970). *Can. J. Chem.*, **48**, 2853
323. Greenwood, N. N. and Ruddick, J. N. R. (1967). *J. Chem. Soc. A*, 1679
324. Herber, R. H. and Cheng., H-S. (1969). *Inorg. Chem.*, **8**, 2145
325. Clausen, C. A. and Good, M. L. (1970). *Inorg. Chem.*, **9**, 817
326. Brown, T. L., McDugle, W. G. and Kent, L. G. (1970). *J. Amer. Chem. Soc.*, **92**, 3645
327. Brown, T. L. and Kent, L. G. (1970). *J. Phys. Chem.*, **74**, 3572
328. Brill, T. B., Hugus, Z. Z. and Schreiner, A. F. (1970). *J. Phys. Chem.*, **74**, 2999
329. Graybeal, J. D., McKown, R. J. and Ing., S. D. (1970). *J. Phys. Chem.*, **74**, 1814
330. Singh, B. P., Pandey, A. N. and Singh, H. S. (1970). *Indian J. Pure Appl. Phys.*, **8**, 193
331. Rao, D. V. R., Thakuo, S. N. and Rai, D. K. (1970). *Proc. Indian Acad. Sci. A*, **71**, 42
332. Awasthi, M. N. and Mehta, M. L. (1969). *Z. Naturforsch A*, **24**, 2029
333. Debeau, M. (1969). *Spectrochim. Acta*, **25A**, 1311
334. Debeau, M. and Poulet, H. (1969). *Spectrochim. Acta*, **25A**, 1553
335. Wharf, I. and Shriver, D. F. (1969). *Inorg. Chem.*, **8**, 914
336. Nakatani, M., Takahashi, Y. and Ouchi, A. (1969). *J. Inorg. Nucl. Chem.*, **31**, 3330
337. Fratiello, A., Peak, S., Schuster, R. E. and Davis, D. D. (1970). *J. Phys. Chem.*, **74**, 3730
338. McFarlane, W. and Wood, R. J. (1969). *Chem. Commun.*, 262
339. Haman, K., Hesse, L., Kleman, L., Kocher, C., McKinley, S. and Young, A. (1969). *Inorg. Chem.*, **8**, 1054
340. Wharf, I. and Onyszchuk, M. (1970). *Can. J. Chem.*, **48**, 2250
341. Cradock, S., Harland, P. W. and Thynne, J. C. J. (1970). *Inorg. Nucl. Chem. Lett.*, **6**, 425
342a. Dean, P. A. W. and Evans, D. F. (1970). *J. Chem. Soc. A*, 2569
342b. Hajek, B. and Benda, F. (1970). *Collect. Czech. Chem. Commun.*, **35**, 2494
342c. Portier, J., Menil, F. and Grannec, J. (1969). *Compt. Rend. C*, **269**, 327
343. Wharf, I., Lobos, J. Z. and Onyszchuk, M. (1969). *Can. J. Chem.*, **91**, 2787
344. Fitzsimmons, B. W., Seeley, N. J. and Smith, A. W. (1969). *J. Chem. Soc. A*, 143
345. Homann, R. and Hoppe, R. (1969). *Z. Anorg. Allgem. Chem.*, **368**, 271
346. Day, P. and Hall, I. D. (1970). *J. Chem. Soc. A*, 2679
347. Bloom, H. and Mackay, C. J. (1970). *Aust. J. Chem.*, **23**, 1523
348. Easteal, A. J. and Hodge, I. M. (1970). *J. Phys. Chem.*, **74**, 730
349. Touberg-Jensen, N. (1969). *J. Chem. Phys.*, **50**, 559
350. Poskozim, P. S. and Stone, A. L. (1970). *J. Inorg. Nucl. Chem.*, **32**, 1391
351. Poskozim, P. S. and Stone, A. L. (1969). *J. Organometal. Chem.*, **16**, 314
352. Cotton, J. D. and Peachey, R. M. (1970). *Inorg. Nucl. Chem. Lett.*, **6**, 727

353. Wharf, I. and Shriver, D. F. (1970). *J. Inorg. Nucl. Chem.*, **32**, 1831
354. Nunes, T. L. (1970). *Inorg. Chem.*, **9**, 1325
355. Bergh, A. A. and Haight, G. P. (1969). *Inorg. Chem.*, **8**, 189
356. Nunes, T. L. and Powell, R. E. (1970). *Inorg. Chem.*, **9**, 1912
357. Daugherty, N. A. and Schiefelbein, B. (1969). *J. Amer. Chem. Soc.*, **91**, 4328
358. Schiefelbein, B. and Daugherty, N. A. (1970). *Inorg. Chem.*, **9**, 1716
359. Elizarova, G. L. and Matvienko, L. G. (1970). *Russ. J. Inorg. Chem.*, **15**, 823
360. Poulson, F. R. and Rasmussen, S. E. (1970). *Acta Chem. Scand.*, **24**, 150
361. Sevast'yanova, T. N. and Karpenko, N. V. (1969). *Russ. J. Inorg. Chem.*, **14**, 1645
362a. Donaldson, J. D. and Otenz, R. (1969). *J. Chem. Soc. A*, 2696
362b. Bergerhoff, G. and Goost, L. (1970). *Acta Crystallogr. B*, **26**, 19
363. Clark, R. J. H., Maresea, L. and Smith, P. J. (1970). *J. Chem. Soc. A*, 2687
364a. Donaldson, J. D. and Senior, B. J. (1969). *J. Inorg. Nucl. Chem.*, **31**, 881
364b. Truka, J. (1970). *Czech. J. Phys.*, **20**, 908
365a. McDonald, J. R., Scherr, V. M. and McGlynn, S. P. (1969). *J. Chem. Phys.*, **51**, 1723
365b. Fedorov, V. A., Samsonova, N. P. and Mironov, V. E. (1969). *Russ. J. Inorg. Chem.*, **14**, 1721
365c. Chamberlain, B. R. and Moses, W. (1969). *J. Chem. Soc. A*, 354
366. Vandrish, G. and Onyszchuk, M. (1970). *J. Chem. Soc. A*, 3327
367. Beattie, I. R., Milne, M., Webster, M., Blayden, H. E., Jones, P. J., Killean, R. C. G. and Lawrence, J. L. (1969). *J. Chem. Soc. A*, 482
368. Beattie, I. R., McQuillan, G. P., Rule, L. and Webster, M. (1963). *J. Chem. Soc.*, 1514
369. Fowles, G. W. A., Rice, D. A. and Walton, R. A. (1969). *Spectrochim. Acta*, **25A**, 1035
370. Pichungina, E. K. and Glybovskaya, V. A. (1969). *Zh. Strukt. Khim.*, **12**, 132
371. Paul, R. C., Aggaruval, V. K., Ahluwalia, S. C. and Navalu, S. P. (1970). *Inorg. Nucl. Chem. Lett.*, **6**, 487
372. Beattie, I. R. and Ozin. G. A. (1970). *J. Chem. Soc. A*, 370
373. Carty, A. J., Hinsperger, T., Mihichuk, L. and Sharma, H. D. (1970). *Inorg. Chem.*, **9**, 2573
374. Huggins, K. G., Parrett, F. W. and Patel, H. A. (1969). *J. Inorg. Nucl. Chem.*, **31**, 1209
375. Webster, M. and Blayden, H. E. (1969). *J. Chem. Soc. A*, 2443
376. Slavinskaya, R. A., Litvyak, I. G. and Levchenko, L. (1969). *Zh. Obshch. Khim.*, **39**, 487
377. Jain, S. C. and Rivest, R. (1970). *J. Inorg. Nucl. Chem.*, **32**, 1117
378. Jain, S. G. and Rivest, R. (1970). *J. Inorg. Nucl. Chem.*, **32**, 1579
379. Jain, S. C. and Rivest, R. (1969). *Can. J. Chem.*, **47**, 2209
380. Nicholls, D., Warburton, B. A. and Wilkinson, D. H. (1970). *J. Inorg. Nucl. Chem.*, **32**, 1075
381. Michaelson, C. E., Dyer, D. S. and Ragsdale, R. O. (1970). *J. Inorg. Nucl. Chem.*, **32**, 833
382. McGregor, W. R. and Bridgland, B. E. (1969). *J. Inorg. Nucl. Chem.*, **31**, 3325
383. Yeats, P. A., Sams, J. R. and Aubke, F. (1970). *Inorg. Chem.*, **9**, 740
384. Paul, R. C. and Chadha, S. L. (1969). *J. Inorg. Nucl. Chem.*, **31**, 1679
385. Maksyutin, Y. K., Makvidin, V. P., Guryanova, E. N. and Semin, G. K. (1970). *Isv. Acad. Nauk SSSR Ser. Khim*, **7**, 1634
386. Ichiba, S., Mishimu, M. and Negita, H. (1969). *Bull. Chem. Soc., Jap.*, **42**, 1486
387. Vasiliva, V. N., Yatskovskaya, M. A. and Mendvedev, S. S. (1969). *Dokl. Acad. Nauk SSSR*, **187**, 797
388. Moore, C. Z. and Nelson, W. H. (1969). *Inorg. Chem.*, **8**, 138
389. Serpone, N. and Fay, R. C. (1969). *Inorg. Chem.*, **8**, 2379
390. Moore, C. Z. and Nelson, W. H. (1969). *Inorg. Chem.*, **8**, 143
391. Belousova, E. M. and Seifullina, I. I. (1970). *Zh. Neorg. Khim.*, **15**, 455
392. Belousova, E. M., Seifullina, I. I. and Stasenko, I. V. (1970). *Zh. Obshch. Khim.*, **40**, 815
393. Belousova, E. M., Seifullina, I. I. and Reznichenko, V. N. (1970). *Zh. Strukt. Khim.*, **11**, 545
394. Marcarovici, C. G. and Volasniuc-Bivou, M. (1969). *Rev. Roum. Chim.*, **14**, 1231
395. Paul, R. C. and Chadha, S. L. (1969). *Aust. J. Chem.*, **22**, 1381
396. Jain, S. C. and Rivest, R. (1969). *J. Inorg. Nucl. Chem.*, **31**, 399
397. Paul, R. C. and Chadha, S. L. (1969). *J. Inorg. Nucl. Chem.*, **31**, 2753
398. Matsubayashi, G. and Tanaka, T. (1969). *J. Inorg. Nucl. Chem.*, **31**, 1963
399. Farona, M. F., Grasselli, J. G., Grossman, H. and Ritchey, W. M. (1969). *Inorg. Chim. Acta*, **3**, 495

400. Hart, D. M., Rolfes, P. S. and Kessinger, J. M. (1970). *J. Inorg. Nucl. Chem.*, **32**, 469
401. Ouchi, A., Takeuchi, T. and Taminaga, I. (1970). *Bull. Chem. Soc., Jap.*, **43**, 2840
402. Paul, R. C., Singal, H. R. and Chadha, S. L. (1970). *J. Inorg. Nucl. Chem.*, **32**, 3205
403. Kogan, V. A., Sokolov, V. P. and Osipov, O. A. (1970). *Zh. Obshch. Khim.*, **40**, 322.
404. Van den Bergen, A., Cozens, R. J. and Murray, K. S. (1970). *J. Chem. Soc. A*, 3060
405. Biradar, N. S., Kulkarni, V. H. and Sirmakadam, N. N. (1970). *Indian J. Chem.*, **8**, 838
406. Ali, K. M., Cunningham, D., Frazer, M. J., Donaldson, J. D. and Senior, B. J. (1969). *J. Chem. Soc. A*, 2836
407. Poller, R. C. and Ruddich, J. N. R. (1969). *J. Chem. Soc. A*, 2273
408. Barker, M., Swift, P., Cunningham, D. and Frazer, M. J. (1970). *Chem. Commun.*, 1338
409. Shelepina, V. L., Osipov, D. A., Shelepin, O. E., Garnovskii, A. D. and Orlova, L. V. (1969). *Zh. Neorg. Khim.*, **14**, 3276
410. Kogan, V. A., Sokolov, V. P. and Osipov, O. A. (1970). *Zh. Obshch. Khim.*, **40**, 833
411. Paul, R. C., Gupta, S. C., Ahluwalia, S. C. and Parkash, R. (1969). *Indian J. Chem.*, **7**, 631
412. Slavinskaya, R. A., Livyak, I. G., Levchenko, L. V., Sumavokova, T. N. and Batyrova, N. D. (1969). *Zh. Obshch. Khim.*, **39**, 481
413a. Shelepina, V. L., Shelepin, A. E. and Osipov, O. A. (1969). *Zh. Vses. Khim. Obshchest.*, **14**, 586
413b. Shelepina, V. L., Shelepin, A. E. and Osipov, O. A. (1969). *Russ. J. Inorg. Chem.*, **14**, 748
414. Mikulski, C. M., Karayaunis, N. M., Minkiewicz, J. V., Pytlewski, L. L. and Labes, M. M. (1969). *Inorg. Chim. Acta*, **3**, 523
415. Dick, V. J. and Maurer, A. (1969). *Rev. Roum., Chim.*, **14**, 1603
416. Ashley, P. J. and Torrible, E. G. (1969). *Can. J. Chem.*, **47**, 2583
417. Weiss, J. (1966). *Fortschr. Chem. Forsch.*, **5**, 635
418. Gerding, H. and Stufkens, D. J. (1969). *Rev. Chim. Miner.*, **6**, 795
419. Deic, A. and Lyubimov, G. A. (1969). *Latv. PSR Zinat. Acad. Vestis, Kim. Ser.*, **4**, 502
420. Christie, K. O. and Schack, C. J. (1970). *Inorg. Chem.*, **9**, 2296
421a. Greenwood, N. N. and Ruddich, J. N. R. (1967). *J. Chem. Soc. A*, 1679
    b. Mullins, M. A. and Curran, C. (1967). *Inorg. Chem.*, **6**, 2017
    c. Philip, J., Mullins, M. A. and Curran, C. (1968). *Inorg. Chem.*, **7**, 1895
    d. Fitzsimmons, B. W., Seeley, N. J. and Smith, A. W. (1968). *Chem. Commun.*, 390
422. Poller, R. C., Ruddich, J. N. R., Thevarasa, M. and McWhinnie, W. R. (1969). *J. Chem. Soc. A*, 2327
423. Hendricker, D. G. (1969). *Inorg. Chem.*, **8**, 2328
424. Tanaka, T., Matsubayashi, G., Shimizu, A. and Matsuo, S. (1969). *Inorg. Chim. Acta*, **3**, 187
425. Harada, T. (1970). *Bull. Chem. Soc., Jap.*, **43**, 266
426. Jaura, K. L., Khurana, N. S. and Verma, V. K. (1970). *Indian J. Chem.*, **8**, 186
427. Jaura, K. L., Chander, K. and Sharma, K. K. (1970). *Z. Anorg. Allgem. Chem.*, **376**, 303
428. Kitching, W., Moore, C. J. and Doddrell, D. (1969). *Aust. J. Chem.*, **22**, 1149
429. Isaacs, N. W. and Kennard, C. H. L. (1970). *J. Chem. Soc. A*, 1257
430. Fitzsimmons, B. W. (1970). *J. Chem. Soc. A*, 3235
431. Kimara, T., Veki, T., Yasaoka, N., Kasai, N. and Kakudo, M. (1969). *Bull Chem. Soc. Jap.*, **42**, 2479
432. Matsabayashi, G., Nishii, N. and Tanaka, T. (1969). *Bull. Chem. Soc. Jap.*, **42**, 2369
433. Faraglia, G., Maggio, F., Cefalu, R., Bosco, R. and Barbieri, R. (1969). *Inorg. Nucl. Chem. Lett.*, **5**, 177
434. Hill, J. C., Drago, R. S. and Herber, R. H. (1969). *J. Amer. Chem. Soc.*, **91**, 1644
435. Parish, R. V. and Platt, R. H. (1969). *J. Chem. Soc. A*, 2145
436. Davies, A. G., Smith, L. and Smith, P. J. (1970). *J. Organometal. Chem.*, **23**, 135
437. Mullins, F. R. (1970). *Can. J. Chem.*, **48**, 1677
438. Park, J. J., Collins, D. M. and Hoard, J. L. (1970). *J. Amer. Chem. Soc.*, **92**, 3636
439. Jaura, K. L., Chandler, K. and Sharma, K. K. (1970). *Z. Anorg. Allgem. Chem.*, **375**, 107
440. Van den Berghe, E. V., Verdonck, L. and Van der Kelen, G. P. (1969). *J. Organometal. Chem.*, **16**, 497
441. Srivastava, T. N. and Agarwal, M. P. (1970). *J. Inorg. Nucl. Chem.*, **32**, 3416
442. Holloway, J. H., McQuillan, G. P. and Ross, D. S. (1969). *J. Chem. Soc. A*, 2505
443. Fitzsimmons, B. W., Owusu, A. A., Seeley, N. J. and Smith, A. W. (1970). *J. Chem. Soc. A*, 935

444. McGlinchey, M. J., Odom, J. D., Reynoldson, T. and Stone, F. G. A. (1970). *J. Chem. Soc. A*, 31
445. Knyazev, E. A. and Klebanov, M. S. (1970). *Russ. J. Inorg. Chem.*, **15**, 10
446. Bonchov, T., Khristov, D., Burin, K. and Kostadincheva, B. (1968). *God. Sofii. Univ. Fiz. Fak.*, **61**, 115
447. Nagita, H., Okuda, T. and Mishima, M. (1969). *Bull. Chem. Soc. Jap.*, **42**, 2509
448. Belousova, E. M., Seifullina, I. I. and Bobrovskaya, M. M. (1970). *Russ. J. Inorg. Chem.*, **15**, 508
449. Belousova, E. M. and Seifullina, I. I. (1970). *Russ. J. Inorg. Chem.*, **15**, 235
450. Belousova, E. M. and Seifullina, I. I. (1970). *Russ. J. Inorg. Chem.*, **15**, 299
451. Weingarten, H. and Wager, J. S. (1970). *Chem. Commun.*, 854
452. Yeats, P. A., Poh, B. L., Ford, B. F. E., Sams, J. R. and Aubke, F. (1970). *J. Chem. Soc. A*, 2188
453. Peacock, R. D. and Wilson, I. L. (1969). *J. Chem. Soc. A*, 2030
454. Roesky, H. W. and Dietl, M. (1970). *Z. Anorg. Allgem. Chem.*, **376**, 230
455. Manoussakis, G. E. and Tossidis, J. A. (1969). *Inorg. Nucl. Chem. Lett.*, **5**, 733
456. Frydnych, R. (1967). *Chem. Ber.*, **100**, 3588
457. Frydnych, R. (1970). *Chem. Ber.*, **103**, 327
458. Donaldson, J. D. and Nicholson, D. G. (1970). *Inorg. Nucl. Chem. Lett.*, **6**, 151
459. Donaldson, J. D. and Nicholson, D. G. (1970). *J. Chem. Soc. A*, 145
460. Kulishov, V. I., Bokii, N. G., Strachkov, Y. T., Nefedov, O. M., Kolesnikov, S. P. and Perl'mutter, B. L. (1970). *Zh. Strukt. Khim.*, **11**, 71
461. Orlyanskaya, A. K., Katseva, G. N., Streletes, N. L. and Gyunne, E. A. (1969). *Russ. J. Inorg. Chem.* **14**, 199
462. Cassidy, J. E., Moser, W., Donaldson, J. D., Jelen, A. and Nicholson, D. G. (1970). *J. Chem. Soc. A*, 173
463. Sumarokova, T. N. and Surpina, D. E. (1970). *Rocz. Chem.*, **44**, 947
464. Brini-Fritz, M., Geistel, M. M. and Pousse, A. (1969). *Compt. Rend. C*, **268**, 2040
465. Sumavokova, T. N. and Surpina, D. E. (1969). *Isv. Acad. Nauk SSSR Ser. Khim.*, **19**, 16
466. Doskey, M. A. and Curran, C. (1969). *Inorg. Chim. Acta*, **3**, 169
467. Shlenskaya, V. I., Burgukov, A. A. and Moryakova, L. N. (1969). *Russ. J. Inorg. Chem.*, **14**, 255
468. Wannagat, U. and Rabet, F. (1970). *Inorg. Nucl. Chem. Lett.*, **6**, 155
469. Fenton, D. E., Gould, R. R., Harrison, P. G. Harvey, T. B., Omietanski, G. M., Sze, K. C.-T. and Zuckerman, J. J. (1970). *Inorg. Chim. Acta*, **4**, 235
470. Hoeft, J., Lovas, F. J., Tiemann, E. and Torring, T. (1970). *J. Chem. Phys.*, **53**, 2736
471. Raymonda, J. W., Muenter, J. S. and Kemplerer, W. A. (1970). *J. Chem. Phys.*, **52**, 3458
472. Yamdagni, R. (1970). *J. Mol. Spectrosc.*, **33**, 531
473. Hoeft, J., Lovas, F. J., Tiemann, E., Tisher, R. and Torring, T. (1969). *Z. Naturforsch. A*, **24**, 1217
474. Ogden, J. S. and Ricks, M. J. (1970). *J. Chem. Phys.*, **52**, 352
475. Ogden, J. S. and Ricks, M. J. (1970). *J. Chem. Phys.*, **53**, 896
476. Baggio, E. M. and Sonnino, T. (1970). *J. Chem. Phys.*, **52**, 3786
477. Bollades, B. I., Perepech, K. V., Seregin, P. P. and Shipatov, V. T. (1970). *Isv. Acad. Nauk SSSR Neorg. Mater.*, **6**, 818
478. Scott, J. F. (1970). *Phys. Rev. B*, **3**, 3488
479. Scott, J. F. (1970). *J. Chem. Phys.*, **53**, 852
480. Murty, A. N. and Curl, R. F. (1969). *J. Mol. Spectrosc.*, **30**, 102
481. Malinin, G. V. and Tolmachev, Y. M. (1969). *Russ. J. Inorg. Chem.*, **14**, 159
482. Malinin, G. V. and Tolmachev, Y. M. (1969). *Russ. J. Inorg. Chem.*, **14**, 1529
483. Kusnetsov, Y. P., Petrov, E. S. and Vakrusheva, A. I. (1969). *Isv. Sib. Otd. Acad. Nauk SSSR Ser. Khim.*, 63
484. Komarova, T. N., Kindeeva, V. P. and Belonogova, L. N. (1969). *Isv. Vyssh. Zaved. Khim.*, **12**, 550
485. Ivashentev, Y. I. and Ivantsova, V. I. (1969). *Russian, J. Phys. Chem.*, **43**, 505
486. Kogan, E. A. and Evdokimov, D. Y. (1969). *Russian J. Inorg. Chem.*, **14**, 418
487. Lyakh, O. D., Sheka, I. A., Perfil'ev, A. I. (1969). *Russian J. Inorg. Chem.*, **14**, 420
488. Shevyakina, V. K., Shpirt, M. Y. and Blavatnik, V. M. (1969). *Russian J. Inorg. Chem.*, **14**, 598
489. Tarte, P. and Cahay, R. (1970). *Compt. Rend. C*, **271**, 777

490. Fang, J., Townes, W. D. and Robinson, P. D. (1969). *Z. Kristallogr., Kristalgeometre, Kristalphys., Kristallchem.*, **130,** 185
491. Voellenkle, H., Wittmann, A. and Nowotny, H. (1970). *Monatsh. Chem.*, **101,** 46
492. Lazarev, A. N. and Kolesova, V. A. (1970). *Isv. Acad. Nauk SSSR Neorg. Mater.*, **6,** 1445
493. Verkhovskii, V., Kuz'min, E. A., Ilyukhin, V. V. and Belov, N. V. (1970). *Dokl. Acad. Nauk SSSR*, **190,** 91
494. Choisnet, J., Deschauvres, A. and Raveau, B. (1970). *Compt. Rend. C*, **270,** 1003
495. Grebenshikov, R. G. and Shurvinskaya, A. K. (1969). *Isv. Acad. Nauk SSSR Neorg. Mater*, **5,** 987
496. Vinke, H., Voellenkle, H. and Nowotny, H. (1970). *Monatsh. Chem.*, **101,** 275
497. Tarte, P. and Prendhomme, J. (1970). *Spectrochim. Acta*, **26A,** 2207, and (1970). *Compt. Rend. B.*, **270,** 474
498. Lazarev, A. N. and Ignat'ev, I. S. (1970). *Opt. Spektrosk.*, **28,** 971
499. Tenisheva, T. F., Lazarev, A. N., Bondor, I. A. and Petrova, M. A. (1970). *Isv. Acad. Nauk SSSR Neorg. Mater.*, **6,** 766
500. Smolin, Y. I. (1970). *Kristallografiya*, **15,** 47
501. Tromel, M. (1969). *Z. Anorg. Allgem. Chem.*, **371,** 237
502. Tromel, M. and Hauck, J. (1969). *Z. Anorg. Allgem. Chem.*, **368,** 248
503. Hauck, J. (1969). *Z. Naturforsch. B*, **24,** 1067
504. Hauck, J. (1970). *Z. Naturforsch. B*, **25,** 109
505. Seeger, K. and Hoppe, R. (1970). *Z. Anorg. Allgem. Chem.*, **375,** 255
506. Hoppe, R. and Seeger, K. (1970). *Z. Anorg. Allgem. Chem.*, **375,** 264
507. Hebecker, C., Hoppe, R. and Kreuzburg, G. (1970). *Z. Anorg. Allgem. Chem.*, **375,** 270
508. Moller, B. (1970). *Z. Anorg. Allgem. Chem.*, **376,** 144
509. Tromel, M. and Hauck, J. (1969). *Z. Anorg. Allgem. Chem.*, **368,** 160
510. Christensen, A. N. (1970). *Acta Chem. Scand.*, **24,** 1287
511. Morengstern-Gadarau, I. and Michel, A. (1969). *Compt. Rend. C*, **271,** 1313
512. Scholder, R., Rade, D. and Schwarz, H. (1969). *Z. Anorg. Allgem. Chem.*, **364,** 113
513. Gatehouse, B. M. and Lloyd, D. J. (1969). *Chem. Commun.*, 727
514. Sevryukov, N. N., Salikova, G. E. and Dolganev, V. P. (1969). *Russian J. Inorg. Chem.*, **14,** 13
515. Ribes, M., Philippot, E. and Maurin, M. (1970). *Compt. Rend. C*, **270,** 1873
516. Ribes, M. and Maurin, M. (1970). *Rev. Chim. Miner.*, **7,** 75
517. Ribes, M. (1969). *Compt. Rend. C.*, **269,** 695
518. Nanobashvili, E. M., Putkaradze, N. V., Gordzholadze, L. A. and Sharasheuidze, V. (1969). *Isv. Acad. Nauk SSSR Neorg. Mater.*, **5,** 1659
519. Krebs, B., Pohl, S. and Schiwy, W. (1970). *Angew. Chem. Int. Ed. Engl.*, **9,** 897
520. Leblanc, A., Danot, M. and Rouxel, J. (1969). *Bull. Soc. Chim. France*, 87
521. Renaud, M., Poidatz, E. and Chaix, J.-E. (1969). *Can. J. Chem.*, **48,** 2061
522. Rabenan, A. and Rau, H. (1969). *Z. Anorg. Allgem. Chem.*, **369,** 295
523. Krebs, B. (1970). *Z. Naturforsch. B*, **25,** 223
524. Sterzel, W. and Horn, J. (1970). *Z. Anorg. Allgem. Chem.*, **376,** 254
525. Scholder, R., Malle, K. G., Triebskorn, B. and Schwarz, H. (1969). *Z. Anorg. Allgem. Chem.*, **364,** 41
526. Spiro, T. G., Templeton, D. H. and Zalkin, A. (1969). *Inorg. Chem.*, **8,** 856
527. Maroni, V. A. and Spiro, T. G. (1967). *J. Amer. Chem. Soc.*, **89,** 45
528. Spiro, T. G., Maroni, V. A. and Quicksall, C. O. (1969). *Inorg. Chem.*, **8,** 2524
529. Bulliner, P. A. and Spiro, T. G. (1970). *Spectrochim. Acta*, **26A,** 1641
530. Ramamurthy, P., Secco, E. A. and Badri, M. (1969). *Can. J. Chem.*, **91,** 2616
531. Avduevskaya, K. A. and Mironova, V. S. (1969). *Russ. J. Inorg. Chem.*, **14,** 1073
532. Everest, D. A. (1953). *J. Chem. Soc.*, 660
533. Davies, C. G., Donaldson, J. D. and Simpson, W. B. (1969). *J. Chem. Soc. A*, 417
534. Yellin, W. and Cilley, W. A. (1969). *Spectrochim. Acta*, **25A,** 879
535. Frydrych, R. and Lohoff, K. (1969). *Chem. Ber.*, **102,** 4070
536. Huber, F. and El-Meligy, M. S. A. (1969). *Z. Anorg. Allgem. Chem.*, **367,** 154
537. Huber, F. and El-Meligy, M. S. A. (1969). *Chem. Ber.*, **102,** 872
538. Izhekova, O. V., Kudra, O. K. and Suprunchuk, V. I. (1969). *Russ. J. Inorg. Chem.*, **14,** 1075
539. Engel, G. (1970). *Naturvissenschaften*, **57,** 355
540. Golub, A. M., Wen, N. C. and Grigovenko, F. F. (1969). *Russ. J. Inorg. Chem.*, **14,** 607
541. Brooker, M. H., Irish, D. E. and Boyd, G. E. (1970). *J. Chem. Phys.*, **53,** 1083

542. Newkick, A. E. and Hughes, V. B. (1970). *Inorg. Chem.*, **9**, 401
543. Pavlinova, A. V. and Dem'yanchuk, L. S. (1969). *Russ. J. Inorg. Chem.*, **14**, 500
544. Potts, D. and Walker, A. (1969). *Can. J. Chem.*, **47**, 1621
545. Ferraro, J. R., Potts, D. and Walker, A. (1970). *Can. J. Chem.*, **48**, 711
546. Addison, C. C. and Simpson, W. B. (1965). *J. Chem. Soc.*, 598
547. Srivastava, T. N. and Agarwal, M. P. (1970). *Indian J. Chem.*, **8**, 652
548. Durand-Henchoz, S. and Masdupuy, E. (1970). *Compt. Rend. C*, **270**, 1408
549. Nazarenko, V. A., Vinarova, L. I. and Lebedeva, N. V. (1969). *Russ. J. Inorg. Chem.*, **14**, 365
550. Lebedeva, N. L. and Egorova, A. L. (1969). *Russ. J. Inorg. Chem.*, **14**, 368
551. Shagisultanova, G. A., Kurnevich, G. I., Vishnevskii, V. B. and Bogdanova, I. V. (1970). *Russ. J. Inorg. Chem.*, **15**, 333
552. Meilleur, R. and Benoit, R. L. (1969). *Can. J. Chem.*, **47**, 2569
553. Nazarenko, V. A., Lebedeva, N. V. and Vinarova, L. I. (1970). *Russ. J. Inorg. Chem.* **15**, 331
554. Dean, P. A. W., Evans, D. F. and Phillips, R. F. (1969). *J. Chem. Soc. A,* 363
555. Yadava, K. L. and Pandey, U. S. (1970). *J. Inorg. Nucl. Chem.*, **32**, 1737
556. Yadava, K. L. and Pandey, U. S. (1969). *J. Inorg. Nucl. Chem.*, **31**, 2343
557. Petillon, F., Youinou, M. and Guerchais, J. (1969). *Bull. Soc. Chim. Fr.*, 4293
558. David, J., Laurent, Y. and Lang, J. (1970). *Bull. Soc. Fr. Mineral Crystallogr.*, **93**, 153
559. Poller, R. C., Ruddick, J. N. R. and Spillman, J. A. (1970). *Chem. Commun.*, 680
560. Petridis, D., Mullins, F. P. and Curran, C. (1970). *Inorg. Chem.*, **9**, 1270
561. Cefalu, R., Bosco, R., Bonati, F., Maggio, F. and Barbieri, R. (1970). *Z. Anorg. Allgem. Chem.*, **376**, 180
562. Kitching, W., Kumar Das, V. G. and Moore, C. J. (1970). *J. Organometal. Chem.*, **22**, 399
563. Poller, R. C. and Toley, D. L. B. (1969). *J. Inorg. Nucl. Chem.*, **31**, 2973
564. Hayes, J. W., Le Fèvre, R. J. W. and Radford, D. V. (1970). *Inorg. Chem.*, **9**, 400
565. Yeats, P. A., Ford, B. F. E., Sams, J. R. and Aubke, F. (1969). *Chem. Commun.*, 791
566. Dubac, J., Manuel, G. and Mazerolles, P. (1970). *Compt. Rend. C*, **271**, 465
567. Mehrota, R. C. (1967). *Inorg. Chim. Acta Rev.*, **1**, 99
568. Massol, M., Satgé, J., Rivière, P. and Barrau, J. (1970). *J. Organometal. Chem.*, **22**, 599
569. Voronkov, M. G. and Romadans, J. (1970). *Ortkytiya, Izobret., Prom., Obraztsy, Tovarnye Znaki,* **47**, 24
570. Sara, A. N. and Tangböl, K. (1970). *J. Inorg. Nucl. Chem.*, **32**, 3199
571. Mathur, S., Ouahi, R., Mathur, V. K., Mehrotra, R. C. and Maire, J. C. (1969). *Indian J. Chem.*, **7**, 284
572. Dube, G. (1969). *Z. Chem.*, **9**, 316
573. Strauss, I., Zueva, G. Y., Andreeva, L. V. and Maijs, L. (1969). *Isv. Acad. Nauk SSSR Ser. Khim.*, 1729
574. Mehrotra, R. C. and Bachlas, B. P. (1970). *J. Organometal. Chem.*, **22**, 121
575. Mehrotra, R. C. and Bachlas, B. P. (1970). *J. Organometal. Chem.*, **22**, 129
576. Hoenigschmid-Grossich, R. and Amberger, E. (1969). *Chem. Ber.*, **102**, 3589
577. Mitchell, T. N. and Neumann, W. P. (1970). *J. Organometal. Chem.*, **22**, C25
578. Kokunov, Y. V. and Buslaev, Y. A. (1970). *Russ. J. Inorg. Chem.*, **15**, 147
579. Paul, R. C., Makhini, H. S., Singh, P. and Chadha, S. L. (1970). *Z. Anorg. Allgem. Chem.*, **377**, 108
580. Paul, R. C., Singh, P., Chadha, S. L. and Makhini, H. S. (1970). *J. Inorg. Nucl. Chem.*, **32**, 2141
581. Harrison, P. G. and Zuckerman, J. J. (1969). *Chem. Commun.*, 321
582. Peruzzo, V., Plazzogna, G. and Tagliavini, G. (1970). *J. Organometal. Chem.*, **24**, 347
583. Peruzzo, V., Plazzogna, G. and Tagliavini, G. (1969). *J. Organometal. Chem.*, **18**, 89
584. Plazzogna, G., Peruzzo, V. and Tagliavini, G. (1969). *J. Organometal. Chem.*, **16**, 500
585. Hester, R. E. (1970). *J. Organometal. Chem.*, **23**, 123
586. Poder, C. and Sams, J. R. (1969). *J. Organometal. Chem.*, **19**, 67
587. Deacon, G. B. and Felder, P. W.(1970). *Aust. J. Chem.*, **23**, 1359
588. Ford, F. E., Liengme, B. V. and Sams, J. R. (1969). *J. Organometal. Chem.*, **19**, 53
589. Ford, F. E. and Sams, J. R. (1970). *J. Organometal. Chem.*, **21**, 345
590. Donaldson, J. D. and Filmore, E. J. (1969). *Spectrochim. Acta,* **25A**, 339
591. Jelen, A. and Lindquist, O. (1969). *Acta Chem. Scand.*, **23**, 3071

592. Bunting, J. W. and Thong, K. M. (1970). *Can. J. Chem.*, **48,** 1654
593. Steward, O. W. and Dziedzic, J. E. (1969). *J. Organometal. Chem.*, **16,** P5
594. Elbourne, R. G. P. and Buchanan, G. S. (1970). *J. Inorg. Nucl. Chem.*, **32,** 3559
595. Elbourne, R. G. P. and Buchanan, G. S. (1970). *J. Inorg. Nucl. Chem.*, **32,** 493
596. Dube, S. S. and Dhindsa, S. S. (1969). *J. Indian Chem. Soc.*, **66,** 838
597. Ivanova, E. D. and Migal, P. K. (1969). *Russ. J. Inorg. Chem.*, **14,** 1723
598. Ghosh, N. N. and Dasgupta, M. (1970). *Z. Anorg. Allgem. Chem.*, **375,** 315
599. Tanabe, T., Kimura, K. and Takamoto, S. (1969). *Nippon Kagaku Zasshi,* **90,** 598
600. Bernard, M. A. and Borel, M. M. (1969). *Bull. Soc. Chim. Fr.* **9,** 3064
601. Harrison, P. G. and Zuckerman, J. J. (1969). *Inorg. Nucl. Chem. Lett.*, **5,** 545
602. Krishnan, V. and Zingara, R. A. (1969). *Inorg. Chem.*, **8,** 2337
603. Friedel, M. K., Hoskins, B. F., Martin, R. L. and Mason, S. A. (1970). *Chem. Commun.*, 400
604. Choi, C. S. and Boutin, H. P. (1969). *Acta Crystallogr. B*, **25,** 982
605. Maciejewski, M., Layko, J. and Werezynski, J. (1970). *Bull. Acad. Pol. Sci. Ser. Sci. Chim.*, **18,** 205
606. Baranova, N. N. (1969). *Russ. J. Inorg. Chem.*, **14,** 1717
607. Fromage, F. and Fiorina, S. (1969). *Compt. Rend. C,* **268,** 1764
608. Andreeva, V. N. and Limar, T. F. (1970). *Russ. J. Inorg. Chem.*, **15,** 1077
609. Philippot, E. and Maurin, M. (1969). *Rev. Chim. Miner.*, **6,** 901
610. Watters, K. L. and Risen, W. M. (1969). *Inorg. Chim. Acta Rev.*, **3,** 129
611. Kolobova, N. E., Antonova, A. B. and Anisimov, K. N. (1969). *Russ. Chem. Rev.*, **38,** 822
612. King, R. B. (1970). Accounts Chem. Res., **3,** 417
613. Kenworthy, J. G. and Myatt, J. (1970). *Chem. Commun.*, 447
614. Davison, A. and Ellis, J. S. (1970). *J. Organometal. Chem.*, **23,** Cl
615. Keppie, S. A. and Lappert, M. F. (1969). *J. Organometal. Chem.*, **19,** P5
616. Cradwick, E. M. and Hall, D. (1970). *J. Organometal. Chem.*, **25,** 91
617. George, T. A. (1970). *Chem. Commun.*, 1632
618a. Mays, M. J. and Pearson, S. M. (1969). *J. Chem. Soc. A,* 136
   b. Harrison, P. G., Zuckerman, J. J., Long, T. V., Poeth, T. P. and Willeford, B. R. (1970). *Inorg. Nucl. Chem. Lett.*, **6,** 627
619. Wynter, C. and Chandler, L. (1970). *Bull. Chem. Soc. Jap.*, **43,** 2115
620. Gay, R. S. and Graham, W. A. G. (1969). *Inorg. Chem.*, **8,** 1561
621. Gapotchenko, N. I., Alekseev, N. V., Antonova, A. B., Anisimov, K. N., Kolobova, N. E., Ronova, I. A. and Struchkov, Y. T. (1970). *J. Organometal. Chem.*, **23,** 525
622. Kilbourn, B. T., Blundell, T. L. and Powell, H. M. (1965). *Chem. Commun.*, 444
623. Booth, M. R., Cardin, D. J., Carey, N. A. D., Clark, H. C. and Streenathan, B. R. (1970). *J. Organometal. Chem.*, **21,** 171
624. King, R. B. and Korenowski, T. F. (1969). *J. Organometal. Chem.*, **17,** 95
625. Nesmeyanov, A. N., Kolobova, N. E., Anisimov, K. N. and Skripkin, V. V. (1969). *Isv. Acad. Nauk SSSR Ser. Khim.*, 2859
626. Andrianov, V. G., Martynov, V. P., Anisimov, K. N., Kolobova, N. E. and Skripkin, V. V. (1970). *Chem. Commun.*, 1252
627. Elder, M. and Hall, D. (1969). *Inorg. Chem.*, **8,** 1424
628. Elder, M. (1969). *Inorg. Chem.*, **8,** 2703
629. Nesmeyanov, A. N., Kolobova, N. E., Anisimov, K. N. and Denisov, F. S. (1970). *Dokl. Acad. Nauk SSSR,* **192,** 813
630. Barrett, P. F. and Sun, K. K. W. (1970). *Can. J. Chem.*, **48,** 3300
631. Mays, M. J. and Prater, B. E. (1969). *J. Chem. Soc. A,* 2525
632. Bancroft, G. M., Mays, M. J. and Prater, B. E. (1969). *Discuss. Faraday Soc.*, **47,** 136
633. Bryan, R. F., Greene, P. T., Melson, G. A., Stokeley, P. F. and Manning, A. R. (1969). *Chem. Commun.*, 720
634. Greene, P. T. and Bryan, R. F. (1970). *J. Chem. Soc. A,* 1696
635. Melson, G. A., Stokeley, P. F. and Bryan, R. F. (1970). *J. Chem. Soc. A,* 2247
636. Greene, P. T. and Bryan, R. F. (1970). *J. Chem. Soc. A,* 2261
637. Burlitch, J. M. and Ulmer, S. W. (1969). *J. Organometal. Chem.*, **19,** P21
638. Casey, M. and Manning, A. R. (1970). *Chem. Commun.*, 674
639. Delbeke, F. T., Claeys, E. G., Van der Kelen, G. P. and Eeckhault, Z. (1970). *J. Organometal. Chem.*, **25,** 213
640. Bichler, R. E. J., Booth, M. R. and Clark, H. C. (1970). *J. Organometal. Chem.*, **24,** 145

641. Bichler, R. E. J. and Clark, H. C. (1970). *J. Organometal. Chem.*, **23**, 427
642. Bonati, F. and Minghetti, G. (1969). *J. Organometal. Chem.*, **16**, 332
643. Jehn, W. (1969). *J. Organometal. Chem.*, **16**, 419
644. Knox, S. A. R. and Stone, F. G. A. (1969). *J. Chem. Soc. A*, 2559
645. Watkins, S. F. (1969). *J. Chem. Soc. A*, 1552
646. Elder, M. and Hall, D. (1970). *J. Chem. Soc. A*, 254
647. Howard, J., Knox, S. A. R., Stone, F. G. A. and Woodward, P. (1970). *Chem. Commun.*, 1477
648. Knox, S. A. R. and Stone, F. G. A. (1970). *J. Chem. Soc. A*, 3147
649. Patmore, D. J. and Graham, W. A. G. (1968). *Inorg. Chem.*, **7**, 771
650. Kummer, R. and Graham, W. A. G. (1968). *Inorg. Chem.*, **7**, 1208
651. Pomeroy, R. K., Elder, M., Hall, D. and Graham, W. A. G. (1969). *Chem. Commun.*, 381
652. Moss, J. R. and Graham, W. A. G. (1969). *J. Organometal. Chem.*, **18**, P24
653. Knox, S. A. R., Mitchell, C. M. and Stone, F. G. A. (1969). *J. Organometal. Chem.*, **16**, P67
654. Burlitch, J. M. (1969). *J. Amer. Chem. Soc.*, **91**, 4562
655. Burlitch, J. M. (1969). *J. Amer. Chem. Soc.*, **91**, 4563
656. Graybeal, J. D., Ing, S. D. and Hsu, M. W. (1970). *Inorg. Chem.*, **9**, 678
657. Spencer, D. D., Kirsch, J. L. and Brown, T. L. (1970). *Inorg. Chem.*, **9**, 235
658. Nesmeyanov, A. N., Semin, G. K., Bryukhova, E. V., Anisimov, K. N., Kolobova, N. E. and Khandozhko, V. N. (1969). *Isv. Acad. Nauk SSSR Ser. Khim.*, **9**, 1936
659. Speiss, H. W. and Sheline, R. K. (1970). *J. Chem. Phys.*, **58**, 3036
660. Ball, R., Bennett, M. J., Brooks, E. H., Graham W. A. G., Hoyano, J. and Illingworth, S. M. (1970). *Chem. Commun.*, 592
661. Stalick, J. K. and Ibers, J. A. (1970). *J. Organometal Chem.*, **22**, 213
662. Fieldhouse, S. A., Freeland, B. H. and O'Brien, R. J. (1969). *Chem. Commun.*, 1297
663. Bennett, M. J., Brooks, W., Elder, M., Graham, W. A. G., Hall, D. and Kummer, R. (1970). *J. Amer. Chem. Soc.*, **92**, 208
664. Glockling, F. and Wilbey, M. D. (1969). *Chem. Commun.*, 286
665. Glockling, F. and Wilbey, M. D. (1970). *J. Chem. Soc. A*, 1675
666. Camia, M., Lachi, M. P., Benzoni, L., Zanzottera, C. and Tacchi Ventari, M. (1970). *Inorg. Chem.*, 251
667. Lappert, M. F. and Travers, N. F. (1970). *J. Chem. Soc. A*, 3303
668. Glockling, F. and Hill, G. C. (1970). *J. Organometal. Chem.*, **22**, C48
669. Glockling, F., McGregor, A., Schneider, M. L. and Shearer, H. M. M. (1970). *J. Inorg. Nucl. Chem.*, **32**, 3103
670. Kruck, T. and Herber, B. (1969). *Angew. Chem. Int. Ed. Engl.*, **8**, 679
671. Sakakibara, M., Takahashi, Y., Sakai, S. and Ishii, Y. (1969). *Inorg. Nucl. Chem. Lett.*, **5**, 427
672. Mason, R. and Whimp, P. O. (1969). *J. Chem. Soc. A*, 2709
673. Kingston, J. V. and Scollary, G. R. (1970). *Chem. Commun.*, 362
674. Hooton, K. A. (1969). *J. Chem. Soc. A*, 680
675. Clemmit, A. F. and Glockling, F. (1970). *Chem. Commun.*, 705
676. Deganello, G., Carturan, G. and Uguagliati, P. (1969). *J. Organometal. Chem.*, **17**, 179
677. Akhtar, M. and Clark, H. J. (1970). *J. Organometal. Chem.*, **22**, 233
678. Novikov, G. V., Trukhtanov, V. A., Khrushch, A. P., Shilov, A. E. and Gol'danski, V. I. (1969). *Dokl. Acad. Nauk SSSR*, **189**, 1294
679. Chatt, J., Eaborn, C. and Kapoor, P. N. (1970). *J. Organometal. Chem.*, **23**, 109
680. Bonati, F. and Minghetti, G. (1970). *J. Organometal. Chem.*, **24**, 251
681. Chatt, J. and Mingos, D. M. P. (1969). *J. Chem. Soc. A*, 1770
682. Abel, E. W. and Moorhouse, S. (1970). *J. Organometal. Chem.*, **24**, 687
683. Abel, E. W. Keppie, S. A., Lappert, M. F. and Moorhouse, S. (1970). *J. Organometal. Chem.*, **22**, C31
684. Fenton, D. and Zuckermann, J. J. (1969). *Inorg. Chem.*, **8**, 1771
685. Harrison, P. G. and Zuckermann, J. J. (1969). *J. Amer. Chem. Soc.*, **91**, 9886